Spice Bioactive Compounds

Nature offers us spices, which are a significant part of healthy and nutritious foods. The presence of abundant bioactive compounds in these spices makes them interesting from a scientific and health perspective. Extracts obtained from spice materials possess many health benefits and are rich sources of antioxidants, which suppress reactive oxygen species.

Spice Bioactive Compounds: Properties, Applications, and Health Benefits collects such information together in one book, presenting all necessary features related to spices and their properties. Exploring the most recent research related to the extraction, isolation, encapsulation, identification, and characterization of bioactive compounds present in spices, this book also covers the health element of spices and its utilization as a treatment for various disorders.

Key Features:

- Discusses about 14 different spices and their salient features
- Presents the novel technologies used in the extraction, isolation, and identification of bioactive compounds from spices
- Explores the utilization of spices for culinary use in food

Industries such as the food and pharmaceutical industries have great interest in the use of bioactive compounds for the production of drugs and functional foods. Written by experts in their field, this book will be useful to anyone in either industry, as well as those who have an interest in the use of such bioactive compounds for the production of drugs and functional foods.

Spice Bioactive Compounds

Properties, Applications, and Health Benefits

Edited by
Sajad Ahmad Wani
Ajay Singh
Pradyuman Kumar

CRC Press
Taylor & Francis Group
Boca Raton London New York

CRC Press is an imprint of the
Taylor & Francis Group, an **informa** business

First edition published 2023
by CRC Press
6000 Broken Sound Parkway NW, Suite 300, Boca Raton, FL 33487-2742

and by CRC Press
4 Park Square, Milton Park, Abingdon, Oxon, OX14 4RN

CRC Press is an imprint of Taylor & Francis Group, LLC

ISBN: 978-1-032-06291-4 (hbk)
ISBN: 978-1-032-10447-8 (pbk)
ISBN: 978-1-003-21538-7 (ebk)

DOI: 10.1201/9781003215387

Typeset in Times
by KnowledgeWorks Global Ltd.

Contents

Preface

Nature offered us spices which are a significant part of healthy and nutritious foods. The presence of abundant bioactive compounds in these spices makes them interesting from health and scientific points of view. The extracts obtained from spice materials possess a lot of health benefits. These extracts are rich sources of antioxidants, which suppress reactive oxygen species and related diseases and disorders. Different industries, like the food industry and pharmaceutical industry, show interest towards the use of bioactive compounds for the production of drugs and functional foods. Therefore, collecting such information in a book is a great achievement.

Spice Bioactive Compounds: Properties, Applications, and Health Benefits comprehensively explores the properties of bioactive compounds present in spices, and the utilization of spices for culinary use in food. This book has 16 chapters primarily focusing on spice bioactive compounds from selected spices. It covers the most recent research related to the extraction, isolation, encapsulation, identification, and characterization of bioactive compounds present in the spices. It covers the novel methodologies used in the extraction, isolation, encapsulation, identification, and characterization of bioactive compounds of the spices. It also includes the relationship of bioactive compounds of spices with health. It also includes the antibacterial, antiviral, and antifungal properties of spices.

Contributors

Shilpi Ahluwalia
Dr SSBUICET, Panjab University
Chandigarh, India

Misbah Ajaz
Department of Basic Sciences and Humanities
Sher-e-Kashmir University of Agricultural Sciences and Technology
Srinagar, Jammu & Kashmir, India

Mona S. Alwahibi
Department of Botany and Microbiology
College of Science
King Saud University
Riyadh, Saudi Arabia

Prajya Arya
Department of Food Engineering and Technology
Sant Longowal Institute of Engineering and Technology
Longowal, Punjab, India

Vikas Bansal
Department of Food Technology
School of Engineering & Technology
Jaipur National University
Jagatpura, Jaipur, India

Suradeep Basak
University Institute of Biotechnology
Chandigarh University
Mohali, Punjab, India

Aziza Begum
Dairy Chemistry Division
ICAR-National Dairy Research Institute
Karnal, Haryana, India

Anuradha Dutta
Department of Foods and Nutrition
Govind Ballabh Pant University of Agriculture & Technology
Pantnagar, Udham Singh Nagar, Uttarakhand, India

Mohamed S. Elshikh
Department of Botany and Microbiology
College of Science, King Saud University
Riyadh, Saudi Arabia

Umar Farooq
Department of Botany
Govt. Degree College Shopian
Jammu & Kashmir, India

Nawaz Ahmad Ganie
Department of Vegetable Science
Sher-e-Kashmir University of Agricultural Sciences and Technology
Srinagar, Jammu & Kashmir, India

Prerna Gupta
School of Agriculture
Lovely Professional University
Jalandhar-Phagwara, Punjab, India

Swati Haldar
Drug Discovery and Development Division
Patanjali Research Institute
Haridwar, Uttarakhand, India

Navneet Kaur
Department of Food Processing and Technology
Sri Guru Granth Sahib World University
Fatehgarh Sahib, India

Isha Kaushik
Department of Food Science & Technology
Chaudhary Charan Singh Haryana Agricultural University
Hisar, Haryana, India

Ashwani Kumar Khajuria
Government Degree College for Women
Kathua, Jammu & Kashmir, India

Srishti Khurana
KU Leuven – Department of Chemical Engineering
BioTeC+ - Chemical & Biochemical Process Technology & Control
Gebroeders De Smetstraat
Gent, Belgium

Jasmeet Kour
Department of Food Science and Technology
Padma Shri Padma Sachdev
Govt. PG College for Women Gandhi Nagar
Jammu, Jammu & Kashmir, India

Kawaljeet Kour
Govt. PG College for Women Gandhi Nagar
Jammu, Jammu & Kashmir, India

Pradyuman Kumar
Department of Food Engineering and Technology
Sant Longowal Institute of Engineering and Technology
Longowal, Punjab, India

Viney Kumar
Molecular Endocrinology Laboratory
Department of Biotechnology, Indian Institute of Technology, Roorkee
Uttarakhand, India

Wasim Yousuf Mir
Division of Entomology
Sher-e-Kashmir University of Agricultural Sciences and Technology
Srinagar, Jammu & Kashmir, India

Abhirup Mitra
Department of Agriculture
Invertis University, Bareilly
Uttar Pradesh, India

Jessica Pandohee
Centre for Crop and Disease Management
School of Molecular and Life Sciences
Curtin University
Bentley, Australia

Indira Dey Paul
Haldia Institute of Technology
Haldia, West Bengal, India

Poornima
Dairy Technology Division
ICAR-National Dairy Research Institute
Karnal, Haryana, India

Karnika Prakash
Centre for Rural Development
Indian Institute of Technology
New Delhi, India

Bhairav Prasad
Vidya Jyoti Institute of Higher Education
Mohali, Punjab, India

Asif Mohiuddin Rather
Division of Vegetable Science
Sher-e-Kashmir University of Agricultural Sciences and Technology
Srinagar, Jammu & Kashmir, India

Winny Routray
Department of Food Process Engineering
NIT Rourkela, Odisha, India

Vandana Sablania
Department of Food and Nutrition and Food Technology
Institute of Home Economics, University of Delhi
New Delhi, India

Mukul Sain
Dairy Engineering Division
ICAR-National Dairy Research Institute
Karnal, Haryana, India

Sangeeta
Guru Nanak College
Budhlada, Mansa, Punjab, India

Shelja Sareen
Department of Biotechnology
B.B. K. D.A.V. College for Women
Amritsar, Punjab, India

Happy Dev Sharma
Department of Vegetable Science
Dr YS Parmar University of Horticulture and Forestry
Nauni, Solan (HP), India

Kanchan Sharma
Department of Home Science
University of Jammu
Jammu, Jammu & Kashmir, India

Renu Sharma
Department of Applied Sciences
Bhai Gurdas Degree College
Sangrur, Punjab, India

Vipin Sharma
Department of Vegetable Science
Dr YS Parmar University of Horticulture and Forestry
Nauni, Solan (HP), India

Ajay Singh
Department of Food Technology
Mata Gujri College, Fatehgarh Sahib
Punjab, India

Amritpal Singh
Department of Chemistry
Mata Gujri College
Fatehgarh Sahib, Punjab, India

Kanchan Suri
Department of Food Science and Technology
Guru Nanak Dev University
Amritsar, Punjab, India

Shweta Suri
Department of Foods and Nutrition
Govind Ballabh Pant University of Agriculture & Technology
Pantnagar, Udham Singh Nagar, Uttarakhand, India

Anju Tanwar
Department of Botany
Government Post Graduate College
Ambala Cantt, India

Lisa F. M. Lee Nen That
School of Science
RMIT University
Bundoora, Australia

Javed Ahmad Wagay
Department of Botany
Govt. Degree College Shopian
Jammu & Kashmir, India

Sajad Ahmad Wani
Department of Food Technology
Islamic University of Science and Technology
Srinagar, Jammu & Kashmir, India

Upasana Yadav
Department of Nutrition and Dietetics
Faculty of Allied Health Sciences
Manav Rachna International Institute of Research and Studies
Faridabad, Haryana, India

Vidhu Yadav
Department of Food and Nutrition and Food Technology
Institute of Home Economics, University of Delhi
New Delhi, India

Syeda Zahra
Department of Pharmacognosy
Shifa College of Pharmaceutical Sciences
Shifa Tameer-e-Millat University
Islamabad, Pakistan

About the Editors

Dr. Sajad Ahmad Wani is currently working as D. S. Kothari Post-doctoral Fellow in the Department of Food Technology, Islamic University of Science and Technology, Kashmir, India. He has completed his Master's Degree in Food Technology from IUST, Awantipora, J&K, India, and Ph.D. from Sant Longowal Institute of Engineering and Technology, Punjab, India. He has also qualified National Eligibility Test (2015) (NET) conducted by Indian Council of Agricultural Research (ICAR) in December 2015. He has published more than 35 research/review articles, ten book chapters, and one book. He has also written articles for *Food & Beverage News* (India's first magazine for food and beverage industry). He has attended 50 national and 20 international conferences, seminars, and workshops throughout the world. Dr. Wani has participated in various faculty-development programmes. Dr. Wani serves as an Editorial Board Member of a number of national and international journals. He is Potential Peer Reviewer of reputed international journals related to food science and technology that belong to popular publishing houses viz. Elsevier, Taylor & Francis, Wiley, Springer, and others. He is also working as a member of various associations such as IFT, IFERP, International Association for Agricultural Sustainability, AFSTI, Asia Society of Researchers, and Asian Council of Science Editors. Dr. Wani is the recipient of the prestigious D. S. Kothari Post-doctoral Fellowship and the Maulana Azad National Fellowship (MANF-2013–14) from University Grants Commission, New Delhi, India.

Dr. Ajay Singh is presently working in the Department of Food Technology of Mata Gujri College, an autonomous body (established in 1957) in Punjab, India. He has obtained his Bachelor's of Science in Life Sciences from Kurukshetra University, Kurukshetra, India; Master's in Food Science & Technology from CDLU, Sirsa, India; and Doctorate in Food Engineering from SLIET, Longowal. He has qualified ICAR NET-2014 too. Dr. Singh is also experienced in High Pressure Processing (HPP) impact on physico-chemical characterization of buffalo milk from NDRI and IIT Kharagpur collaborated project. Besides, he has also qualified Rheology diploma-Rotation and Oscillations from Anton Paar (Austria), and has working experience in the same field on to the MCR-52 (Anton-Paar) model which is worth mentioning. He has attended and participated in various conferences and workshops of national and international repute. He has been DST-sponsored speaker for an oral talk at an international conference in Chicago, USA. Numerous reviews, research works, and popular articles of national and international journals have been assigned to his credit along with lists of online MOOCS course (IIT Kanpur & COL, Canada) awarded with distinction. In addition, he has demonstrated TEQIP-sponsored workshop as a resource person in Kurukshetra University, Kurukshetra. He has been also awarded with 2nd prize in MOFPI-sponsored international conference at JNU, New Delhi. Recently, he has submitted a bilateral project (Indo-Uzbekistan) on formula optimization as smoothie for severely acute malnourished (SAM) children to DST, Government of India.

Dr. Pradyuman Kumar is working as a professor in the Department of Food Engineering and Technology, Sant Longowal Institute of Engineering and Technology, Longowal, Punjab, India. Dr. Kumar has a teaching experience of more than 20 years and about 15 years of research experience. He has published more than 100 research/review articles, and a number of book chapters as well as books. He has also written articles for *Food & Beverage News* (India's first magazine for food and beverage industry). He has attended more than 100 national and international conferences, seminars, and workshops throughout the world. Dr. Kumar serves as an Editorial Board Member of numerous national and international journals. He is Potential Peer Reviewer of reputed international journals related to food science and technology that belong to popular publishing houses viz. Elsevier, Taylor & Francis, Wiley, Springer, and others. He is also working as a member of various associations such as IFT, AFSTI, Asia Society of Researchers, and Asian Council of Science Editors.

CHAPTER 1

Introduction to Spices and Their Bioactive Components

Jasmeet Kour, Prerna Gupta, Syeda Zahra, Vikas Bansal, Renu Sharma, Misbah Ajaz, and Ashwani Kumar Khajuria

CONTENTS

1.1 INTRODUCTION

Since times immemorial, a plethora of herbs as well as spices are exploited in the fortification of foods in the form of several preservatives, flavour enhancers, and therapeutic ingredients (El-Sayed and Youssef 2019). A survey by World Health Organization has reported that around 70–80% of the population across the globe is dependent on modern medicinal world in the form of herbal origin as the prime healthcare (Chan 2003). There has been a significant upsurge in the import of spices and herbs in spice producing countries from 1999 to 2017 which is credited to the escalating application of spices and herbs derived naturally as clean label ingredients imparting typical aroma, flavour

DOI: 10.1201/9781003215387-1

as well as colour (Embuscado 2019). Since antiquity, the addition of spices and culinary herbs to food was done for enhancing flavour and other organoleptic attributes related to food (Yashin et al. 2017; Wani and Kumar 2018). These herbs and spices are loaded abundantly with several bioactives having variable chemical composition playing a pivotal role in food and medicinal world (Siddique et al. 2020). Amid various sources of these bioactives, the most prominent ones being plants, spices, and several by-products produced by agriculture apart from marine world which have also proven their mettle in an impeccable way (Ibañez et al. 2012).

Owing to the abundant literature available on the utilization of several significant herbs and spices in the field of medicine and cooking sector, a big storehouse of information can be procured from peer-reviewed sources (Siddique et al. 2020). Apart from fruits and vegetables, spices and herbs provide us with additional sources of antioxidants of natural origin (Yashin et al. 2017). The classification of spices and herbs can be done on the basis of flavour/taste and taxonomical traits. These spices and herbs belong to the class Angiospermae or the flowering plants depending on the taxonomic classification (Embuscado 2019). Being a treasury of phenolic compounds, these herbs and spices have widespread use in pharmaceutical industries as well as additives in food products due to their functional properties (Embuscado 2015).

There is no dearth in literature which reports that spices along with herbs, primarily rosemary, oregano, sage, and thyme which are subjected to super critical fluid extraction are abundant in phenolics and exhibit antioxidant activity equivalent to that of butyl hydroxy anisole (Babovic et al. 2010). There are several herbs and spices from which certain phytochemicals such as polyphenols, flavonoids, flavonols, tannins, and terpenoids can be extracted which have the potential to diminish the onset of tumour development by retarding the phenomena of carcinogenesis, tumour angiogenesis while encouraging tumour cell differentiation and apoptosis (Valdés et al. 2015). Spices owe their antioxidant activity to their chemical organization which in turn depends on the polyphenolic content and bioactive compounds (Yashin et al. 2017).

The utilization of medicinal plants, aromatic herbs and spices, being an important component of the human diet has been carried out for thousands of years traditionally in the field of medicine and as well as in uplifting the sensory attributes of foods such as flavour, colour, and aroma. The recommendation given by various studies has highlighted the use of herbs and spices on human health owing to their immune modulating, antimutagenic, anti-inflammatory, and antioxidative properties (Conn 1995).

Many herbs and spices hold their significance in exerting beneficial effects for human health primarily including triggering of digestion, and several other pharmacological properties such as anti-inflammatory, antimicrobial, and anticarcinogenic ones which are mainly attributed to polyphenolic compounds (Valdes et al. 2015). The antioxidants in spices reduce oxidative stress, which is the outcome of free radical mechanism of cells and tissues triggered by ultraviolet, gamma, and X-rays as well as other devastating environmental conditions (Yashin et al. 2017).

The extraction of valuable medicinal ingredients from herbs and spices is carried out to selectively separate bioactives keeping quality and quantity intact (Siddique et al. 2020). There are various plants which have been reported to contain potential antioxidants out of which herbs and spices are the prominent ones when it comes to natural antioxidants (Valdes et al. 2015). The unending data reporting the ever-increasing evidence demonstrating the efficacy of these spices and herbs in human health has elevated the attention of research world and pharma industries as well (Sánchez-Camargo 2020).

1.2 PROPERTIES

Spices have specific flavour, taste (pungency, bitterness, and sweetness), and colour (red, green, and yellow). All spices such as mace, cinnamon, tarragon, basil, dill, nutmeg, fennel, parsley, marjoram, anise, cardamom, cumin, and mint have flavouring properties. Mustard, red pepper, ginger,

Table 1.1 Biological Properties of Spices

S. No.	Biological Properties	Spices
1	Antioxidant	Corianders, Cumin, Cinnamon, Nutmeg, Long pepper, Mustard (Samarth et al. 2008), Chili (Lee et al. 1995), Anise (Ruberto et al. 2000), Tamarind (Martinello et al. 2006), and Mint (Kanatt et al. 2007).
2	Antidiabetic	Corianders (Gray and Flatt 1999), Fenugreek (Wani and Kumar 2018), larger cardamom, Curry leaf (Kesari et al. 2005), and Tamarind (Maiti et al. 2004).
3	Anti-inflammatory	Turmeric, Ginger (Vendruscolo et al. 2006), Fenugreek (Wani and Kumar 2018), Small cardamom, Long pepper, Carom seeds (Thangam and Dhananjayan 2003), Anise, and Saffron (Hosseinzadeh and Younesi 2002)
4	Antinociceptive	Ginger (Vendruscolo et al. 2006), Fenugreek, Nutmeg, Clove, Pepper, Anise, Fennel, and Saffron (Hosseinzadeh and Younesi 2002)
5	Antimicrobial	Turmeric, Onion, Cinnamon, Clove, Long pepper, Anise (Ruberto et al. 2000), Fennel, Tamarind (Doughari 2006), and Mint (Marin et al. 2006)
6	Hypotensive	Corianders, Ginger, and Mint (Lahlou et al. 2002)
7	Antipyretic	Ginger, Fenugreek (Chatterjee and Pakrashi 1991), Small cardamom, Nutmeg, Asafoetida, Anise, Fennel, and Tamarind (Izquierdo et al. 2007)
8	Anticancer	Fenugreek (Chatterjee and Pakrashi 1991), Cumin, Clove, Asafoetida (Saleem et al. 2001), Fennel, and Mustard (Qiblawi and Kumar 1999)
9	Antihyperlipidemic	Corianders, Turmeric, Garlic, Fenugreek, Nutmeg, and Tamarind (Martinello et al. 2006)
10	Antihepatotoxic	Turmeric, Long pepper (Chatterjee and Pakrashi 1991), Anise, Fennel, and Curry leaf
11	Antiulcerative	Fennel (Ali and Blunden 2003) and Nutmeg
12	Antitussive	Onion, Ginger, Garlic, Large cardamom (Chatterjee and Pakrashi 1991), Chili, Asafoetida, and Anise
13	Digestive (Carminative, anti-dyspepsia and Stomachic)	Cumin, Ginger, Chili, Garlic, Small cardamomum, Large cardamomum, Cinnamon, Nutmeg, Clove (Chatterjee and Pakrashi 1991), Pepper, and Carom seeds
14	Diuretic	Garlic, Onion, and Anise
15	Antiarthritic	Turmeric (Chatterjee and Pakrashi 1991).
16	Anti-diarrheal	Carom seeds
17	Stimulant	Cumin, Chili, Garlic, Onion, Large cardamom (Chatterjee and Pakrashi 1991), Clove, Pepper, Carom seeds, and Asafoetida
18	Antifertility	Corianders
19	Antiproliferative	Corianders
20	Anthelmintic	Garlic

Source: Yashin et al. (2017).

Japanese pepper, pepper, and horse radish have pungent smell. Paparika, turmeric, and saffron have colourant property.

Spices also increase appetite, mask taste, help improve texture, and help in preservation. Caraway, rosemary, sage, clove, bay leaves, onion, thyme, coriander, and garlic have masking properties. Spices also have carminative, analgesic, blood purifier, hepatoprotective, antipyretic, and hypolipidemic properties (Gupta 2010). Table 1.1 depicts manifold biological properties of spices.

1.3 SPICES AND THEIR BIOACTIVE COMPOUNDS

Spices are defined as dry plant parts such as leaves, bark, roots, rhizomes, seeds, flowers, and buds which impart specific strong taste and smell to food. Term spice is usually derived from species. Spice is usually a plant made up of gum, protein, ash, sugar, fibre, fat, and essential oil. Spices are usually used in small quantity and have a strong taste (Li 2006). Spices are mostly extracted from those plants which are cultivated in hot or tropical environment. Since the ancient times, spices are the most common food complements in the world especially in India, China, and other Asian countries. In Rome, some spices were used in cooking to show their financial status and then in Medieval and Renaissance era until people developed a strong taste for foods full of spices (Gupta 2010).

Spices, because of their flavouring and medicinal properties, play a significant role in the pharmaceutical and food industries. Nowadays, spices and herbs are mainly used for medicinal purposes like prevention of coughs and colds (Carlsen et al. 2010), treatment of heart, diabetes, inflammation, and chronic degenerative diseases (Kaefer and Milner 2008). Use of spices for medicinal purpose is because of their antimicrobial, anti-inflammatory, antioxidant, and nutritional properties. Spices because of their specific odour, taste, and colour are main ingredients of our food. Many preservation methods are in use nowadays in order to enhance the life of spices as most spices are seasonal. For preservation and for adding more worth to spices, different drying and extraction techniques are used. Table 1.2 depicts various spices and their chemical constituents.

1.3.1 Extraction

Spices are composed of different bioactive compounds, thus their extraction is necessary in order to gain maximum benefits of spices both in food and medicinal industries. The basic principle of spice extraction is separation of bioactive compounds on the basis of their solubility. Bioactive compounds are usually distributed between two immiscible phases which result in separation of these components according to their solubility in respective phase. Traditionally, bioactive components of spices are extracted by two methods, solid–liquid extraction and liquid–liquid extraction, depending on the physical condition of spices (Siddique et al. 2020). In liquid–liquid extraction, the bioactive components are partitioned between two immiscible liquids, polar (such as water) and non-polar (chloroform). In solid–liquid extraction, the solid phase containing the bioactive compounds is extracted with solvent.

Nowadays, different new techniques are used which have less environmental hazards and are more effective and economical. Different modern and traditional extraction methods include extractions assisted by enzymes, ultrasound and microwave, supercritical fluid, and pressurized liquid (Sánchez-Camargo et al. 2020). Combinations of these techniques are also used for extracting bioactive compounds from spices.

1.3.2 Encapsulation of Various Bioactive Compounds in Spices

A process of enclosing the substances (food ingredients and bioactive compounds) in an inert and stable material to protect and enhance the shelf life and to prevent degradation from unwanted reactions is encapsulation (Kundu and Gupta 2021). Spices can be used as a reusable, naturally decomposed and important component of polyphenols with rich antioxidant and antimicrobial properties (Valdes et al. 2015). They are enriched with a number of bioactive compounds such as phenolic acids, terpenoids, flavonoids, etc., present in different plant parts.

Spices like cinnamon, clove, rosemary, oregano, and thyme have strong antimicrobial property because of the active compounds present in them (Bajwa et al. 2019). The major drawback with spices is the instability of functional compounds which hinders the delivery of sufficient health benefits to the consumers. Various factors like hydrogen ion concentration, luminescence, temperature, free radicals, chemical nature and concentration, presence of oxygen and solvents affect their stability and solubility (Bodade and Bodade 2020 and Carocho et al. 2015). Therefore, in order to control the release of flavour, odour, and colour and oxidative reactions as well as lessening the use of chemicals/or irradiation/or enzymatic alleviations, encapsulation in a more stable form is required (Garavand et al. 2021). Spray drying along with encapsulation helps in the containment of volatile

Table 1.2 Chemical Composition of Spices (Seasonings) and Culinary Herbs

Spices and Herbs	Important Chemical Constituents
Cloves	Eugenol, acetyleugenol, isoeugenol, pinene, sesquiterpene, flavonoids, gallic acid, vanillin, and phenolic acids
Cinnamon	Eugenol, terpineol, limonenetannins, proanthocyanidins, linalool, pinene, safrole, catechins, benzaldehyde, and methyleugenol
Cardamon	Limonene, terpinolene, 1,8-cineole, myrcene, quercetin, caffeic acid, kaempferol, pelargonidin, and luteolin
Coriander	Linalool, geraniol, borneol, terpineol, pinene, cumene, kaempferol, quercetin, terpinene, caffeic, vanillic, n-coumaricacids, ferulic, rutin, pyrogallol, and tocopherols
Saffron	Crocins (water-soluble carotenoids), flavonoids, safranal, gallic, ferulic, n-catechuic, caffeic, salicylic, syringic, and vanillic acids
Turmeric	Curcumins, vanillic acid, essential oils, carotene, eugenol, caffeic, protocatechuic, syringic, p-coumaric, and ascorbic acid
Ginger	Gingerol, zingiberon, turmeric, paradol, geranial, geraniol, borneol, linalool, zingerol, and camphene
Anise	Camphene, linalool, pinene, acetanisole, eugenol, rutin, apigenin-7-glucoside, trans-, cis-anetholes, luteolin-7-glucoside, and isoorientin
Caraway	Monoterpenes, aromatic aldehydes, sesquiterpene, terpenol, terpene esters, limonene, terpenon, terpenal, safranal, quercetin, kaempferol, ferulic, tannins, p-coumaric, caffeic, and chlorogenic acids
Fenugreek	Sesquiterpenes, terpenes, and aromatic aldehydes
Black pepper	Piperine, sarmentine, pinene, terpenes, limonene, camphene, isoquercetin, and piperidine
Oregano	Apigenin, luteolin, quercetin, myricetin, eriodictyol, diosmetin, carvacrol, thymol, caffeic, rosmarinic, protocatechuic, and p-coumaric acid
Basil	Apigenin, quercetin, catechins, rutin, anthocyanins, kaempferol, terpinene, limonene, eugenol, carvacrol, geraniol, tannins, safrole, menthol, ursolic, rosmarinic, and p-coumaric acids
Bay leaf	Cinnamtannin, 1,8-cineole
Dill	Quercetin, myricetin, catechins, kaempferol, carvone, isorhamnetin, and limonene
Garlic	Allicin, diallylsulfide, diallyltrisulfide, diallyldisulfide, S-allyl cysteine, and allylisothiocyanate
Horseradish	Allylisothiocyanate, phenyl methyl isothiocyanate, asparagine, and sinigrin
Allspice	Quercetin, eugenol, pimentol, and gallic acid
Marjoram	Limonene, terpinene, pinene, ferulic, apigenin, p-cumene, caffeic, sinapinic, syringic, rosmarinic, 4-hydroxybenzoic, and vanillic acids
Mustard	Carotene, allylisothiocyanate, isorhamnetin, kaempferol glucoside, and isorhamnetin-7-O-glucoside
Fennelflower	Pinene, thymohydroquinone, thymoquinone, thymol, p-cumene, nigellidine, carvacrol, nigellicine, and hederin
Onion	Quercetin, apigenin, dipyridyldisulfide, quercetin-4-glucoside,andrutin
Parsley	Apigenin, caffeic acid, kaempferol, luteolin, quercetin, and myricetin
Red pepper	Capsaicin, lutein, carotene, tocopherol, quercetin, capsanthin, and ascorbic acid
Peppermint	Menthol, limonene, menthone, eriocitrin, isomenthone, rutin, hesperidin, luteolin, apigenin, tocopherols, carotenes, caffeic, chlorogenic, and rosmarinic acid
Rosemary	Carnosol, rosmanol, pinene, limonene, geraniol, apigenin, luteolin, naringin, ursolic, vanillic, rosmarinic, and caffeic acids
Sage	Geraniol, carnosol, pinene, limonene, catechins, saponin, apigenin, carnosine, luteolin, rosmarinic, vanillic, and caffeic acids
Nutmeg	Catechins, caffeic acid, myricetin, lignans, and orgentin
Myrtle	Anthocyanins, limonene, pinene, myrtocommulone, ellagic, gallic acids, myricetin-3-O-rhamnoside, and myricetin-3-O-galactoside
Lavender	Limonene, apigenin, quercetin, rosmarinic, ferulic, kaempferol glucoside, p-coumaric, and caffeic acid

compounds and better retention of flavour and aroma in their active state for extended period of time so that they can be utilized in their fresh inherent form. Such compounds can be easily incorporated into any product formulations and are much suited for water-dispersible systems with lesser germ count. These products have an organised release in the human body as well as in the food products wherein they are used with better solubility, stability, and bioavailability (Sari et al. 2015).

The encapsulation of bioactive compounds with spray drying process is found to have a shelf life of up to 12 months and is heat stable as well. Other volatile and oxidizable compounds like 2-phenylethyl alcohol, phenylpropanoid (cinnamaldehydetrans (E) isomer), and β-ionone from various extracts were also encapsulated by using spray drying technique with a polar water-soluble group attached to a water-insoluble hydrocarbon chain, i.e., amphiphilic methylcellulose (MC) as an emulsifier in the inner wall material and alginate (ALG)/methylcellulose (MC) as complex outer wall materials to increase the firmness of core materials. It was observed that the delivery of core matter was more in the pre-stage of storage rather than afterwards. This was due to the hurdling effect of capsule-block of tiny microcapsules that can prevent the straight and negative effects of environment, controlling the main release thereby prolonging the shelf life of wall materials (Zohu et al. 2018). Microencapsulation by freeze dying process is also one of the new techniques that has been implemented on clove (*Syzygium aromaticum (L.)* by Cortes-Rojas et al. (2014). The main focus was to encapsulate the volatile and poor water-soluble compounds, i.e., eugenol and eugenyl acetate. Another study to restore eugenol oil by the process of coacervation using gelatin-sodium alginate was attempted with a major application to restore the antioxidants present in it (Shinde and Nagarsenker 2011). Rodrigues et al. (2014) has also reported the use of micro and nano encapsulation of salt with its greater advantage and use in meat and meat product industry, bakery and dairy industry, as well as preserved food materials. Substances like proteins, fats, carbohydrates, and natural and synthetic gums, as well as other additional biopolymer, can be used for encapsulating flavours. Different materials successfully used for encapsulation of various spices are sodium caseinate, maltodextrin, gum Arabic, sodium salt, and polyvinyl acetate matrix (Chatterjee and Bhattacharjee 2015 and Pan et al. 2014). The advantages of encapsulating bioactive compounds in spices are as follows:

- Reduction in modification of the core material by the external conditions of the surrounding environment as well as the variation in body pH (Zhu 2017)
- Easy handling and distribution
- Maintaining functionality, integrity, and activeness and promptness of the bioactive compounds by their appropriate release in the target organ
- Prevention from microbial and chemical damage
- Overall improvisation in the organoleptic quality parameters
- Increase in the water solubility of bioactive compounds

1.4 CHARACTERIZATION OF VARIOUS BIOACTIVE COMPOUNDS IN SPICES

The use of spice as a preservative, colourants, and flavour enhancer is known since ancient times. They have many significant physiological effects on the body by being rich in anticarcinogenic and free radical scavenging activities, aiding in digestion by releasing stimulants needed for digestion, hypolipidemics, and by fighting against foreign bodies thereby possessing antibacterial, anti-inflammatory, and antiviral activity. Compounds isolated from spices are well known for their functional properties because of which they are preferably used in food, pharma, and medicinal services (Juana et al. 2012). These functional properties are generally associated with the role and type of phenolic compounds and the presence of terpenes and or monoterpenes and sesquiterpenes (Capecka et al. 2005).

Many studies have been put forward to explain this phenomenon. Some of these are as follows: Kim et al. (2003) have reported that eugenol slows down the proliferation of HT-29 human colon cancer cells and the mRNA expression of cyclooxygenase-2 (COX-2). Hwang et al. (2009) suggested that eugenol imports the apoptotic signal *via* ROS formation, thereby lowering or preventing anti-apoptotic protein Bcl-2 level following apoptotic cell death in human promyelocytic leukaemia cells (HL-60). Phenolic compounds like carnosic acid and carnosol present in rosemary and sage showed anti-inflammatory activity via the production of pro-inflammatory compounds such as leukotrienes (Poeckel et al. 2008). Apigenin, a flavonoid, showed reduced inflammation in human body. This inhibitory action is related to NF-κB activation (Xu et al. 2008). One of the major phenolic compounds of significant importance is rosmarinic acid isolated from mint, sweet basil, oregano, rosemary, sage, and thyme. Among the various types oregano, sage, rosemary, mint, sweat basil, and thyme contain highest amounts of rosmarinic acid with an approximate concentration of 2562.7, 2186.1, 1908.5, 1086.1, and 681.1 mg/100 g of oven weight, respectively (Yashin et al. 2017). Rosmarinic acid has been evaluated to possess strong antioxidant activity and oxygen radical absorption capacity (ORAC) (Zheng and Wang 2001). The therapeutic effects of spices against various cancer-causing cells, inflammations, and gastrointestinal diseases are well known since historic times. Curcumin, apigenin, luteolin, quercetin, thymoquinone, and isothiocyanate have proven anticarcinogenic effect (Randhawa and Alghamdi 2011 and Aguilera et al. 2010). Some of the major bioactive compounds and their composition with reported therapeutic effect found in spices are depicted in Table 1.3.

1.5 ROLE OF BIOACTIVE COMPOUNDS IN FOOD PRESERVATION

The origin of spices is known in Egypt where it was considered as a valuable trade item and thereafter the importance and beneficial effects of spices spread globally. As per Food and Agriculture Organisation (FAO) (2011), India is the leader in global spice production, followed by Bangladesh, Turkey, China, Pakistan, Iran, Nepal, etc. In India, since long, spices of numerous varieties have been growing at large scale, so these have become an inherent part of their lifestyle, customs, traditions, and above all, their kitchen. The reason is their self-belief of various therapeutic and medicinal characteristics that spices have which has been proven scientifically also in the modern world. Various spices contain an infinite number of bioactive compounds and their role in food preservation can be well understood owing to their antioxidant and antimicrobial activity, ability to prevent food oxidation, and food spoilage.

The main bioactive compounds present in various spices include different forms of alkaloids, flavonoids, polypeptides, polyphenols, and quinines (Negi 2012). Some natural compounds with bioactive potential act as stress relievers like caffeine in tea or coffee or some are responsible for a "hot" sensation in the mouth like gingerols in ginger. Several researchers across the globe have made efforts to understand the role of bioactive compounds in food preservation and the same efforts have been made to summarize those findings in brief here. The details about the individual spices have been documented in the subsequent chapters.

1.5.1 Bioactive Compounds as Natural Antioxidants

Antioxidants are substances that neutralize free radicals or prevent oxidation of other components. Spices are reported to be rich sources of various antioxidants which include phenolic substances, flavonoids, tannins, sulphur-containing compounds, alkaloids, phenolic diterpenes, and vitamins (Peter 2004; Charles 2013; Yesiloglu et al. 2013). The antioxidant activity of these bioactives is reported to cure various diseases.

Table 1.3 Therapeutic Effects of Some Biologically Active Spices Along with the Active Compounds Present

Spices	Biologically Active Compound	Potential Beneficial Effects	Flavonoid Concentration mg/100 g	Potential Mechanism for Disease Prevention	References
Ginger	Gingerol, Zerumbone	Antioxidant, Anti-inflammatory	33.60	Cardiovascular diseases, including heart attack, reduction in platelet aggregation, against alcohol abuse, reduction in the oxidative response of macrophages	Gorinstein et al. (2010), Shati and Elsaid (2009), Vasanthi and Rarameswari (2010)
Garlic	Allicin, S-allyl cysteine	Antioxidant	3.61	Cardiovascular diseases including heart attack, renal diseases, inhibit angiotensin-converting enzyme, decrease platelet aggregation, prostate cancer, colon cancer Prevent lipid peroxidation of oxidized erythrocytes	Elbarbry et al. (2009), Mahmoud et al. (2012), Rahman (2001), Kim et al. (2011)
Red onion	Quercetin, Polyphenols, Flavonols, Flavonoids	Antioxidant	35.31	Reduce platelet aggregation, reduce cholesterol level, enhance blood fibrinolytic activity, Gastric carcinoma	USDA (2014), Hwang et al. (2009), Zohu et al. (2018)
Fennel	Thymoquinone	Antioxidant, Anticarcinogenic	84.50	Renal diseases	Khan et al. (2009)
Mustard	Allyl isothiocyanate	Anticarcinogenic, Antiallergin, Antidiabetic	62.90	Reduced cholesterol levels, rectal carcinoma, bladder cancer	Mills and Koren (2007), Yuan et al. (2011), Kim et al. (2003)
Sage	1,8-cineole, camphor, α-thujone, β-thujone, borneol, and viridiflorol	Antioxidant, Anti-inflammatory	17.90	Against common infections, gastrointestinal and urinary tract infections	Stavinoha, R.C.
Turmeric	Cucurmin, carotene, protocatechuic	Antioxidative, Anti-inflammatory,	549.2 mg quercetin equivalent/g of freeze-dried crude extract (mg QE/g CE) for 80% acetone extraction and 0.6 mg QE/g CE for water extraction	Prevention against common cough and cold, cardiovascular diseases, and neurodegenerative diseases	Yashin et al. (2017) Sepahpour et al. (2018)
Black Pepper	Piperine	Anti-inflammatory	983.82 ± 8.19 mg CE/100 g	Gastrointestinal diseases	Rubio et al. (2013) Mehmood and Gilani (2010)
Rosemary	Rosmarinic acid	Anti-inflammatory	9,075 ± 0,002 mg QE/g of dry material for flavonoids	Cortisol level reduction in saliva	Atsumi and Tonosaki (2007)

Flavonoids can scavenge free radicals and form complexes with catalytic metal ions that render them inactive. Spices and herbs such as sage, oregano, and rosemary have been demonstrated to be rich sources of antioxidants. Food lipids and oils can be protected against oxidative destruction by antioxidants. Antioxidant substances are known for enhancing the shelf life as well as maintaining the nutritional quality of food products as they prevent rancidity and hinder the formation of toxic oxidative products. Natural antioxidants contained in spices help to minimize oxidative stress.

Curcuminoids have capacity to scavenge or neutralise free radicals by inhibiting oxidative enzymes such as cytochrome P-450 and quenching the accessible oxygen responsible for oxidative processes (Majeed 1995). The exceptional antioxidant characteristics of curcumins are reported be highly effective for preventing lipid peroxidation in comparison to β-tocopherols, pine bark extract (*Pinus sylvestris L.*), grape seeds extract (*Vitis vinifera L.*) or the normally employed artificial antioxidant, BHT (Sreejayan and Rao 1994; Majeed 1995). The strong antioxidant activity might be attributed to their phenolic structure. Turmeric is also recognized for strong antioxidant activity than garlic and onion for preventing oxidative rancidity in salted, cooked fish (Ramanathan and Das 1993).

1.5.2 Bioactives as Antimicrobial Agents

Essential oils have been utilized for fungicidal, anti-parasitical, bactericidal, therapeutic, insecticidal, and viricidal purpose since the Middle Ages. Spices are usually employed to preserve fresh meat and meat products that are susceptible to microbial contamination. The long chain (C_6–C_{10}) alcohols and aldehydes in cilantro essential oil were efficient against *L. monocytogenes*. The inhibitory strength of fractions was generally exceeded in comparison to crude oils (Delaquis et al. 2002).

Erturk (2006) tested 11 spices extracts namely peppermint (*Mentha piperita L. emend. Huds*), lemon balm, Sicilian sumac (*Rhus coriaria L.*), bay, black pepper, carnation (*Dianthus caryophyllus L.*), cade (*Juniperus oxycedrus L.*), bladder senna (*Colutea arborescens L.*), cumin (*Cuminum cyminum L.*) and chilli, briar (*Erica arborea*), in vitro for the antibacterial activity against two gram negative (*E. coli and P. aeruginosa*) and three gram positive bacteria (*S. aureus, Bacillus subtilis and S. epidermidis*). The antifungal activity against *C. albicans* and *A. niger* was also examined. The extracts of cade (*Juniperus oxycedrus L.*), carnation (*Dianthus caryophyllus L.*), bay, and bladder senna (*Colutea arborescens L.*) displayed greater inhibitory action against *A. niger* and *C. albicans* in comparison to standard antifungal nystatin, a synthetic chemical drug.

Antimicrobial and antioxidant activities of essential oils extracted from basil were explored by Hussain et al. (2008). From the chemical composition analysis, the occurrence of linalool (56.7–60.6%), epi-α-cadinol (8.6–11.4%), α-bergamotene (7.4–9.2%), and alpha cadinene (3.2–5.4%) was reported. Among all the constituents, linalool is found to chief component of basil. The extracts along with linalool demonstrated significant antimicrobial and antioxidant activity against *E. coli, S. aureus, Pasteurella multocida, B. subtilis*, and pathogenic fungi such as A. *niger, Mucor mucedo, Fusarium solani, Botryodiplodia theobromae*, and *Rhizopus solani*.

Thymol, an active constituent of thyme, ajowan (*Trachyspermum ammi Sprague*), and basil (*Ocimum spp*) were reported to be antifungal and antibacterial. The bioactive component, allyl isothiocyanate, contained in mustard, was described to be very effective on numerous food microbes (Chacon et al. 2006).

Thus, apart from exotic capabilities of spicing up the meal to provide taste and flavour, spices have been utilized in India for hundreds of years to heal ailments. India's rich tradition has streamlined the utilizations of spices in foodstuffs, beverages, and medication for both preventive and therapeutic purposes, which has been well documented. The latest research reports demonstrated that spices exhibit an essential role in healthcare without adding substantial calories or causing harmful side effects through eco-friendly green chemistry.

Several antioxidants (both water-soluble and lipid-soluble) are present naturally in spices like turmeric, garlic, ginger, black cumin, black pepper, clove, and red pepper. Gingerol (Ginger) is responsible for the inhibition of hydroxyl radicals (Stoilova et al. 2007). Thymoquinone (black cumin) has an antioxidant activity that is higher than butylated hydroxyanisole and butylated hydroxytoluene (Kapoor et al. 2010). Similarly, cinnamaldehyde (cinnamon) scavenges the peroxide radicals and this ability can be comparable to synthetic antioxidants (Shahid et al. 2018). Moreover, Eugenol (Clove) also has antioxidant activity against hydroxyl radicals and act as an iron chelator (Ghadermazi et al. 2017).

The antimicrobial activity of spices is effective against various microorganisms. Some of the examples include cinnamon, ginger, clove, and coriander. Cinnamaldehyde (cinnamon) is effective against *Salmonella enterica, Staphylococcus aureus,* and *Listeria monocytogenes* (Shan et al. 2009) while Eugenol (clove) has antimicrobial activity against food spoilage pathogens like *Salmonella enterica, Staphylococcus aureus, Bacillus subtilis, Escherichia coli, Campylobacter jejuni, Listeria monocytogenes,* and *Saccharomyces cerevisiae* (Singh 2018). Similarly, clove oil affects the growth of some of the fungus-like *Rhizoctonia solani, Cladosporium herbarum, Penicillium glabrum, P. expansum, Fusarium verticilloides,* and *A. niger* (Veluti et al. 2004). Moreover, coriander leaves are effective against *Escherichia coli, Bacillus subtilis,* and *Salmonella typhi* (Elgayyar et al. 2001), and ginger inhibits the growth of *Staphylococcus aureus* and *Streptococcus pyogenes* (Sikrodia et al. 2018).

1.5.3 Bioactive Compounds as Food Oxidation Inhibitor

Many times, food loses its integrity due to oxidation phenomenon and eventually nutritional quality and also harms human health. The problem can be resolved by using bioactive compounds which ensure the stability of nutritional quality of food. Synthetic bioactive compounds like butylated hydroxyanisole (BHA), butylated hydroxytoluene (BHT), and tertiary butylated hydroquinone (TBHQ) are supposed to have undesirable effects on human health in terms of allergy and headache at initial stages and cancer and dermatitis at extreme stages (Bondi et al. 2017). So, in this context, spices containing natural bioactive (phenolic) compounds are the best option for arresting food oxidation.

1.5.4 Bioactive Compounds in the Prevention of Food Spoilage

Every year tonnes of food are spoiled due to microbial proliferation during storage. The use of natural bioactive compounds inhibits or delays the process of spoilage although the efficacy of bioactive compound depends upon the chemical composition of these compounds, concentration of compounds used, and initial microbial load. Some of the food products, treated with these bioactive compounds, have been summarized as follows:

Meat and its products are consumed in large quantities across the globe and its perishability makes it difficult to keep fresh and stable for a longer period of time. The application of thyme and cinnamon essential oil significantly reduces the risk of *L. monocytogenes* in meat products (Dussault et al. 2014) while the storage life of bologna and mortadella sausage can be increased by using rosemary and oregano essential oil, respectively (Viuda-Martos et al. 2010). Similarly, garlic and oregano oil was found effective against the proliferation of *Salmonella spp., L.* monocytogenes, and *S. aureus* (Garcia-Diez et al. 2016).

The bioactive compounds can be directly applied to milk and milk products like cheese and yogurt. The use of mint oil in low-fat yogurt against *Salmonella enteritidis* surprised the researchers. Similarly, different combinations of the cinnamon, bay, and thyme essential oil in low-fat yogurt against *L. monocytogenes* and full-fat yogurt against *Salmonella enteritidis* were found effective by Smith-Palmer et al. (2001). The intrinsic (pH and oxygen) and extrinsic (temperature, time, and relative humidity) factors significantly affect the growth of pathogenic microorganisms, and to prevent this, various chemical preservation techniques are used by food industries. In recent times, the use of smart packaging (incorporation of bioactive compounds in packaging material) by food manufacturers helped a lot in arresting the growth of pathogenic microorganisms. Raybaudi-Massilia et al. (2009) prepared an alginate-based edible coating by incorporating active compounds of herbs and spices and tested against *E. coli* by applying it on fresh-cut apple. Earlier in the year 2008, Raybaudi-Massilia used the combination of malic acid with different concentration of essential oil of lemongrass, palmarosa, and cinnamon incorporated in the same alginate-based edible coating and tested against *S. enteritidis* in fresh-cut melon. In both cases, fresh-cut apple and melon, the proliferation of bacteria reduced, and shelf life increased to a significant level. Similarly, Campos et al. (2016) used another combination of sage and thyme essential oil in the headspace of the package of strawberries and found good results against the growth of fungi.

The incorporation of allyl isothiocyanate in a sachet of sliced mozzarella cheese was found effective against the growth of *Staphylococcus* species and psychrotrophic bacteria (Pires et al. 2009). Ojagh et al. (2010) experimented with cinnamon and chitosan essential oil-based coating on trout fillet and witnessed the increase in shelf life from 10 days to 16 days.

1.6 QUALITY ISSUES AND SPECIFICATIONS OF SPICES

Spices are valuable commodities for which European Union (EU) and United States of America (USA) had formulated and implemented required standards at international level. These standards are also influenced by specifications set by major spice exporting countries. Cleanliness, ash level, acid insoluble ash, volatile oil content, moisture content, water activity, pesticide level, mycotoxin level, bulk density, and particle size are some common tests that contribute to setting up the range of international standards. With this view, International Standards Organization (ISO) had established many testing protocols, and examples of some standards include moisture (ISO 939), volatile oil (ISO 6571), and total ash (ISO 928). Some other tests which are spice-specific include piperine level, colour values, capsaicin content, Scoville heat units, and curcumin content.

According to the American Standards, cleanliness specifications have been fixed for every spice in terms of the permitted quantity of foreign matter, insects, excreta, and any visible mould. Salmonella is the major pathogen of public health concern in spice quality so fumigation with ethylene oxide and irradiation processes are followed with specified doses to maintain the quality of spices. In recent years, EU had banned ethylene oxide treatment and replaced it with heat treatment, i.e., high-pressure steam.

1.7 HEALTH BENEFITS OF BIOACTIVES FROM VARIOUS SPICES

Spices and herbs have been utilized in variety of ways throughout history. Since the ancient periods, spices have been basically utilized to improve aroma and organoleptic characteristics of food. Spices have also been broadly utilized as additives and medication. The bioactive components, which are the significant natural flavouring ingredients of numerous spices, contribute a crucial role in healthcare. Besides the food applications, the various components contained in spices have been extensively studied for their nutraceutical properties by Srinivasan (2005) and Singh and Laishram (2011). Spices have been accounted for to have various health benefits which include anti-thrombotic, gastroprotective, anti-rheumatic, lipid-lowering activity, anti-sclerotic, anticarcinogenic, and anti-inflammatory (Naidu and Thippeswamy 2002; Shaik et al. 2006; Dearlove et al. 2008; Assayed 2010; Babili et al. 2011).

1.7.1 Anti-Inflammatory and Antiarthritic

Inflammation is the key cause and aggravator of a variety of acute and chronic pathological disorders, such as cancer, photoaging, and diabetes. The inflammatory reactions cause transcriptional elements and proinflammatory cytokines to be activated, which can prompt an unresolved inflammatory reaction, insulin signalling suppression, and a high risk of cardiovascular diseases. Several studies have been conducted to identify foods possessing bioactive compounds with anti-inflammatory properties that are protective. Thus, using bioactive compounds found in herbs and spices that have anti-inflammatory and antioxidant characteristics can help to reduce inflammation. As a result, anti-inflammatory phytochemicals could be a valuable ally for preventing chronic diseases caused by inflammatory processes (Teodoro 2019).

Turmeric oil, obtained from rhizome of turmeric (*Curcuma longa L.*), displayed a reduction in inflammatory puffiness which when persuaded by Freund's adjuvant was prominent in comparison to control demonstrating antiarthritic and anti-inflammatory activity in albino rats (Chandra and Gupta 1972). The anti-inflammatory action was linked to retardation of hyaluronidase and trypsin activity and formaldehyde anabolism (Tripathi et al. 1973). Gupta and Modh (1969) found that 0.04 mL/kg turmeric oil inhibited formaldehyde in a way that was equal to 5 mg/kg of hydrocortisone, a synthetic anti-inflammatory drug.

Curcumin, a constituent contained in yellow pigment of turmeric rhizome, is reported for prevention of cyclooxygenase, cytokines, lipoxygenase, and steroidal hormones discharge. Curcumin possesses anti-inflammatory properties via stabilizing the lysosomal sheath, causing oxidative phosphorylation decoupling and significant O_2 radical scavenging action. A dosage of 100–200 mg per kilogram of body weight has shown adequate anti-inflammatory characteristics in numerous animal studies and appeared to have low deleterious impacts on human systems. Banji et al. (2011) described the reduced hepatotoxicity and oxidative stress by collective influence of curcumin (30 and 100 mg/kg) and methotrexate (1 mg/kg), a synthetic drug utilized for treating rheumatoid arthritis. Thus, turmeric being a nontoxic food can be used as functional health food (Mae et al. 2007).

Mace lignan obtained from nutmeg (*Myristica fragrans Houtt.*) is known for suppressing the intracellular ROS formation in glutamate-stimulated neuronal hippocampus cells and also reducing COX-2 expression in lipopolysaccharide (LPS) stimulated primary microglial cells (Jin et al. 2005). Their findings revealed that mace lignan repressed phosphorylation of mitogen-activated protein kinases and the disintegration of inhibitory-kappa B and enhanced nuclear factor-κB in LPS stimulated BV-2 microglial cells. Therefore, mace lignan possesses anti-inflammatory characteristics (Ma et al. 2009).

1.7.2 Neuroprotective Effects

Neurodegenerative ailments such as Alzheimer's disease, Parkinson's disease, and multiple sclerosis results in gradual neuronal loss accompanied with cognitive and sensory impairment (Mattson 2004). These disorders have recently been linked to a variety of multifactorial etiologies, as well as social and financial issues (Saxena and Caroni 2011).

Anti-inflammatory drugs have also been recommended as a way to slow down the progression of such degenerative diseases (Khazdair et al. 2019). Several research studies reported that non-steroidal anti-inflammatory drugs may reduce the risk of Alzheimer's disease (McGeer et al. 1996). The continuing consumption of non-steroidal anti-inflammatory medicines has been linked to gastrointestinal, liver, and renal toxicity. Curcumin is an alternative to non-steroidal anti-inflammatory medicines that have been evaluated at low (160 ppm) and high (5000 ppm) doses on oxidative destruction and inflammation. Curcumin at small and higher dosages dramatically reduced oxidised proteins and interleukin-1, a proinflammatory cytokine raised in the mice brain (Lim et al. 2001). The neuroprotective potency of black cumin (*Nigella sativa L.*) was examined against beta amyloid cells. The harmful consequences of amyloid have been linked to neurodegeneration in Alzheimer's disease (Ismail et al. 2008).

1.7.3 Cardiovascular Protection

Dietary factors exhibit an important role in the prevention of some degenerative ailments, such as cardiovascular diseases. There is currently a surge in global interest in identifying therapeutic plants which are pharmacologically effectual and have minimal or no adverse effects, with the goal of using them in preventative medicine. Spices constitute a significant part of human nutrition in all over the world. A number of research findings concerning the possible health benefits of certain herbs and spices, particularly in terms of preventing cardiovascular diseases, have been reported (Rastogi et al. 2017). Literature studies revealed that spices are rich sources of antioxidants and are effective inhibitors of inflammation and tissue damage by elevated blood sugar and circulating lipids. The phenolic substances contained in spices are known to inhibit the synthesis of chemicals that contribute towards metabolic diseases. Because of their lower calorie content and lesser cost, they are considered to be reliable sources of antioxidants and other potential bioactive components in diet (Muthulakshmi et al. 2009).

Epidemiologic studies have specified that consumption of fruits, herbs, and spices, especially garlic, may reduce the risk of cardiovascular disorders (Mykolas 2009). Cardiovascular diseases are associated with numerous issues such as elevated serum cholesterol, increased low density lipoprotein oxidation as well as platelet aggregation, hypertension, and smoking (Rahman and Lowe 2006). Garlic has been reported to hinder lipid synthesis enzymes, decline platelet aggregation, reduce blood pressure, inhibit lipid peroxidation of oxidized erythrocytes and low density lipoprotein, enhance antioxidant activity, and obstruct angiotension-converting enzyme (Rahman 2001). The key component of garlic, allicin, was found to greatly reduce the production of fatty streaks in the aortic sinus (Abramovitz et al. 1999).

According to Li (2006), an antioxidant-rich spice blend has the potential to improve heart healthiness. Consumption of highly fatted foods might result in enhanced plasma levels of malondialdehyde and also cytotoxic and genotoxic compounds that are linked to stress-related heart ailment and cancer. Research studies showed that combination of oregano (*Oregano mmajorana L.*), clove (*Eugenia caryophyllata Thunb*), cinnamon (*Cinnamomum zeylanicum Blume*), ginger (*Zingiber officinale Rosc.*), and turmeric spices lowers the level of malondialdehyde substantially during cooking of meat products.

Coriander (*Coriandrum sativum*) has been recognized as a conventional medicine meant for treating cholesterol and diabetes. The coriander seeds have a substantial hypolipidemic activity. The total cholesterol and triglycerides levels declined considerably in the animal tissues that received coriander seeds. A substantial rise in methyl glutaryl CoA reductase and plasma lecithin cholesterol acyl transferase (LCAT) activity was also noticed in the treated animals. The increased activity of plasma LCAT seemed to contribute towards its hypocholesterolemic effect by increasing hepatic bile acid production and cholesterol breakdown to faecal bile acids and neutral sterols (Chithra and Leelamma 1997). Thrombosis, a common complication of cardiovascular disorder, can be disastrous if platelets aggregation occurs in the narrow lumen of arteries, obstructing blood movement to the heart. Leaf spice extracts have been studied for their antiplatelet action since they are high in natural antioxidant compounds. Water extracts of curry and coriander leaves were shown to inhibit platelet aggregation in humans, when tested over varied concentrations (Suneetha and Krishnakantha 2005).

1.7.4 Weight Management

Obesity is one of the world's serious health issues, resulting from an imbalance between calorie intake and energy consumption (Williamson et al. 1993). Obesity has been linked to a number of diseases, including hypertension, dyslipidemia, diabetes, and heart diseases (Singla et al. 2010; De Miguel-Etayo et al. 2013). Medicines and bariatric operations are key strategies in the prevention and treatment of obesity, but they come with many drawbacks such as greasy stools, flatulence, diarrhoea, and reduced absorption of fat soluble vitamins (Padwal and Majumdar 2007). A number of naturally occurring bioactive compounds found in common species such as ginger, cinnamon, saffron, rosemary, garlic, pepper, and turmeric have been shown to exhibit weight-loss properties (Ninomiya et al. 2004; Joo et al. 2013; Misawa et al. 2015). Bioactive substances derived from these spices have been shown to diminish lipid accretion in adipose tissue by regulating the functionalities of fatty acid synthase, acyl-CoA carboxylase, and glycerol-3-phosphate dehydrogenase enzymes, related to lipogenesis (Lu et al. 2018).

Cinnamon extract is known to have beneficial effects in blood glucose management (Anderson and Polansky 2002). The analytical examinations reported that the most important constituents of cinnamon are cinnamic acid, eugenol, cinnamaldehyde, copane, cinnamyl acetate, and camphor (Jayaprakasha and Rao 2011). These bioactive substances have the potential to improve glucose absorption, insulin sensitivity, and anti-obesity effects (Baker et al. 2008). Turmeric is recognized to offer therapeutic weight loss owing to cholesterol reduction and altered body configuration. This dietary supplement is useful in treating obesity, comprising jointly weight loss and decrease in weight attainment (Wei and Xu 2003).

1.8 BIOACTIVE COMPONENTS OF SPICES IN ACTIVE FOOD PACKAGING SYSTEMS

Literature studies revealed that antioxidants contained in various herbs and spices exhibit numerous health beneficial effects. Because of their favourable effects on human health, the utility of these chemicals has increased in food type applications (Viuda-Martos et al. 2010). Herbs and spices have been recognized for enhancing or improving the flavour of food and for their preservation capabilities since ancient times (Vallverdú-Queralt et al. 2014). Several common herbs and spices have been shown to offer health benefits, including anti-inflammatory, digestive stimulant, antibacterial, and anticarcinogenic properties, that are accredited to polyphenols found in their chemical constitution. Beneficial impacts of bioactive substances acquired from plants and spices can be implemented not only into food but in packaging materials as well for enhancing the food shelf life.

In current years, there has been growing demand for natural, eco-friendly, and renewable resources. Furthermore, food producers and consumers have raised their expectations for processed food quality, especially in terms of extending shelf life while retaining the nutritional qualities. Packaging technologies have advanced considerably in the recent times, with the goal of meeting the demand for long-lasting processed foods by adding antioxidant and antimicrobial substances in the packaging material. These ingredients are designed to be discharged into the foodstuffs in a controlled manner. In order to meet the demands of packing technologies, spices and herbs have been recognized as renewable and biodegradable sources of chemical substances such as polyphenolics having considerable antioxidant and antimicrobial characteristics (Valdes et al. 2015).

Literature studies showed that packaging of fat enriched foods is a specific problem as oxidation of lipids due to bacterial progression is the basic reason of food decomposition. Fatty acids enriched foods are more susceptible towards oxidative degradation, which results in the development of off-flavours and renders food products unfit for human consumption (Sanchez-Bel et al. 2011; Valdes et al. 2015).

The incorporation of antioxidants to the packaging material is considered to be promising strategy for improving food quality. However, the usage of artificial antioxidant substances in the packaging systems is prohibited due to their adverse effects. The alternative approach involving natural antioxidants, predominantly polyphenols, derived from spices, plants, and agricultural by-products is among the present-day hot issues in food packaging research. Within this framework, natural extracts, having significant antioxidant capacity, have been integrated into plastic packing materials in order to maintain the nutritional quality of food products (Silva-Weiss et al. 2013; Valdés et al. 2014). Li et al. (2014) prepared gelatin films by incorporating natural antioxidants obtained from grape seed proanthocyanidins and polyphenols, green tea, gingko leaf, and ginger for food packaging uses.

Antimicrobial and antioxidant packaging materials are receiving consideration from food industries due to the consumer preference for nominally processed and preservative-less food items.

The isomeric forms such as carvacrol and thymol, obtained from oregano, have gained popularity as a result of their natural origins, broad antibacterial and antifungal spectrum against a variety of pathogens, significant antioxidant capacity, and thermal constancy when employed for packaging applications (Kuorwel et al. 2011). Keeping in view the versatile properties of these bioactive compounds, Ramos et al. (2012) developed polypropylene-based antimicrobial packaging sheets by incorporating carvacrol and thymol, representing their potential to be acquainted as active additives. Low-density polythene, in combination with carvacrol and organo-modified montmorillonite, has been proposed as a promising antimicrobial packaging material (Shemesh et al. 2015). Thus, amalgamation of natural active components to polymeric substances and specifically to biopolymers has revealed the great potential of these combinations to mend designated properties in food packaging materials and to guard the packed foodstuffs in a more effective manner.

1.9 CONCLUSION

Over the years, spike has been noticed in interest and demand to utilize these spices which are natural and biodegradable resources. These play a crucial role in fortification of foods in the form of preservatives. These also act as food additives which are used to enhance the organoleptic properties of food apart from increasing the shelf life. Apart from this, these spices also play a significant role in increasing interest in their application in active packaging systems. The antioxidative potential exhibited by various spices is an indicative of its chemical constitution which depends upon the concentration of polyphenolic and other active compounds. The maximum proportion of these compounds has shown anticarcinogenic properties apart from being therapeutic agents. The major phenolics present in spices having antioxidant potential comprise of flavonoids and phenolic acids. More research needs to be conducted to unfold the impact of concentration on treatment, various reaction of humans to different phytochemicals as well as employing nutrigenomics to understand how the compounds in the diet (e.g., from spices and herbs) affect health.

REFERENCES

Abramovitz, D., S. Gavri, D. Harats. et al. 1999. Allicin-induced decrease in formation of fatty streaks (atherosclerosis) in mice fed a cholesterol-rich diet. *Coronary Artery Disease* 10(7): 515–519.

Aguilera, P., M. E. Chanez-Cardenas, A. Ortiz-Plata. et al. 2010. Aged garlic extract delays the appearance of infarct, an effect likely conditioned by the cellular antioxidant systems. *Phytomedicine* 17: 241–247.

Ali, B. H. and G. Blunden. 2003. Pharmacological and toxicological properties of Nigella sativa. *Phytotherapy Research: An International Journal Devoted to Pharmacological and Toxicological Evaluation of Natural Product Derivatives* 17(4): 299–305.

Anderson, R. A. and M. M. Polansky. 2002. Tea enhances insulin activity. *Journal of Agricultural and Food Chemistry* 50(24): 7182–7186.

Assayed, M. E. 2010. Radioprotective effects of black seed (*Nigella sativa*) oil against hemopoietic damage and immunosuppression in gamma-irradiated rats. *Immunopharmacology and Immunotoxicology* 32(2): 284–296.

Babili, F. E., J. Bouajila, J. P. Souchard. et al. 2011. Oregano: chemical analysis and evaluation of its antimalarial, antioxidant, and cytotoxic activities. *Journal of Food Science* 76(3): C512–C518.

Babovic, N., S. Djilas, M. Jadranin. et al. (2010). Supercritical carbon dioxide extraction of antioxidant fractions from selected Lamiaceae herbs and their antioxidant capacity. *Innovative Food Science & Emerging Technologies* 11(1): 98–107.

Bajwa, S., D. Anang, T. Dew, and N. Ahmed. 2019. Antimicrobial properties of bioactive compounds of Indian Spices and Herbs in food: *IFST Journal article (Sept 2019)*. Antimicrobial properties of spices.

Baker, W. L., G. Gutierrez-Williams, C. M. White. et al. 2008. Effect of cinnamon on glucose control and lipid parameters. *Diabetes Care* 31(1): 41–43.

Banji, D., J. Pinnapureddy, O. J. Banji. et al. 2011. Synergistic activity of curcumin with methotrexate in ameliorating Freund's Complete Adjuvant induced arthritis with reduced hepatotoxicity in experimental animals. *European Journal of Pharmacology* 668(1–2): 293–298.

Bodade, R. and A. Bodade. 2020. Microencapsulation of bioactive compounds and enzymes for therapeutic applications. In *Biopolymer-Based Formulations*, ed. K. Pal, I. Banerjee, P. Sarkar, D. Kim, W. P. Deng, N. K. Dubey, and K. Majumder, 381–404. USA: Elsevier.

Bondi, M., A. Laukova, S. de Niederhausern. et al. 2017. Review on Natural Preservatives to Improve Food Quality and Safety. *Journal of Food Quality* 1–3.

Carlsen, M. H., B. L. Halvorsen, K. Holte. et al. 2010. The total antioxidant content of more than 3100 foods, beverages, spices, herbs and supplements used worldwide. *Nutrition Journal* 9(1), 1–11.

Campos, T., S. Barreto, R. Queiros. et al. 2016. Conservacao de morangos com utilizacao de oleos essenciais. *AGROTEC* 18: 90–96.

Capecka, E., A. Mareczek, and M. Leja. 2005. Antioxidant activity of fresh and dry herbs of some Lamiaceae species. *Food Chemistry* 93: 223–226.

Carocho, M., P. Morales, and I. C. Ferreira. 2015. Natural food additives: Quo vadis? *Trends in Food Science and Technology* 45(2): 284–295.

Chatterjee, D. and P. Bhattacharjee. 2015. Use of eugenol-lean clove extract as a flavoring agent and natural antioxidant in mayonnaise: product characterization and storage study. *Journal of Food Science and Technology* 52: 4945–4954. https://doi.org/10.1007/s13197-014-1573-6

Chacon, P. A., R. A. Buffo, and R. A. Holley. 2006. Inhibitory effects of microencapsulated allyl isothiocyanate (AIT) against *Escherichia coli* O157: H7 in refrigerated, nitrogen packed, finely chopped beef. *International Journal of Food Microbiology* 107(3): 231–237.

Chan, K. 2003. Some aspects of toxic contaminants in herbal medicines. *Chemosphere* 52(9): 1361–1371.

Chandra, D. and S. S. Gupta. 1972. Anti-inflammatory and anti-arthritic activity of volatile oil of Curcuma longa (Haldi). *The Indian Journal of Medical Research* 60(1): 138–142.

Charles, D. J. 2013. *Antioxidant properties of spices, herbs and other sources*. Springer: New York, Heidelberg Dordrecht, London.

Chatterjee, A. and Pakrashi, S. C. 1991. *Treatise on Indian medicinal plants*. Publications & Information Directorate.

Chen, I. N., C. C. Chang, C. C. Ng. et al. 2008. Antioxidant and antimicrobial activity of *Zingiberaceae* plants in Taiwan. *Plant Foods for Human Nutrition* 63(1): 15–20.

Chithra, V. and S. Leelamma. 1997. Hypolipidemic effect of coriander seeds (*Coriandrumsativum*): mechanism of action. *Plant Foods for Human Nutrition* 51(2): 167–172.

Conn, E. E. 1995. The world of phytochemicals. In *Phytochemicals and Health*, ed. D. L. Gustine and H. E. Flores, 1–14. Rockville, Maryland.

Cortes-Rojas, D. F., C. R. F. Souza, and W. P. Oliveira. 2014. Encapsulation of eugenol rich clove extract in solid lipid carriers. *Journal of Food Engineering* 127: 34–42.

De Miguel-Etayo, P., L. A. Moreno, and I. Iglesia. 2013. Body composition changes during interventions to treat overweight and obesity in children and adolescents; a descriptive review. *Nutricionhospitalaria* 28(1): 52–62.

Dearlove, R. P., P. Greenspan, D. K., Hartle, Swanson, R. B. and Hargrove, J. L. 2008. Inhibition of protein glycation by extracts of culinary herbs and spices. *Journal of Medicinal Food* 11(2): 275–281.

Delaquis, P. J., K. Stanich, B. Girard, and G. Mazza. 2002. Antimicrobial activity of individual and mixed fractions of dill, cilantro, coriander and eucalyptus essential oils. *International Journal of Food Microbiology* 74(1–2): 101–109.

Doughari, J. H. 2006. Antimicrobial activity of Tamarindusindica Linn. *Tropical Journal of Pharmaceutical Research* 5(2): 597–603.

Dussault, D., K. D. Vu, and M. Lacroix. 2014. *In vitro* evaluation of antimicrobial activities of various commercial essential oils, oleoresin and pure compounds against food pathogens and application in ham. *Meat Science* 96(1): 514–520.

Elbarbry, F., S. Gazarin, and A. Shoker. 2009. The protective effect of thymoquinone, an antioxidant andanti-inflammatory agent, against renal injury: A review. *Saudi Journal of Kidney Disease Transplant* 20: 741–752.

Elgayyar, M., F. A. Draughon, and D. A. Golden. 2001. Antimicrobial activity of essential oils from plants against selected pathogenic and saprophytic microorganisms. *Journal of Food Protection* 64(7): 1019–1024.

El-Sayed, S. M. and A. M. Youssef. 2019. Potential application of herbs and spices and their effects in functional dairy products. *Heliyon* 5(6): e01989.

Embuscado, M. E. 2019. Bioactives from culinary spices and herbs: a review. *Journal of Food Bioactives* 6.

Embuscado, M. E. 2015. Spices and herbs: Natural sources of antioxidants – a mini review. *Journal of Functional Foods* 18: 811–819. https://doi.org/10.1016/j.jff.2015.03.005

Erturk, O. 2006. Antibacterial and antifungal activity of ethanolic extracts from eleven spice plants. *Biologia* 61(3): 275–278.

Food and Agriculture Organisation (FAO). 2011. Production of spices by countries

Garavand, F., N. Eghbal, M. Nooshkam. et al. 2021. Salt, spices, and seasonings formulated with nano/micro-encapsulated ingredients. In *Application of Nano/Microencapsulated Ingredients in Food Products*. 435–467, Academic press.

Garcia-Diez, J., J. Alheiro, A. L. Pinto. et al. 2016. Behaviour of food-borne pathogens on dry cured sausage manufactured with herbs and spices essential oils and their sensorial acceptability. *Food Control* 59: 262–70.

Ghadermazi, R., J. Keramat, and S. A. H. Goli. 2017. Antioxidant activity of clove *(Eugenia caryophyllata* Thunb), oregano (*Oringanum vulgare* L) and sage (*Salvia officinalis* L) essential oils in various model systems. *International Food Research Journal* 24(4): 1628–1635.

Gorinstein, S., H. Leontowicz, M. Leontowicz. et al. 2010. The influence of raw and processed garlic and onions on plasma classical and non-classical atherosclerosis indices: Investigations *in vitro* and *in vivo. Phytotherapy Research* 24: 706.

Gray, A. M. and P. R. Flatt. 1999. Insulin-releasing and insulin-like activity of the traditional anti-diabetic plant Coriandrumsativum (coriander). *British Journal of Nutrition* 81(3): 203–209.

Gupta, S. S. and P. R. Modh. 1969. Anti-inflammatory effect of Curcuma longa in relation to inhibition of histamine release and mast cells. *Indian Journal of Pharmacology* 1: 22–25.

Gupta, M. 2010. Pharmacological properties and traditional therapeutic uses of important Indian spices: A review. *International Journal of Food Properties* 13(5) 1092–1116.

Hosseinzadeh, H. and H. M. Younesi. 2002. Antinociceptive and anti-inflammatory effects of Crocus sativus L. stigma and petal extracts in mice. *BMC Pharmacology* 2(1): 1–8.

Hussain, A. I., F. Anwar, and S. T. H. Sherazi. 2008. Chemical composition, antioxidant and antimicrobial activities of basil (*Ocimumbasilicum*) essential oils depends on seasonal variations. *Food Chemistry* 108(3): 986–995.

Hwang, I. K., C. H. Lee, K. Y. Yoo. et al. 2009. Neuroprotective effects of onion extract and quercetin against ischemic neuronal damage in the gerbil hippocampus. *Journal of Medicinal Food* 12(5): 990–995.

Ibañez, E., M. Herrero, J. A. Mendiola. et al. (2012). Extraction and characterization of bioactive compounds with health benefits from marine resources: macro and micro algae, cyanobacteria, and invertebrates. In *Marine Bioactive Compounds: Sources, Characterization and Applications*, ed. M. Hayes, 585–598. New York: Springer US.

Ismail, N., M. Ismail, and L. A. Latiff. 2008. Black cumin seed (*nigella sativalinn.*) oil and its fractions protect against beta amyloid peptide-induced toxicity in primary cerebellar granule neurons. *Journal of Food Lipids* 15(4): 519–533.

Izquierdo, T., F. Garcia-Tamayo, C. Soto. et al. 2007. A Tamarindusindica. Linn Pulp Polysaccharide Inhibits Fever In Vivo. And IL-1β Release by Murine Peritoneal Exudate Cells. *Pharmaceutical Biology* 45(1): 22–30.

Jayaprakasha, G. K. and L. J. M. Rao. 2011. Chemistry, biogenesis, and biological activities of Cinnamomumzeylanicum. *Critical Reviews in Food Science and Nutrition* 51(6): 547–562.

Jin, D. Q., C. S. Lim, and J. K. Hwang. 2005. Anti-oxidant and anti-inflammatory activities of macelignan in murine hippocampal cell line and primary culture of rat microglial cells. *Biochemical and Biophysical Research Communications* 331(4): 1264–1269.

Joo, H., C. T. Kim, and I. H. Kim. 2013. Anti-obesity effects of hot water extract and high hydrostatic pressure extract of garlic in rats fed a high-fat diet. *Food and Chemical Toxicology* 55: 100–105.

Juana, F. L., P. A. J. Angel, and V. L. Angel. 2012. Beneficial health effects of bioactive compounds present in spices and aromatic herbs. In *Studies in Natural Products Chemistry*, ed. Atta-ur-Rahman, volume 37, 115–134. Elsevier.

Kaefer, C. M. and J. A. Milner. 2008. The role of herbs and spices in cancer prevention. *The Journal of Nutritional Biochemistry* 19(6): 347–361.

Kanatt, S. R., R. Chander, and A. Sharma. 2007. Antioxidant potential of mint (Menthaspicata L.) in radiation-processed lamb meat. *Food Chemistry* 100(2): 451–458.

Kapoor, I. P., G. Singh, C. S. De Heluani, M. P. De Lampasona and C. A. Catalan. 2010. Chemistry and antioxidant activity of essential oil and oleoresins of black caraway (*Carum bulbocastanum*) fruits. *Journal of the Science of Food and Agriculture* 90: 385–90.

Kesari, A. N., R. K. Gupta, and G. Watal. 2005. Hypoglycemic effects of Murrayakoenigii on normal and alloxan-diabetic rabbits. *Journal of Ethnopharmacology* 97(2): 247–251.

Khan, A., G. Zaman, and R. A. Anderson. 2009. Bay leaves improve glucose and lipid profile of people with type 2 diabetes. *Journal of Clinical Biochemistry and Nutrition* 44: 52–56.

Khazdair, M. R., A. Anaeigoudari, M. Hashemzehi. et al. 2019. Neuroprotective potency of some spice herbs, a literature review. *Journal of Traditional and Complementary Medicine* 9(2): 98–105.

Kim, S. S., O. J. Oh, H. Y. Min. et al. 2003. Eugenol suppresses cyclooxygenase-2 expression in lipopolysaccharide-stimulated mouse macrophage RAW264.7 cells. *Life Science* 73: 337–348.

Kim, I., J. Y. Kim, and Y. J. Hwang. 2011. The beneficial effects of aged black garlic extract on obesity and hyperlipidemia in rats fed a high-fat diet. *Journal of Medicinal Plants Research* 5(14): 3159–3168.

Kuorwel, K. K., M. J. Cran, K. Sonneveld. et al. 2011. *Journal of Food Science* 76: R164–R177.

Kundu, P. and P. Gupta. 2021. Advanced methods of encapsulation. In *Handbook of Research on Food Processing and Preservation Technologies, Vol. 2.* Innovations in Agriculture and Biological Engineering, ed. M. R. Goyal, P. Birwal, and M. Sharma. Apple Academic Press.

Lahlou, S., P. J. C. Magalhães, and R. F. L. Carneiro-Leão. 2002. Involvement of nitric oxide in the mediation of the hypotensive action of the essential oil of Mentha × villosa in normotensive conscious rats. *Planta Medica* 68(08): 694–699.

Lee, Y., L. R. Howard, and B. Villalon. 1995. Flavonoids and antioxidant activity of fresh pepper (Capsicum annuum) cultivars. *Journal of Food Science* 60(3): 473–476.

Li, T. S. C. 2006. The range of medicinal herbs and spices. In *Handbook of Herbs and Spices*, ed. K. V. Peter, 113–125. Elsevier.

Li, J. H., J. Miao, and J. L. Wu. 2014. Preparation and characterization of active gelatin-based films incorporated with natural antioxidants. *Food Hydrocolloids* 37: 66–173.

Lim, G. P., T. Chu, F. Yang. et al. 2001. The curry spice curcumin reduces oxidative damage and amyloid pathology in an Alzheimer transgenic mouse. *Journal of Neuroscience* 21(21): 8370–8377.

Lu, M., Y. Cao, J. Xiao. et al. 2018. Molecular mechanisms of the anti-obesity effect of bioactive ingredients in common spices: a review. *Food & Function* 9(9): 4569–4581.

Ma, J., Y. K. Hwang, W. H. Cho. et al. 2009. Macelignan attenuates activations of mitogen-activated protein kinases and nuclear factor kappa B induced by lipopolysaccharide in microglial cells. *Biological and Pharmaceutical Bulletin* 32(6): 1085–1090.

Mae, T., T. Nishiyama, H. Kishida. et al. 2007. Agent for preventing/ameliorating life style related diseases containing turmeric essential oil component. European Patent Application, Intl. Nat. Pub No. WO 2005/115366, EP1752144A1, 1–12.

Maiti, R., D. Jana, U. K. Das. et al. 2004. Antidiabetic effect of aqueous extract of seed of Tamarindusindica in streptozotocin-induced diabetic rats. *Journal of Ethnopharmacology* 92(1): 85–91.

Majeed, M. 1995. *Curcuminoids: Antioxidant Phytonutrient.* New Jersey, USA: NutriScience Publishers Inc.

Mahmoud, M. F., A. A. Diaai, and F. Ahmed. 2012. Evaluation of the efficacy of ginger, arabic gum and boswelliain acute and chronic renal failure. *Renal Failure* 34: 73–82.

Marin, P. D., J. Glamoclija, M. Soković. et al. 2006. Antifungal activity of the essential oil of Mentha x piperita. *Pharmaceutical Biology* 44(7).

Martinello, F., S. M. Soares, J. J. Franco. et al. 2006. Hypolipemic and antioxidant activities from Tamarindusindica L. pulp fruit extract in hypercholesterolemic hamsters. *Food and Chemical Toxicology* 44(6): 810–818.

Mattson, M. P. 2004. Metal-catalyzed disruption of membrane protein and lipid signaling in the pathogenesis of neurodegenerative disorders. *Annals of the New York Academy of Sciences* 1012(1): 37–50.

McGeer, P. L., M. Schulzer, and E. G. McGeer. 1996. Arthritis and anti-inflammatory agents as possible protective factors for Alzheimer's disease: a review of 17 epidemiologic studies. *Neurology* 47(2): 425–432.

Mills, E. and G. Koren. 2007. From type 2 diabetes to antioxidant activity: A systematic review of the safety andefficacy of common and cassia cinnamon bark. *Canadian Journal of Physiology and Pharmacology* 85: 837–847.

Misawa, K., K. Hashizume, M. Yamamoto. et al. 2015. Ginger extract prevents high-fat diet-induced obesity in mice via activation of the peroxisome proliferator-activated receptor δ pathway. *The Journal of Nutritional Biochemistry* 26(10): 1058–1067.

Muthulakshmi, V., V. Vijayakumar, M. Vasanthkumar. et al. 2009. Food is medicine and medicine is food- A Siddha perspective. In *Functional Foods for Chronic Diseases (Vol. 4)*, ed. D. M. Martirosyan, and N. Abate, 274–317. USA: D &AInc/FF Publishing.

Mykolas, A. 2009. New and Popular cardiovascular medicines in the Baltic states. In *Functional Foods for Chronic Diseases (Vol. 4)*, ed. D. M. Martirosyan, and N. Abate, 234–241. USA: D & AInc/FF Publishing.

Naidu, K. A. and N. B. Thippeswamy. 2002. Inhibition of human low density lipoprotein oxidation by active principles from spices. *Molecular and Cellular Biochemistry* 229(1): 19–23.

Ninomiya, K., H. Matsuda, H. Shimoda. et al. 2004. Carnosic acid, a new class of lipid absorption inhibitor from sage. *Bioorganic & Medicinal Chemistry Letters* 14(8): 1943–1946.

Negi, P. S. 2012. Plant extracts for the control of bacterial growth: Efficacy, stability and safety issues for food application. *International Journal of Food Microbiology* 156: 7–17.

Ojagh, S. M., M. Rezaei, S. H. Razavi. et al. 2010. Effect of chitosan coatings enriched with cinnamon oil on the quality of refrigerated rainbow trout. *Food Chemistry* 120: 193–198.

Padwal, R. S. and S. R. Majumdar. 2007. Drug treatments for obesity: orlistat, sibutramine, and rimonabant. *The Lancet* 369(9555): 71–77.

Pan, K., H. Chen, P. D. Michael. et al. 2014. Thymol nanoencapsulated by sodium caseinate: Physical and antilisterial properties. *Journal of Agricultural and Food Chemistry* 62(7): 1649–1657.

Peter, K. V. 2004. *Handbook of Herbs and Spices* (Volume 2). Cambridge, UK: Woodhead Publishing Limited.

Pires, A. C. S., N. F. F. Soares, N. J. Andrade. et al. 2009. Increased preservation of sliced mozzarella cheese by antimicrobial sachet incorporated with allylisothiocyanate. *Brazilian Journal of Microbiology* 40: 1002–1008.

Poeckel, D., C. Greiner, M. Verhoff. et al. 2008. *Biochemical Pharmacology* 76(1): 91–97.

Qiblawi, S. and A. Kumar. 1999. Chemopreventive action by an extract from Brassica compestris (varsarason) on 7, 12-dimethylbenz (A) anthracene induced skin papillomagenesis in mice. *Phytotherapy Research* 13(3): 261–263.

Rahman, K. 2001. Historical perspective on garlic and cardiovascular disease. *The Journal of Nutrition* 131(3): 977S–979S.

Rahman, K. and G. M. Lowe. 2006. Garlic and cardiovascular disease: a critical review. *The Journal of nutrition* 136(3): 736S–740S.

Ramanathan, L. and N. P. Das. 1993. Natural products inhibit oxidative rancidity in salted cooked ground fish. *Journal of Food Science* 58(2): 318–320.

Ramos, M., A. Jiménez, M. Peltzer. et al. 2012. Characterization and antimicrobial activity studies of polypropylene films with carvacrol and thymol for active packaging. *Journal of Food Engineering* 109(3): 513–519.

Randhawa, M. A. and M. S. Alghamdi. 2011. Anticancer activity of Nigella sativa(blackseed)—A review. *American Journal of Chinese Medicine* 39: 1075–1091.

Rastogi, S., M. Mohan Pandey, and A. K. Singh Rawat. 2017. Spices: Therapeutic potential in cardiovascular health. *Current Pharmaceutical Design* 23(7): 989–998.

Raybaudi-Massilia, R. M., J. Mosqueda-Melgar, A. Sobrino-Lopez. 2009. Use of malic acid and other quality stabilizing compounds to assure the safety of fresh-cut "Fuji" apples by inactivation of Listeria Monocytogenes, Salmonella Enteritidis and Escherichia coli O157, H7. *Journal of Food Safety* 29: 236–252.

Rodrigues, J. F., G. Junqueira, C. S. Goncalves. et al. 2014. Elaboration of garlic and salt spice with reduced sodium intake. *Agrarian Sciences* 86(04).

Ruberto, G., M. T. Baratta, S. G. Deans. et al. 2000. Antioxidant and antimicrobial activity of Foeniculumvulgare and Crithmum maritimum essential oils. *Planta Medica* 66(08): 687–693.

Rubió, L., M. J. Motilva, and M. P. Romero. 2013. Recent advances in biologically active compounds in herbs and spices: a review of the most effective antioxidant and anti-inflammatory active principles. *Critical Reviews in Food Science and Nutrition* 53: 943–953.

Saleem, M., A. Alam, and S. Sultana. 2001. Asafoetida inhibits early events of carcinogenesis: A chemopreventive study. *Life Sciences* 68(16): 1913–1921.

Samarth, R. M., M. Panwar, M. Kumar. et al. 2008. Evaluation of antioxidant and radical-scavenging activities of certain radioprotective plant extracts. *Food Chemistry* 106(2): 868–873.

Sánchez-Camargo, A. P., L. Montero, J. A. Mendiola. et al. 2020. Novel Extraction Techniques for Bioactive Compounds from Herbs and Spices. *Herbs, Spices and Medicinal Plants: Processing, Health Benefits and Safety* 95–128.

Sanchez-Bel, P., I. Egea, M. T. Pretel. et al. 2011. Roasting and packaging in nitrogen atmosphere protect almond var. Guara against lipid oxidation. *Food Science and Technology International* 17(6): 529–540.

Sari, T. P., B. Mann, R. Kumar. et al. 2015. Preparation and characterization of nanoemulsion encapsulating curcumin. *Food Hydrocolloids* 43: 540–546.

Saxena, S. and P. Caroni. 2011. Selective neuronal vulnerability in neurodegenerative diseases: from stressor thresholds to degeneration. *Neuron* 71(1): 35–48.

Sepahpour, S., J. Selamat, M. Y. Abdul Manap, A. Khatib, and A. F. Abdull Razis. (2018). Comparative analysis of chemical composition, antioxidant activity and quantitative characterization of some phenolic compounds in selected herbs and spices in different solvent extraction systems. *Molecules*, 23(2): 402.

Shaik, Y. B., M. L. Castellani, A. Perrella. et al. 2006. Role of quercetin (a natural herbal compound) in allergy and inflammation. *Journal of Biological Regulators and Homeostatic Agents* 20(3–4): 47–52.

Shati, A. A. and F. G. Elsaid. 2009. Effects of water extracts of thyme (Thymus vulgaris) and ginger (*Zingiberofficinale Roscoe*) on alcohol abuse. *Food Chemistry and Toxicology* 47: 1945–1949.

Shemesh, R., D. Goldman, M. Krepker. et al. 2015. LDPE/clay/carvacrol nanocomposites with prolonged antimicrobial activity. *Journal of Applied Polymer Science* 132(2): 41261.

Silva-Weiss, A., M. Ihl, P. D. A. Sobral. et al. 2013. Natural additives in bioactive edible films and coatings: functionality and applications in foods. *Food Engineering Reviews* 5(4): 200–216.

Shahid, M. Z., H. Saima, A. Yasmin. et al. 2018. Antioxidant capacity of cinnamon extract for palm oil stability. *Lipids in Health and Disease* 17(1): 1–8.

Shan, B., Y. Z. Cai, J. D. Brooks and Corke H. 2009. Antibacterial and antioxidant effects of five spice and herb extracts as natural preservatives of raw pork.

Shinde, U. and M. Nagarsenker. 2011. Microencapsulation of eugenol by gelatin-sodium alginate complex coacervation. *Indian Journal of Pharmaceutical Sciences* 73(3): 311.

Siddique, M. A. B., K. Tzima, D. K. Rai. et al. 2020. Conventional extraction techniques for bioactive compounds from herbs and spices. *Herbs, Spices and Medicinal Plants: Processing, Health Benefits and Safety* 69–93.

Sikrodia, R., D. Chhabra, S. D. Audarya. et al. 2018. Efficacy of antimicrobial activity of aqueous extract of garlic (*Allium sativum*) and ginger (*Zingiber officinale*) against different bacterial species. *International Journal of Current Microbiology and Applied Sciences* 7: 3021–3025.

Singh, R. 2018. Antimicrobial activity of selected medicinal plants cinnamon and clove. *International Journal of Research in Pharmacy and Pharmaceutical Sciences* 3(5): 48–50.

Smith-Palmer, A., J. Stewart, and L. Fyfe. 2001. The potential application of plant essential oils as natural foods preservatives in soft cheese. *Food Microbiology* 18: 463–470.

Srinivasan, K. 2005. Role of spices beyond food flavoring: Nutraceuticals with multiple health effects. *Food Reviews International* 21(2): 167–188.

Stoilova, I., A. Krastanv, A. Stoyanova. et al. 2007. Antioxidant activity of a ginger extract (*Zingiber officinale*). *Food Chemistry* 102: 764–770.

Singh, A. and N. Laishram. 2011. Properties of essential oils: A review. *Indian Perfumer* 55: 23–27.

Singla, P., A. Bardoloi and A. A. Parkash. 2010. Metabolic effects of obesity: a review. *World journal of diabetes* 1(3): 76.

Sreejayan, N. and M. N. A. Rao. 1994. Curcuminoids as potent inhibitors of lipid peroxidation. *Journal of Pharmacy and Pharmacology* 46: 1013–1016.

Srinivasan, K. 2005. Role of spices beyond food flavoring: Nutraceuticals with multiple health effects. *Food Reviews International* 21(2): 167–188.

Suneetha, W. J. and T. P. Krishnakantha. 2005. Antiplatelet activity of coriander and curry leaf spices. *Pharmaceutical Biology* 43(3): 230–233.

Teodoro, A. J. 2019. Bioactive compounds of food: their role in the prevention and treatment of diseases. *Oxidative Medicine and Cellular Longevity* 1–4.

Thangam, C. and R. Dhananjayan. 2003. Antiinflammatory potential of the seeds of Carum Copticum Linn. *Indian Journal of Pharmacology* 35(6): 388.

Tripathi, R. M., S. S. Gupta, and D. Chandra. 1973. Anti-trypsin and antihyaluronidase activity of volatile oil of Curcuma longa (Haldi). *Indian Journal of Pharmacology* 5: 260–261.

Valdés, A., A. C. Mellinas, M. Ramos. et al. 2015. Use of herbs, spices and their bioactive compounds in active food packaging. *RSC Advances* 5(50): 40324–40335.

Valdés, A., A. C. Mellinas, M. Ramos Garrigós et al. 2014. Natural additives and agricultural wastes in biopolymer formulations for food packaging. *Frontiers in Chemistry* 2: 1–10.

Vallverdú-Queralt, A., J. Regueiro, M. Martínez-Huélamo et al. 2014. A comprehensive study on the phenolic profile of widely used culinary herbs and spices: rosemary, thyme, oregano, cinnamon, cumin and bay. *Food Chemistry* 154: 299–307.

Vasanthi, H. R. and R. P. Rarameswari. 2010. Indian spices for healthy heart—An overview. *Current Cardiology Reviews* 6: 274–279.

Veluti, A., V. Sanchis, A. J. Ramos, and S. Marin. 2004. Effect of essential oils of cinnamon, clove, lemongrass, oregano and palmarosa on growth and fumonisin B1 production by *Fusarium verticillioides* in maize. *Journal of the Science of Food and Agriculture* 84(10): 1141–1146.

Vendruscolo, A., I. Takaki, L. E. Bersani-Amado. et al. 2006. Antiinflammatory and antinociceptive activities of zingiberofficinale roscoe essential oil in experimental animal models. *Indian Journal of Pharmacology* 38(1): 58.

Viuda-Martos, M., Y. Ruiz Navajas, E. Sánchez Zapata. et al. 2010. Antioxidant activity of essential oils of five spice plants widely used in a Mediterranean diet. *Flavour and Fragrance journal* 25(1): 13–19.

Viuda-Martos, M., Y. Ruiz-Navajas, J. Fernandez-Lopez. et al. 2010. Effect of added citrus fibre and spice essential oils on quality characteristics and shelf-life of mortadella. *Meat Science* 85(3): 568–76.

Wani, S. A. and Kumar, P. 2018. Fenugreek: A review on its nutraceutical properties and utilization in various food products. *Journal of the Saudi Society of Agricultural Sciences* 17(2): 97–106.

Wei, K. and X. Xu. 2003. Herbal composition and method for controlling body weight and composition. United States Patent 6541046, USA.

Williamson, D. F., J. A. R. F. Madans, R. F. Anda. et al. 1993. Recreational physical activity and ten-year weight change in a US national cohort. *International Journal of Obesity and Related Metabolic Disorders: Journal of the International Association for the Study of Obesity* 17(5): 279–286.

Xu, L., Zhang, L., Bertucci, A. M., Pope. et al. 2008. Apigenin, a dietary flavonoid, sensitizes human T cells for activation-induced cell death by inhibiting PKB/Akt and NF-κB activation pathway. *Immunology Letter* 121: 74–83.

Yashin, A., Y. Yashin, X. Xia. 2017. Antioxidant activity of spices and their impact on human health: A review. *Antioxidants* 6(3): 70.

Yesiloglu, Y., H. Aydin, and I. Kilic. 2013. In vitro antioxidant activity of various extracts of ginger (*Zingiberofficinale L.*) seed. *Asian Journal of Chemistry* 25(7): 3573–3578.

Yuan, H., M. Zhu, W. Guo. et al. 2011. Mustard seeds attenuate azoxymethane-induced colon carcinogenesis. *Redox Report* 16: 38–44.

Zheng, W. and S. Y. Wang. 2001. Antioxidant activity and phenolic compounds in selected herbs. *Journal of Agriculture and Food Chemistry* 49: 5165–5170.

Zhu, F. 2017. Encapsulation and delivery of food ingredients using starch based systems, *Food Chemistry* 229: 542–552.

Zohu, Y., X. Q. Zin, and S. Zhang. et al. 2018. Microencapsulation of spices with Methylcellulose/alginate/methylcellulose and the components stability during storage. IOP Conference Series: *Materials Science and Engineering* 382(2): 022006.

CHAPTER 2

Spices and Their Uses in Foods

Nawaz Ahmad Ganie, Sajad Ahmad Wani, Javed Ahmad Wagay, Umar Farooq, Wasim Yousuf Mir, and Asif Mohiuddin Rather

CONTENTS

2.1 INTRODUCTION

Spices are the pungent, coloured, or aromatic dried plant stuffs used mainly for seasoning the food. They are obtained from plant parts such as fruits, seeds, roots, flowers, and bark. Spices can be differentiated from herbs as dried part of plants other than leaves. Herbs could be fresh or dried leaves. They have strong flavours while herbs have subtle flavours. Spices have different origin and characteristic flavours. Spices are specially used in the preparation of food and culinary. They are used to flavour the food since ancient times. They can be used in the ground, broken or whole form to impart specific aroma, flavour, and taste to foods. The main purpose of spices in food is seasoning instead of nutrition. The abundance of phytochemicals in spices is responsible for their preservative nature in foods. They are rich sources of antioxidants and therefore have been used as emulsion extract, ground or whole spice in foods. Hence, they have been used in fatty foods to control lipid oxidation. To protect the food from degradation, off-flavour because of oxidation, utilization of spices is a need of an hour for the foods which are highly sensitive to oxidation. Spices are important sources of natural antioxidants. The following compounds are some of the natural antioxidants such as phenolics, phospholipids, carotenoids, vitamin C, and protein-based compounds (Choe and Min 2009; Wani and Kumar 2018).

DOI: 10.1201/9781003215387-2

Utilization of spices in food depends upon the properties of foods. Most of the foods are included in the diet because of understanding their certain characteristics of health, which are supported by food (Jacobs and Steffen 2003). For validating the health claim associated with food, developments are in progress through evidence-based backgrounds (Contor and Asp 2004). Currently, recommendations are given to foods, which are rich in bioactive compounds (Kris-Etherton et al. 2002), for example, spices and herbs. Gradually, scientific evidences can be expected to support the benefits of spices so far as health and disease protection is concerned.

2.2 SPICE ORIGIN AND PRODUCTION

Spices origin has been known to date back to ancient times in Egypt as recorded in the Bible as valuable trade items. The use of spices moves from Egypt through the Middle East and spread to the Mediterranean and Europe. For so many years, Arab middlemen control the spice trade. Spices trade later spread to India, China, and Indonesia and in the 17th century to America. Spices could be of indigenous or exotic origin. Indigenous species are those that actually originated in a particular region while exotic are those are imported from the other region. Some spices are of temperate plants while from are from the tropical region.

2.2.1 Indian Origin

Indians are using spices from ancient times for seasoning and culinary purposes. Indigenous spices such as turmeric and cardamom were cultivated in India in the eighteenth century in the estates of Babylon (Duke 2002). Sushruta, an ancient surgeon in about the fourth century BC, has utilized certain aromatic plants and white mustard in bed sheets to avoid malignant spirits. He has also wrapped sesame over wounds to act as antiseptic. Spices and herbs are included in the medical literature of Sushruta II in the 2nd century and Charaka.

Use of spices such as pepper, turmeric, ginger, cardamom, and cinnamon as healing agent has been noticed in Sushruta II. Spices like mustard seed, cumin, black pepper, ginger, and cardamom have been used as ancient herbal medicines. They (cardamom and cloves) have been used in ayurvedic medicines to aid in digestion.

2.2.2 Area and Production

Global largest exporter, consumer, and producer is India. They share an output of about 44% and 36% in world's spice market. During past decade, steady rise in a spice production and area in India was reported. India dominates in cardamom and is the biggest producer of turmeric and ginger leads in production and area. Out of the total spices production, Indian households consume about 70–75 per cent either in whole form or in value added form (powder or masalas), 5–10 per cent is utilized by oleoresins, pharmaceuticals and cosmetics companies, 15–20 per cent is exported and the remaining 5 per cent goes for seed purposes (Anonymous 2021).

Among the 109 varieties listed by the International Organization for Standardization, India produces about 75 varieties and makes half of the world market in spices. In the financial year 2020, India exported spices cost US$3.62 billion. However, in the financial year 2021, top importers of India were United States, United Arab Emirates, United Kingdom, Canada, Australia, and Hong Kong. From April 2020 to March 2021, total spices exported from India were US$4.00 billion and in March 2021, US$448.77 million was exported (APEDA 2021).

2.3 POTENTIAL APPLICATIONS OF SPICES IN FOOD

Spices have been used in foods as colorant, flavouring agent, and as a preservative since ancient times. Mainly spices have been used as natural flavouring agent (Shobana and Akhilender 2000), colouring agent (Ravindran et al. 2006), as antioxidant, and antimicrobial (Ceylan and Fung 2004) in foodstuff. Spices and their potential uses in various food materials are shown in Table 2.1. Mainly the application of spice is in food industry; however, they have also been used in several other industries such as medicine (Shan et al. 2007), nutraceuticals, cosmetics, and perfumery industry (Peter and Shylaja 2012).

Table 2.1 Spices and Potential Uses in Food

Spices	Food Uses
Trigonella foenum-graecum (Fenugreek)	Its green leafy vegetable and seeds are used for seasoning purpose. It has been also used in pickles, extruded products.
Cuminum cyminum (Cumin)	Its seeds are used in cuisines to add flavour.
Coriandrum sativum (Coriander)	Leaves of coriander as well as seeds are used to add flavour to foodstuff.
Allium sativum (Garlic)	Garlic has a pungent odour used for cooking purpose.
Elettaria cardamomum (Cardamom)	Cardamom is used in most of the Indian and other sweet dishes to give a good flavour and smell. It has been also used in Kashmir to produce a special tea.
Curcuma longa (Turmeric)	Turmeric has been used in several products such as beverages, cereals, milk products, bakery products, eggs, fish, meat, ice cream, confectionery, and pickles.
Piper nigrum (Pepper)	Pepper is the most widely used spice. It has been used in various types of foods as flavouring agent. In food industry because of the presence of piperine, antioxidant activity, antimicrobial terpenoids such as α-terpinene-4-ol, *p*-cymene, limonene, α-pinene, β-pinene and δ^3-carene, pepper has been used.
Myristica (Nutmeg)	Nutmeg has been used to garnishing the food in powder form and also in the preparation of a mixture of spices.
Brassica nigra (Mustard)	Mustard leaves are used in cooking, salads, its seeds are also used to add flavour to the food material. It has been used as seasoning.
Ferula assa-foetida (Asafoetida)	Asafoetida has been used as digestive aid and as spice in foods as well as in pickles. It has been widely used in Indian foods.
Zingiber officinale (Ginger)	Ginger has been used as a spice. In China leaves of ginger have been used to add flavour to food. Oil obtained from ginger has been used as flavouring agent in soft drink, in bakery items, in confectionary products, sauces, pickles. It has been used as preservatives.
Crocus sativus (Saffron)	Saffron has been used in food as a spice to add flavour, aroma, and yellow colour. It has been used in a number of foods such as cheese, chicken and meat, rice, mayonnaise, and cakes.
Syzygium aromaticum (Clove)	In India dried clove has been used in tea. It has also been used to add flavour to chewing gum, candies, cookies, pastries, meat products, pickles, curries, sandwiches, chocolate drinks, cakes and puddings. It is the common spice used in cuisines of China, India, Russia, Greece, and Scandinavia.
Cinnamomum verum (Cinnamon)	*Cinnamomum verum* has been used mostly as spice to add flavour to different varieties of foods, breakfast cereals, sweets, tea and other traditional recipes. Essential oil, cinnamaldehyde and other components including eugenol derived from the cinnamon has been found to be responsible ingredient to develop flavour and aroma.
Capsicum annuum (Chilli)	It is one of the main components used in food for the development of flavour.

Sources: Sachan et al. (2018), Lakshmi (2014), Solymosi et al. (2015), Karsha and Lakshmi (2010), Tipsrisukond et al. (1998), Kapoor et al. (2014), Nakatani et al. (1986), Shahrajabian et al. (2019), Roden (1975), Basker and Negbi (1983), Milind and Deepa (2011), and Wani and Kumar (2016), Wani et al. (2021).

Foods without spices are not preferred by consumers. Globally spices are mostly used in India. They are used extensively for cooking in India. Not only in India, but also in many other countries has been used in food as a flavouring agent and as a colorant. Some of the important properties of spices that reflect its use in foods are discussed below.

2.4 PROPERTIES OF SPICES FOR THEIR POTENTIAL APPLICATION IN FOOD

2.4.1 Flavouring Properties

Spices are used mainly for flavouring the food. They have a strong flavour and therefore a pinch of spice enough to produce flavour. Dried spices are generally used for seasoning. The main responsible flavouring ingredient of spice is the Essential oils and oleoresins. Extraction of essential oils is generally done by cold pressing, resin tapping, absolute oil extraction, solvent extraction, expression, and steam distillation. Besides used in foods, they are also used in scents, cosmetics, soaps, and other flavouring products for different end-use application (Sunder 2016). Table 2.2 shows the plant parts used for flavouring foods.

Oleoresins are the semi-solid extract from spices possessing both volatile as well as non-volatile flavour compounds. They provide mouthfeel and aroma characteristic with strong flavour. Oleoresins are generally extracted using solvent. Spices are the sources of natural flavouring and they can be used either as such or in processed form (Ganapathy 2012).

2.4.2 Colouring Properties

Spices are the natural sources of colorants. Some spices provide both colour as well as flavour to the food material such as annatto, parsley, turmeric, chilli powder, and saffron. Colouring components in spices can be either water or oil soluble. Spices owe their colouring properties because of some chemical compounds. Main compounds accountable for the food colour are β-carotene, neoxanthin, and lutin (Bartley and Scolnik 1995). Some of the distinctive colouring compounds in

Table 2.2 Plant Parts Used for Flavouring Foods

S. No.	Spices	Plant Part Used
1	*Elettaria cardamomum*	Seeds
2	*Zingiber officinale*	Rhizomes
3	*Cinnamomum verum*	Bark
4	*Aframomum danielli*	Seeds
5	*Aframomun melegueta*	Seeds
6	*Pimenta dioica*	Berry
7	*Syzygium aromaticum*	Flower buds
8	*Myristica*	Seeds, arils
9	*Piper nigrum*	Whole fruit, seed
10	*Amomum subulatum*	Pod and the seeds
11	*Trigonella foenum-graecum*	Seeds and green leaves
12	*Cuminum cyminum*	Cumin oil
13	*Carum carvi*	Oil
14	*Coriandrum sativum*	Leaves and seeds
15	*Foeniculum vulgare*	Fennel seeds
16	*Nigella sativa*	Seeds

Source: Fasoyiro (2014).

Table 2.3 Various Colouring Components of Different Spices

Spice	Colouring Component	Colour
Bixa Orellana	Bixin	Golden yellow
	Norbixin	Orange yellow
Crocus sativus	Crocin	Yellowish orange
	Crocetin	Dark red
	Beta Carotene	Reddish orange
Petroselinum crispum	Chlorophyll	Green
	Lutein	Dark red
	Neoxanthin	Orange yellow
	Violaxanthin	Orange
Curcuma longa	Curcumin	Orange yellow
Capsicum annuum	Carotenoids:	
	Capsanthin	Dark red
	Violaxanthin	Orange
	Cryptoxanthin	Red
	Capsorbin	Purplish red
	Beta Carotene	Reddish orange
	Lutein	Dark red
	Zeaxanthin	Yellow

Source: Raghavan (2006).

spices are curcumin (turmeric), bixin (annatto), capsanthin (chilli pepper), carotenoids (paprika), and crocin (saffron) (Ravindran et al. 2006). Sometimes overall colouring of food is the joint effect of two or more colouring compounds of spices.

Colouring provided by spices generally ranges from yellow, orange, and red. Spices have been used as natural colorant in foods. Using spices as a natural source of colorant in food is advantageous to avoid synthetic colorants. Table 2.3 shows the different colouring components of spices.

2.4.3 Nutritional Properties

Spices are not only used for their sensory acceptability, but they are also enriched with vitamins, minerals, antioxidants, essential oils, and phytonutrients. Spice seeds are good sources of carbohydrate, protein, and fat. They play an important role in nutrition, because of the presence of both micro and macro nutrients. Ogunka-Nnoka and Mepha (2008) conducted research on the chemical composition of some spices from Nigeria. These were shown to be good sources of nutrients. Even most spices are abundant sources of important minerals such as zinc, magnesium, phosphorus, and calcium. Spices are considered as functional foods because they provide not only potential health benefit but also are rich sources of nutrients.

2.4.4 Antioxidant Properties

Antioxidants are either synthetic or natural. They inhibit the process of oxidation (Shahidi and Zhong 2010). There is a great demand of natural antioxidants because of the safety concern of synthetic antioxidants and less utilization of food additives in foodstuffs. Spices are rich sources of natural antioxidants and thus help in food preservation by inhibiting the process of lipid oxidation in foods. One of the advantage of using spices as a source of natural antioxidant is that that are healthy and they can be easily assimilated in our digestive system. On the other hand, synthetic antioxidant can have harmful effect on the consumer's body and do not contribute to extra

Table 2.4 Antioxidant Compounds in Spices (Embuscado 2015)

Spice	Antioxidant Compounds
Clove	Phenolic acids (gallic acid), flavonol glucosides, phenolic volatile oils (eugenol, acetyl eugenol), tannins
Ginger	Shogoal, gingerol, zingerone
Nutmeg	Myristphenone, phenolic volatile oils, phenolic acid (caffeic acid), flavanols (catechin)
Red pepper	Beta-carotene fruit, beta-sitosterol plant, caffeic acid campesterol, camphene fruit, capsaicin fruit, capsanthin fruit, chlorogenic-acid fruit, eugenol fruit, gamma-terpinene fruit, hesperidin fruit, myristic acid
Turmeric	Curcumin, 4-hydroxycinnamoylmethane
Sesame seeds	Sesaminol, α-tocopherol, sesamol
Black pepper	Kaempferol, rhamnetin, quercetin
Cumin	Cuminal, γ-terpinene, pinocarveol, linalool, 1-methyl-2-(1methylethyl)benzene, carotol

Source: From various sources.

nutritional advantage. Table 2.4 shows some antioxidant components in spices. To protect the food from degradation because of oxidation, it is important to add antioxidants as one of the ingredient to the foodstuff. Spices have abundances of antioxidants and can have safe usage. Therefore, using spices in food can protect the food from degradation.

Rancidity is the undesirable flavour, appearance, and texture and nutritional loss of fat, caused by oxidation/hydrolysis of fat (Smith 1991). Antioxidant substances retard deterioration, rancidity, and discoloration (Halliwell and Gutteridge 1984).

2.4.5 Preservative Properties

Food spoilage signifies to an unalterable change to food, in which foodstuff becomes inedible and its quality is lost. Many factors have been found to responsible for such changes these include physical and biological. In spite of having modern methods available such as use of preservatives, drying, pasteurization, and freezing, it seems food spoilage hazards cannot be eradicated completely (Gutierrez et al. 2009). The most important issue of food spoilage is the lipid oxidation. Therefore, antioxidants have been used by food industries to stop food spoilage (Stoilova et al. 2007). However, as discussed above there is a safety concern with respect to the use of synthetic antioxidants. Therefore, it is highly recommended to use spices as a source of natural antioxidants (Hyldgaard et al. 2012). Spices possess antioxidant properties, primarily because of the presence of phenolic components. The antioxidant property is due to their scavenging activity, metal chelation, oxygen quenching, and increasing the activities of enzyme antioxidants (Rubió et al. 2013).

A study revealed CO_2 extract of ginger at a temperature of 37 and 80°C in inhibiting oxidation of lipid by in vitro activity as compared to butylated hydroxytoluene (Stoilova et al. 2007). Besides, black pepper and pimento extracts decrease the creation of acrylamide to about 50 and 75%, in an ideal blend of simulating potato matrix at 180°C for about 20 min. The formation of acrylamide was restricted to 50% with the eugenol, a key compound of pimento essential oil (Ciesarova et al. 2008). Several other antioxidants in different spices include marjoram ginger (gingerol), black pepper (piperine), oregano, thyme (thymol), oregano, thyme, marjoram (carvacrol), turmeric (curcumin), red chilli (capsaicin), and dill (quercetine) (Srinivasan 2014; Przygodzka et al. 2014; Rubió et al. 2013). Spices have been reported to have antimicrobial activity. Food spoilage can be also due to microbial growth. The antimicrobial activity of spices has been found to retard/delay food spoilage by inhibiting the growth of microorganisms as well as pathogenic microorganisms (Tajkarimi et al. 2010). Hence, preserves the food.

2.5 CONCLUSION

Consumer's interest for using natural ingredients in foodstuff has been increasing due to the health concern for synthetic ingredients. Beneficial health effects of spices are encouraging them to be used as an important ingredient of our daily diet, beyond their main job of providing flavour and taste to our daily food. Intake of spices has been proved to be safe. The use of spices or their extracts in the development of a functional food product should be supported. The detrimental effects of lipid oxidation on foodstuff make the food undesirable and inedible to be used for consumption. With respect to this, spices have been used as colouring, flavouring, and preservative agents since time immemorial. They possess powerful antioxidants, which inhibits the processing of lipid oxidation or reduces the inception of rancidity in foodstuffs.

REFERENCES

Anonymous, 2021. Present status of production in India, www.commodityindia.com. (Accessed August, 2021)

APEDA 2021. Agricultural and Processed Food Products Export Development Authority. https://apeda.gov.in/apedawebsite/ (Accessed September, 2021)

Bartley, G. E., Scolnik, P. A. 1995. Plant carotenoids: Pigments for photoprotection, visual attraction, and human health. *Plant Cell* 7 (7): 1027–1038.

Basker, D., Negbi, M. 1983. Uses of saffron. *Economic Botany* 37 (2), pp. 228–236.

Ceylan, E., Fung, D. Y. C. 2004. Antimicrobial activity of spices. *Journal of Rapid Methods and Automation in Microbiology* 12: 1–55.

Choe, E., Min, D. B. 2009. Mechanisms of antioxidants in the oxidation of foods. *Comprehensive Reviews in Food Science and Food Safety* 8: 345–358.

Ciesarova, Z., Suhaj, M., Horváthová, J. 2008. Correlation between acrylamide contents and antioxidant capacities of spice extracts in a model potato matrix. *Journal of Food Nutrition and Research* 47: 1–5.

Contor, L., Asp, N. G. 2004. Process for the assessment of scientific support for claims on foods (PASSCLAIM) phase two: moving forward. *European Journal of Nutrition* 43 (2): ii3–ii6.

Duke, J. A. 2002. *Handbook of Medicinal Spices*. Boca Raton: CRC Press/Taylor & Francis.

Embuscado, M. E. 2015. Spices and herbs: Natural sources of antioxidants – a mini review, *Journal of Functional Foods*. doi: 10.1016/j.jff.2015.03.005

Fasoyiro, S. B. 2014. The value and utilization of spice plants in tropical Africa. *Journal of Agricultural and Food Information* 15: 109–120.

Ganapathy, V. 2012. Spice Oleoresins. Food and Beverage Magazine.

Gutierrez, J., Barry-Ryan, C., Bourke, P. 2009. Antimicrobial activity of plant essential oils using food model media: efficacy, synergistic potential and interaction with food components. *Food Microbiology* 26, 142–150. doi: 10.1016/j.fm.2008.10.008

Halliwell, B., Gutteridge, J. M. C. 1984. Oxygen toxicity, oxygen radicals, transition metal and diseases. *Biochemistry Journal* 219, 1–14.

Hyldgaard, M., Mygind, T., Meyer, R. L. 2012. Essential oils in food preservation: mode of action, synergies, and interactions with food matrix components. *Frontiers in Microbiology* 3: 12. doi: 10.3389/fmicb.2012. 00012

Jacobs, D. R., Steffen, L. M. 2003. Nutrients, foods, and dietary patterns as exposures in research: a framework for food synergy. *American Journal of Clinical Nutrition* 78: 508S–513S.

Kapoor, I. P. S., Singh, B., Singh, S., Singh, G., 2014. Essential oil and oleoresins of black pepper as natural food preservatives for orange juice. *Journal of Food Process and Preservation* 38 (1): 146–152.

Karsha, P. V., Lakshmi, O. B. 2010. Antibacterial activity of black pepper (Piper nigrum Linn.) with special reference to its mode of action on bacteria. *Indian Journal of Natural Product and Resources* 1 (2): 213–215.

Kris-Etherton, P. M., Hecker, K. D., Bonanome, A., et al. 2002. Bioactive compounds in foods: their role in the prevention of cardiovascular disease and cancer. *The American Journal of Medicine* 113 (Suppl 9B): 71S–88S.

Lakshmi, G. 2014. Food coloring: the natural way. *Research Journal of Chemical Science* 4: 87–96.

Milind, P., Deepa, K. 2011. Clove: A champion spice. *International Journal of Research in Ayurveda & Pharmacy* 2 (1): 47–54.

Nakatani, N., Inatani, R., Ohta, H., Nishioka, A., 1986. Chemical constituents of peppers (Piper spp.) and application to food preservation: naturally occurring antioxidative compounds. *Environmental Health Perspectives* 67: 135–142.

Ogunka-Nnoka, C. U., Mepha, H. D. 2008. Proximate composition and anti-nutrient contents of some common spices in Nigeria. *Food Science Journal* 2: 62–67.

Peter, K. V., Shylaja, M. R. 2012. Introduction to herbs and spices: Definitions, trade and applications. In: *The Handbook of Herbs and Spices*, 2nd ed., vol. 1, pp. 1–24. Woodhead Publishing, Cambridge, UK.

Przygodzka, M., Zielinska, D., Ciesarová, Z., Kukurová, K., Zielinski, H. 2014. Comparison of methods for evaluation of the antioxidant capacity and phenolic compounds in common spices. *LWT-Food Science and Technology* 58: 1–6. doi: 10.1016/j.lwt.2013.09.019

Raghavan, S. 2006. *Forms, Functions, and Applications of Spices from: Handbook of Spices, Seasonings, and Flavorings*. Boca Raton: CRC Press/Taylor & Francis.

Ravindran, P. N., Nirmal Babu, K., Shiva, K. N. 2006. Genetic resources of spices and their conservation. In: *Advances in Spices Research*, ed. P. N. Ravindran, K. Nirmal Babu, K. N. Shiva, and K. Johny, 63–91. Agrobios.

Roden, C. 1975. *A Book of Middle Eastern Food. Penguin*, Harmondsworth, Middlesex.

Rubió, L., Motilva, M. J., Romero, M. P. 2013. Recent advances in biologically active compounds in herbs and spices: a review of the most effective antioxidant and anti-inflammatory active principles. *Critical Reviews in Food Science and Nutrition* 53: 943–953. doi: 10.1080/10408398.2011.574802

Sachan, A. K. R., Kumar, S., Kumari, K., Singh, D. 2018. Medicinal uses of spices used in our traditional culture: Worldwide. *Journal of Medicinal Plants Studies* 6 (3): 116–122.

Shahidi, F., Zhong, Y. 2010. Novel antioxidants in food preservation and health promotion. *European Journal of Lipid Science and Technology* 112: 930–940.

Shahrajabian, M. H., Sun, W., Cheng, Q. 2019. Pharmacological uses and health benefits of ginger (Zingiber officinale) in traditional Asian and ancient Chinese medicine, and modern practice. *Notulae Scientia Biologocae* 11 (3): 309–319. doi: 10.15835/nsb11310419

Shan, B., Cai, Y., Brooks, J. D., Corke, H. 2007. The in vitro antibacterial activity of dietary spice and medicinal herb extracts. *International Journal of Food Microbiology* 117: 112–119.

Shobana, S., Akhilender, K. 2000. Antioxidant activity of selected Indian spices. Prostagladinins, *Leukotrienes and Essential Fatty Acids* 62: 107–110.

Smith, J. 1991. *Food Additives Users Handbook*. Academic Press, Glasgow. pp 5–46.

Solymosi, K., Latruffe, N., Morant-Manceau, A., Schoefs, B. 2015. "1 – Food colour additives of natural origin," In *Colour Additives for Foods and Beverages*. ed. M. J. Scotter, 3–34, Oxford: Woodhead Publishing.

Srinivasan, K. 2014. Antioxidant potential of spices and their active constituents. *Critical Review in Food Science and Nutrition* 54: 352–372. doi: 10.1080/10408398.2011.585525

Stoilova, I., Krastanov, A., Stoyanova, A., Denev, P., Gargova, S. 2007. Antioxidant activity of a ginger extract (*Zingiber officinale*). *Food Chemistry* 102: 764–770. doi: 10.1016/j.foodchem.2006.06.023

Sunder, S. 2016. *Spices as Flavouring Substances*. LinkedIn Corporation.

Tajkarimi, M. M., Ibrahim, S. A., Cliver, D. O. 2010. Antimicrobial herb and spice compounds in food. *Food Control* 21, 1199–1218. doi: 10.1016/j.foodcont.2010.02.003

Tipsrisukond, N., Fernando, L. N., Clarke, A. D., 1998. Antioxidant effects of essential oil and oleoresin of black pepper from supercritical carbon dioxide extractions in ground pork. *Journal of the Agriculture and Food Chemistry* 46 (10): 4329–4333.

Wani, S. A., Kumar, P. 2018. Fenugreek: A review on its nutraceutical properties and utilization in various food products. *Journal of the Saudi Society of Agricultural Sciences* 17 (2): 97–106.

Wani, S. A., Alwahibi, M. S., Elshikh, M. S. et al. 2021. Sensory, functional characteristics and in vitro digestibility of snacks supplemented with non-traditional ingredient raw and processed fenugreek. *International Journal of Food Science and Technology*. https://doi.org/10.1111/ijfs.15441

Wani, S. A., Kumar, P. 2016. Development and parameter optimization of health promising extrudate based on fenugreek oat and pea. *Food Bioscience* 14 (1): 34–40.

CHAPTER **3**

Bioactive Compounds in Garlic

Jasmeet Kour, Kawaljeet Kour, Vikas Bansal, Sangeeta, Renu Sharma, Kanchan Sharma, Mohamed S. Elshikh, and Mona S. Alwahibi

CONTENTS

3.1 INTRODUCTION

Garlic (*Allium sativum L.*) is considered as one of the most significant vegetables across the globe credited with the total harvesting area of 1.437.690 ha with around 24.255.303 tonnes of dry bulbs being produced on an annual basis (FAO 2013). Hailing from the family *Liliaceae*, it is abundantly rich in sulphur compounds imparting characteristic flavour and taste yielding enormous beneficial effects comprising of water (65%), carbohydrate (28%), 2.3%, organosulfur compounds (2%), arginine (1.2%), and fibre (1.5%) as well (Santhosha et al. 2013). Garlic is an herb pertaining to the genus and family of *Allium* and *Liliaceae*, respectively with the bulbs comprising of 6–35 smaller bulblets (Kuete 2017). Figure 3.1 shows the garlic bulb. Tsao et al. (2003) reported that fresh garlic and vacuum dried garlic powder were effective against *Staphylococcus aureus, Streptococcus, Streptococcum hemolyticus, Llebsiella pneumonae, Proteus vulgaris, E. coli, Bacillus subtilis, Salmonella Enteritidis, B. mycoides, and Serratia marcescens.*

Various studies have reported garlic possessing strong antioxidant properties with one of these evaluating the antioxidant value of both raw and cooked form with this finding that the raw garlic was found to reveal stronger antioxidant activity (1,1-diphenyl-2-picrilhydrazyl (DPPH) radical scavenging assay, 2,2′-Azino-bis(3-ethyl benzothiazoline-6-sulfonic acid) (ABTS) radical scavenging assay, and ferric ion reducing antioxidant power (FRAP) assay) (Martins et al. 2016). Garlic (*Allium sativum L.*) is a common spice loaded with manifold bioactives with respect to organic

DOI: 10.1201/9781003215387-3

Figure 3.1 Garlic bulb.

sulfides, phenolics, and polysaccharides exhibiting various health benefits (Diretto et al. 2017). There is a synergistic relationship between various compounds in garlic and garlic's chemical complexity yielding preparations with varying efficacy and safety (Amagase 2006).

3.2 BIOACTIVE COMPOUNDS IN GARLIC

The extraction of bioactive compounds from various agricultural sources depends upon the effectiveness and efficiency of the method chosen for the extraction of a particular compound. Conventional methods of extraction are time-consuming and are more costly than novel ones (solvent extraction, maceration, hydro or steam distillation, squeezing or cold pressing, etc.). The former ones need high quality solvent and in large amounts and have low extraction rates (Azmir et al. 2013). On the other hand, novel extraction methods such as ultrasound-assisted extraction (UAE), enzyme-assisted extraction (EAE), and microwave-assisted extraction (MAC) are widely used now-a-days for extraction of bioactive compounds as these are fast and efficient (Tiwari 2015; Aires 2017; Chemat and Khan 2011). Various other novel extraction methods used are microwave-ultrasonic-assisted enzymatic extraction method (MUAEE), ultrasonic microwave-assisted extraction (UMAE), microwave assisted-enzymatic extraction (MAEE), and ultrasound-assisted enzymatic extraction (UAEE). MUAEE is a combination of three extraction techniques (UAE, MAE, and EAE) employed successively or simultaneously (Yin et al. 2018; Wu et al. 2015). Similarly, UAEE is a combination of EAE and UAE, and MAEE is a combination of EAE and MAE extraction methods (Cheng et al. 2015).

The first isolation and characterization of allicin was performed in 1940s by Cavallito and Bailey. They used 95% ethanol for extraction and the product obtained was colourless and unstable oil with 2.5% solubility in water. The product thus obtained was allicin. The study also reported its antibiotic activity against gram-positive and gram-negative bacteria (Cavallito and Bailey 1944). Alliin gets converted into allicin on enzymatic reaction in the presence of enzyme allinase. The highest content of allicin was reported in freshly crushed garlic (Koca and Tasci 2016) having its main metabolites, diallyl sulphide, diallyl disulphide, diallyl trisulfide, tertrasulfides, and sulphur dioxide. Analysis of thiosulfinates was done (Leontiev et al. 2018) by HPLC method using a Jasco HPLC system with a UV-2077 multichannel detector and a PU-980 pump rontosil Kromaplum column (150 × 4.6 nm, 5 μm) at 25°C was used for the separation and the UV detector used was operated 254 nm. Flow rate was 1 ml/min. The mobile phase consisted of water (A) and methanol. The gradient used was: 56% A (0 min), 53% A (10 min), 7% A (15 min), 7% A (30 min), 56% A (31 min), 56% (35 min). The study resulted that most of the thiosulfinates tested were more stable than allicin. It was also observed that allicin when kept in open for 1 hr showed its activity by diffusion through the gas phase. Also the thiosulfinated tested in the study showed higher antimicotic activity than antibacterial activity.

Cavalittto and Bailey (1944) isolated allicin from ground garlic bulbs (4 kg) treated with 5 lt. of 95% ethanol stirred for 30 minutes. The mixture was then filtered by suction method which yields 5200 cc of solution. The extract was then concentrated under reduced pressure (15–20 mm) till most of the alcohol got removed. The distillate obtained was discarded and the distillation process was continued at 10–15 mm pressure. The aqueous distillate was collected while maintaining the volume in the distillation flask up to 500 cc by addition of water from a dropping funnel. The process was continued till the final solution contained less than 10 units per cc of active principle compound. In the end about 9 lt. of aqueous distillate was obtained. This is followed by extraction of one-third of the product obtained with 500 cc of ethyl ether. Then extraction was done with 4 times with ether (300 cc). The ether extracts thus obtained were mixed with the ether removed under reduced pressure. The residue collected was composed of water and oil, which was then shaked thoroughly with 250 cc of water and 10 cc of Skelly solve B. The aqueous solution obtained after separation and filtration was frozen and stored in dry ice after separation and filtration, until desired for future work. The pure product thus obtained was isolated and extracted from aqueous concentrate four times, each with one-fifth the volume of ether. The ether extract was mixed and cooled in a dry ice-bath and ice-crystals obtained were filtered. The ether present in it was then removed under reduced pressure and the residual oil was exposed to vacuum (0.5 mm) or less for 30 min at room temperature. Approximate 6 gm of the oil was obtained from 4 kg of garlic.

Bat-Chen et al. (2013) extracted allicin from freshly peeled garlic cloves (10 g) crushed with 100 ml of water and 10 g NaCl for 5 min. followed by the addition of diethyl ether (100 ml) with vigorous mixing. The solution is then left to stand for 15 min. until two phases appeared. The organic phase present was separated by evaporation under reduced pressure at room temperature. Double distilled water is then added to the residue and was mixed followed by the filtration of solution using Whatman no. 1 filter paper. The purified allicin solution was then stored at 4°C.

Chen et al. (2017) prepared garlic extract from 10 gm of raw garlic using crushing method with deionized water (20 ml) at different pH values (3.0, 5.0, 7.0, 9.0, and 10.7 adjusted with buffer), methanol or ethanol and kept for 2 hr at room temperature followed by centrifugation at 10,000 rpm for 20 mon at 4°C, and filtration of supernatant through membrane filters (0.22 μm). It was then stored at 4°C in dark until required for further use. To identify antioxidants in aged garlic extract (AGE) Wang et al. (2016) conducted a study using Gas Chromatography-Mass Spectrometry and Liquid Chromatography-Mass Spectrometry. A new method of formulation of AGE was developed having high antioxidant activity compared to that formed by the common commercial method. Sliced garlic was soaked in 10% ethanol solution for 90 days at room temperature followed by filtration and concentration at 40°C under reduced pressure. AGE was dissolved in distilled water and fractionated by petroleum ether. Followed by fractionation with ethyl acetate and n-butanol successively. HPLC and Gas chromatography-mass spectrometry was employed to analyse the antioxidant activity of AGE. This novel method took less time than the commercial process of preparation of AGE. Ikram et al. (2018) also conducted a similar study and used gas chromatography-mass spectrometry for characterisation of bioactives compounds present in garlic. Another study conducted isolation of allicin using Agilent Technologies 1200 series device. Zorbax XDB-C-18 (6.2 × 150 mm, 5 μm) column was used to done the separations. The mobile phase consisted of water and acetonitrile in the ratio of 80:20, v/v having flow rate 4 ml/min and injection volume 70 μl at temperature 25°C (Ilic et al. 2012).

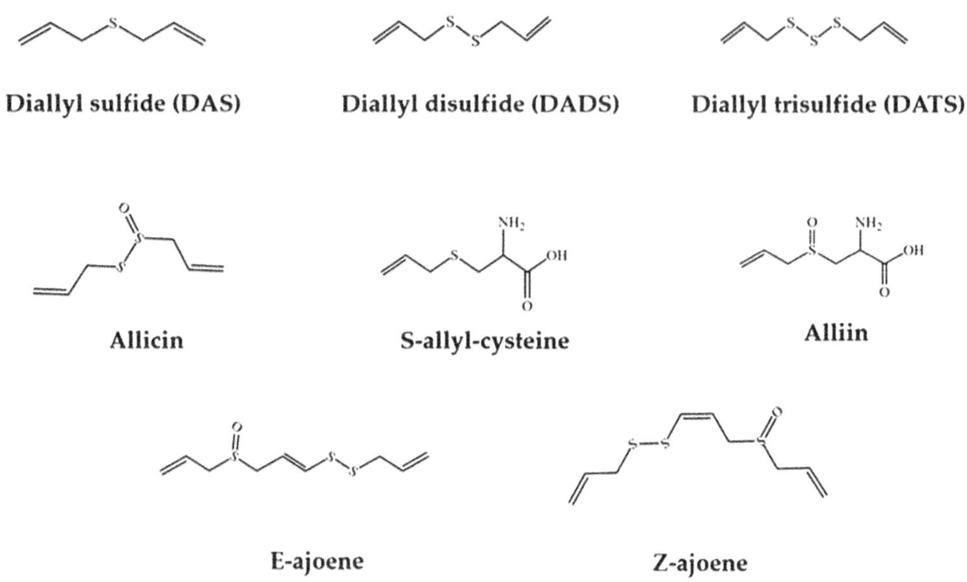

Figure 3.2 The chemical structures of the main organosulfur compounds in garlic.

In a comparative study of organosulfur compounds and antioxidant activity of Allium species Kim et al. (2018) used ORAC (oxygen radical absorbance capacity) and DPPH and ABTS assay to evaluate its antioxidant capacities. In a recent study conducted by Zaini et al. (2020) reviewed various methods of extraction of allicin from garlic. It was reported that common extraction techniques (UAE-ultrasonic assisted extraction, PLE- pressurized liquid extraction, supercritical carbon dioxide method and solvent extraction, etc.) of allicin has many drawbacks since these require organic solvent, longer times and also require enzymatic extraction process. The product obtained from the solvent extraction method was low in purity and also possess toxicity. It was suggested that SWE (sub critical water extraction) technique has better options for extraction of allicin against the conventional methods as it yields product with high purity and uses water in place of organic solvent. In SWE, both the enzymatic and extraction processes occur in one go.

Li et al. (2021) reviewed the characterization of garlic and its related species from available literature and found that many researchers performed characterisation of allicin using HPLC-UV and UPLC-UV detectors methods whereas few of them used Infrared spectroscopy (Wang et al. 2013; Thomas and Parkin 1994). Other methods like supercritical fluid chromatography-mass spectrometry, HPLC with diode array detection was also used for this purpose (Rybak et al. 2004; Jiang et al. 2014). It was also provided that besides these methods HPLC-multiple mass spectrometry without derivatization (Arnault et al. 2003; Horie and Yamashita 2006), capillary electrophoresis (CE), high performance thin layer chromatography (HPTLC) and raman spectroscopy has also been used after characterisation of allin and allicin compounds from garlic (Wang et al. 2013). Figure 3.2 shows the structure of main organosulphur compounds in garlic.

3.3 ENCAPSULATION OF THE GARLIC BIOACTIVE COMPOUNDS

Encapsulation is the process of coating or entrapping either one or a mixture of compounds within another carrier material (Saifullah et al. 2019). The entrapped/coating components are called as actives, fill, core material, payload, or internal phase, whereas carrier materials are commonly known as wall material, carrier, shell, or membrane. Encapsulation in case of garlic and other spices is mainly used for the stabilization of their bioactive components and to protect them against various damaging factors (environmental) like light, extreme pH, oxygen, moisture content, high shear mixer, and heat; or unfavourable digestive tract conditions like stomach acidic conditions, intestine pH (near neutral) (Shishir et al. 2018), etc. Encapsulation allows the bioactive compounds to achieve appropriate concentration effect at the site of action; maintains their (bioactive complex) chemical integrity or cell viability; and prevents chemical modification of active substance during storage, prevent harmful/undesired compounds formation (Garti and McClements 2012).

In a broad term, three main categories are reported for encapsulation techniques: physicochemical (ionic gelation, coacervation, liposome encapsulation/emulsification); mechanical (spray drying/cooling/chilling, fluid bed coating, supercritical fluid and extrusion); and chemical techniques (interfacial/in situ polymerization and molecular inclusion) (Majeed et al. 2015; Marisa Ribeiro et al. 2019). Among these all encapsulation techniques, the most widely used methods for encapsulation of garlic bioactive compounds are: molecular inclusion, complex coacervation, spray drying, nanoliposomes, nanoemulsions, and nanophytosomes (Tavares, Santos and Norena 2021).

3.3.1 Molecular Inclusion (Cyclodextrins)

Encapsulation of bioactive compounds by molecular inclusion using wall material cyclodextrins has various advantages like trapping of poorly soluble, temperature/chemically sensitive bioactive components, improving antioxidant properties and many more (Yu et al. 2019; Ozkan et al. 2019) but has some limitations too, like high cost, segregation and low payload problems, etc. (Estevinho and Rocha 2017). In spite of all these issues, successful use of cyclodextrins has been reported for the encapsulation of garlic bioactive compounds. Tavares and Norena (2021) have reported the successful use of β-cyclodextrin as wall materials in the encapsulation of garlic essential oil. Encapsulation of garlic essential oil to retain its organosulphurous compounds has been done by using different cyclodextrins (γ, α, and β) as wall materials and γ-cyclodextrin was reported to be the most effective wall material (Yu et al. 2019). Garlic essential oil microparticles obtained by using β-cyclodextrin has been reported to be as potent pesticide against *Aedesalbopictus* (mosquito) (Junnila et al. 2015) and cyclodextrin-based garlic essential oil microparticles have excellent inhibition capacity of *Staphylococcus aureus* (Piletti et al. 2019).

3.3.2 Lipid-Based Nano-Encapsulation

Garlic bioactive compounds have been reported to be encapsulated by various methods like nanoemulsions, nanophytosomes, and nanoliposomes (Fernandes et al. 2021). Garlic extract was encapsulated by using phosphatidylcholine/oleic acid mixture (liposomes with bilayer) and reported to have antifungal and inhibition activity against various fungi (Pinilla, Thys and Brandelli 2019). Garlic extract has been encapsulated by phosphatidylcholine nanovesicles, and the resulted nanoliposome exhibited antimicrobial activity against different *Listerial strains* and reported to have biopreservative potential in food preservation. Garlic extract has been encapsulated by using liposome bilayer (phosphatidylcholine) and reported to have antimicrobial effect (against *Listeria monocytogenes, Escherichia coli, Salmonella enteritidis*, and *Staphylococcus aureus*) (Pinilla and Brandelli 2016). Nazari et al. (2019) encapsulated garlic essential oil in nanophytosomes, and reported its potential to have natural preservative activity (antimicrobial) in samples of yogurt. Raeisi et al. (2019) reported that nanocapsules of garlic essential oil have potential use in food as well as pharmaceutical industries because of their physicochemical and thermal stability. Raeisi et al. (2021) reported that nanocapsules of garlic essential oil and fish oil can be used for fortification and shelf life extension of products.

3.3.3 Coacervation

Simple and complex coacervation are the two methods of encapsulation using coacervation (Saifullah et al. 2019). Various studies reported the use of complex coacervation for garlic extract encapsulation followed by spray drying (Tavares and Norena 2019) or freeze drying (Tavares et al. 2019) and reported their use as wall materials to protect bioactive compounds of garlic extract. The microparticles obtained by freeze drying were reported to have high water solubility, amorphous nature, mesopores/macropores on their surface, and high phenolics, which enables their use in food products application (Tavares and Norena 2021).

3.3.4 Spray Drying

Various researchers reported that spray drying is the common method of encapsulation but has various limitations (Carvalho et al. 2016; Tavares and Norena 2019). The spray dried garlic extract encapsulated powder has been reported to have indentation and collapse on the surface may be due to contraction/shrinkage of the particles because of quick drying, and also have volatile loss (Tavares and Norena 2019); concavities and thiosulfinate degradation have also been reported in spray dried encapsulated garlic extract powder (Kinalski and Norena 2019). Among the all techniques of encapsulation, spray drying has been reported to the least suitable method of encapsulation for garlic bioactive compounds may be because of the high operating temperature of spray-drying process which degrades the heat sensitive bioactive components. The freeze-drying method reported to a good alternative over spray drying as it found to be protective for heat sensitive compounds of garlic extract and maintaining their biological activity.

However, the available data on functional foods based on encapsulated product of bioactive compounds from garlic is very limited but various studies demonstrate the potential use encapsulated garlic products in various food applications. The use of garlic essential oil-based nanophytosomes in yogurt has no significant effects on its textural and colour properties as reported by Nazari et al. (2019). Bread fortified with microcapsules of garlic essential oil was reported to have better sensory and physical characteristics than normal garlic bread (Narsaiah et al. 2019). Bread fortified with garlic essential oil and flax seed-based microcapsules was reported to have various health beneficial effects (Kairam, Kandi and Sharma 2021). Various studies demonstrated that the use of garlic essential oil and extract in encapsulated form (nanoparticles) have shelf-life extension, sensory quality improving, food preservation, antimicrobial, antibacterial, antifungal and in turn bio-preserving effect in various food products (Raeisi et al. 2021; Pinilla and Brandelli 2016). However, the encapsulation of garlic essential oil and garlic extract have been demonstrated by different studies, but the selection of suitable encapsulation technique and proper wall material selection still have many challenges ahead to produce biologically functional nano or micro particles without affecting nutritional value and enable their best possible use in different food products with different purposes.

Garlic bioactive compounds in encapsulated form when incorporated as functional food components have reported to have various health promoting effects and improve the performance of human health and prevent various life threatening disease. So the use and demand of products enriched with encapsulated garlic bioactive is regularly increasing in food industry.

3.4 UTILIZATION OF GARLIC

Garlic (*A. sativum*) has been recognized for its medicinal food and mystical values since ancient times. Various researchers have studied the meditative value of garlic for treating several diseases (Valente et al. 2014). In the present days, it is widely used as flavouring agent and food preservative.

This valued culinary spice is utilized as one of the ingredients in cooking various vegetables and meat dishes as it improves the aroma and flavour of foods. The various processed forms of garlic including dehydrated garlic powder, essential oil, aged garlic extract, oil macerate can be used as an ingredient in food processing (Wiczkowski 2011). The flavour of foods and beverages is reported to be improved with the incorporation of garlic powder, fresh garlic, and garlic oil (Cupp 1999). The garlic by-products can be obtained through various processes such as low temperature or freeze drying, distillation, maceration in oil, hydroalcoholic short extraction, and hydroalchoholic long maceration (Staba et al. 2001; Wiczkowski 2011). The waste products including stem, peels, straw, and unwanted bulbs obtained during garlic harvesting and processing can be employed as source of antioxidants, for purification of water as well as for production of bioenergy (Kallel et al. 2014; Kallel and Chaabouni 2017).

The waste obtained from garlic in raw or processed form can be utilized as biosorbent. The potential of garlic peels to remove methylene blue from aqueous solutions was evaluated by Hameed and Ahmad (2009). They presented that there is increase in adsorption of methylene blue on increasing pH from 4 to 6 while there is no change in adsorption in the pH range of 6 to 12. The utilization of acid treated garlic peels as a sorbent medium for removal of quinolone antibiotics excreted by animals and human beings and have negative impact on human wellbeing and aquatic ecosystems was presented by Zhao et al. (2017). The adsorption capacity of garlic peels treated with nitric acid showed higher adsorption efficiency in comparison to untreated peels. The greater affinity of acid modified peels for quinolone antibiotics might be attributed to the increase in acidic oxygen groups.

Garlic (*Allium sativum* L.) is recognized for its various pharmacological attributes such as antiarthritic, antimicrobial, antitumor, antithrombotic, hypolipidemic, and hypoglycaemic activities. Literature revealed that active metabolites contained in garlic are known for killing of cancerous cells and thus provides strong protection against cancer. The occurrence of various bioactive molecules such as diallyl trisulfide, allicin, diallyl disulfide, diallyl sulfide, and allyl mercaptan with anticancer activity was reported by many researchers. The impact of various products derived from garlic, their phytoconstituents and nanoformulations on skin, ovarian, breast, gastric, colorectal, oral, liver, and pancreatic cancers have been evaluated by Mondal et al. (2021). The current study explores the potential of various phytoconstituents contained in garlic to inhibit the development of different stages of cancer as well as the potential of utilizing them as potent anticancer pharmaceuticals.

Since thousand years, garlic is used to provide protection against various infections such as deafness, earaches, diarrhoea, leprosy, constipation and parasitic diseases. In Europe and Indian, garlic incorporated medicines have been widely utilized to cure stomach ache, fever, cold, hay fever, cough, and asthma. Literature showed the potency of garlic to prevent cardiovascular diseases also (Tattelman 2005). Several herbalists suggested the therapeutic potential of garlic oil to treat ear infections (Kemper 2000). Garlic on mixing with honey can be used against rheumatism. It can also be applied externally for the treatment of ringworm. Asthma can be treated by boiling garlic with sugar and vinegar. Garlic possesses several properties for treating skin problems. The antioxidants present in garlic help the people to stay young by replenishment of skin and tissues. It is also beneficial in the treatment of acne and pimples.

Pang et al. (2014) utilized wastewater produced by garlic processing in a lotus pond. Lotus pond wetlands possess significant capability to eradicate organic pollutants and reduce chemical oxygen demand, biochemical oxygen demand and substances in suspended form in garlic processing wastewater. The study showed stability of pH of lotus pond on discharge of wastewater. The water quality in the lotus pond garlic processing wastewater irrigation did not pollute the pond sediment and did not produce undesirable effect on the growth of lotus roots. The lotus roots quality as well as output was improved by 3.0–8.3% with the utilization of garlic processing wastewater.

The antimicrobial potential of garlic peel was studied by Ifesan et al. (2014). Results of the study indicated that garlic extracts, when inoculated on cooked beef, significantly reduced the number of

Staphylococcus aureus, *Escherichia coli,* and *Proteus vulgaris*. The antibacterial properties might be due to the interaction of garlic sulphur compounds naming allicin, with thiolfunctional groups of microbial enzymes such as trypsin and other proteases (Bakri and Douglas 2005). The antimicrobial activity of garlic extracts against gram-positive bacteria was also tested by Kallel et al. (2014). In their study, phenolic substances were obtained in good yield from garlic skins using ethanol or water solvents individually or in combination of methanol/water (50:50), and an ethanol/water mixture (50:50) for 45 min at room temperature. Due to polar nature and higher solubility for phenolic components, methanol was considered to be good solvent for extracting phenolic compounds. The methanolic extracts and its mixture with water (50/50 methanol/water) curbed the growth of *Pseudomonas aeruginosa* and *Klebsiella pneumonia*. The other uses of garlic processing wastes might include the production of dietary fibres, cellulose, sugars, microfibers, and nanocrystals (El Mashad et al. 2019).

3.5 HEALTH BENEFITS OF GARLIC

There is an ever ending dependence of human beings on medicinal plants which have the potential to combat various diseases due to their therapeutic value (Ramaa et al. 2006). Garlic is anticancerous owing to its ability to accumulate selenium resisting cancer as well as essential for the production of antioxidant enzyme i.e. glutathione peroxidase (Santhosha et al. 2013). It has widespread applications by mankind in different cultures since years in food industry as a principal condiment and as folk medicine apart from exhibiting various prominent biological activities owing to its rich content (Ugwu and Suru 2016). Table 3.1 shows the potential health benefits of garlic and its components.

Table 3.1 Potential Health Benefits of Garlic and Its Components (Shang et al. 2019)

Physiological Benefit	Chemical Components	Main Effects	References
Antioxidant Activity	Saponins	Guarding cells from oxidative stress	Liu et al. (2018)
Anti-inflammatory Activity	Ethyl linoleate	Decreasing the production of NO and prostaglandin E-2	Park et al. (2014)
	Garlic 14-kDa protein	Obstructing the inflammatory mediators for example IL-1β TNF-α and NO	Zamani Taghizadeh Rabe et al. (2015)
	Allicin	Schistosome infection induced inflammatory reaction is being guarded	Metwally et al. (2018)
		Improving osteoarthritis	Dehghani et al. (2018)
Antimicrobial Activity	Allicin	Obstructing bacteria growth	Fratianni et al. (2016)
	Garlic oil	Obstructing bacteria growth	Guo (2014)
		In the stomach the *Helicobacter pylori* is being inhibited	Zardast et al. (2016)
Modulating Immune System	Polysaccharides/ Fructan	Employing immunomodulatory Effect	Li et al. (2017)
	Garlic oil	T-cell subtype CD4+ and serum total immunoglobulin concentration of rats are being normalized T-helper 1/ T-helper 2 response are being stabilized by the combination of levamisole and garlic oil	Hassouna et al. (2015), Mohamed et al. (2016)

(Continued)

Table 3.1 Potential Health Benefits of Garlic and Its Components (Shang et al. 2019) (*Continued*)

Physiological Benefit	Chemical Components	Main Effects	References
Cardiovascular Protection	Polyphenols	Myocardial arteries are relaxed just before and after ischemia-reperfusion in rat Stopping the reduction in heart contractility	García-Villalón et al. (2016)
	Alliin	The activity of captopril on preventing ACE and high blood pressure is increased	Asdaq and Inamdar (2010)
		The cholesterol and lipid metabolism is being controlled	Ha et al. (2015)
	Allyl methyl sulfide/Allyl methyl sulfoxide	Decrease heart hypertrophy remodelling induced by isoproterenol	Khatua et al. (2017)
Anticancer Activity	Lipid bioactive compounds	Cell proliferation is being prevented Inducing cell cycle arrest Inducing apoptosis	Bagul et al. (2015)
	Allicin	Cell proliferation is being prevented	Mansingh et al. (2018)
	Z-ajoene	Population growth of cancer stem cells is being inhibited	Jung et al. (2014)
	SAC	Cell proliferation is being prevented Stopping the Stimulation of G1/S-phase cell cycle Increasing apoptosis Cell migration is being reduced	Xu et al. (2014)
	Alliin	Cell apoptosis is being regulated	Mansingh et al. (2018)
		Tumour growth is being inhibited	Tabari and Ebrahimpour (2014)
Hepatoprotective Activity	Garlic oil	Guarding liver	Ko et al. (2016)
	DADS	Guarding mice from the liver diseases caused by long-term high-fat food	Lai et al. (2014)
	LAFGE	Without adverse effects, the levels of gamamyl glutamyl transpeptias and alanine aminotransferase are improved	Kim et al. (2017)
Digestive System Protection	Allicin	The ulcerative colitis were decreased, which were induced by dextran sulphate sodium	Shi et al. (2017)
Anti-Diabetic Activity	LAFGE	Weight loss was observed Mesenteric adipose tissue mass, etroperitoneal and epididymal were decreased	Lee et al. (2016)
	Garlic oil	Stabilizing the effect of high-fat food on the adipose tissue weight and body mass	Yang et al. (2018)
Neuroprotection	FruArg	Decreasing neuroinflammation	Zhou et al. (2014)
	Z-ajoene	Stopping I/R-induced delayed neuronal death and gliosis region of the hippocampus	Yoo et al. (2014)
Renal Protection	DATS	Guarding the kidney from the damage caused by oxidation induced by As	Miltonprabu et al. (2017)

3.6 CONCLUSION

Herbs and spices have been proven to enhance the overall acceptability of foodstuffs apart from improving health. Out of various spices, garlic is the one which comprises of significant bioactives prominently sulphur compounds and minerals as well. Apart from this, garlic is also a store house for phenolic compounds. The extracts of garlic and various isolated compounds have been reported to exhibit broad biological activities. The anticancerous activity of phytochemicals derived from garlic comprises of alteration in the permeability of mitochondria as well as increasing the antioxidative and proapoptotic potential. One of the best techniques to safeguard these bioactives in garlic is the encapsulation technology by creating barriers giving them protection against adverse environment. Nevertheless future studies need to be conducted in order to analyse the interactions of garlic based bioactives in the human body system based on in-vivo preclinical trials.

REFERENCES

Aires, A. 2017. Phenolics in foods: Extraction, analysis and measurements. In *Phenolic compounds: Natural sources, importance and applications*, ed. M. Soto-Hernández, M. Palma-Tenango, and R. García-Mateos, UK: InTech.

Amagase, H. 2006. Clarifying the real bioactive constituents of garlic. *The Journal of Nutrition* 136 (3): 716S–725S.

Arnault, I., J. P. Christides, N. Mandon. et al. 2003. High-performance ion-pair chromatography method for simultaneous analysis of alliin, deoxyalliin, allicin and dipeptide precursors in garlic products using multiple mass spectrometry and UV detection. *Journal of Chromatography* 991 (1): 69–75. doi: 10. 1016/S0021-9673(03)00214-0

Asdaq, S. M., and M. N. Inamdar. 2010. Potential of garlic and its active constituent, S-allyl cysteine, as antihypertensive and cardioprotective in presence of captopril. *Phytomedicine* 17 (13): 1016–1026.

Azmir, J., I. S. M. Zaidul, M. M. Rahman. et al. 2013. Techniques for extraction of bioactive compounds from plant materials: a review. *Journal of Food Engineering* 117 (4): 426–36. doi: 10.1016/j.jfoodeng.2013.01.014

Bagul, M., S. Kakumanu, and T. A. Wilson. 2015. Crude garlic extract inhibits cell proliferation and induces cell cycle arrest and apoptosis of cancer cells in vitro. *Journal of Medicinal Food* 18 (7): 731–737.

Bakri, I. M., and C. W. I. Douglas. 2005. Inhibitory effect of garlic extract on oral bacteria. *Archives of Oral Biology* 50 (7): 645–651.

Bat-Chen, W., T. Golan, I. Peri. et al. 2013. Allicin purified from fresh garlic cloves induces apoptosis in colon cancer cells via Nrf2. *Nutrition and Cancer* 62 (7): 947–957. doi: 10.1080/01635581.2010.509837

Carvalho, I. T., B. N. Estevinho, and L. Santos. 2016. Application of microencapsulated essential oils in cosmetic and personal healthcare products – a review. *International Journal of Cosmetic Science* 38 (2): 109–119.

Cavallito, C. J., and J. H. Bailey. 1944. Allicin, the antibacterial principle of Allium sativum L.: Isolation, physical properties and antibacterial action. *Journal of the American Chemical Society* 66 (11): 1950–1951.

Chemat, F., and M. K. Khan. 2011. Applications of ultrasound in food technology: Processing, preservation and extraction. *Ultrasonics Sonochemistry* 18: 813–835.

Chen, C., C. H. Liu, J. Cai. et al. 2017. Broad-spectrum antimicrobial activity, chemical composition and mechanism of action of garlic (Allium sativum) extracts. *Food Control* 1–17.

Cheng, Z., H. Song, Y. Yang. et al. 2015. Optimization of microwave-assisted enzymatic extraction of polysaccharides from the fruit of SchisandrachinensisBaill. *International Journal of Biological Macromolecules* 76: 161–168.

Cupp, M. J. 1999. Herbal remedies: adverse effects and drug interactions. *American Family Physician* 59 (5): 1239–1245.

Dehghani, S., E. Alipoor, A. Salimzadeh, et al. 2018. The effect of a garlic supplement on the pro-inflammatory adipocytokines, resistin and tumor necrosis factor-alpha, and on pain severity, in overweight or obese women with knee osteoarthritis. *Phytomedicine* 48: 70–75.

Diretto, G., A. Rubio-Moraga, J. Argandona. et al. 2017. Tissue-specific accumulation of sulfur compounds and saponins in different parts of garlic cloves from purple and white ecotypes. *Molecules* 22: 1359.

Estevinho, B. N., and F. Rocha 2017. A key for the future of the flavors in food industry: Nanoencapsulation and microencapsulation. In *Nanotechnology applications in food Flavor, Stability, Nutrition and Safety*, ed. A. M. Grumezescu and A. E. Oprea, 1–19. Academic Press.

El Mashad, H. M., R. Zhang, and Z. Pan. 2019. Onion and garlic. In *Integrated Processing Technologies for Food and Agricultural by-Products*. 273–296. Academic Press.

FAO. 2013. *Production and trade statistics*. FAO, Rome, Italy.

Fernandes, F., M. Dias-Teixeira, C. Delerue-Matos. 2021. Critical review of lipid-based nanoparticles as carriers of neuroprotective drugs and extracts. *Nanomaterials* 11 (3): 563.

Fratianni, F., R. Riccardi, P. Spigno et al. 2016. Biochemical characterization and antimicrobial and antifungal activity of two endemic varieties of garlic (Allium sativum L.) of the campania region, southern Italy. *Journal of Medicinal Food* 19 (7): 686–691.

García-Villalón, A. L., S. Amor, L. Monge et al. 2016. In vitro studies of aged black garlic extract enriched in S-allylcysteine and polyphenols with cardioprotective effects. *Journal of Functional Foods* 27: 189–200.

Garti, N., and D. J. McClements. 2012. *Encapsulation technologies and delivery systems for food ingredients and nutraceuticals*. Elsevier.

Guo, Y. 2014. Experimental study on the optimization of extraction process of garlic oil and its antibacterial effects. *African Journal of Traditional, Complementary and Alternative Medicines* 11 (2): 411–414.

Ha, A. W., T. Ying, and W. K. Kim. 2015. The effects of black garlic (Allium satvium) extracts on lipid metabolism in rats fed a high fat diet. *Nutrition Research and Practice* 9 (1): 30–36.

Hameed, B. H., and A. A. Ahmad. 2009. Batch adsorption of methylene blue from aqueous solution by garlic peel, an agricultural waste biomass. *Journal of Hazardous Materials* 164 (2–3): 870–875.

Hassouna, I., H. Ibrahim, F. Abdel Gaffar, I. El-Elaimy, and H. Abdel Latif. 2015. Simultaneous administration of hesperidin or garlic oil modulates diazinon-induced hemato-and immunotoxicity in rats. *Immunopharmacology and Immunotoxicology* 37 (5): 442–449.

Horie, H., and K. i. Yamashita. 2006. Non-derivatized analysis of methiin and alliin in vegetables by capillary electrophoresis. *Journal of Chromatography* 1132 (1–2): 337–339. doi: 10.1016/j.chroma.2006.09.018

Ifesan, B. O. T., E. A. Fadipe, and B. T. Ifesan. 2014. Investigation of antioxidant and antimicrobial properties of garlic peel extract (Allium sativum) and its use as natural food additive in cooked beef. *Journal of Scientific Research and Reports* 711–721.

Ikram, R., K. H. Low, N. B. Hashim. et al. 2018. Characterization of sulfur-compounds as chemotaxonomic markers in the essential oils of allium species by solvent-free microwave extraction and gas chromatography–mass spectrometry. *Analytical Letters* 2–12. doi: 10.1080/00032719.2018.1479411

Ilic, D., V. Nikolic, M. Stankovic. et al. 2012. Transformation of synthetic allicin: The influence of ultrasound, microwaves, different solvents and temperatures, and the products isolation. *The Scientific World Journal* 1–7. doi: 10.1100/2012/561823

Jiang, X. M., Y. M. Lu, and C. P. Tan. 2014. Combination of aqueous two-phase extraction and cation-exchange chromatography: New strategies for separation and purification of alliin from garlic powder. *Journal of Chromatography B* 957: 60–67.

Jung, Y., H. Park, H. Y. Zhao. et al. 2014. Systemic approaches identify a garlic-derived chemical, Z-ajoene, as a glioblastoma multiforme cancer stem cell-specific targeting agent. *Molecules and Cells* 37 (7): 547–553.

Junnila, A., E. E. Revay, G. C. Müller. et al. 2015. Efficacy of attractive toxic sugar baits (ATSB) against Aedesalbopictus with garlic oil encapsulated in beta-cyclodextrin as the active ingredient. *ActaTropica* 152: 195–200.

Kairam, N., S. Kandi, and M. Sharma. 2021. Development of functional bread with flaxseed oil and garlic oil hybrid microcapsules. *Lebensmittel-Wissenschaft & Technologie* 136: 110300.

Kallel, F., and S. E. Chaabouni. 2017. Perspective of garlic processing wastes as low-cost substrates for production of high-added value products: A review. *Environmental Progress & Sustainable Energy* 36 (6): 1765–1777.

Kallel, F., D. Driss, F. Chaari. et al. 2014. Garlic (*Allium sativum* L.) husk waste as a potential source of phenolic compounds: Influence of extracting solvents on its antimicrobial and antioxidant properties. *Industrial Crops and Products* 62: 34–41.

Kemper, K. J. 2000. Garlic (*Allium Sativum*), the Longwood Herbal Task Force and the center for Holistic Pediatric Education and Research 1–49.

Khatua, T. N., R. M. Borkar, S. A. Mohammed. et al. 2017. Novel sulfur metabolites of garlic attenuate cardiac hypertrophy and remodeling through induction of Na+/K+-ATPase expression. *Frontiers in Pharmacology* 8: 18.

Kim, H. N., S. G. Kang, Y. K. Roh, M. K. Choi, and S. W. Song. 2017. Efficacy and safety of fermented garlic extract on hepatic function in adults with elevated serum gamma-glutamyl transpeptidase levels: A double-blind, randomized, placebo-controlled trial. *European Journal of Nutrition* 56 (5): 1993–2002.

Kim, S., D. B. Kim, W. Jin. et al. 2018. Comparative studies of bioactive organosulphur compounds and antioxidant activities in garlic (Allium sativum L.), elephant garlic (Allium ampeloprasum L.) and onion (Allium cepa L.). *Natural Product Research* 32 (10): 1193–1197.

Kinalski, T. and C. P. Z. Norena. 2019. Effect of spray drying encapsulation of garlic extract on inulin and thiosulfinates contents. *Journal of Food Measurement and Characterization* 13 (3): 2438–2447.

Ko, J. W., S. H. Park, I. C. Lee. et al. 2016. Protective effects of garlic oil against 1, 3-dichloro-2-propanol-induced hepatotoxicity: role of CYP2E1 and MAPKs. *Molecular & Cellular Toxicology* 12 (2): 185–195.

Koca, I., and B. Tasci. 2016. Garlic as a functional food. *Acta Horticulturae* 1143: 139–146.

Kuete, V. 2017. Allium sativum. In *Medicinal spices and vegetables from Africa Therapeutic Potential against Metabolic, Inflammatory, Infectious and Systemic Diseases*, ed. V. Kuete, 363–377. Academic Press.

Lai, Y. S., W. C. Chen, C. T. Ho. et al. 2014. Garlic essential oil protects against obesity-triggered nonalcoholic fatty liver disease through modulation of lipid metabolism and oxidative stress. *Journal of Agricultural and Food Chemistry* 62 (25): 5897–5906.

Lee, H. S., W. C. Lim, S. J. Lee. et al. 2016. Antiobesity effect of garlic extract fermented by Lactobacillus plantarum BL2 in diet-induced obese mice. *Journal of Medicinal Food* 19 (9): 823–829.

Leontiev, R., N. Hohaus, C. Jacob. et al. 2018. A Comparison of the antibacterial and antifungal activities of thiosulfnate analogues of allicin. *Scientific Reports* 1–19.

Li, J., Y. Dadmohammadi, and A. Abbaspourrad. 2021. Flavor components, precursors, formation mechanisms, production and characterization methods: garlic, onion, and chili pepper flavors, *Critical Reviews in Food Science and Nutrition* 1–23.

Li, M., Y. X. Yan, Q. T. Yu, et al. 2017. Comparison of immunomodulatory effects of fresh garlic and black garlic polysaccharides on RAW 264.7 macrophages. *Journal of Food Science* 82 (3): 765–771.

Liu, J., W. Guo, M. Yang. et al. 2018. Investigation of the dynamic changes in the chemical constituents of Chinese "laba" garlic during traditional processing. *RSC Advances* 8 (73): 41872–41883.

Majeed, H., Y.-Y. Bian, B. Ali. et al. 2015. Essential oil encapsulations: Uses, procedures, and trends. *RSC Advances* 5 (72): 58449–58463.

Mansingh, D. P., N. Dalpati, V. K. Sali, and A. H. R. Vasanthi. 2018. Alliin the precursor of allicin in garlic extract mitigates proliferation of gastric adenocarcinoma cells by modulating apoptosis. *Pharmacognosy Magazine* 14 (55): S84–S91.

Marisa Ribeiro, A., B. N. Estevinho, and F. Rocha. 2019. Microencapsulation of polyphenols - the specific case of the microencapsulation of SambucusNigra L. extracts - A review. *Trends in Food Science & Technology* 105: 454–467.

Martins, N., S. Petropoulos, and I. C. Ferreira. 2016. Chemical composition and bioactive compounds of garlic (Allium sativum L.) as affected by pre-and post-harvest conditions: A review. *Food Chemistry* 211: 41–50.

Metwally, D. M., Al-Olayan, E. M., Alanazi, M., Alzahrany, S. B., and Semlali, A., 2018. Antischistosomal and anti-inflammatory activity of garlic and allicin compared with that of praziquantel in vivo. *BMC Complementary and Alternative Medicine* 18 (1): 1–11.

Miltonprabu, S., N. C. Sumedha, and P. Senthilraja. 2017. RETRACTED: Diallyl trisulfide, a garlic polysulfide protects against As-induced renal oxidative nephrotoxicity, apoptosis and inflammation in rats by activating the Nrf2/ARE signaling pathway. *International Immunopharmacology* 50: 107–120.

Mohamed, E. H., A. A. A. Baiomy, Z. S. Ibrahim, and M. M. Soliman. 2016. Modulatory effects of levamisole and garlic oil on the immune response of Wistar rats: Biochemical, immunohistochemical, molecular and immunological study. *Molecular Medicine Reports* 14 (3): 2755–2763.

Mondal, A., S. Banerjee, and S. Bose. 2021. Garlic constituents for cancer prevention and therapy: From phytochemistry to novel formulations. *Pharmacological Research* 105837.

Narsaiah, K., M. Sharma, K. Sridhar. et al. 2019. Garlic oil nanoemulsions hybridized in calcium alginate microcapsules for functional bread. *Agricultural Research* 8 (3): 356–363.

Nazari, M., B. Ghanbarzadeh, H. SamadiKafil. et al. 2019. Garlic essential oil nanophytosomes as a natural food preservative: Its application in yogurt as food model. *Colloid and Interface Science Communications* 30: 100176.

Ozkan, G., P. Franco, I. De Marco. et al. 2019. A review of microencapsulation methods for food antioxidants: Principles, advantages, drawbacks and applications. *Food Chemistry* 272: 494–506.

Pang, J., X. L. Feng, and X. F. Wang. 2014. Purification and utilization of garlic processing wastewater in lotus pond wetlands. *Water Science and Engineering* 7 (4): 395–402.

Park, S. Y., Seetharaman, R., Ko, M. J. et al. 2014. Ethyl linoleate from garlic attenuates lipopolysaccharide-induced pro-inflammatory cytokine production by inducing heme oxygenase-1 in RAW264. 7 cells. *International Immunopharmacology* 19 (2): 253–261.

Piletti, R., M. Zanetti, G. Jung. et al. 2019. Microencapsulation of garlic oil by β-cyclodextrin as a thermal protection method for antibacterial action. *Materials Science and Engineering: C* 94: 139–149.

Pinilla, C. M. B., and A. Brandelli. 2016. Antimicrobial activity of nanoliposomes co-encapsulating nisin and garlic extract against Gram-positive and Gram-negative bacteria in milk. *Innovative Food Science & Emerging Technologies* 36: 287–293.

Pinilla, C. M. B., R. C. S. Thys, and A. Brandelli. 2019. Antifungal properties of phosphatidylcholine-oleic acid liposomes encapsulating garlic against environmental fungal in wheat bread. *International Journal of Food Microbiology* 293: 72–78.

Raeisi, S., S. M. Ojagh, P. Pourashouri. et al. 2021. Shelf-life and quality of chicken nuggets fortified with encapsulated fish oil and garlic essential oil during refrigerated storage. *Journal of Food Science and Technology* 58: 121–128.

Raeisi, S., S. M. Ojagh, S. Y. Quek. et al. 2019. Nano-encapsulation of fish oil and garlic essential oil by a novel composition of wall material: Persian gum-chitosan. *Lebensmittel-Wissenschaft & Technologie* 116: 108494.

Ramaa, C. S., A. R. Shirode, A. S. Mundada. et al. (2006). Nutraceuticals-an emerging era in the treatment and prevention of cardiovascular diseases. *Current Pharmaceutical Biotechnology* 7 (1): 15–23.

Rybak, M. E., E. M. Calvey, and J. M. Harnly. 2004. Quantitative determination of allicin in garlic: supercritical fluid extraction and standard addition of alliin. *Journal of Agricultural and Food Chemistry* 52(4): 682–687.

Saifullah, M., M. R. I. Shishir, R. Ferdowsi. et al. 2019. Micro and nano encapsulation, retention and controlled release of flavor and aroma compounds: A critical review. *Trends in Food Science & Technology* 86: 230–251.

Santhosha, S. G., P. Jamuna, and S. N. Prabhavathi. 2013. Bioactive components of garlic and their physiological role in health maintenance: A review. *Food Bioscience* 3: 59–74.

Shang, Ao, S. Y. Cao, X. Y. Xu. et al. 2019. Bioactive compounds and biological functions of garlic (*Allium sativum* L.). *Foods* 8: 246; doi: 10.3390/foods8070246

Shi, L., Q. Lin, X. Li. et al. 2017. Alliin, a garlic organosulfur compound, ameliorates gut inflammation through MAPK-NF-κB/AP-1/STAT-1 inactivation and PPAR-γ activation. *Molecular Nutrition & Food Research* 61 (9): 1601013.

Shishir, M. R. I., L. Xie, C. Sun. et al. 2018. Advances in micro and nano-encapsulation of bioactive compounds using biopolymer and lipid-based transporters. *Trends in Food Science & Technology* 78: 34–60.

Staba, E. J., L. Lash, and J. E. Staba. 2001. A commentary on the effects of garlic extraction and formulation on product composition. *The Journal of Nutrition* 131 (3): 1118S–1119S.

Tabari, M. A. and S. Ebrahimpour. 2014. Effect of aged garlic extract on immune responses to experimental fibrosarcoma tumor in BALB/c mice. *Indian Journal of Cancer* 51 (4): 609–613.

Tattelman, E. 2005. Health effects of garlic. *American Family Physician* 72 (1): 103–106.

Tavares, L. and C. P. Z. Norena. 2019. Encapsulation of garlic extract using complex coacervation with whey protein isolate and chitosan as wall materials followed by spray drying. *Food Hydrocolloids* 89: 360–369.

Tavares, L. and C. P. Z. Norena. 2021. Characterization of the physicochemical, structural and thermodynamic properties of encapsulated garlic extract in multilayer wall materials. *Powder Technology* 378: 388–399.

Tavares, L., H. L. B. Barros, J. C. P. Vaghetti. et al. 2019. Microencapsulation of garlic extract by complex coacervation using whey protein isolate/chitosan and gum Arabic/chitosan as wall materials: Influence of anionic biopolymers on the physicochemical and structural properties of microparticles. *Food and Bioprocess Technology* 12 (12): 2093–2106.

Tavares, L., L. Santos, and C. P. Z. Norena. 2021. Bioactive compounds of garlic: A comprehensive review of encapsulation technologies, characterization of the encapsulated garlic compounds and their industrial applicability. *Trends in Food Science & Technology* 114: 232–244.

Thomas, D. J., and K. L. Parkin. 1994. Quantification of alk(en)yl-L-cysteine sulfoxides and related amino acids in Alliums by high-performance liquid chromatography. *Journal of Agricultural and Food Chemistry* 42 (8): 1632–1638. doi: 10.1021/jf00044a010

Tiwari, B. K. 2015. Ultrasound: A clean, green extraction technology. *TrAC Trends in Analytical Chemistry* 71:100–9. doi: 10.1016/j.trac.2015.04.013.

Tsao, S. M., C. C. Hsu, and M. C. Yin. 2003. Garlic extract and two diallyl sulphides inhibit methicillin-resistant Staphylococcus aureus infection in BALB/cA mice. *Journal of Antimicrobial Chemotherapy* 52 (6): 974–980.

Ugwu, C. E., and S. M. Suru. 2016. The functional role of garlic and bioactive components in cardiovascular and cerebrovascular health: what we do know. *Journal of Biosciences and Medicines* 4 (10): 28–42.

Valente, C., G. Aboua, and S. S. Du Plessis. 2014. Garlic and its effects on health with special reference to the reproductive system. In *Antioxidant-Antidiabetic Agents and Human*, ed. O. Oguntibeju, 259–277. IntechOpen, doi: 10.5772/57191

Wang, X., H. P. Liu, C. C. Zeng. et al. 2013. Infrared and Raman spectral analysis of alliin and methiin. *Spectroscopy and Spectral Analysis* 33 (6): 1570–1574. doi: 10.3964/j.issn.1000- 0593(2013)06-1570-05

Wang, X., Y. Yang, R. Liu, Z. Zhou, and M. Zhang. 2016. Identification of antioxidants in aged garlic extract by gas chromatography-mass spectrometry and liquid chromatography-mass spectrometry. *International Journal of Food Properties* 19 (2): 474–483. doi: 10.1080/10942912.2015.1033548

Wiczkowski, W. 2011. Garlic and onion: Production, biochemistry, and processing. *Handbook of Vegetables and Vegetable Processing* 1: 625–642.

Wu, D., T. Gao, H. Yang, Y. Du. et al. 2015. Saneous microwave/ultrasonic-assisted enzymatic extraction of antioxidant ingredients from NitrariatangutorunBobr. juice by-products. *Industrial Crops and Products* 66: 229–238.

Xu, Y. S., J. G. Feng, D. Zhang. et al. 2014. S-allylcysteine, a garlic derivative, suppresses proliferation and induces apoptosis in human ovarian cancer cells in vitro. *Acta Pharmacologica Sinica* 35 (2): 267–274.

Yang, C., L. Li, L. Yang. et al. 2018. Anti-obesity and Hypolipidemic effects of garlic oil and onion oil in rats fed a high-fat diet. *Nutrition & Metabolism* 15 (1): 1–8.

Yin, C., X. Fan, Z. Fan, D. Shi, and H. Gao. 2018. Optimization of enzymes-microwave-ultrasound assisted extraction of Lentinusedodes polysaccharides and determination of its antioxidant activity. *International Journal of Biological Macromolecules*. 111: 446–454. doi: 10.1016/j.ijbiomac.2018.01.007

Yoo, D. Y., W. Kim, S. M. Nam. et al. 2014. Neuroprotective effects of Z-ajoene, an organosulfur compound derived from oil-macerated garlic, in the gerbil hippocampal CA1 region after transient forebrain ischemia. *Food and Chemical Toxicology* 72: 1–7.

Yu, J., H. Z. Castada, X. Huang. et al. 2019. Comparison of encapsulation of garlic oil with α-, β-, and γ-cyclodextrin using selected ion flow tube-mass spectrometry (SIFT-MS). *Journal of Food Processing and Preservation* 43 (2): e13865.

Zaini, A. S., N. R. Putra, Z. Idham. et al. 2020. Mini Review: Extraction of Allicin from Allium sativum using Subcritical Water Extraction. *IOP Conf. Series: Material and Science Engeneering*, 932. doi: 10. 1088/1757-899X/932/1/012023 932 012023

Zamani Taghizadeh Rabe, S., T. Ghazanfari, Z. Siadat, M. Rastin, S. Zamani Taghizadeh Rabe, and M. Mahmoudi. 2015. Anti-inflammatory effect of garlic 14-kDa protein on LPS-stimulated-J774A. 1 macrophages. *Immunopharmacology and Immunotoxicology* 37 (2): 158–164.

Zardast, M., K. Namakin, J. E. Kaho, and S. S. Hashemi. 2016. Assessment of antibacterial effect of garlic in patients infected with Helicobacter pylori using urease breath test. *Avicenna Journal of Phytomedicine* 6 (5): 495–501.

Zhao, Y., W. Li, J. Liu. et al. 2017. Modification of garlic peel by nitric acid and its application as a novel adsorbent for solid-phase extraction of quinolone antibiotics. *Chemical Engineering Journal* 326: 745–755.

Zhou, H., Z. Qu, V. V. Mossine, D. L. Nknolise. et al. 2014. Proteomic analysis of the effects of aged garlic extract and its FruArg component on lipopolysaccharide-induced neuroinflammatory response in microglial cells. *PLoS One* 9 (11): e113531.

CHAPTER 4

Bioactive Compounds in Cinnamon

Swati Haldar, Anju Tanwar, Ajay Singh, Geeta Mehra, Mukul Sain, and Viney Kumar

CONTENTS

DOI: 10.1201/9781003215387-4

4.1 INTRODUCTION

Over the past few decades, consumers are very particular regarding their eating habits and very conscious about nutritional and healthy performance for the opted food (Menrad 2003), which has consequently led to unprecedented increase in popularity for functional foods. Functional foods are known for their medicinal benefits in improving physical and mental health, reducing risks for various illnesses, and in many cases curing ailments (Bigliardi and Galati 2013). There are observable variations among the regulatory definitions of functional foods. Academy of Nutrition and Dietetics defines functional foods as whole, fortified, enriched and/or enhanced foods with potential health benefits, when taken at effective levels as part of any regular diet. The Institute of Food Technologists recognizes functional foods as "Foods and food components that provide a health benefits beyond basic nutrition (for the intended population)" (Crowe and Francis 2013). Ministry of Health, Labour and Welfare, Japan (Ministry of Health Labour and Welfare), understands functional foods as "FOSHU" which expands to Food for Specified Health Uses. As the name suggests, "FOSHU" includes foods with constituents having potential applications for health benefits. "FOSHU" has official approval to lay claims on physiological effects in humans, such as, maintaining healthy limits of blood pressure or blood cholesterol. There was no official or legal definition of functional foods or specific regulations in Europe; however, in the work of Stein and Rodríguez-Cerezo (Stein and Rodriguez-Cerezo 2008), functional food is defined as "food that is taken as part of the usual diet and has beneficial effects that go beyond traditional nutritional effects." So, summarizing these classifications, providing health benefits is the necessary and sufficient condition to be a functional food. Regular food can be upgraded to functional food by fortifying or enriching them with ingredients or components with direct health benefits. In order to keep pace with the steeply growing functional food market, the past two decades have witnessed functional foods to develop at an incredible speed (Bigliardi and Galati 2013). A plethora of functional foods, embedded with lists of bioactive constituents viz; minerals, vitamins, phytosterols, probiotics, and antioxidants came into existence during this period (Bröring 2010). However, functional foods from natural sources such as, herbs and spices are increasingly being preferred over the synthetic ones, that were prevalent earlier (Embuscado 2015). Since ancient times, spices and herbs have been great favourites as natural culinary endowments to make food flavourful, curiously, with full knowledge of their nutritional values documented in the traditional medicinal texts and lexicons of eastern hemisphere, primarily, India and China. In recent times, the bioactive components of these functional foods have drawn much attention from the food scientists across the globe (Vallverdú-Queralt et al. 2014).

In this chapter, we have focused on cinnamon, a spice known for its unique flavour. Cinnamon (commonly known as 'dalchini' in India and "Kwai" in China) dates back to nearly 2800 BC. The generic name Cinnamomum has its roots in the Greek word kinnamomon, meaning "sweet wood" (Charles 2013). Cinnamon belongs to Lauraceae family, and has been a part of almost every kitchen as a flavouring agent for a long time. In traditional Indian medicinal system, cinnamon has been a part of various remedial formulations for respiratory and digestive ailments. Cinnamon is rich in structurally diverse phytocompounds (Udayaprakash et al. 2015). The genus *Cinnamomum* comprises of nearly 250 phenotypically and genotypically distinct species (Tung et al. 2008), most of which still awaited for their compositional and functional characterization. However, *Cinnamomum zeylanicum* (Ceylon Cinnamon, True Cinnamon), *Cinnamomum verum* (Mexican cinnamon), *Cinnamomum burmanni* (Indonesian Cinnamon, Korintje Cinnamon, Padang Cassia), *Cinnamomum loureiroi* (Saigon Cinnamon, Vietnamese Cassia, Vietnamese Cinnamon), and *Cinnamomum aromaticum* (Chinese Cinnamon, Cassia Cinnamon) are the popular and economically important species in the world (Table 4.1) (Sivapriya and John 2016), with widespread use in food and pharmaceutical industries.

Table 4.1 Characteristics of the Popular Cinnamon Species (Sivapriya and John 2016)

	Cinnamomum zeylanicum, Cinnamomum verum	*Cinnamomum burmanni*	*Cinnamomum loureiroi*	*Cinnamomum aromaticum*
Local name	Ceylon Cinnamon, True Cinnamon, Mexican Cinnamon	Indonesian Cinnamon, Korintje Cinnamon, Padang Cassia	Saigon Cinnamon, Vietnamese Cinnamon, Vietnamese Cassia	Chinese Cinnamon, Cassia Cinnamon
Country	Sri Lanka, India, brazil, Caribbean, Madagascar	Indonesia	Vietnam	China
Height	32–49 ft.	20–22 ft.	35–50 ft.	32–49 ft.
Taste	Mild sweet	Spicy	Spicy and Sweet	Spicy and bitter
Colour	Light to medium reddish brown	Dark reddish brown	Dark reddish brown	Dark reddish brown
Good	Delicate taste, Fragnant smell	Cheap, Spicy, Cinnamon flavour, Smells strong	Strong smell, Spice taste	Cheap, Spicy, Strong smell
Bad	Ultra-low coumarin, Expensive	High coumarin level	Very high coumarin level	High coumarin level

4.2 PLANT HIERARCHY, HABITAT, AND CHEMICAL COMPOSITION

Cinnamon is a tropical evergreen tree indigenous to Sri Lanka, South eastern India, Indonesia, South America, and the West Indies (Vangalapati et al. 2012). Phylogenetically, genus Cinnamon belongs to family Lauraceae of order Laurales within division Magnoliophyta (Sangal 2011). Cinnamon trees grow to a height of 10–15 m. However, cultivated cinnamon trees are grown as bush to a height of not more than 3 m. The stems are continuously pruned to facilitate bushy growth for new shoots with concomitant higher bark/quills production. The thick scabrous brown barks of the trunk and branches are the sources of the actual spice (Vangalapati et al. 2012). The peeled barks are rolled into striated pipes called quills. Besides spice, bark along with leaves and roots yields cinnamon oils enriched in different phytocompounds, such as, cinnamaldehyde (in the bark oil), camphor (in the root-bark oil) and eugenol (in the leaf oil) (Yeh et al. 2014). The fruits and flowers of cinnamon are rich in trans-cinnamyl acetate (Jayaprakasha and Rao 2011). Cinnamon has significant medicinal values, which are documented in the 6000 year old traditional Indian system of medicine, Ayurveda (Sangal 2011).

The moisture content of cinnamon bark is 9.9%. It is most enriched in carbohydrates, which comprise of 22.6%, followed by 20.3% fibre, 4.65% protein, 3.55% total ash, 1.6% calcium, 0.4% potassium, 0.05% phosphorous, 0.04% iron and 0.01% sodium. It also contains different types of vitamins such as, vitamins B1, B2, C (0.14, 0.21, 39.8 mg/100 g) and niacin (1.9 mg/100 g) (Thomas and Kuruvilla 2012). The cinnamon nutritional composition is given in Table 4.2.

4.3 CULTIVATION, HARVESTING, PROCESSING, PACKAGING, AND QUALITY ISSUES

4.3.1 Cultivation

The annually worldwide production of cinnamon is about 27,000–35,000 tons. Large portion of cinnamon is cultivated in Sri Lanka, Seychelles, China, Madagascar and Maldives, whereas, India and Vietnam are enlisted as small-scale producers of cinnamon. Cinnamon grows well between 20 to 30°C, with an annual precipitation from 1250 to 2500 mm. In India, Kerala, Karnataka

Table 4.2 Nutritive Value of Ground Cinnamon (Sivapriya and John 2016)

Cinnamon, Ground (13 Kcal)	
2.00 tsp/ (5.20 gm)	**Glycemic Index** -Very Low
Nutrient	**Amount**
Basic Macronutrient	**Calories (g)**
Protein	0.21
Carbohydrates	4.19
Fat	0.06
Dietary Fibre	2.76
Calories	12.8
Micronutrients	**Calories (mg)**
Calcium	52.10
Iron	0.43
Magnesium	3.12
Phosphorous	3.33
Potassium	22.41

coastal areas and different parts of Tamil Nadu produce high quality cinnamon bark. Besides peninsular India, cinnamon also grows in Andaman and Nicobar and Lakshadweep islands (Saeed et al. 2018).

4.3.2 Harvesting

The cinnamon bark is harvested after the second or even third year of sowing. A gap of minimum 12 to 18 months is given, before two consecutive harvests. The branches retrieved by peeling of the exterior bark and then dislodge the inside bark which is more useful then outer. The tree is snipped from centre parts of the shoot during the wet season. A thin consistent brown colour bark was retrieved from shoot with a length and diameter of 1.0–1.25 m and 1.25 cm, respectively. Red to green colour conversion of young leaves indicates the appropriate timing to snipe of the stem and also an indication of free flow of sap between the wood and bark. There are two main periods for harvesting, first one comes around May and the second one in November. The appropriate timing for peeling of shoot is chosen based on the circulation of sap between the corky layer and root (Thomas and Kuruvilla 2012).

4.3.3 Processing

Wet barks are processed further immediately after harvest. Quill yields are typically around 60–125 kg/ha after the first harvest. Plants aged between 10 and 12 years yield around 225–300 kg per hectare of quills. Compared to smaller and shattered parts, the bigger fragments of quills fetch higher prices. Immediate drying of the barks is ensured within 4–6 h of harvest to control pest infestation (Thomas and Kuruvilla, 2012).

4.3.4 Packaging

Polypropylene is used for cinnamon packaging over polythene, because the latter cannot prevent the diffusion of flavour volatile components. Quills were chopped into ~10 cm long parts and packed into moisture-resistant sealed polypropylene bags for sale (Thomas and Kuruvilla 2012).

4.3.5 Quality Issues

The quality of the finished product depends on the bark thickness, appearance, flavour and aroma. Soil quality and environmental factors strongly affect bark quality. The bark acquired from mid branches is better than that which acquired from the outside of the shoot.

The grade of cinnamon was evaluated, on the grounds of their appearance, proportions and fragrance features of volatile oil. Cinnamon of good quality must not be thicker than thick paper with light brown in colour and at the time of broken gives a fracture sound, after chewing it will melt and sweeten blowing in the mouth (Thomas and Kuruvilla 2012).

4.4 BIOLOGICAL ACTIVE COMPONENTS

Although the final phytochemical composition of cinnamon is affected by the geographical location of its cultivation and post-harvest downstream processing conditions, gum, cinnamaldehyde, tannin, mannitol, essential oils and coumarins are its primary phytoconstituents (Thomas and Kuruvilla 2012). Cinnamon is an attractive spice because of its taste, but it may also be of pharmaceutical interest (Mohammad and Gharibzahedi 2018).

Secondary metabolites comprise a huge number of chemical compounds, which are not essential for the growth and life of the producing plant, synthesized through metabolism of primary metabolites and having numerous health promoting effects. Most of them are dietetically neutral, but they usually have a positive effect on human health. Besides, being a flavourant for many types of cuisines, cinnamon is traditionally recognized for its medicinal values, particularly, as natural antioxidants (Mohammad and Gharibzahedi 2018). The phytochemical composition of cinnamon oil varies significantly depending on the plant part, it has been derived from, and is shown in Figure 4.1. The medicinal

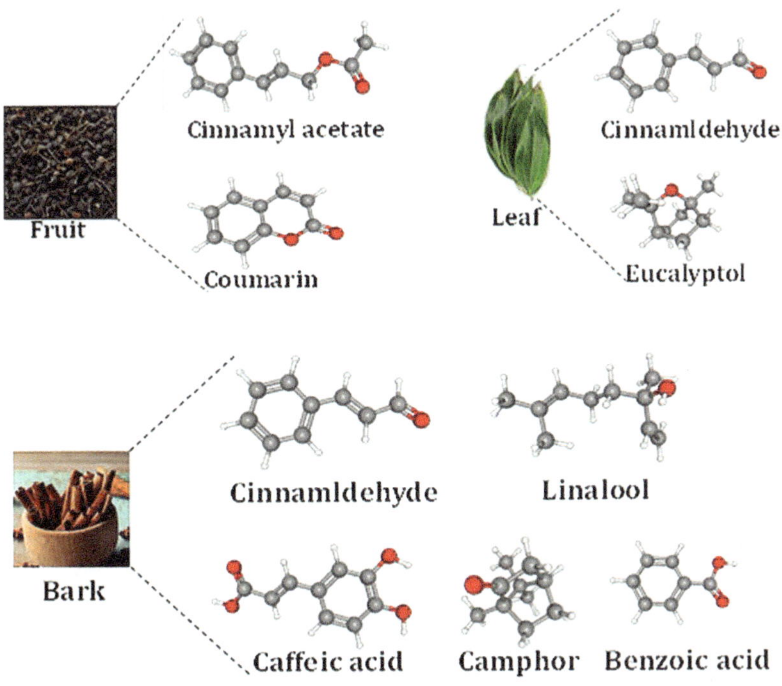

Figure 4.1 Major phytocompounds present in volatile cinnamon oils from different parts of the plant.

Table 4.3 List of Phytocompounds Present in Different Part of Various Cinnamon Species (Muhammad and Dewettinck 2017)

Sr. No.	Species	Part	Bioactive Phytocompound	References
1.	***Cinnamomum burmanni***	Leaf	Cyclohexane methanol, trans-cinnamaldehyde, trans-cinnamyl acetate, eugenol, caryophyllene, D-borneol, eucalyptol, guaiol, (-)-Spathulenol	(Kumar et al. 2019)
		Fruit	Anthocyanin, proanthocyanidine, camphene, linalool, terpineol, pinene, caryophyllene, sylvestrene, myrcene, citral, fenchol	(Kumar et al. 2019)
		Shoot	Camphene, linalool, terpineol, pinene, caryophyllene, sylvestrene, myrcene, citral, fenchol, nerolidol	(Kumar et al. 2019)
2.	***Cinnamomum aromaticum***	Bark	Benzene, trans-cinnamaldehyde, 1,3-dimethyl, camphene, linalool, benzaldehyde, camphor, benzenepropanol, trimethyl-eugenol, acetophenone, copaene, geranyl acetate, sinapic acid, vanillin, *p*-coumaric acid, caffeic acid, cinnamic acid, tannic acid, syringic acid, kaempferol	(Helal et al. 2014; Jiao et al. 2013; Kumar et al. 2012)
		Leaf	*o*-methoxy-cinnamaldehyde, 3-methoxy-1,2-propanediol, benzenemethanol, coumarin ethyl alcohol, 2-nitroethanol, glycerine, cinnamyl alcohol, dodecane, isopropyl acetate	(Tung et al. 2008)
3.	***Cinnamomum verum***	Bark	Trans-cinnamaldehyde, benzene, benzaldehyde, styrene, camphene, acetophenone, linalool, camphor, benzenepropanol, trimethyl-eugenol, copaene, caryophyllene, geranyl acetate	(Kumar et al. 2012)
4.	***Cinnamomum zeylanicum***	leaf	Eugenol, α-pinene, α-thujene, α-phellandrene, 1,8-cineole, *p*-cymene, β-caryophyllene, germacrene-d, aromadendrene, bicyclogermacrene, spathulenol, δ-cardinene, α-amorphene, viridiflorol, docosane, chlorogenic acid, catechin, caffeic acid, e-cinnamaldehyde, benzenemethanol, linalool, borneol, α-terpineol, *cis*-isoeugenol, vanilene, neohexane, *p*-vinylbenzohydrazide, 3-methylbenzothiophene, coumarine	(Cheng et al. 2012; Jiao et al. 2013; Tung et al. 2008)
		Bark	Trans-cinnamaldehyde, δ-cardinene, α-amorphene, α-copaene, β-caryophyllene, coumarine, β-bisabolene, α-muurolene, cadina (1,2)-4, diene, *o*-methoxycinnamaldehyde, tetracosane, cubenol	(Cheng et al. 2012)
5.	***Cinnamomum loureiroi***	Bark	Trans-cinnamaldehyde, benzene, 1,3-dimethyl, camphene, styrene, linalool, acetophenon, camphor, trimethyl-eugenol, cinnamyl acetate, geranyl acetate, cardiophyllene, anthraquinone, borneol, tannin, saponin, phellandrene, limonene	(Kumar et al. 2012, 2019)

effects of cinnamon are due to polyphenols, like, phenolic acids (coumarin, proanthocyanidin), cinnamaldehyde and eugenol. Volatile oils also contribute significantly towards the health benefits associated with cinnamon. Volatile oils from different parts of cinnamon tree, like, bark, leaves and roots, although share the same spectrum of monoterpene hydrocarbons but differ in their main phytoconstituent, cinnamic acid and type A-Procyanidin polymer (TAPP) are the other major phytocompounds derived from cinnamon (Błaszczyk et al. 2021). A collection of the phytocompounds along with the corresponding part of the cinnamon tree, they are derived from is provided in Table 4.3.

4.5 SEPARATION, EXTRACTION, AND PURIFICATION OF PHYTOCOMPOUNDS FROM CINNAMON

The type and amount of phytocompounds and volatile oils present in cinnamon extracts not only depend on the technique of extraction employed, but also depends on the species, part (bark, leaves, roots or fruits), and time of harvest with respect to the growth of the of cinnamon tree. For example, volatile oils can be obtained through single-phase extraction (Khasanah et al. 2017) and upper and middle segments of the bark is the best raw material for extracting cinnamon oil while a mature tree yields maximum cinnamon trans-aldehyde (Błaszczyk et al. 2021). Likewise, other factors that potentially affect the yield of phytocompounds are selection of extraction solvents, adopted process of extraction and variation among physical tactics involved in it such as; time, temperature, and pressure. Therefore, maximum yield of phytocompounds in cinnamon oil or its other extracts can be achieved only through meticulous optimization of the extraction procedure (Adarsh et al. 2020; Cha et al. 2019; Othman et al. 2020). So, depending on the target phytocompound/s needs to be extracted, different types of preparative techniques are used, and thus, there is no one particular procedure to be claimed as the best. Rather, success of an extraction technique is contextual and is largely dictated by the final product aimed for.

Among the traditional procedures of extracting essential oils from cinnamon include hydro-distillation, organic solvent extraction and steam distillation. These processes use solvents, like, ethanol, methanol and chloroform (Adarsh et al. 2020). The recently developed super-critical fluid extraction method using microwaves, ultrasounds or carbon dioxide at a very low temperature and under very high pressure have added new dimensions to isolation of oils from cinnamon and other herbal sources (Chen et al. 2020). As analysed on gas chromatography-mass spectrometry (GC-MS), major phytochemical yields from cinnamon through supercritical carbon dioxide extraction are β-caryophyllene, cinnamic aldehyde, longifolene, and β-amyrin (Li et al. 2018). Despite unprecedented advance in the field of extraction chemistry, however, phytocompounds isolated through water extraction still remain the safest and most compatible with human body. Ultraperformance liquid chromatography-high-resolution mass spectrometry (UPLC-HRMS) analysis of aqueous cinnamon extract has shown the presence of biologically active phytocompounds, like, camphor, L(-)-carnitine, and rosavin (Tang et al. 2020). Just like there is a decent collection of separation techniques to choose from when it comes to extracting phytocompounds from cinnamon, a fairly good choice of techniques exists for chemical analysis of the extracts. These include liquid chromatography-mass spectrometry (LC-MS), high-performance liquid chromatography (HPLC), gas chromatography coupled with flame ionization detection (GC-FID) and gas chromatography-mass spectrometry (GC-MS) (da Silva et al. 2018). The most recent analytical technique of choice is gas chromatography with time-of-flight mass spectrometry (GCxGC-TOFMS) for impeccable accuracy during separation and characterization of essential oil components, especially those present in complex samples (Silva et al. 2020). The current state of knowledge about active compounds in different parts of cinnamon plant, their separation, and purification method is summarized in Table 4.4.

4.6 APPLICATION OF CINNAMON IN FOOD INDUSTRY

Cinnamon is an unrivalled culinary flavourant in a variety of recipes for rice, meat, desserts, baked items, spiced drinks and syrups, cereals and fruits across several cuisines around the globe. Some of these applications required cinnamon sticks whereas, others demand it in powdered form. Naturally, the most prominent application of cinnamon lies in the food industry. However, the phytocompounds present in cinnamon make it much more than just a food flavourant, by imparting several chemical properties that make it a highly attractive raw material in food industries. For example, the anti-fungal and anti-bacterial properties of cinnamon bark oil is very effective against meat spoilage. Hence, the popular practice of spicing up meat dishes in warmer climates with cinnamon to spice up meat dishes in warmer climates. Thus, cinnamon is used as a natural food preservative (Shan et al. 2007). The most common uses of cinnamon in the food industries are discussed here.

4.6.1 Anti-Microbial Activity

Cinnamon prevents food spoilage from infectious microorganism (Jessica Elizabeth et al. 2017). Anti-bacterial activity of cinnamon against *Staphylococcus aureus*, *Bacillus subtilis*,

Table 4.4 Separation/Purification and Analytical Methods to Isolate Phytocompounds from Different Parts of Cinnamon Tree (Błaszczyk et al. 2021)

Part	Separation Technique	Analytical Method	Determined Compound	References
Fruit	Hydrodistillation	GC-MS	Linalool	(Kaul et al. 2003)
Leaf	Hydrodistillation	GCxGC-TOFMS, GC-FID, GC-MS,	4-Hydroxy-4-methyl-2-pentanone, noipnone, p-mentha-1,5-dien-8-ol, methyl eugenol, verbenone, spathulenol, camphene, mustakone, sabinene, furanoid, linalool, e-cinnamaldehyde, borneol, e-caryophyllene, e-cinnamyl acetate, camphor, α-terpineol	(Farias et al. 2020; Jayaprakasha and Rao 2011; Silva et al. 2020)
	Supercritical extraction	GC-MS	Linollol, eugenol	(Jayaprakasha and Rao, 2011)
	Maceration	GC-MS	Trans-caryophyllene, 1,4,8-cycloundecatriene, eugenol, glycerol triacetate, oleic acid amide, benzyl benzoate	(Khasanah et al. 2017)
Bark	Hydrodistillation	GC-FID, GC-MS	Linalool, methyl eugenol, limonene, terpineol, sabinene, α-pinene, α-terpinene, α-selinene, trans-cinnamylacetae, cinnamaldehyde, coumarin acid, 1,4-benzenedicarboxylic acid, 9-octadecenoic acid, 1,8-cineole	(Abdelwahab et al. 2017; Kallel et al. 2019)
	Distillation	GC-MS, HPLC	Benzene propanol, eucalyptol, trans-cinnamaldehyde, eugenol, coumarin, α-muurolene	(Gotmare and Tambe 2019); Wong et al. 2014)
	Supercritical CO_2 extraction	GC	Cinnamaldehyde, eugenol	(Masghati and Ghoreishi 2018)
	Subcritical extraction	GC-MS **(for n-butanol extract),**	**n-butanol extract** e-cinnamaldehyde, benzaldehyde, styrene, camphene, linalool, β-pinene, eugenol, methyl cinnamate, coumarin	(Liang et al. 2019)
		HPLC-MS **(for ethanol extract)**	**ethanol extract** Cinnamyl alcohol, coumarin, cathechin, procyanidinetrimer	
	Soxhelt extraction	HPLC	Cinnamaldehyde	(Wong et al. 2014)
	Sonohydrodistillation	GC-MS	Cinnamaldehyde	(Modi et al. 2019)
	Solvent extraction	UPLC-HRMS	Protocathechuic acid, catechin, 3,4-dihydroxybenzaldehyde, epicatechin, cinnamic acid, procyanidin B2	(Muhammad et al. 2021)
	Pulse-electric field extraction	Folin-ciocalteu method	Phenol	(Pashazadeh et al. 2020)

Pseudomonas aeruginosa, Salmonella typhi, and *Escherichia coli* is well studied. Likewise, cinnamon also noted for its anti-fungal activity, particularly against different species of *Aspergillus*, like *A. ochraceus, A. flavus, A. niger, and A. terreus, Fusarium*, like, *F. moniliforme and F. graminearum*, and *Penicillium, such as, P. viridicatum and P. citrinum* (Cheng et al. 2012; Goñi et al. 2009). Essential oil extracted from *Cinnamon zeylanicum*is reported to inhibit growth of some species of *Enterococus, Streptococcus, Enterobacter, Acinetobacter, Proteus, Klebsiella, Clostridium, Mycobacterium, Candida* and *Listeria* even at a very low concentration of 500 ppm, with minimum inhibitory and minimum bactericidal concentrations, ranging from 200–400 and 400–800 mg/L, respectively (Miyashita and Sadzuka 2013). Thus, there are several reports evincing the anti-microbial efficacy of cinnamon and essential oils from cinnamon in food industry. For example, 5% cinnamon, when added to skim milk pasteurized with pulsed electric field, reduced *Salmonella typhimurium* contamination by 52% (Pina-Pérez et al. 2012). Similarly, cinnamon was more effective in inhibiting the growth of *Helicobacter pylori* in white wine, than green tea and raspberry (Lin et al. 2005). Likewise, essential oils from cinnamon over a concentration range of 0.005-0.5% proved to be more efficient in preventing *Streptococcus thermophiles* contamination in yogurt than those isolated from mint, clove and cardamom (Bayoumi 1992). The volatile compounds in essential oil from cinnamon were observed to disrupt hyphae formation, and consequently, inhibited sporulation in *A. flavus* (Manso et al. 2013). At molecular level, essential oils from cinnamon inhibited DNA aflatoxin-induced pathogenicity not only by reducing the DNA binding ability of aflatoxin but also scavenging the reactive oxygen species (ROS) produced by this toxin (Alpsoy, 2010). It was observed that cinnamon efficiently inhibited the uncontrolled expansion of *L. monocytogenes* in meat products and having a stronger inhibitory capacity than thyme and garlic (Dussault et al. 2014). Alginate and agar bilayer sheets coated with essential cinnamon oil (2 g/100 ml) demonstrated anti-bacterial properties against *L. monocytogenes* in scraped shrimp (Arancibia et al. 2014). Anti-microbial efficacy in fresh goods (tomatoes and peppers) of essential cinnamon oils is also documented (Tzortzakis, 2009). Thus, it is established beyond doubt is that cinnamon and its essential oil are very potent anti-microbial agents, safe for being used in foods to prevent microbial growth-induced spoilage. The most advantageous attribute of essential oils from cinnamon, observed so far, is that no anti-microbial resistance is developed against them, thus, making these oils one of the best options for treating drug resistant varieties of microbial pathogens. Naturally, these merits the significant contribution cinnamon offers towards maintaining food safety.

4.6.2 Cinnamon as a Flavourant

Cinnamon is globally recognized food flavourant. Both cinnamon bark and leaf oils are used to season different types of foods, like, baked goods, confectionary, soft drinks, savoury snacks and sauces and pickles. Oleoresin and ex-eugenol are the main phytocompounds imparting cinnamon its flavouring properties for food preparations. Besides, cinnamon is also used to flavour various pharmaceutical, dental and tobacco preparations. Cinnamon buds are equally flavourful as its bark and leaves. Additionally, cinnamon is a popular deodorant or masking agent in American food industry (Thomas and Kuruvilla 2012).

4.6.3 Cinnamon as a Colourant

Cinnamon is one of the popular sources of natural colourants for food. It has become a standard practice to blend different species of cinnamon to impart the right amount of colour and flavour to the baked and confectionary food items. Besides food items, cinnamon is also a colourant for soaps (Thomas and Kuruvilla 2012). Cinnamon has mouth refreshing effects and helps to fight bad breadth because of which it is also used to flavour chewing gums (Jakhetia et al. 2010).

4.6.4 Cinnamon in Cosmetics and Encapsulation

Regarding the cosmetic sector, Herman *et al.* showed that standard essential cinnamon oil has very strong anti-bacterial activity with 2.5% of cosmetic emulsion concentration. In disk-diffusion assay, the test for anti-bacterial activity was reported to be higher than the test for methylparaben, employed as a positive control. Inhibition zone diameters scored between 24 and 44 mm for the essential cinnamon oil, and for methylparabenes it was between 9 and 8 mm (Herman et al. 2013). A further potential application of cinnamon essential oil for anti-microbial activity was documented, that showed the nanoparticles made from poly-DL-lactide-co-glycolide (PLGA), a nontoxic biocompatible polymer, increasingly employed in the healthcare market, can also be used in food sector for supplies of food matrices. The authors also documented the anti-bacterial activity against *S. typhimurium* and *L. monocytogenes* in response to cinnamon extract loaded nanoparticles (Hill et al. 2013). Thus, nanoencapsulation could be a promising way to supply immobilized anti-bacterial compounds to infectious micro-organisms present in food products without any substantial damage on the textural properties.

4.7 CINNAMON IN TRADITIONAL AND MODERN MEDICINES

For thousand years, cinnamon has been the best remedy for toothaches urinary tract infections and stomach irritations, hence, employed by several historic cultures to treat diarrhoea, arthritis and various menstrual disorders. Traditional medicines have been using cinnamon as an adjuvant in carminative and stomachic medicines to treat anorexia, inflammation, vomiting and tubercular ulcers (Thomas and Kuruvilla 2012). Small quantities of cinnamon applied to the forehead, helps to relief neuralgic headaches. Traditionally, cinnamon is used as a rubefacient against the stings of poisonous insects. The huge number of therapeutic applications for cinnamon reflects the widespread admiration of folk herbalists for its restorative virtues. In several traditional systems of medicine, cinnamon has been used as a medicinal agent, the most prominent ones being Ayurveda, the Indian traditional medicine and Unani, Perso-Arabic traditional medicine.

In Ayurveda, cinnamon is recognized as a medicine as well as a spice. Several Ayurvedic formulations include cinnamon bark such as, Caturjata Churna, Astanga Lavana Churna, Talisadya Churna, Sudarsana Churna, Sitopaladi Churna, Khadirarishta, Chandraprabha vati, Lavanbhaskara Churna, and Pippayadyasava. Ayurveda recommends the formulations with cinnamon against a variety of ailments of respiratory systems, like, bronchitis, colds, congestion, lung inflammation, dysfunctional digestive system, having diarrhoea, dysentry, indigestion, bloating, liver problems, nausea and vomiting, and reproductive and uro-genital indispositions, like, menorrhagia and difficult uterine contractions during labour. Ayurvedic formulations prescribed for relieving physical malaise, like, oedema and muscle tension contains cinnamon. In Ayurveda, cinnamon is considered a very good agent to promote metabolic and heart strengthening, and ease mental affliction, like, melancholy (Thomas and Kuruvilla 2012). Like Ayurveda, Unani medicine uses cinnamon as, a cardiac stimulant, cephalic tonic and for treating coughs.

Although, like traditional medicine, cinnamon does not have an established usage in modern medicine, nevertheless, there are several studies reporting its potential medicinal applications. These studies are gradually accumulating into a body of evidence that can irrefutably used as foundation for integrating cinnamon into the prescriptions of modern medicine. Cinnamon fortifies colon health, inevitably, reducing the risks of colon cancer (Wondrak et al. 2010). Besides, cinnamon is also a potent coagulant (Hossein et al. 2013). As mentioned earlier, essential oils from cinnamon are efficient anti-bacterial (Gende et al. 2008; Hili et al. 1997) and anti-fungal (Wang et al. 2005) agents, besides, being potent antioxidant (Mancini-Filho et al. 1998; Mathew and Abraham, 2006; Shobana and Naidu 2000), and anti-diabetic (Kim et al. 2006; Onderoglu et al. 1999; Subash Babu et al. 2007). Anti-inflammatory (Chao et al. 2005; Tung et al. 2008), nematicidal (Kong et al. 2007; Park et al. 2005), and anti-cancer (Koppikar et al. 2010; Kwon et al. 2010; Lu et al. 2010) properties of cinnamon are also reported. Cinnamon is also very effective against inflammation and rheumatism (Wagner et al. 1986). Methoxyhydroxy chalcone polymer present in cinnamon, if becomes a part of a daily diabetic diet, can effectively reduce the blood glucose level. No wonder, cinnamon is now being widely recognized as a herbal remedy in Europe and the USA (Gustafson 2015).

4.8 MEDICINAL PROPERTIES OF NATURALLY OCCURRING COMPOUNDS IN CINNAMON

The plethora of medicinal properties associated with cinnamon is attributable to the phytocompounds present in it. In order to find safer alternatives to modern medicines, the rapid rise in chronic and metabolic ailments over the past four to five decades has led to a large number of investigations to identify phytocompounds present in plants commonly used in traditional systems of medicine (Aludatt et al. 2016). Cinnamon is one such remedial plant recognized by several traditional systems of medicines around the world (Elliott and Brimacombe, 1987; Gupta, 2010).

Till now, numerous studies have been conducted to identify the various phytocompounds present in cinnamon and their medicinal effects. A brief summary of pharmacological activities of the phytocompounds identified in cinnamon is provided in Table 4.5. Thereafter, the studies leading to the identification of important medicinal effects of cinnamon are recollected in the subsequent paragraphs under corresponding sub-headings.

4.8.1 Anti-Inflammatory Activity

Release of cytokines, like, interleukin-1β (IL-1β), interleukin-6 (IL-6) and tumour necrosis factor-α (TNF-α) lead to inflammation, an innate immunological response to infection, to protect uninjured tissues and remove the damaged ones. Infectious agents trigger inflammation. For example, the bacterial toxin, lipopolysaccharide (LPS), induces pro-inflammatory cytokine response by macrophage activation and production of TNF-α and IL-1β (Miguel, 2010). There are designated immunogenic cells responsible for the synthesis of specific cytokines, such as, IL-1β is produced in macrophages, monocytes, endothelial and fibroblasts cells, whereas TNF-α is mainly a product of monocytic/macrophagic lineage. There is close molecular cross-talk between inflammation, oxidative stress and cancer. TNF-α and IL-1β, triggers the activation of the nuclear factor NF-κB, which in turn responsible for cancer development (Karin 2009; Lawrence 2009; Miguel 2010). Besides cytokines, prostaglandins, and nitric oxide (NO), produced due to the activities of cyclooxygenase (COX), and NO synthase, respectively (Miguel 2010) cause inflammation. In the following paragraphs, we will refer to the studies that have provided evidence and insight into the moderation of inflammatory pathways by different phytocompounds present in cinnamon.

Several studies have reported for anti-inflammatory activity of cinnamon and its essential oils (Chao et al. 2005; Tung et al. 2008, 2010). Orally administered water extract of *C. cassia* administered orally for 6 consecutive days at 20–100 mg/kg of body weight could significantly reduce the serum level of TNF-α in the LPS injected mice. This anti-inflammatory activity was attributed to

Table 4.5 Potential Health Benefit of Cinnamon and Its Constituents (Muhammad and Dewettinck 2017)

Species	Major Compound	Medicinal Traits	References
Cinnamomum burmannii	Proanthocyanidin, Cinnamic acid, Type-A-procyanidine polyphenol	Anti-diabetic, Anti-mutagenic effect	(Cao et al. 2010; Couturier et al. 2011; Prasad et al. 2009)
Cinnamomum aromaticum	Polyphenol, Cinnamaldehyde	Anti-inflammatory effect, Anti-tumour effect, Digestive system protection, Anti-diabetic effect, Anti-viral effect	(Hong et al. 2012; Kwon et al. 2009; Tabak et al. 1999; Yan et al. 2015; Yeh et al. 2013)
Cinnamomum zeylanicum	Cinnamaldehyde, Type-A-procyanidine polyphenol, Type-B-procyanidine polyphenol	Anti-hyperglycemic effect, Anti-Alzheimer effect, Anti-asthmatic effect, Anti-inflammatory and Anti-arthritic effects, Anti-diabetic effect	(Kandhare et al. 2017; Muthenna et al. 2014; Peterson et al. 2009; Subash Babu et al. 2007; Vetal et al. 2013)

the polyphenol compounds present in the extract (Hong et al. 2012). *C. cassia* ethanolic extract at 100 μg/mL was found to effectively reduce NO production, prostaglandin E2 and TNF-α secretion levels in LPS-activated cells (Yu et al. 2012). The essential oils from *C. osmophloeum* twig, containing 15.89% L-bornyl acetate and 4.07% cinnamaldehyde, effectively decreased NO production from LPS-stimulated macrophages at a concentration of 50 μg/mL. Cinnamaldehyde is implicated in this activity (Lin et al. 2008). Oral dose of 8 mg/kg body weight of Type-A procyanidin polyphenol (TAPP) acts as an anti-inflammatory agent (Ping et al. 2010). In fact, TAPP is associated with inhibition of lysyl oxidase (LOX) activity, COX enzyme activity, leukotrienes synthesis and pro-inflammatory cytokines production (TNF-α and IL-1β) (Miguel 2010).

Results from different independent studies collectively implicate the potential of cinnamon to inhibit NF-κB and COX-2 (isoform of COX), by interfering with the signalling among the transcription factors responsible for regulating their expressions (Mueller et al. 2010). Cinnamaldehyde is one of the major phytoconstituent of cinnamon, act as multilevel attenuators of NF-κB function through activation of peroxisome proliferator activated receptors (PPARs) (Huang et al. 2011). PPARs can interfere with the transcription activating capacity of the NF-κB complex or they can upregulate the suppressors of NF-κB activation and nuclear translocation or they might attenuate the DNA binding ability of NF-κB (Miguel, 2010). Yet another phytoconstituent, 2-hydroxycinnamaldehyde isolated from *C. cassia* bark was found to inhibit NO production by inhibiting NF-κB activation in B cells (Lee et al. 2005). NF-κB-mediated activation of Src/spleen-tyrosine kinase (Src/Syk) was observed in case of ethanolic extract of *C. cassia*, thus, indicating associated anti-inflammatory effects (Youn et al. 2008; Yu et al. 2012). Therefore, taken together, we could summarize that cinnamon is capable of targeting inflammation at multiple steps.

4.8.2 Anti-Tumour and Anti-Cancer Activity

The strong correlation between inflammation and cancer development is evident from the fact that IL-1β and TNF-α activated NF-κB production leads to cancer development. Thus, the anti-inflammatory property of cinnamon, by extension, serves as anti-tumour and anti-cancer activities. The second mode of preventing cancer development by cinnamon is through inhibition of COX pathway, that facilitates angiogenesis and metastasis (Muhammad and Dewettinck 2017). *C. cassia* water extract at 100 μg/mL and 0.5 mg/mL concentrations could respectively, block NF-κB activation (Azab et al. 2011) and prevent growth of hematologic tumour cells in male mice (Kwon et al. 2009). COX-2 and HIF-1α expressions were significantly reduced in the tumour tissues. Treatment for 24 and 48 h with 25 μg/mL essential oils isolated from *C. zeylanicum*, contain 68.95% cinnamaldehyde, reduced viability of cancerous H-ras active rat fibroblastic cells, 5RP7 by 10 and 70%, respectively (Unlu et al. 2010). These data indicate the interference of essential oils, which concomitantly induce apoptosis and control cell proliferation and differentiation with ras signal transduction.

Polyphenols and cinnamaldehyde present in cinnamon essential oil are implicated in prevention of cervical cancer, by reducing the migration of cancer cells in the cervix by inducing apoptosis through mitochondrial membrane potential depolarization (Koppikar et al. 2010). Procyanidins, present in the water extracts of cinnamon exhibited anti-angiogenic property associate with their ability to inhibit vascular endothelial growth factor subtype 2 (VEGFR2) kinase activity (Lu et al. 2010). CB403, the chemical synthesized from the cinnamaldehyde-derivative, 2-hydroxycinnamaldehyde have anti-tumour and anti-proliferative properties in *in vitro* as well as *in vivo* studies, thereby, indicating the potential of cinnamon as an anti-cancer agent (Jeong et al. 2003). Cinnamic acid, another phytoconstituent of cinnamon, was also observed to inhibit the production of TNFα-induced interleukin-8 (IL-8) and activity of NF-κB in A375 cells, thereby, proving its anti-cancer potential (Cabello et al. 2009). Taken together, we see that various phytocompounds present in cinnamon are capable of preventing cancer by targeting different pathways.

4.8.3 Anti-Diabetic Activity

A set of metabolic illnesses resulting to high blood glucose (blood sugar) levels is clinically recognized as diabetes. High blood glucose levels are detrimental to several organs leading to other complications, like, nephropathy, cardiovascular diseases, neuropathy, retinopathy, strokes, dementia, depression, and sexual dysfunction (Forbes and Cooper 2013).

Studies on the effect of cinnamon on diabetes are fewer in number as compared to those on its other medicinal activities. Nevertheless, the collection is sufficient to provide insight in to the anti-diabetic effect of cinnamon and identify the phytochemicals present in it, which are associated with the activity. In streptozocine-induced diabetic rats, anti-diabetic properties of cinnamon bark were observed two decades ago (Onderoglu et al. 1999) and even before that "insulin-potentiating factor" (IPF) was isolated from cinnamon (Khan et al. 1990). A reduction in blood glucose levels and increase in serum insulin of type-2 diabetic animals was observed in response to treatment with *C. cassia* extract at 200 mg lyophilized extract/kg mice body weight (Kim, Hyun and Choung 2006). Likewise, in terms of decreased insulin sensitivity, *C. burmanii* water extract with a meal of 20 g/kg comprising more than 5 percent TAPP per gram cinnamon extract increased hepatic glycogen synthesis. This implicated its ability to maintaining homeostasis due to high dietary consumption when fat or carbohydrates are overweight (Couturier et al. 2011). Anti-diabetic potential of cinnamon is attributable, since it increases insulin release, decreases intestinal glucose absorption, promotes glycogen production and triggers PPAR-γ (Mueller and Jungbauer 2009). Additionally, the polyphenols in cinnamon can regulate glucose metabolism and repair pancreatic beta cells (Li et al. 2013). In patients with type 2 diabetic mellitus and pre-diabetic conditions, cinnamon was useful for decreasing diastolic or systolic blood pressure (Akilen et al. 2013). Type-A procyanidine polymers (TAPP) isolated from cinnamon were found to be insulin-like molecules (Anderson et al. 2004). Subsequently, naphthalenemethyl ester, a new compound from hydroxycinnamic acid derivatives, was identified, which could lower blood glucose levels (Kim Hyun and Choung 2006). Several polyphenols such as, catechin (1.9%), rutin (90.0672%), quercetin (0.172%), isorhamnetin (0.103%) and kaempferol (0.016%) purified from aqueous extract of cinnamon through high performance liquid chromatography (HPLC) exhibited insulin-like activity (Yang, Li, and Chuang 2012; Bin Li et al. 2008; Cao, Polansky and Anderson 2007). The aqueous extract of cinnamon also decreased the absorption of alanine, vital for gluconeogenesis, in the rat intestine (Kreydiyyeh, Usta and Copti 2000). With this body of evidence, we can recognize the anti-diabetic effect of cinnamon.

4.8.4 Anti-Hypertriglyceridemia

Feeding cinnamon reduced plasma triglycerides and high density lipoprotein (HDL) cholesterol in rats (Kim, Hyun and Choung 2006; Rahman et al. 2013). Similar effects were also observed in case of humans (Khan et al. 2003). Such improvements in lipid profile due to cinnamon can be attributed to its phytoconstituent, cinnamaldehyde that can increase the activity of lecithin cholesterol acyl transferase (Subash Babu, Prabuseenivasan, and Ignacimuthu 2007). Besides, polyphenols in cinnamon are also likely to influence lipid metabolism (Li et al. 2013) by effectively inhibiting hepatic lipid peroxidation (Lee et al. 2003) and consequently, protecting from concomitant cytotoxic and genotoxic effects (Esterbauer, Muskiet and Horrobin 1993).

4.8.5 Anti-Viral Activity

Traditional herbal drugs and pure herbal extracts are always regarded as a great resource for the formation of new anti-virals agents over the years (Yang et al. 2020). In order to suppress viral infections, the possible anti-virus medicines must target each step of the viral cycle of life, for example, viral entry, reproduction, assembly, and release (Lin et al. 2015).

A unique cinnamon extract, IND02, which is a rich source of pentamer polyphenols and type A procyanidin trimer, has anti-viral activity at very low concentration (μM) against CCR5 and CXCR4 receptor to prevent entry of the viruses (Connell et al. 2016). Cluster of differentiation antigen 4 (CD4), Heparan sulphate (HS), and co-receptor (CCR5/CXCR4) were found as hot spots on cell surface for attachment and entry of the HIV and Hepatitis C virus (HCV). Connel *et al.*, demonstrated experimentally using a technique called "surface plasmon resonance-based binding assay" that HIV-1 glycoprotein binding with cellular CD4, HS and a monoclonal antibody (mAb17b) was inhibited in presence of cinnamon extract, IND02. Anti-HIV-1 activity was recorded by employing β-galactosidase activity in a "multinuclear activation of indicator" assay and surprisingly, the results showed that IND02-trimer was found to prevent HIV-1 infection (Connell et al. 2016). Calland *et al.*, demonstrated that as compared to other natural anti-HCV molecules, like epigallocatechin-3-gallate (ECGC) and silicone, cinnamon extract IND02 appears to be a more potent entry inhibitor of Hepatitis C virus (HCV) at very low concentration (EC50 of 2.06 μM). Overall results of this study showed the ability of cinnamon extract as a barrier against entry of varieties of viruses (Calland et al. 2012; Fauvelle et al. 2017). Thus, cinnamon extract could be a very promising candidate for preventing viral attachment and reducing viral load.

4.8.6 Prevent Neurological Disorders

Implications of cinnamon and its phytoconstituents and their derivatives in the managing neurological disorders are compelling. isolated From cinnamon's water-soluble extract, procyanidin type-A trimer (trimer 1) was isolated and reducing cell swelling by affecting intracellular calcium (Ca^{2+}) effective in efflux and alleviated oxygen glucose deprivation-induced detrimental effects on glutamate uptake through moderation of mitochondria (Panickar et al. 2012). Sodium benzoate, one of the metabolites present in cinnamon modulated mevalonate to upregulate PD protein 7 (PARK7), one of the main neuro-protective proteins (Bonifati, Oostra and Heutink 2004) (Brahmachari, Jana, and Pahan, 2009; Khasnavis and Pahan, 2012). Neurotrophin-3 (NT-3) and neurotropic factors BDNF (brain-derived neurotropic factors) were upregulated by cinnamon and its metabolite sodium benzoate in the mouse central nervous system (Jana et al. 2013). CEppt, the natural compound isolated from cinnamon extract was observed to reduce the formation of toxic β-amyloid polypeptide (Aβ) oligomers and thereby, prevented its toxicity on neuronal pheochromocytoma (PC12) cells (Frydman-Marom et al. 2011). Thus, CEppt resolved the reduced permanence, deficiencies in locomotion, and completely eradicated the tetrameric species of Aβ in the brain. This, in turn reduced plaques and improved the cognitive performance in transgenic mice models of Parkinson's Disease (Frydman-Marom et al. 2011). The aqueous extract of *C. zeylanicum* not only reduced filament formation and tau aggregation, but also facilitated complete fragmentation of tau filaments in Alzheimer's disease brain (Peterson et al. 2009), thereby, indicating the potential of cinnamon in the treatment of Alzheimer's disease.

4.8.7 Prevent Cardiovascular Diseases

The evidence in support of cinnamon as a potent cardiovascular fortifying agent are quite encouraging. C. cassia extracted 2-methoxycinnamaldehyde (2-MCA), reduced the vascular cell adhesion molecule-1 (VCAM-1) expression in endothelial cells triggered by TNFα and suggesting an ameliorative function in reperfusion/ischaemia damage as the result of hemeoxygenase-1 (HO-1) induction (Hwa et al. 2012). Cinnamaldehyde and cinnamic acid, from *C. cassia* are effective against myocardial ischemia (Song et al. 2013). The lignan, cinnamophilin isolated from *C. philippinensis* exhibited thromboxane A2 (TXA2) receptor blocking activity in rats and guinea pigs (Yu, Wu and Teng 1994), implicating inhibition of thromboxane receptor-mediated vascular smooth muscle cell proliferation and by extension potential in preventing vascular diseases and atherosclerosis

(Ko et al. 1995). Through impediment of Ca^{2+} release and Ca^{2+} influx, cinnamaldehyde has been observed to have vasodilatory function (Xue et al. 2011). Therefore, the phytoconstituents of cinnamon are adept in preventing cardiac malfunctioning.

4.9 BIOACCESSIBILITY AND BIOAVAILABILITY

Bioaccessibility refers to the amount of a component liberated from the food matrix that is available for absorption in the small intestine (Durak, Gawlik-Dziki and Pecio 2014). A thorough knowledge of the bioaccessibility of functional foods are essential for the understanding of their bioactivities (Rodríguez-Roque et al. 2015). The bioaccessibility and bioavailability of substances are influenced by the compound's starting concentration, the dietary matrix, and gastrointestinal parameters (Tagliazucchi et al. 2012). Components of cinnamon and cinnamon itself have received limited research into their bioavailability and bioaccessibility. The bioaccessibility of cinnamon's total polyphenol was approximately 79%, which was measured through a digestion model (Helal et al. 2014).

4.10 TOXICITY

The Bundesinstitut for riskobewertung (BFR's) recommendation holds good for the intake of high coumarin-containing Cassia Cinnamon. Since, the Tolerable Daily Intake (TDI) of coumarin is 0.1 mg/kg of body weight, therefore consumers who frequently eat large quantities of cinnamon as a spice are advised to opt for the low coumarin ceylon cinnamon (BFR Fact sheet 2012).

Species differences in toxico-kinetics and metabolism of coumarin in humans and rodents were observed, as reported by the SCF (1999). For example, in humans, enzyme CYP2A6 converts coumarin to 7-Hydroxycoumarin for downstream detoxification. Contrastingly, in rats and mice coumarin is bioactivated to 3,4-coumarin epoxide as the first in biotransformation reaction, which is then rapidly converted to *o*-hydroxyphenyl acetaldehyde (*o*-HPA) (Sivapriya and John 2016).

4.11 CLINICAL STUDIES ON CINNAMON

Due to significant restorative potential of cinnamon, several clinical investigations were conducted. A total 26 clinical trials including six studies, 14 completed studies, two recruited studies, two terminated studies and two active studies are listed in the Clinical Trials Govt. database. Most of these studies still under progress (Table 4.6) (Kumar, Kumari, and Mishra, 2019).

4.12 CONSUMER ACCEPTANCE

Spices are the traditional flavourant for foods (Durak, Gawlik-Dziki and Pecio 2014). Asian cultures, like Chinese, Indonesian, Arab, Korean, Turkish, and Malaysian recognize wide spread use of cinnamon in their traditional cuisines (Elshafie et al. 2012; Helal et al. 2014; Nor et al. 2012; Song et al. 2009; Suryati et al. 2014; Tagliazucchi et al. 2012; Yeğin and Üren 2008). Egypt has a popular traditional cinnamon bark powder and water drink, which is made in 200 mL of water by mixing of 4 g cinnamon powder, below its boiling point (Helal et al. 2014). In some European countries, cinnamon-flavoured foods are gaining popularity, where, *C. cassia* is a greater favourite than *C. verum* (Ballin and Sørensen 2014; Blahová and Svobodov 2012; Fotland et al. 2012; Seo,

Table 4.6 List of Completed and Undergoing Clinical Trial for Cinnamon (Kumar, Kumari and Mishra 2019)

Clinical Trials	Title
NCT01301521	Cinnamon trial-lifestyle intervention plus water-soluble cinnamon extract on lowering blood glucose in pre-diabetics
NCT03778099	The effect of cinnamon on ovulation induction in women with polycystic ovary syndrome
NCT01302743	Cinnamon bark, water-soluble cinnamon extract, and metformin for treatment of type 2 DM
NCT01847053	Bioavailability study of cinnamon in healthy subjects
NCT00331279	The effect of cinnamon extract on insulin resistance parameters in polycystic ovary syndrome: a pilot study
NCT00445354	Randomized controlled clinical trial of cinnamon to lower haemoglobin A1c
NCT00371800	The effect of cinnamon on HbA1c among adolescents with type I diabetes
NCT01483118	Cinnamon extract on menstrual cycles in polycystic ovary syndrome (PCOS)
NCT00237640	Effect of cinnamon on glucose and lipid levels in non-insulin dependent type 2 diabetes mellitus
NCT00951639	Cassia cinnamon for glucose uptake in young women
NCT0084898	Is there a metabolic effect of cinnamon on HbA1c, blood pressure and serum lipids in type 2 diabetes mellitus?
NCT00970541	Effect of cinnamon extract on insulin resistance in polycystic ovary syndrome
NCT03061916	Cinnamon and ginger in comparison to chlorhexidine gluconate 0.2% on oral *Streptococcus mutants*
NCT03219411	Effects of Cinnamon supplementation on glucose metabolism in patients with pre-diabetes
NCT02942056	The effect of cinnamon cassia on diabetes control and cardio metabolic risk factors in adults with type 2 diabetes mellitus
NCT00479973	The anti-diabetic and cholesterol-lowering effects of cinnamon and cassia bark

Buschhüter and Hummel 2009). A commercial biscuit made in mass production in Germany contains about 1% cinnamon (Bundesinstitut für Risikobewertung 2006). Flavoured cinnamon meat in northern Greece is a famous traditional food (Hanlidou et al. 2004). Cinnamon roll is, a favourite American dessert (Slining, Mathias and Popkin 2013).

4.13 CONCLUSION

Cinnamon has been an integral part of several ancient cultures in capacities of both spice and medicine. Cinnamon possesses many bioactive chemicals with pharmaceutical characteristics. The knowledge of scientists and general population about its use is crucial, since it can be served as an option, which is affordable and readily available as compared to other costly, often ineffective, conventional medicinal products. Presence of various bioactive compound in cinnamon made it to be effective against different types of disease (like cancer, inflammation, diabetes, microbial infection, abdominal disorder, nerves disorder, asthma, urinary infection, bronchitis, arthritis, blood pressure and anaemia), in food industry (as anti-microbial, as colourant, as flavourant and encapsulation of essential oil for slow release of volatile anti-microbial chemical on food) and in cosmetic sector. Thus, cinnamon as a multi-targeting medicinal entity fits well into the current 'one drug-multitarget' paradigm of modern medicine. We have summarized the pre-existing studies on pharmacological activities of cinnamon in Figure 4.2. We have tried to provide a comprehensive status quo of cinnamon as a spice and medicine in this chapter by covering all aspects relevant to both the identities of this herb. Despite identification and isolation of several phytocompounds from cinnamon, still much remains unexplored in terms of safety, toxicity and rigorous validation for medicinal aspects of individual phytocompounds, and of course, better insight into mechanism of its action as medicinal herb.

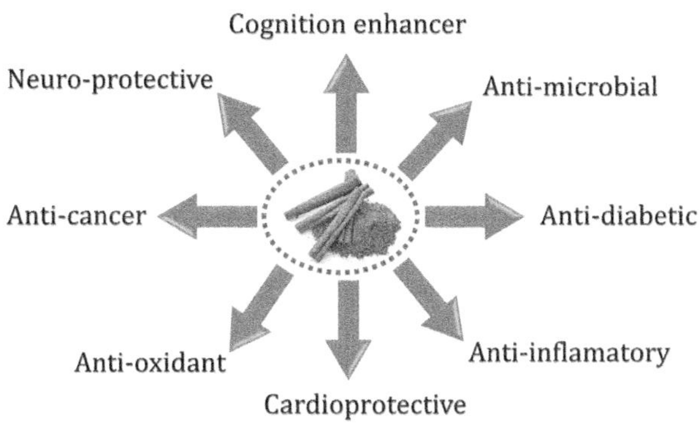

Figure 4.2 Various pharmacological activity of cinnamon.

REFERENCES

Abdelwahab, Siddig Ibrahim, Abdalbasit Adam Mariod, Manal Mohamed Elhassan Taha, Faridah Qamaruz Zaman, Adil Hassan Ahmed Abdelmageed, Shamsul Khamis, Yasodha Sivasothy, and Khalijah Awang. 2017. "Chemical Composition and Antioxidant Properties of the Essential Oil of Cinnamomum Altissimum Kosterm. (Lauraceae)." *Arabian Journal of Chemistry* 10 (1). King Saud University: 131–135. doi:10.1016/j.arabjc.2014.02.001.

Adarsh, A., C. Bharath, B. M. Kanthesh, and N. Raghu. 2020. "Phytochemical Screening and Antimicrobial Activity of 'Cinnamon Zeylanicum.'" *International Journal of Pharmaceutical Research and Innovation* 13: 22–33. doi:10.13140/RG.2.2.33937.04962.

Akilen, Rajadurai, Zeller Pimlott, Amalia Tsiami, and Nicola Robinson. 2013. "Effect of Short-Term Administration of Cinnamon on Blood Pressure in Patients with Prediabetes and Type 2 Diabetes." *Nutrition* 29 (10). Elsevier Inc.: 1192–1196. doi:10.1016/j.nut.2013.03.007.

Alpsoy, Lokman. 2010. "Inhibitory Effect of Essential Oil on Aflatoxin Activities." *African Journal of Biotechnology* 9 (17): 2474–2481. doi:10.4314/ajb.v9i17.

Aludatt, Muhammad H., Taha Rababah, Mohammad N. Alhamad, Sana Gammoh, Khalil Ereifej, Ayman Johargy, Stan Kubow, Ali M. Almajwal, and Mervat Rawashdeh. 2016. "Optimization of Phenolic Content, Antioxidant, and Inhibitory Activities of Aglucosidase and Angiotensin Converting (AC) Enzymes from Zingiber Officinale Z." *International Journal of Food Properties* 19 (6): 1303–1316. doi: 10.1080/10942912.2015.1063066.

Anderson, Richard A., C. Leigh Broadhurst, Marilyn M. Polansky, Walter F. Schmidt, Alam Khan, Vincent P. Flanagan, Norberta W. Schoene, and Donald J. Graves. 2004. "Isolation and Characterization of Polyphenol Type-A Polymers from Cinnamon with Insulin-like Biological Activity." *Journal of Agricultural and Food Chemistry* 52 (1): 65–70. doi:10.1021/jf034916b.

Arancibia, M., B. Giménez, M. E. López-Caballero, M. C. Gómez-Guillén, and P. Montero. 2014. "Release of Cinnamon Essential Oil from Polysaccharide Bilayer Films and Its Use for Microbial Growth Inhibition in Chilled Shrimps." *LWT - Food Science and Technology* 59 (2P1): 989–995. doi:10.1016/j.lwt.2014.06.031.

Azab, Khaled Sh, Abdel Halem A. Mostafa, Ehab M.M. Ali, and Mohamed A.S. Abdel-Aziz. 2011. "Cinnamon Extract Ameliorates Ionizing Radiation-Induced Cellular Injury in Rats." *Ecotoxicology and Environmental Safety* 74 (8). Elsevier: 2324–2329. doi:10.1016/j.ecoenv.2011.06.016.

Ballin, Nicolai Z., and Ann T. Sørensen. 2014. "Coumarin Content in Cinnamon Containing Food Products on the Danish Market." *Food Control* 38 (1). Elsevier Ltd: 198–203. doi:10.1016/j.foodcont.2013.10.014.

Bayoumi, S. 1992. "Bacteriostatic Effect of Some Spices and Their Utilization in the Manufactureof Yoghurt." *Chemie, Mikrobiologie, Technologie Der Lebensmittel* 14 (1): 21–26.

BFR Fact sheet. 2012. "Cassia Cinnamon with High Coumarin Contents to Be Consumed in Moderation - BfR." *The Federal Institute for Risk Assessment (*BFR*).* http://www.bfr.bund.de/en/press_information/2012/26/cassia_cinnamon_with_high_coumarin_contents_to_be_consumed_in_moderation-131836.html.

Bigliardi, B., and F. Galati. 2013. "Innovation Trends in the Food Industry: The Case of Functional Foods." *Trends in Food Science and Technology* 31 (2). Elsevier Ltd: 118–129. doi:10.1016/j.tifs.2013.03.006.

Blahová, Jana, and Zdeňká Svobodov. 2012. "Assessment of Coumarin Levels in Ground Cinnamon Available in the Czech Retail Market." *The Scientific World Journal* 2012. doi:10.1100/2012/263851.

Błaszczyk, N., A. Rosiak, and J. Kałuzna-Czaplinska. 2021. "The Potential Role of Cinnamon in Human Health." *Forests* 12 (648): 1–17. doi:https://doi.org/ 10.3390/f12050648.

Bonifati, Vincenzo, Ben A. Oostra, and Peter Heutink. 2004. "Linking DJ-1 to Neurodegeneration Offers Novel Insights for Understanding the Pathogenesis of Parkinson's Disease." *Journal of Molecular Medicine* 82 (3): 163–174. doi:10.1007/s00109-003-0512-1.

Brahmachari, Saurav, Arundhati Jana, and Kalipada Pahan. 2009. "Sodium Benzoate, a Metabolite of Cinnamon and a Food Additive, Reduces Microglial and Astroglial Inflammatory Responses." *The Journal of Immunology* 183 (9): 5917–5927. doi:10.4049/jimmunol.0803336.

Bröring, S. 2010. "Innovation Strategies for Functional Foods and Supplements—Challenges of the Positioning between Foods and Drugs." *Food Science & Technology (IFST), Sheffield* 7 (8): 111–123. doi:10.1616/1476-2137.15996.

Bundesinstitut für Risikobewertung. 2006. "High Daily Intakes of Cinnamon: Health Risk Cannot Be Ruled Out." *BfR Health Assessment No. 044/2006, Germany: Bundesinstitut Für Risikobewertung* (044): 1–15.

Cabello, Christopher M., Warner B. Bair, Sarah D. Lamore, Stephanie Ley, Alexandra S. Bause, Sara Azimian, and Georg T. Wondrak. 2009. "The Cinnamon-Derived Michael Acceptor Cinnamic Aldehyde Impairs Melanoma Cell Proliferation, Invasiveness, and Tumor Growth." *Free Radical Biology and Medicine* 46 (2). Elsevier Inc.: 220–231. doi:10.1016/j.freeradbiomed.2008.10.025.

Calland, Noémie, Jean Dubuisson, Yves Rouillé, and Karin Séron. 2012. "Hepatitis C Virus and Natural Compounds: A New Antiviral Approach?." *Viruses* 4 (10): 2197–2217. doi:10.3390/v4102197.

Cao, Heping, Donald J. Graves, and Richard A. Anderson. 2010. "Cinnamon Extract Regulates Glucose Transporter and Insulin-Signaling Gene Expression in Mouse Adipocytes." *Phytomedicine* 17 (13). Elsevier GmbH.: 1027–1032. doi:10.1016/j.phymed.2010.03.023.

Cao, Heping, Marilyn M. Polansky, and Richard A. Anderson. 2007. "Cinnamon Extract and Polyphenols Affect the Expression of Tristetraprolin, Insulin Receptor, and Glucose Transporter 4 in Mouse 3T3-L1 Adipocytes." *Archives of Biochemistry and Biophysics* 459 (2): 214–222. doi:10.1016/j.abb.2006.12.034.

Cha, Jaeyoon, Chong Tai Kim, Tae Eun Kim, and Yong Jin Cho. 2019. "Optimization of Subcritical Extraction Process for Cinnamon (Cinnamomum Cassia Blume) Using Response Surface Methodology." *Food Science and Biotechnology* 28 (6). Springer Singapore: 1703–1711. doi:10.1007/s10068-019-00616-6.

Chao, L. K., K.-F. Hua, H.-Y. Hsu, S.-S. Cheng, J.-Y. Liu, and T. S. Chang. 2005. "Study on the Antiinflammatory Activity of Essential Oil from Leaves of Cinnamomum Osmophloeum." *Journal of Agricultural and Food Chemistry* 53 (18): 7274–7278. doi:10.1021/jf051151u.

Charles, D. J. 2013. *Antioxidant Properties of Spices, Herbs and Other Sources.*

Chen, Guanghui, Fengrui Sun, Shougui Wang, Weiwen Wang, Jipeng Dong, and Fei Gao. 2020. "Enhanced Extraction of Essential Oil from Cinnamomum Cassia Bark by Ultrasound Assisted Hydrodistillation." *Chinese Journal of Chemical Engineering.* doi:10.1016/j.cjche.2020.08.007.

Cheng, Diana M., Peter Kuhn, Alexander Poulev, Leonel E. Rojo, Mary Ann Lila, and Ilya Raskin. 2012. "In Vivo and in Vitro Antidiabetic Effects of Aqueous Cinnamon Extract and Cinnamon Polyphenol-Enhanced Food Matrix." *Food Chemistry* 135 (4). Elsevier Ltd: 2994–3002. doi:10.1016/j.foodchem.2012.06.117.

Connell, Bridgette Janine, Sui Yuan Chang, Ekambaranellore Prakash, Rahima Yousfi, Viswaraman Mohan, Wilfried Posch, Doris Wilflingseder, et al. 2016. "A Cinnamon-Derived Procyanidin Compound Displays Anti-HIV-1 Activity by Blocking Heparan Sulfate- and Co-Receptor- Binding Sites on Gp120 and Reverses T Cell Exhaustion via Impeding Tim-3 and PD-1 Upregulation." *PLoS ONE* 11 (10): 1–18. doi:10.1371/journal.pone.0165386.

Couturier, Karine, Bolin Qin, Cecile Batandier, Manar Awada, Isabelle Hininger-Favier, Frederic Canini, Xavier Leverve, Anne Marie Roussel, and Richard A. Anderson. 2011. "Cinnamon Increases Liver Glycogen in an Animal Model of Insulin Resistance." *Metabolism: Clinical and Experimental* 60 (11). Elsevier B.V.: 1590–1597. doi:10.1016/j.metabol.2011.03.016.

Crowe, K. M., and C. Francis. 2013. "Position of the Academy of Nutrition and Dietetics: Functional Foods." *Journal of the Academy of Nutrition and Dietetics* 113 (8): 1096–1103. doi:10.1016/j.jand.2013.06.002.

Durak, Agata, Urszula Gawlik-Dziki, and Łukasz Pecio. 2014. "Coffee with Cinnamon - Impact of Phytochemicals Interactions on Antioxidant and Anti-Inflammatory in Vitro Activity." *Food Chemistry* 162. Elsevier Ltd: 81–88. doi:10.1016/j.foodchem.2014.03.132.

Dussault, Dominic, Khanh Dang Vu, and Monique Lacroix. 2014. "In Vitro Evaluation of Antimicrobial Activities of Various Commercial Essential Oils, Oleoresin and Pure Compounds against Food Pathogens and Application in Ham." *Meat Science* 96 (1). Elsevier Ltd: 514–520. doi:10.1016/j.meatsci.2013.08.015.

Elliott, Stephen, and Joseph Brimacombe. 1987. "The Medicinal Plants of Gunung Leuser National Park, Indonesia." *Journal of Ethnopharmacology* 19 (3): 285–317. doi:10.1016/0378-8741(87)90006-7.

Elshafie, M. M., I. A. Nawar, M. A. Algamal, and S. M. Ahmad. 2012. "Evaluation of the Biological Effects for Adding Cinnamon Volatile Oil and TBHQ as Antioxidant on Rats' Lipid Profiles." *Asian Journal of Plant Sciences* 11 (3): 100–108.

Embuscado, M. E. 2015. "Spices and Herbs: Natural Sources of Antioxidants - A Mini Review." *Journal of Functional Foods* 18. Elsevier Ltd: 811–819. doi:10.1016/j.jff.2015.03.005.

Esterbauer, H., F. Muskiet, and D. F. Horrobin. 1993. "Cytotoxicity and Genotoxicity of Lipid-Oxidation Products." *American Journal of Clinical Nutrition* 57 (5 SUPPL.). doi:10.1093/ajcn/57.5.779S.

Farias, Ana Patrícia P., Odair dos S. Monteiro, Joyce Kelly R. da Silva, Pablo Luis B. Figueiredo, Antonia Alice C. Rodrigues, Ildenice N. Monteiro, and José Guilherme S. Maia. 2020. "Chemical Composition and Biological Activities of Two Chemotype-Oils from Cinnamomum Verum J. Presl Growing in North Brazil." *Journal of Food Science and Technology* 57 (9): 3176–3183. doi:10.1007/s13197-020-04288-7.

Fauvelle, Catherine, Melanie Lambotin, Laura Heydmann, Ekambaranellore Prakash, Sunil Bhaskaran, Mohan Vishwaraman, Thomas F. Baumert, and Christiane Moog. 2017. "A Cinnamon-Derived Procyanidin Type A Compound Inhibits Hepatitis C Virus Cell Entry." *Hepatology International* 11 (5). Springer India: 440–445. doi:10.1007/s12072-017-9809-y.

Forbes, Josephine M., and Mark E. Cooper. 2013. "Mechanisms of Diabetic Complications." *Physiological Reviews* 93 (1): 137–188. doi:10.1152/physrev.00045.2011.

Fotland, T., J. E. Paulsen, T. Sanner, J. Alexander, and T. Husøy. 2012. "Risk Assessment of Coumarin Using the Bench Mark Dose (BMD) Approach: Children in Norway Which Regularly Eat Oatmeal Porridge with Cinnamon May Exceed the TDI for Coumarin with Several Folds." *Food and Chemical Toxicology* 50 (3–4). Elsevier Ltd: 903–912. doi:10.1016/j.fct.2011.12.005.

Frydman-Marom, Anat, Aviad Levin, Dorit Farfara, Tali Benromano, Roni Scherzer-Attali, Sivan Peled, Robert Vassar, et al. 2011. "Orally Administrated Cinnamon Extract Reduces β-Amyloid Oligomerization and Corrects Cognitive Impairment in Alzheimer's Disease Animal Models." *PLoS ONE* 6 (1): 1–11. doi:10.1371/journal.pone.0016564.

Gende, Liesel Brenda, Ignazio Floris, Rosalia Fritz, and E. G. U. Martin Javier Aras. 2008. "Antimicrobial Activity of Cinnamon (Cinnamomum Zeylanicum) Essential Oil and Its Main Components against Paenibacillus Larvae from Argentine." *Bulletin of Insectology* 61 (1): 1–4.

Goñi, P., P. López, C. Sánchez, R. Gómez-Lus, R. Becerril, and C. Nerín. 2009. "Antimicrobial Activity in the Vapour Phase of a Combination of Cinnamon and Clove Essential Oils." *Food Chemistry* 116 (4). Elsevier Ltd: 982–989. doi:10.1016/j.foodchem.2009.03.058.

Gotmare, S. and Tambe, E. 2019. "Identification of chemical constituents of cinnamon bark oil by GCMS and comparative study garnered from five different countries." *Global Journal of Science Frontier Research: C Biological Science* 19 (1): 35–42.

Gupta, Mradu. 2010. "Pharmacological Properties and Traditional Therapeutic Uses of Important Indian Spices: A Review." *International Journal of Food Properties* 13 (5): 1092–1116. doi:10.1080/10942910902963271.

Gustafson, Craig. 2015. "Mark Blumenthal: Quality and Efficacy of Herbal Medicines." *Integrative Medicine (Boulder)* 14 (4): 54–59.

Hanlidou, E., R. Karousou, V. Kleftoyanni, and S. Kokkini. 2004. "The Herbal Market of Thessaloniki (N Greece) and Its Relation to the Ethnobotanical Tradition." *Journal of Ethnopharmacology* 91 (2–3): 281–299. doi:10.1016/j.jep.2004.01.007.

Helal, Ahmed, Davide Tagliazucchi, Elena Verzelloni, and Angela Conte. 2014. "Bioaccessibility of Polyphenols and Cinnamaldehyde in Cinnamon Beverages Subjected to in Vitro Gastro-Pancreatic Digestion." *Journal of Functional Foods* 7 (1). Elsevier Ltd: 506–516. doi:10.1016/j.jff.2014.01.005.

Herman, Anna, Andrzej Przemysław Herman, Beata Wanda Domagalska, and Andrzej Młynarczyk. 2013. "Essential Oils and Herbal Extracts as Antimicrobial Agents in Cosmetic Emulsion." *Indian Journal of Microbiology* 53 (2): 232–237. doi:10.1007/s12088-012-0329-0.

Hili, P., C. S. Evans, and R. G. Veness. 1997. "Antimicrobial Action of Essential Oils: The Effect of Dimethylsulphoxide on the Activity of Cinnamon Oil." *Letters in Applied Microbiology* 24 (4): 269–275. doi:10.1046/j.1472-765X.1997.00073.x.

Hill, Laura E., T. Matthew Taylor, and Carmen Gomes. 2013. "Antimicrobial Efficacy of Poly (DL-Lactide-Co-Glycolide) (PLGA) Nanoparticles with Entrapped Cinnamon Bark Extract against Listeria Monocytogenes and Salmonella Typhimurium." *Journal of Food Science* 78 (4). doi:10.1111/1750-3841.12069.

Hong J. W., Yang G. E., Kim Y. B., Eom S. H., Lew J. H., and Kang H. et. al. 2012. "Anti-Inflammatory Activity of Cinnamon Water Extract in Vivo and in Vitro LPS-Induced Models." *BMC Complementary And Alternative Medicine* 12 (20): 237.

Hossein, Nazari, Zohrvand Zahra, Mohammadi Abolfazl, Safaryeh Mahdi, and Khaleghian Ali. 2013. "Effect of Cinnamon Zeylanicum Essence and Distillate on the Clotting Time." *Journal of Medicinal Plants Research* 7 (19): 1339–1343. doi:10.5897/JMPR12.1198.

Huang, Bo, Hai Dan Yuan, Do Yeon Kim, Hai Yan Quan, and Sung Hyun Chung. 2011. "Cinnamaldehyde Prevents Adipocyte Differentiation and Adipogenesis via Regulation of Peroxisome Proliferator-Activated Receptor-γ (PPARγ) and AMP-Activated Protein Kinase (AMPK) Pathways." *Journal of Agricultural and Food Chemistry* 59 (8): 3666–3673. doi:10.1021/jf104814t.

Hwa, Jeong Seok, Yong Chun Jin, Young Soo Lee, Young Shin Ko, Young Min Kim, Lian Yu Shi, Hye Jung Kim, et al. 2012. "2-Methoxycinnamaldehyde from Cinnamomum Cassia Reduces Rat Myocardial Ischemia and Reperfusion Injury in Vivo Due to HO-1 Induction." *Journal of Ethnopharmacology* 139 (2). Elsevier Ireland Ltd: 605–615. doi:10.1016/j.jep.2011.12.001.

Jakhetia, V., R. Patel, P. Khatri, N. Pahuja, S. Garg, A. Pandey, and S. Sharma. 2010. "Cinnamon- a Pharmacological Review." *Journal of Advanced Scientific Research* 1 (2): 19–23.

Jana, Arundhati, Khushbu K. Modi, Avik Roy, John A. Anderson, Richard B. Van Breemen, and Kalipada Pahan. 2013. "Up-Regulation of Neurotrophic Factors by Cinnamon and Its Metabolite Sodium Benzoate: Therapeutic Implications for Neurodegenerative Disorders." *Journal of Neuroimmune Pharmacology* 8 (3): 739–755. doi:10.1007/s11481-013-9447-7.

Jayaprakasha, G. K., and L. J. M. Rao. 2011. "Chemistry, Biogenesis, and Biological Activities of Cinnamomum Zeylanicum." *Critical Reviews in Food Science and Nutrition* 51 (6): 547–562. doi:10.1080/10408391003699550.

Jeong, Ha Won, Dong Cho Han, Kwang Hee Son, Mi Young Han, Jong Seok Lim, Ji Hong Ha, Chang Woo Lee, Hwan Mook Kim, Hyoung Chin Kim, and Byoung Mog Kwon. 2003. "Antitumor Effect of the Cinnamaldehyde Derivative CB403 through the Arrest of Cell Cycle Progression in the G2/M Phase." *Biochemical Pharmacology* 65 (8): 1343–1350. doi:10.1016/S0006-2952(03)00038-8.

Jessica Elizabeth, De La Torre, Fatma Gassara, Anne Patricia Kouassi, Satinder Kaur Brar, and Khaled Belkacemi. 2017. "Spice Use in Food: Properties and Benefits." *Critical Reviews in Food Science and Nutrition* 57 (6): 1078–1088. doi:10.1080/10408398.2013.858235.

Jiao, Lihua, Xin Zhang, Lianqi Huang, Hao Gong, Biao Cheng, Yue Sun, Yixuan Li, Qi Liu, Ling Zheng, and Kun Huang. 2013. "Proanthocyanidins Are the Major Anti-Diabetic Components of Cinnamon Water Extract." *Food and Chemical Toxicology* 56. Elsevier Ltd: 398–405. doi:10.1016/j.fct.2013.02.049.

Kallel, Imen, Bilel Hadrich, Bochra Gargouri, Amina Chaabane, Saloua Lassoued, Radhouane Gdoura, Ahmed Bayoudh, and Ezeddine Ben Messaoud. 2019. "Optimization of Cinnamon (Cinnamomum Zeylanicum Blume) Essential Oil Extraction: Evaluation of Antioxidant and Antiproliferative Effects." *Evidence-Based Complementary and Alternative Medicine* 2019. doi:10.1155/2019/6498347.

Kandhare, Amit D., Urmila M. Aswar, Vishwaraman Mohan, and Prasad A. Thakurdesai. 2017. "Ameliorative Effects of Type-A Procyanidins Polyphenols from Cinnamon Bark in Compound 48/80-Induced Mast Cell Degranulation." *Anatomy and Cell Biology* 50 (4): 275–283. doi:10.5115/acb.2017.50.4.275.

Karin, Michael. 2009. "NF-KappaB as a Critical Link between Inflammation and Cancer." *Cold Spring Harbor Perspectives in Biology* 1 (5): 1–15. doi:10.1101/cshperspect.a000141.

Kaul, Pran N., Arun K. Bhattacharya, Bhaskaruni R. Rajeswara Rao, Kodakandla V. Syamasundar, and Srinivasaiyer Ramesh. 2003. "Volatile Constituents of Essential Oils Isolated from Different Parts of Cinnamon (Cinnamomum Zeylanicum Blume)." *Journal of the Science of Food and Agriculture* 83 (1): 53–55. doi:10.1002/jsfa.1277.

Khan, Alam, Noella A. Bryden, Marilyn M. Polansky, and Richard A. Anderson. 1990. "Insulin Potentiating Factor and Chromium Content of Selected Foods and Spices." *Biological Trace Element Research* 24 (2–3): 183–188. doi:10.1007/BF02917206.

Khan, Alam, Mahpara Safdar, Mohammad Muzaffar Ali Khan, Khan Nawaz Khattak, and Richard A. Anderson. 2003. "Cinnamon Improves Glucose and Lipids of People with Type 2 Diabetes." *Diabetes Care* 26 (12): 3215–3218. doi:10.2337/diacare.26.12.3215.

Khasanah, L. U., Kawiji, P. Prasetyawan, R. Utami, W. Atmaka, G. J. Manuhara, and A. P. Sanjaya. 2017. "Optimization and Characterization of Cinnamon Leaves (Cinnamomum Burmannii) Oleoresin." *IOP Conference Series: Materials Science and Engineering* 193 (1). doi:10.1088/1757-899X/193/1/012021.

Khasnavis, Saurabh, and Kalipada Pahan. 2012. "Sodium Benzoate, a Metabolite of Cinnamon and a Food Additive, Upregulates Neuroprotective Parkinson Disease Protein DJ-1 in Astrocytes and Neurons." *Journal of Neuroimmune Pharmacology* 7 (2): 424–435. doi:10.1007/s11481-011-9286-3.

Kim, Sung Hee, Sun Hee Hyun, and Se Young Choung. 2006. "Anti-Diabetic Effect of Cinnamon Extract on Blood Glucose in Db/Db Mice." *Journal of Ethnopharmacology* 104 (1–2): 119–123. doi:10.1016/j.jep.2005.08.059.

Ko, Feng-Nien -N, Sheu-Meei -M Yu, Ya-Fei -F Kang, and Che-Ming -M Teng. 1995. "Characterization of the Thromboxane (TP-) Receptor Subtype Involved in Proliferation in Cultured Vascular Smooth Muscle Cells of Rat." *British Journal of Pharmacology* 116 (2): 1801–1808. doi:10.1111/j.1476-5381.1995.tb16666.x.

Kong, Jeong Ok, Sang Myung Lee, Yil Seong Moon, Sang Gil Lee, and YoungJoon Ahn. 2007. "Nematicidal Activity of Cassia and Cinnamon Oil Compounds and Related Compounds toward Bursaphelenchus Xylophilus (Nematoda: Parasitaphelenchidae)." *Journal of Nematology* 39 (1): 31–36.

Koppikar, Soumya J., Amit S. Choudhari, Snehal A. Suryavanshi, Shweta Kumari, Samit Chattopadhyay, and Ruchika Kaul-Ghanekar. 2010. "Aqueous Cinnamon Extract (ACE-c) from the Bark of Cinnamomum Cassia Causes Apoptosis in Human Cervical Cancer Cell Line (SiHa) through Loss of Mitochondrial Membrane Potential." *BMC Cancer* 10. doi:10.1186/1471-2407-10-210.

Kreydiyyeh, S. I., J. Usta, and R. Copti. 2000. "Effect of Cinnamon, Clove and Some of Their Constituents on the Na+ -K+ -ATPase Activity and Alanine Absorption in the Rat Jejunum." *Food and Chemical Toxicology* 38 (9): 755–762. doi:10.1016/S0278-6915(00)00073-9.

Kumar, S., R. Kumari, and S. Mishra. 2019. "Pharmacological Properties and Their Medicinal Uses of Cinnamomum: A Review." *Journal of Pharmacy and Pharmacology* 71 (12): 1735–1761. doi:10.1111/jphp.13173.

Kumar, S., S. Sharma, and N. Vasudeva. 2012. "Chemical Compositions of Cinnamomum Tamala Oil from Two Different Regions of India." *Asian Pacific Journal of Tropical Disease* 2 (SUPPL2). Asian Pacific Tropical Medicine Press: S761–S764. doi:10.1016/S2222-1808(12)60260-6.

Kwon, Ho Keun, Ji Sun Hwang, Jae Seon So, Choong Gu Lee, Anupama Sahoo, Jae Ha Ryu, Won K. Jeon, et al. 2010. "Cinnamon Extract Induces Tumor Cell Death through Inhibition of NFκB and AP1." *BMC Cancer* 10. doi:10.1186/1471-2407-10-392.

Kwon, Ho Keun, Won Kyung Jeon, Ji Sun Hwang, Choong Gu Lee, Jae Seon So, Jin A. Park, Byoung Seob Ko, and Sin Hyeog Im. 2009. "Cinnamon Extract Suppresses Tumor Progression by Modulating Angiogenesis and the Effector Function of CD8+ T Cells." *Cancer Letters* 278 (2). Elsevier Ireland Ltd: 174–182. doi:10.1016/j.canlet.2009.01.015.

Lawrence, Toby. 2009. "The Nuclear Factor NF-KappaB Pathway in Inflammation." *Cold Spring Harbor Perspectives in Biology* 1 (6): 1–11. doi:10.1101/cshperspect.a001651.

Lee, Jeong Sun, Seon Min Jeon, Eun Mi Park, Tae Lin Huh, Oh Shin Kwon, Mi Kyung Lee, and Myung Sook Choi. 2003. "Cinnamate Supplementation Enhances Hepatic Lipid Metabolism and Antioxidant Defense Systems in High Cholesterol-Fed Rats." *Journal of Medicinal Food* 6 (3): 183–191. doi: 10.1089/10966200360716599.

Lee, Seung Ho, Sun Young Lee, Dong Ju Son, Heesoon Lee, Hwan Soo Yoo, Sukgil Song, Ki Wan Oh, Dong Cho Han, Byoung Mog Kwon, and Jin Tae Hong. 2005. "Inhibitory Effect of 2′-Hydroxycinnamaldehyde on Nitric Oxide Production through Inhibition of NF-KB Activation in RAW 264.7 Cells." *Biochemical Pharmacology* 69 (5): 791–799. doi:10.1016/j.bcp.2004.11.013.

Li, Hua Bin, Chi Chun Wong, Ka Wing Cheng, and Feng Chen. 2008. "Antioxidant Properties in Vitro and Total Phenolic Contents in Methanol Extracts from Medicinal Plants." *LWT - Food Science and Technology* 41 (3): 385–390. doi:10.1016/j.lwt.2007.03.011.

Li, Jiaoning, Enshuang Guan, Li Chen, Xia Zhang, Lei Yin, Lin Dong, Qiong Pan, Xueyan Fu, and Liming Zhang. 2018. "Optimization for Extraction of an Oil Recipe Consisting of White Pepper, Long Pepper, Cinnamon, Saffron, and Myrrh by Supercritical Carbon Dioxide and the Protective Effects against

Oxygen–Glucose Deprivation in PC12 Cells." *Revista Brasileira de Farmacognosia* 28 (3). Sociedade Brasileira de Farmacognosia: 312–319. doi:10.1016/j.bjp.2018.04.001.

Li, Rong, Tao Liang, Lingyuan Xu, Yongwen Li, Shijun Zhang, and Xiaoqun Duan. 2013. "Protective Effect of Cinnamon Polyphenols against STZ-Diabetic Mice Fed High-Sugar, High-Fat Diet and Its Underlying Mechanism." *Food and Chemical Toxicology* 51 (1). Elsevier Ltd: 419–425. doi:10.1016/j.fct.2012.10.024.

Liang, Ying, Yi Li, Aidong Sun, and Xianjin Liu. 2019. "Chemical Compound Identification and Antibacterial Activity Evaluation of Cinnamon Extracts Obtained by Subcritical N-Butane and Ethanol Extraction." *Food Science and Nutrition* 7 (6): 2186–2193. doi:10.1002/fsn3.1065.

Lin, Chien Tsong, Chi Jung Chen, Ting Yu Lin, Judia Chen Tung, and Sheng Yang Wang. 2008. "Anti-Inflammation Activity of Fruit Essential Oil from Cinnamomum Insularimontanum Hayata." *Bioresource Technology* 99 (18): 8783–8787. doi:10.1016/j.biortech.2008.04.041.

Lin, Wei Ling, Shih Yun Guu, Chan Chuan Tsai, Ekambaranellore Prakash, Mohan Viswaraman, Hsing Bao Chen, and Chuan Fa Chang. 2015. "Derivation of Cinnamon Blocks Leukocyte Attachment by Interacting with Sialosides." *PLoS ONE* 10 (6): 18–20. doi:10.1371/journal.pone.0130389.

Lin, Y. T., Dhiraj Vattem, R. G. Labbe, and Kalidas Shetty. 2005. "Enhancement of Antioxidant Activity and Inhibition of Helicobacter Pylori by Phenolic Phytochemical-Enriched Alcoholic Beverages." *Process Biochemistry* 40 (6): 2059–2065. doi:10.1016/j.procbio.2004.07.019.

Lu, Jianming, Keqiang Zhang, Sangkil Nam, Richard A. Anderson, Richard Jove, and Wei Wen. 2010. "Novel Angiogenesis Inhibitory Activity in Cinnamon Extract Blocks VEGFR2 Kinase and Downstream Signaling." *Carcinogenesis* 31 (3): 481–488. doi:10.1093/carcin/bgp292.

Mancini-Filho, J., A. Van-Koiij, D. A. P. Mancini, F. F. Cozzolino, and R. P. Torres. 1998. "Antioxidant Activity of Cinnamon (Cinnamomum Zeylanicum, Breyne) Extracts." *Bollettino Chimico Farmaceutico* 137 (11): 443–447.

Manso, S., F. Cacho-Nerin, R. Becerril, and C. Nerín. 2013. "Combined Analytical and Microbiological Tools to Study the Effect on Aspergillus Flavus of Cinnamon Essential Oil Contained in Food Packaging." *Food Control* 30 (2). Elsevier Ltd: 370–378. doi:10.1016/j.foodcont.2012.07.018.

Masghati, Shima, and Seyyed M. Ghoreishi. 2018. *Supercritical CO2 Extraction of Cinnamaldehyde and Eugenol from Cinnamon Bark: Optimization of Operating Conditions via Response Surface Methodology. Journal of Supercritical Fluids.* 140. Elsevier B.V. doi:10.1016/j.supflu.2018.06.002.

Mathew, Sindhu, and T. Emilia Abraham. 2006. "Studies on the Antioxidant Activities of Cinnamon (Cinnamomum Verum) Bark Extracts, through Various in Vitro Models." *Food Chemistry* 94 (4): 520–528. doi:10.1016/j.foodchem.2004.11.043.

Menrad, K. 2003. *Market and Marketing of Functional Food in Europe. Journal of Food Engineering* 56.

Miguel, Maria Graça. 2010. "Antioxidant and Anti-Inflammatory Activities of Essential Oils: A Short Review." *Molecules* 15 (12): 9252–9287. doi:10.3390/molecules15129252.

Ministry of Health Labour and Welfare, Japan. "Foods for Specified Health Uses (FOSHU)." Http://Www.Mhlw.Go.Jp/English/Topics/Foodsafety/Fhc/02.Html. http://www.mhlw.go.jp/english/topics/foodsafety/fhc/02.html.

Miyashita, Michiko, and Yasuyuki Sadzuka. 2013. "Effect of Linalool as a Component of Humulus Lupulus on Doxorubicin-Induced Antitumor Activity." *Food and Chemical Toxicology* 53: 174–179. doi:10.1016/j.fct.2012.11.035.

Modi, Piyush I., Jigisha K. Parikh, and Meghal A. Desai. 2019. "Sonohydrodistillation: Innovative Approach for Isolation of Essential Oil from the Bark of Cinnamon." *Industrial Crops and Products* 142 (June). Elsevier: 111838. doi:10.1016/j.indcrop.2019.111838.

Mohammad, S., and T. Gharibzahedi. 2018. "The Preparation, Stability, Functionality and Food Enrichment Ability of Cinnamon Oil-Loaded Nanoemulsion-Based Delivery Systems: A Review." *Nutrafoods* 17: 97–105. doi:10.17470/NF-018-1001-2.

Mueller, Monika, Stefanie Hobiger, and Alois Jungbauer. 2010. "Anti-Inflammatory Activity of Extracts from Fruits, Herbs and Spices." *Food Chemistry* 122 (4). Elsevier Ltd: 987–996. doi:10.1016/j.foodchem.2010.03.041.

Mueller, Monika, and Alois Jungbauer. 2009. "Culinary Plants, Herbs and Spices - A Rich Source of PPARγ Ligands." *Food Chemistry* 117 (4). Elsevier Ltd: 660–667. doi:10.1016/j.foodchem.2009.04.063.

Muhammad, Dimas Rahadian Aji, and Koen Dewettinck. 2017. "Cinnamon and Its Derivatives as Potential Ingredient in Functional Food—A Review." *International Journal of Food Properties* 20 (2). Taylor & Francis: 2237–2263. doi:10.1080/10942912.2017.1369102.

Muhammad, Dimas Rahadian Aji, Emmy Tuenter, Graha Darma Patria, Kenn Foubert, Luc Pieters, and Koen Dewettinck. 2021. "Phytochemical Composition and Antioxidant Activity of Cinnamomum Burmannii Blume Extracts and Their Potential Application in White Chocolate." *Food Chemistry* 340 (December 2019). Elsevier: 127983. doi:10.1016/j.foodchem.2020.127983.

Muthenna, P., G. Raghu, P. Anil Kumar, M. V. Surekha, and G. Bhanuprakash Reddy. 2014. "Effect of Cinnamon and Its Procyanidin-B2 Enriched Fraction on Diabetic Nephropathy in Rats." *Chemico-Biological Interactions* 222. Elsevier Ireland Ltd: 68–76. doi:10.1016/j.cbi.2014.08.013.

Nor, Norazmir Md., Mohd Shazali Md. Sharif, Mohd Salehuddin Mohd Zahari, Hannita Mohd Salleh, Noriza Isha, and Rosmaliza Muhammad. 2012. "The Transmission Modes of Malay Traditional Food Knowledge within Generations." *Procedia - Social and Behavioral Sciences* 50 (July). Elsevier B.V.: 79–88. doi:10.1016/j.sbspro.2012.08.017.

Onderoglu, S., S. Sozer, K. M. Erbil, R. Ortac, and F. Lermioglu. 1999. "The Evaluation of Long-Term Effects of Cinnamon Bark and Olive Leaf on Toxicity Induced by Streptozotocin Administration to Rats." *Journal of Pharmacy and Pharmacology* 51 (11): 1305–1312.

Othman, Zetty Shafiqa, Mohamad Yusof Maskat, and Nur Hasyareeda Hassan. 2020. "Optimization of Cinnamaldehyde Extraction and Antioxidant Activity of Ceylon Cinnamon Extract." *Sains Malaysiana* 49 (5): 995–1002. doi:10.17576/jsm-2020-4905-04.

Panickar, K. S., M. M. Polansky, D. J. Graves, J. F. Urban, and R. A. Anderson. 2012. "A Procyanidin Type A Trimer from Cinnamon Extract Attenuates Glial Cell Swelling and the Reduction in Glutamate Uptake Following Ischemia-like Injury in Vitro." *Neuroscience* 202. Elsevier Inc.: 87–98. doi:10.1016/j.neuroscience.2011.11.051.

Park, Il Kwon, Ju Yong Park, Kyung Hee Kim, Kwang Sik Choi, In Ho Choi, Chul Su Kim, and Sang Chul Shin. 2005. "Nematicidal Activity of Plant Essential Oils and Components from Garlic (Allium Sativum) and Cinnamon (Cinnamomum Verum) Oils against the Pine Wood Nematode (Bursaphelenchus Xylophilus)." *Nematology* 7 (5): 767–774. doi:10.1163/156854105775142946.

Pashazadeh, Bahram, Amir Hossein Elhamirad, Hassan Hajnajari, Parvin Sharayei, and Mohammad Armin. 2020. "Optimization of the Pulsed Electric Field -Assisted Extraction of Functional Compounds from Cinnamon." *Biocatalysis and Agricultural Biotechnology* 23. Elsevier Ltd: 101461. doi:10.1016/j.bcab.2019.101461.

Peterson, Dylan W., Roshni C. George, Francesca Scaramozzino, Nichole E. Lapointe, Richard A. Anderson, Donald J. Graves, and John Lew. 2009. "Cinnamon Extract Inhibits Tau Aggregation Associated with Alzheimer's Disease in Vitro." *Journal of Alzheimer's Disease* 17 (3): 585–597. doi:10.3233/JAD-2009-1083.

Pina-Pérez, M. C., A. Martínez-López, and D. Rodrigo. 2012. "Cinnamon Antimicrobial Effect against Salmonella Typhimurium Cells Treated by Pulsed Electric Fields (PEF) in Pasteurized Skim Milk Beverage." *Food Research International* 48 (2). Elsevier Ltd: 777–783. doi:10.1016/j.foodres.2012.06.027.

Ping, Hua, Guijun Zhang, and Guixing Ren. 2010. "Antidiabetic Effects of Cinnamon Oil in Diabetic KK-Ay Mice." *Food and Chemical Toxicology* 48 (8–9). Elsevier Ltd: 2344–2349. doi:10.1016/j.fct.2010.05.069.

Prasad, K. Nagendra, Bao Yang, Xinhong Dong, Guoxiang Jiang, Haiyan Zhang, Haihui Xie, and Yueming Jiang. 2009. "Flavonoid Contents and Antioxidant Activities from Cinnamomum Species." *Innovative Food Science and Emerging Technologies* 10 (4). Elsevier Ltd: 627–632. doi:10.1016/j.ifset.2009.05.009.

Rahman, Sonia, Halima Begum, Zaida Rahman, Ferdous Ara, Md. Jalaluddin Iqbal, and Abul Kalam Mohammad Yousuf. 2013. "Effect of Cinnamon (Cinnamomum Cassia) as a Lipid Lowering Agent on Hypercholesterolemic Rats." *Journal of Enam Medical College* 3 (2): 94–98. doi:10.3329/jemc.v3i2.16132.

Rodríguez-Roque, María Janeth, Begoña de Ancos, Concepción Sánchez-Moreno, M. Pilar Cano, Pedro Elez-Martínez, and Olga Martín-Belloso. 2015. "Impact of Food Matrix and Processing on the in Vitro Bioaccessibility of Vitamin C, Phenolic Compounds, and Hydrophilic Antioxidant Activity from Fruit Juice-Based Beverages." *Journal of Functional Foods* 14. Elsevier Ltd: 33–43. doi:10.1016/j.jff.2015.01.020.

Saeed, M., A. A. Kamboh, S. F. Syed, D. Babazadeh, I. Suheryani, Q. A. Shah, M. Umar, et al. 2018. "Phytochemistry and Beneficial Impacts of Cinnamon (Cinnamomum Zeylanicum) as a Dietary Supplement in Poultry Diets." *World's Poultry Science Journal* 74 (2): 331–345. doi:10.1017/S0043933918000235.

Sangal, A. 2011. "Role of Cinnamon as Beneficial Antidiabetic Food Adjunct : A Review." *Advances in Applied Science Research* 2 (4): 440–450.

Seo, Han Seok, Dorothee Buschhüter, and Thomas Hummel. 2009. "Odor Attributes Change in Relation to the Time of the Year. Cinnamon Odor Is More Familiar and Pleasant during Christmas Season than Summertime." *Appetite* 53 (2): 222–225. doi:10.1016/j.appet.2009.06.011.

Shan, Bin, Yi Zhong Cai, John D. Brooks, and Harold Corke. 2007. "Antibacterial Properties and Major Bioactive Components of Cinnamon Stick (Cinnamomum Burmannii): Activity against Foodborne Pathogenic Bacteria." *Journal of Agricultural and Food Chemistry* 55 (14): 5484–5490. doi:10.1021/jf070424d.

Shobana, S., and Naidu, K.A. 2000. "Antioxidant Activity of Selected Indian Spices." *Prostaglandins Leukotrienes and Essential Fatty Acids* 62 (2): 107–110. doi:10.1054/plef.1999.0128.

Silva, Maria Leonor Tavares da, Maria Alexandra Sardinha Bernardo, Jaipaul Singh, and Maria Fernanda de Mesquita. 2018. *Beneficial Uses of Cinnamon in Health and Diseases: An Interdisciplinary Approach. The Role of Functional Food Security in Global Health.* Elsevier Inc. doi:10.1016/B978-0-12-813148-0.00033-5.

Silva, Fabiana L., Raquel V.S. Silva, Paola C. Branco, Letícia V. Costa-Lotufo, Cynthia Murakami, Maria C.M. Young, Débora A. Azevedo, and Paulo R.H. Moreno. 2020. "Chemical Composition of the Brazilian Native Cinnamomum Stenophyllum (Meisn.) Vattimo-Gil Essential Oil by GC-QMS and GC × GC-TOFMS, and Its Cytotoxic Activity." *Arabian Journal of Chemistry* 13 (4): 4926–4935. doi:10.1016/j.arabjc.2020.01.012.

Sivapriya, T., and S. John. 2016. "Cinnamon : Bioactive Components and Usage for the Diabetes Patients." *World Journal of Pharmaceutical Research* 5 (3): 475–489.

Slining, Meghan M., Kevin C. Mathias, and Barry M. Popkin. 2013. "Trends in Food and Beverage Sources among US Children and Adolescents: 1989-2010." *Journal of the Academy of Nutrition and Dietetics* 113 (12). Elsevier Inc: 1683–1694. doi:10.1016/j.jand.2013.06.001.

Song, Beom Seok, Jin Gyu Park, Jae Nam Park, In Jun Han, Jong il Choi, Jae Hun Kim, Myung Woo Byun, Sang Wook Kang, Gi Hyuk Choi, and Ju Woon Lee. 2009. "High-Dose Processing and Application to Korean Space Foods." *Radiation Physics and Chemistry* 78 (7–8). Elsevier: 671–674. doi:10.1016/j.radphyschem.2009.03.073.

Song, Fan, Hua Li, Jiyuan Sun, and Siwang Wang. 2013. "Protective Effects of Cinnamic Acid and Cinnamic Aldehyde on Isoproterenol-Induced Acute Myocardial Ischemia in Rats." *Journal of Ethnopharmacology* 150 (1). Elsevier: 125–130. doi:10.1016/j.jep.2013.08.019.

Stein, A. J., and E. Rodriguez-Cerezo. 2008. *Functional Foods in the European Union, a Report Based on the ESTO Study Functional Food in European Union. European Commission.* Seville, Spain. doi:10.2791/21607.

Subash Babu, P., S. Prabuseenivasan, and S. Ignacimuthu. 2007. "Cinnamaldehyde-A Potential Antidiabetic Agent." *Phytomedicine* 14 (1): 15–22. doi:10.1016/j.phymed.2006.11.005.

Suryati, T., M. Astawan, H. N. Lioe, T. Wresdiyati, and S. Usmiati. 2014. "Nitrite Residue and Malonaldehyde Reduction in Dendeng - Indonesian Dried Meat - Influenced by Spices, Curing Methods and Precooking Preparation." *Meat Science* 96 (3). Elsevier Ltd: 1403–1408. doi:10.1016/j.meatsci.2013.11.023.

Tabak, Mina, Robert Armon, and Ishak Neeman. 1999. "Cinnamon Extracts' Inhibitory Effect on Helicobacter Pylori." *Journal of Ethnopharmacology* 67 (3): 269–277. doi:10.1016/S0378-8741(99)00054-9.

Tagliazucchi, Davide, Ahmed Helal, Elena Verzelloni, and Angela Conte. 2012. "The Type and Concentration of Milk Increase the in Vitro Bioaccessibility of Coffee Chlorogenic Acids." *Journal of Agricultural and Food Chemistry* 60 (44): 11056–11064. doi:10.1021/jf302694a.

Tang, Pei Ling, Yee Teng Chen, Jianfeng Qin, Xiaotao Hou, and Jiagang Deng. 2020. "Effect of Cinnamon Bark and Twig Extracts on the Chemical, Physicochemical and Antioxidant Properties of Fermented Milk." *Journal of Food Measurement and Characterization* 14 (4). Springer US: 2271–2281. doi:10.1007/s11694-020-00474-5.

Thomas, J., and K. M. Kuruvilla. 2012. "Cinnamon." *Handbook of Herbs and Spices: Second Edition* 1: 182–196. doi:10.1533/9780857095671.182.

Tung, Yu Tang, Meng Thong Chua, Sheng Yang Wang, and Shang Tzen Chang. 2008. "Anti-Inflammation Activities of Essential Oil and Its Constituents from Indigenous Cinnamon (Cinnamomum Osmophloeum) Twigs." *Bioresource Technology* 99 (9): 3908–3913. doi:10.1016/j.biortech.2007.07.050.

Tung, Yu Tang, Pei Ling Yen, Chun Ya Lin, and Shang Tzen Chang. 2010. "Anti-Inflammatory Activities of Essential Oils and Their Constituents from Different Provenances of Indigenous Cinnamon (Cinnamomum Osmophloeum) Leaves." *Pharmaceutical Biology* 48 (10): 1130–1136. doi:10.3109/13880200903527728.

Tzortzakis, Nikos G. 2009. "Impact of Cinnamon Oil-Enrichment on Microbial Spoilage of Fresh Produce." *Innovative Food Science and Emerging Technologies* 10 (1). Elsevier Ltd: 97–102. doi:10.1016/j.ifset.2008.09.002.

Udayaprakash, N. K., M. Ranjithkumar, S. Deepa, N. Sripriya, Abdulla A. Al-Arfaj, and S. Bhuvaneswari. 2015. "Antioxidant, Free Radical Scavenging and GC-MS Composition of Cinnamomum Iners Reinw. Ex Blume." *Industrial Crops and Products* 69. Elsevier B.V.: 175–179. doi:10.1016/j.indcrop.2015.02.018.

Unlu, Mehmet, Emel Ergene, Gulhan Vardar Unlu, Hulya Sivas Zeytinoglu, and Nilufer Vural. 2010. "Composition, Antimicrobial Activity and in Vitro Cytotoxicity of Essential Oil from Cinnamomum Zeylanicum Blume (Lauraceae)." *Food and Chemical Toxicology* 48 (11). Elsevier Ltd: 3274–3280. doi:10.1016/j.fct.2010.09.001.

Vallverdú-Queralt, A., J. Regueiro, M. Martínez-Huélamo, J. F. Rinaldi Alvarenga, L. N. Leal, and R. M. Lamuela-Raventos. 2014. "A Comprehensive Study on the Phenolic Profile of Widely Used Culinary Herbs and Spices: Rosemary, Thyme, Oregano, Cinnamon, Cumin and Bay." *Food Chemistry* 154 (2014). Elsevier Ltd: 299–307. doi:10.1016/j.foodchem.2013.12.106.

Vangalapati, M., N. Sree Satya, D. V. Surya Prakash, and S. Avanigadda. 2012. "A Review on Pharmacological Activities and Clinical Effects of Cinnamon Species." *Research Journal of Pharmaceutical, Biological and Chemical Sciences* 3 (1): 653–663.

Vetal, Sachin, Subhash L. Bodhankar, Vishwaraman Mohan, and Prasad A. Thakurdesai. 2013. "Anti-Inflammatory and Anti-Arthritic Activity of Type-A Procyanidine Polyphenols from Bark of Cinnamomum Zeylanicum in Rats." *Food Science and Human Wellness* 2 (2). Beijing Academy of Food Sciences.: 59–67. doi:10.1016/j.fshw.2013.03.003.

Wagner, Hildebert, Michael Wierer, and Rudolf Bauer. 1986. "In Vitro Inhibition of Prostaglandin Biosynthesis by Essential Oils and Phenolic Compounds." *Planta Medica* 52 (03): 184–187. doi:10.1055/s-2007-969117.

Wang, Sheng Yang, Pin Fun Chen, and Shang Tzen Chang. 2005. "Antifungal Activities of Essential Oils and Their Constituents from Indigenous Cinnamon (Cinnamomum Osmophloeum) Leaves against Wood Decay Fungi." *Bioresource Technology* 96 (7): 813–818. doi:10.1016/j.biortech.2004.07.010.

Wondrak, Georg Thomas, Nicole F. Villeneuve, Sarah D. Lamore, Alexandra S. Bause, Tao Jiang, and Donna D. Zhang. 2010. "The Cinnamon-Derived Dietary Factor Cinnamic Aldehyde Activates the Nrf2-Dependent Antioxidant Response in Human Epithelial Colon Cells." *Molecules* 15 (5): 3338–3355. doi:10.3390/molecules15053338.

Wong, Y. C., M. Y. Ahmad-Mudzaqqir, and W. A. Wan-Nurdiyana. 2014. "Extraction of Essential Oil from Cinnamon (Cinnamomum Zeylanicum)." *Oriental Journal of Chemistry* 30 (1): 37–47. doi:10.13005/ojc/300105.

Xue, Yong Liang, Hai Xia Shi, Ferid Murad, and Ka Bian. 2011. "Vasodilatory Effects of Cinnamaldehyde and Its Mechanism of Action in the Rat Aorta." *Vascular Health and Risk Management* 7 (1): 273–280. doi:10.2147/VHRM.S15429.

Yan, Yong Ming, Ping Fang, Mu Tian Yang, Ning Li, Qing Lu, and Yong Xian Cheng. 2015. "Anti-Diabetic Nephropathy Compounds from Cinnamomum Cassia." *Journal of Ethnopharmacology* 165. Elsevier: 141–147. doi:10.1016/j.jep.2015.01.049.

Yang, Cheng Hong, Rong Xian Li, and Li Yeh Chuang. 2012. "Antioxidant Activity of Various Parts of Cinnamomum Cassia Extracted with Different Extraction Methods." *Molecules* 17 (6): 7294–7304. doi:10.3390/molecules17067294.

Yang, Yang, Md Sahidul Islam, Jin Wang, Yuan Li, and Xin Chen. 2020. "Traditional Chinese Medicine in the Treatment of Patients Infected with 2019-New Coronavirus (SARS-CoV-2): A Review and Perspective." *International Journal of Biological Sciences* 16 (10): 1708–1717. doi:10.7150/ijbs.45538.

Yeğin, Sirma, and Ali Üren. 2008. "Biogenic Amine Content of Boza: A Traditional Cereal-Based, Fermented Turkish Beverage." *Food Chemistry* 111 (4): 983–987. doi:10.1016/j.foodchem.2008.05.020.

Yeh, Chia Feng, Jung San Chang, Kuo Chih Wang, Den En Shieh, and Lien Chai Chiang. 2013. "Water Extract of Cinnamomum Cassia Blume Inhibited Human Respiratory Syncytial Virus by Preventing Viral Attachment, Internalization, and Syncytium Formation." *Journal of Ethnopharmacology* 147 (2). Elsevier: 321–326. doi:10.1016/j.jep.2013.03.010.

Yeh, T., C. Lin, and S. Chang. 2014. "A Potential Low-Coumarin Cinnamon Substitute: Cinnamomum Osmophloeum Leaves." *Journal of Agricultural and Food Chemistry* 62 (7): 1706–1712. doi:10.1021/jf405312q.

Youn, Hyung S., Jun K. Lee, Yong J. Choi, Shin I. Saitoh, Kensuke Miyake, Daniel H. Hwang, and Joo Y. Lee. 2008. "Cinnamaldehyde Suppresses Toll-like Receptor 4 Activation Mediated through the Inhibition of Receptor Oligomerization." *Biochemical Pharmacology* 75 (2): 494–502. doi:10.1016/j.bcp.2007.08.033.

Yu, Sheu-Meei -M, Tian-Shung -S Wu, and Che-Ming -M Teng. 1994. "Pharmacological Characterization of Cinnamophilin, a Novel Dual Inhibitor of Thromboxane Synthase and Thromboxane A2 Receptor." *British Journal of Pharmacology* 111 (3): 906–912. doi:10.1111/j.1476-5381.1994.tb14824.x.

Yu, Tao, Sabin Lee, Woo Seok Yang, Hyun Jae Jang, Yong Jin Lee, Tae Woong Kim, Sun Young Kim, Jaehwi Lee, and Jae Youl Cho. 2012. "The Ability of an Ethanol Extract of Cinnamomum Cassia to Inhibit Src and Spleen Tyrosine Kinase Activity Contributes to Its Anti-Inflammatory Action." *Journal of Ethnopharmacology* 139 (2). Elsevier Ireland Ltd: 566–573. doi:10.1016/j.jep.2011.11.051.

CHAPTER 5

Bioactive Compounds in Turmeric

Vipin Sharma and Happy Dev Sharma

CONTENTS

5.1 INTRODUCTION

Turmeric, derived from the rhizome of the *Curcuma longa* L. plant, is a very important herb that belongs to the ginger family. Turmeric is employed not only as a main spice in Southeast Asia but also in marriages and other religious ceremonies. Due to its brilliant yellow colour, it is also called the golden spice as well as "Indian saffron." It has a very long history of medicinal use that dates back nearly 4000 years to the Vedic culture in India. Probably by 700 AD it reached China, by 800 AD in East Africa, by 1200 AD in West Africa and in the eighteenth century in Jamaica. In 1280 Marco Polo AD, described turmeric better than a vegetable that exhibited qualities so similar to that of saffron. In Susruta's Ayurvedic Compendium even dating back to 250 BC, an ointment containing turmeric is suggested to alleviate the effects of poisoned food. Even Ayurvedic and Unani systems confirm long history of medicinal use of turmeric in South Asia.

Turmeric rhizomes are widely used throughout the globe as a principal spice and drug; and are the main ingredient of dishes originating from India and Bangladesh for colour, flavour and taste since ancient times. In the Indian subcontinent it is widely used in social and religious ceremonies as diet and a complementary alternative system of medicine (CAM). In Ayurveda, it has been considered as *Dashemani Lekhaniya* (emaciating), *Kusthagna* (anti-dermatosis), and *Visaghna* (anti-poisonous); and is therefore used as folk medicine against various ailments, including gastric, hepatic, gynecological and infectious diseases. It is also utilized in Hindu ritual processes as a topical application before a marriage ceremony on the bodies of the bride and groom whereas females apply paste of turmeric on their forehead during day-to-day rituals. Therefore, it is a sacred belief that turmeric is a holy and totally safe material. It is popularly known as "haldi" (from the Sanskrit word *haridra*) in North India and in the South India it is often called "manjal," (from ancient Tamil

DOI: 10.1201/9781003215387-5

literature). The Latin word *terra merita* (meritorious earth) is the root word for turmeric which is related to the colour of ground turmeric which is similar to a mineral pigment whereas in French it is called *terremerite* and in most languages it is commonly known as "yellow root." Its scientific name is derived from the Latin word *curcuma* and has also been adopted by many cultures. In Sanskrit, it has as many as 53 different available references that have been given names, including *anestha* (not offered for sacrifice or homa), *bhadra* (auspicious or lucky), *bahula* (plenty), *dhirgharaja* (long in appearance), *gandhaplashika* (which produces good smell), *gauri* (to make fair), *gharshani* (to rub), haldi (that draws attention to its bright colour), *haridra* (dear to hari, Lord Krishna), *harita* (greenish), *hemaragi* (exhibits golden colour), *hemaragini* (gives the golden colour), *hridayavilasini* (gives delight to heart, charming), *jayanti* (one that wins over diseases), *jawarantika* (which cures fevers), *kanchani* (exhibits golden colour), *kaveri* (harlot), *krimighni/kashpa* (killer of worms), *kshamata* (capability), *laxmi* (prosperity), *mangalprada* (who bestows auspiciousness), *mangalya* (auspicious), *mehagni* (killer of fat), *nisha* (night), *nishakhya* (known as night), *nishawa* (clears darkness and imparts colour), *patwaluka* (perfumed powder), *pavitra* (holy), *pinga* (reddish-brown), *pinja* (yellow-red powder), *pita/pitika* (that imparts yellow colour), *rabhangavasa* (which dissolves fat), *ranjani* (which gives colour), *ratrimanika* (as beautiful as moonlight), *shifa* (fibrous root), *shobhna* (brilliant colour), *shiva* (gracious), *shyama* (dark coloured), *soubhagaya* (lucky), *survana* (golden colour), *survanavara* (which exhibits golden colour), *tamasini* (beautiful as night), *umavara* (Parvati, wife of Lord Shiva), *vairagi* (who remains free from desires), *varavarnini* (which gives fair complexion), *varna datri* (enhancer of body complexion), *varnini* (which gives colour), *vishagni* (killer of poison), *yamini* (night), *yoshitapriya* (beloved of wife), and *yuvati* (young girl).

India ranks first in the production and export of turmeric in the world, even though the crop is grown in many countries, e.g. China, Caribbean Islands, Central America, Japan, Korea, Malaysia, Myanmar, Pakistan, Philippines, Sri Lanka, Thailand and Vietnam. Officially it is estimated that India alone produces about 80% of the global turmeric. Important states in India that grow turmeric are Andhra Pradesh, Odisha, Tamil Nadu, West Bengal, Assam, Maharashtra and Karnataka. Andhra Pradesh stands first in terms of production and productivity followed by Tamil Nadu. India is also the global leader and exporter of value-added products of turmeric which include curcuminoids, dehydrated turmeric powder, oils and oleoresin (Remadevi, Surendran and Kimura 2007; Sasikumar 2005; China 2005; Jayaprakasha, Rao and Sakariah 2002; Tanaka et al. 2008; Garg, Bansal, Gupta and Kumar 1999). Turmeric, like other spices, is available as whole, ground and oleoresin. Its powder is the main component in curry powders and pastes. In the food industry, it is used mostly to colour and flavour mustard (ASTA 2002). It is also used in chicken bouillon and soups, sauces, gravies and dry seasonings, and also as a colourant in cereals (Tainter and Grenis 2001). Commercially, the volatile oil is extracted from dried rhizomes and leaves of turmeric and has its own significance in concerning to human. Dried rhizomes contain 5–6% and leaves contain about 1.0–1.5% oil. Generally, the oil is extracted by steam distillation. Supercritical extraction using liquid carbon dioxide is a novel technique for the extraction of volatile oil and oleoresin. *ar*-turmerone is responsible for the peculiar turmeric aroma which is also the major aroma principle in the oil. Oleoresin extracted from turmeric is used in brine pickle and also in mayonnaise and relish formulations. It is also used in non-alcoholic beverages, for garnishing and in some ice creams (Perotti 1975). In all of the above cases, it is the main colouring agent which is capable of replacing synthetic colours, e.g. tartrazine, which were used formerly (Govindarajan 1980). Epidemiological studies reveal that intake of turmeric may decrease certain forms of cancer and provide other protective biological effects in humans, which may be attributed to its major constituent – curcumin (Radha, Anoop, Gaddipati and Richkab 2006). So, it is commonly employed as anti-inflammatory, antiangiogermic, antioxidant, anticancerous, etc. The curcuminoids, when administered orally, enter the blood circulation and are present as glucuronides and glucuronide sulphate conjugate forms (Asai and Miyazawa 2000) and hence the physiological effects expressed are due to these conjugates. Oleoresin, the organic extract of turmeric is added as a spice and colouring agent to food items. The latest high-technology process like supercritical CO_2 extraction and molecular distillation are

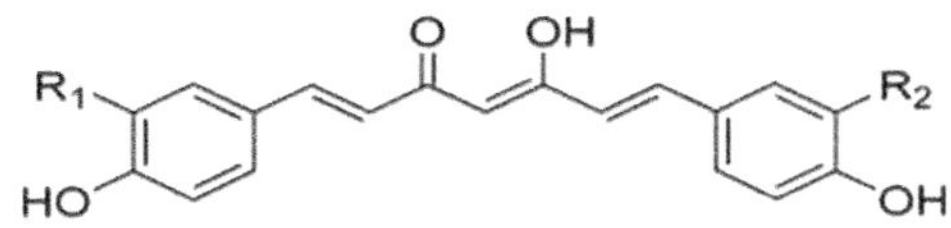

(1) $R_1 = R_2 = OCH_3$
(2) $R_1 = H$; $R_2 = OCH_3$
(3) $R_1 = R_2 = H$

Figure 5.1 Basic structure of Curcumin.

followed to extract and separate turmeric oleoresin of high quality which is essentially used in cooking meat and fish products and certain products such as butter, cheese, pickles and relish formulae. Orange-red coloured turmeric oleoresin is obtained by the solvent extraction of the ground spice with organic solvents like acetone, ethanol and ethylene dichloride for 4–5 hours (Krishnamurthy et al. 1976). Yield of oleoresin ranges from 7.9 to 10.4%. The principal colouring matter that forms one-third of a good quality oleoresin is, Curcumin.

5.2 COLOURING PRINCIPLE

Curcumin: The main colouring principle of turmeric rhizome was isolated in 19th century and named as 'Curcumin'. Roughley and Whiting determined its chemical structure in (Roughley and Whiting 1973) Figure 5.1. Curcumin or curcuminoids concentrate which is basically used as a food colourant, is now included in the list of colours but with a restricted use because it has low acceptable daily intake (ADI) of just 0–1 mg/kg body weight/day. Curcumin can impart a bright yellow colour even at doses of 5–200 ppm.

Curcuminoids (2.5–6%) responsible for the yellow colour are present in the rhizomes. Curcumin (diferuloylmethane) further comprises of Curcumin I (curcumin), Curcumin II (demethoxycurcumin) and Curcumin III (bisdemethoxycurcumin), which are known to be natural antioxidants (Ruby et al. 1995; Figure 5.2). A new curcuminoid called cyclo-curcumin has been isolated from the nematocidally active fraction of turmeric.

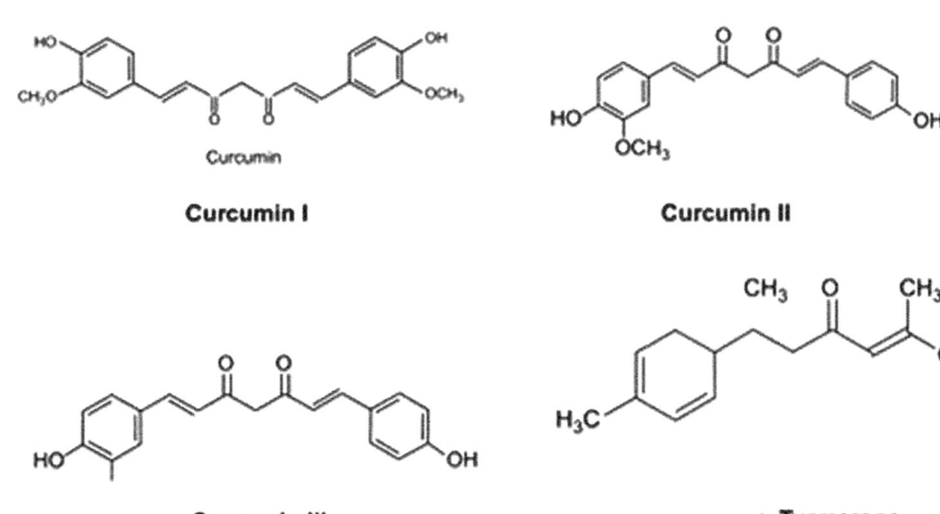

Figure 5.2 Structure of the three main curcuminoids.

The fresh rhizomes contain two pigments and two natural phenolics which possess antioxidant and anti-inflammatory properties. α-phellandrene (1%), sabinene (0.6%), cineol (1%), borneol (0.5%), zingiberene (25%) and sesquiterpines (53%) (Kapoor 1990) are the main constituents of the essential oil (5.8%) obtained by steam distillation of rhizomes of turmeric. The bioactive chemical constituents in turmeric have also been investigated and there are as many as 235 compounds: primarily phenolics and terpenoids that include twenty two diarylheptanoids and diarylpentanoids, eight phenylpropenes as other phenolics and sixty-eight monoterpenes, five diterpenes, three triterpenoids, four sterols, 109 sesquiterpenes, two alkaloids, and fourteen other compounds. Curcuminoids and essential oils (primarily monoterpenes) are the major bioactive constituents showing different bioactivities (Figure 5.3). Vanillins, vanillic acid, quercetin, Calebin-A and other phenolic compounds are also present in the turmeric.

The phenolic compounds present in turmeric rhizome are 1-hydroxy-1, 7-bis (4-hydroxy-3-methoxyphenyl)-(6E)-6-heptene-3, 5-dione; 1-(4-hydroxy-3, 5-dimethoxyphenyl)-7-(4-hydroxy-3-methoxyphenyl)-(1E, 6E)-1, 6-heptadiene-3, 4-dione; 1, 5-bis (4-hydroxy-3-methoxyphenyl)-penta-(1E, 4E)-1, 4-dien-3-one; 1-(4-hydroxy-3-methoxyphenyl-5-(4-hydroxyphenyl)-penta-(1E, 4E)-1, 4-dien-3-one; 1-(4-hydroxy-3-methoxyphenyl)-7-(3, 4-dihydroxyphenyl)-1, 6-heptadiene-3, 5-dione and 1, 7-bis (4-hydroxyphenyl)-1, 4, 6-heptatrien-3-one. The pale yellow to orange-yellow volatile oil (4–6%) obtained from turmeric consists of a number of mono and sesquiterpenes. The sesquiterpenes were named as curcumenone; dehydrocurdione; (4 S, 5 S)-germacrone 4, 5-epoxide; bisabola 3, 10-diene 2-one; arturmerone (Roth, Chandra and Nair 1998); bisacumol; bisacurone; curcumenol; isoprocurcumenol; zedoaronediol; procurcumenol; epiprocurcumenol; germacrone-13-al; 4-hydroxybisabola-2, 10-diene-9-one; 4, 5-dihydrobisabola-2, 10-diene; 4-methoxy- 5-hydroxy-bisabola-2, 10-diene-9-one; 2, 5-dihydroxybisabola-3, 10-diene and procurcumadiol (Ohshiro, Kuroyanagi and Ueno 1990). Some other compounds named as curlone (Kiso, Suzuki, Oshima and Hikino 1983); α-turmerone; β-turmerone; terpinolene; α-phellandrene; curcumadiol; labda-8 (17)-diene-15, 16- dial were also isolated.

5.3 TURMERIC OIL

Dried and cured turmeric generally yields from 1.5 to 5.0% volatile oil. *C. aromatica* is a high volatile oil (4-8%) and low curcuminoids (1.5%) yielder. The oil present in the rhizomes of turmeric is responsible for its aromatic taste and smell. Analysis of the oil, obtained by steam distillation of the powdered rhizome, followed by fractional distillation shows that the components are predominantly a mixture of sesquiterpene ketones and alcohols. A mixture of low-boiling terpenes, *d*-sabinene, α-phellandrene, cineole, borneol and the higher-boiling sesquiterpene, zingiberene, in substantial amounts (25%) have also identified in addition to these major components. Turmeric oil extracted by supercritical carbon dioxide method, followed by further separation and purification of three turmerones, e.g. *ar*-turmerone, α- and β-turmerone gave 71% purity by weight. Subsequent identification by liquid-solid chromatography, NMR qualification and HPLC quantification after purification using a normal phase silica gel 60 column could separate and purify three major turmerones with 86% purity by weight of *ar*-turmerone and 81% purity by weight of α- and β-turmerone. In addition to two new sesquiterpene ketoalcohols-turmeronol A and turmeronol B, five sesquiterpenes, e.g. germacrone-13-al, 4-hydroxybisabola-2, 10-diene-9-one, 4-methoxy-5-hydroxybisabola-2, 10-diene-9-one, 2, 5-dihydroxybisabola-3, 10-diene and pro-curcuma diol, have also been reported from the dried rhizome of turmeric by NMR (^{1}H and ^{13}C) spectroscopy. Bisabolene-type sesquiterpenes are also present in turmeric. Liquid chromatography–electrospray ionization tandem mass spectrometry (LC–ESI-MS/MS) coupled to DAD analysis are the novel online tools for identification of diarylheptanoids in fresh turmeric rhizome extracts. Based on the mass spectra, from both negative and positive mode LC–ESIMS/MS analysis, and supported by their DAD spectra, 19 diarylheptanoids have been identified out of

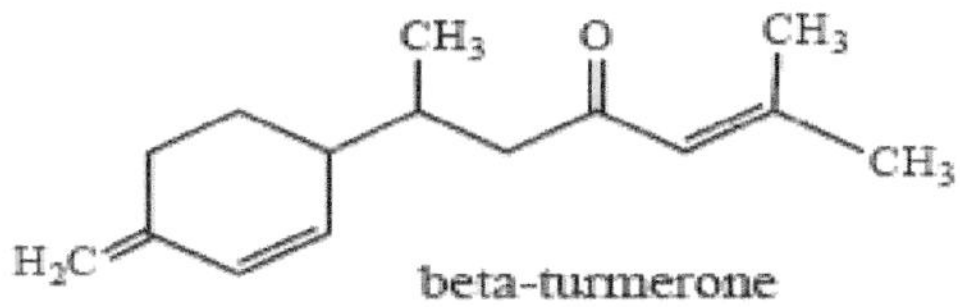
beta-turmerone

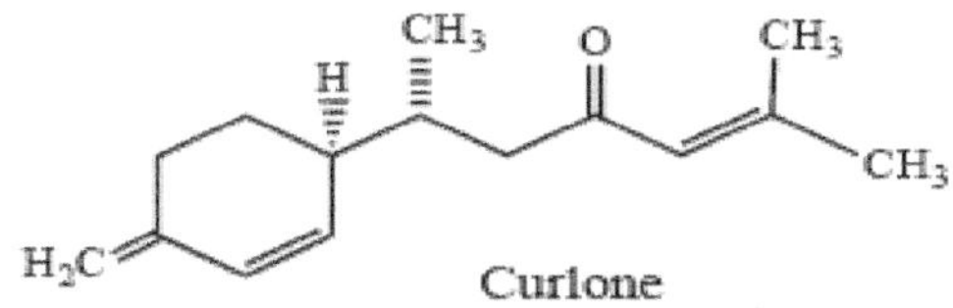
Curlone

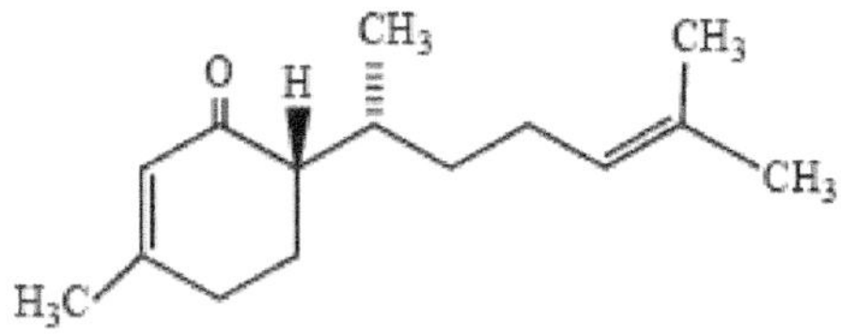
Bisabola-3,10-diene-2-one

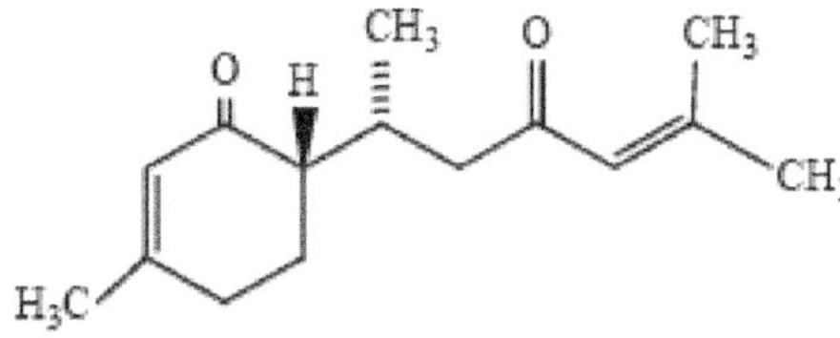
Bisabola-3-10-diene-2,4-dione

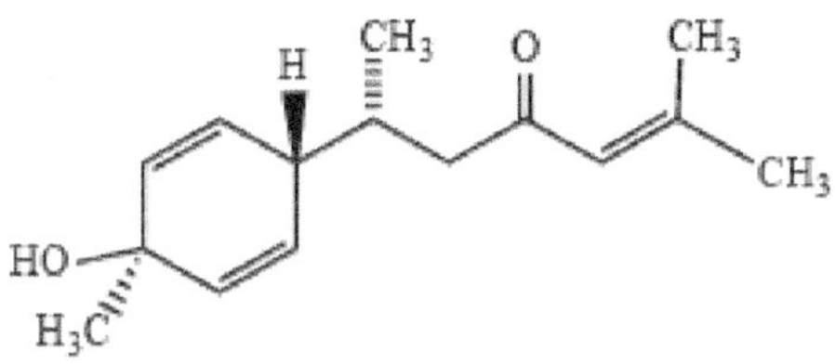
4-Hydroxy-bisbola-2,5,10-triene-9-one

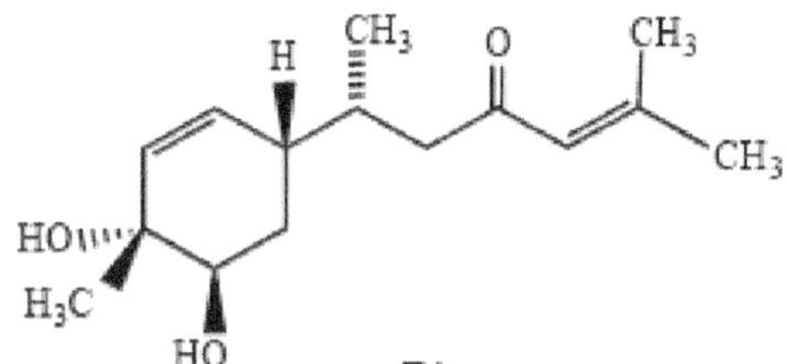
Bisacurone

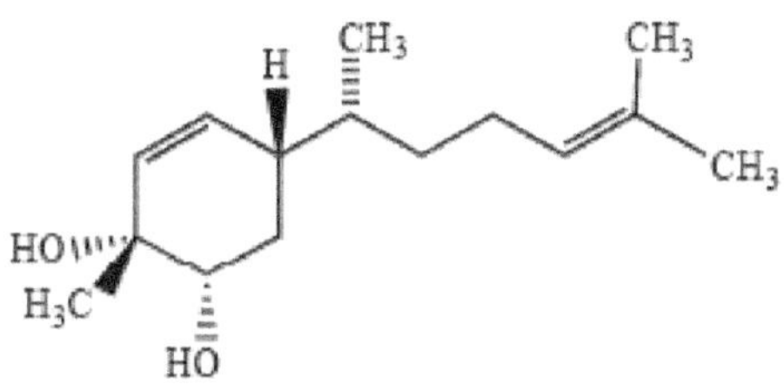
4,5-Dihydroxybisabola-2,10-diene

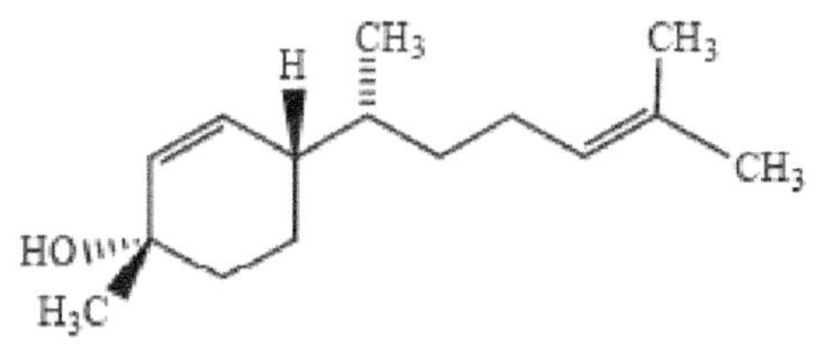
4-Hydroxybisabola-2,10-diene

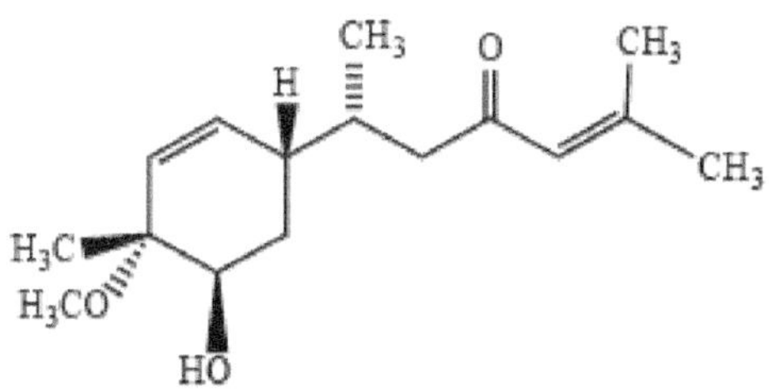
4-Methoxy-5-hydroxy-2,10-dien-9-one

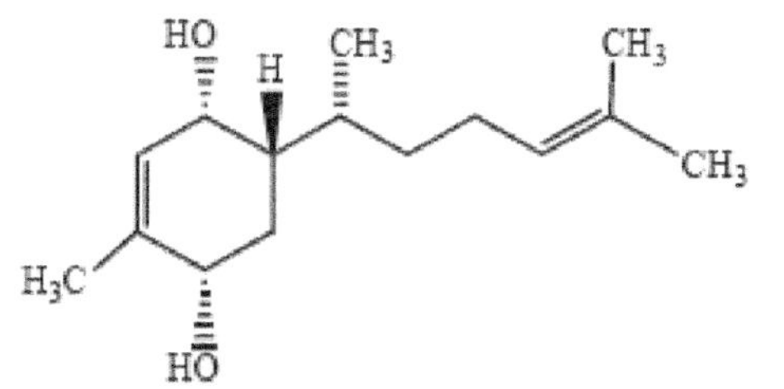
2,5-dihydroxybisabola-3,10-diene

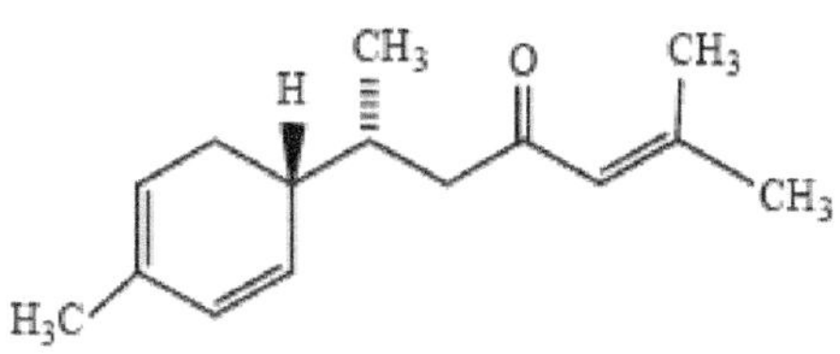
alpha-Turmerone

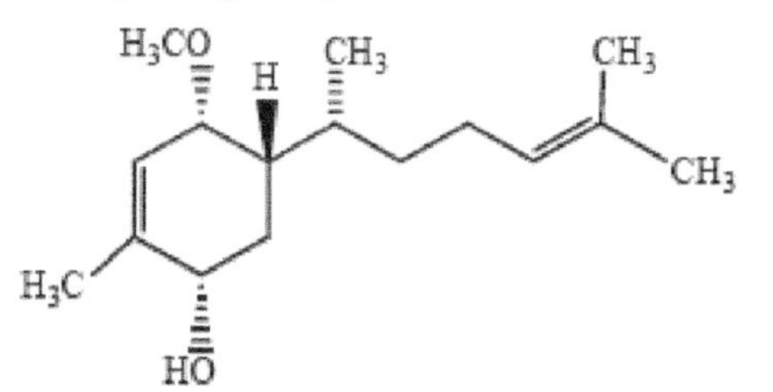
2-Hydroxy-5-methoxybisabola-3,10-diene

Figure 5.3 Molecular structures of some of the bioactive chemical constituents in turmeric. (*Continued*)

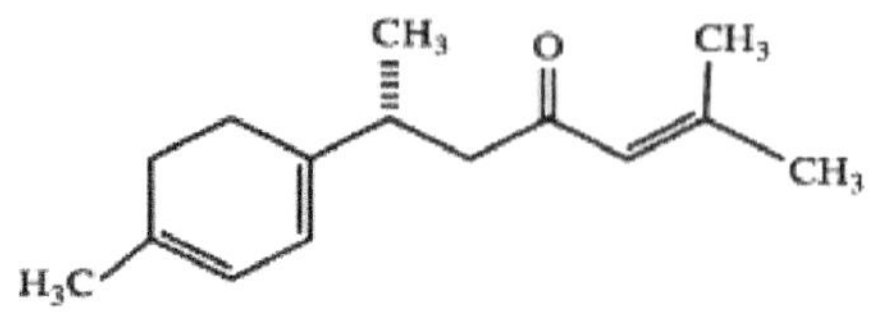

a-Alantone

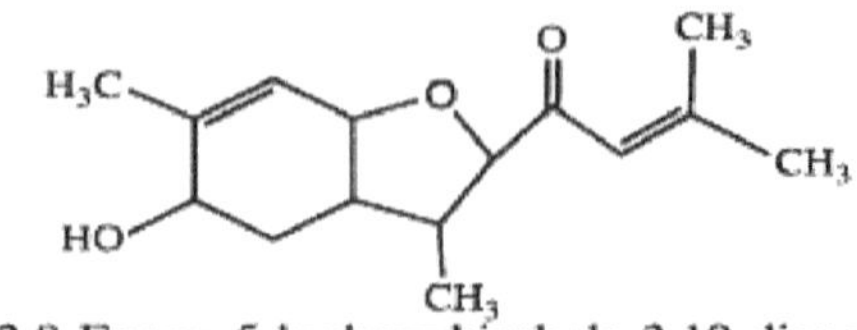

2,8-Epoxy-5-hydroxybisabola-3,10-diene-9-one

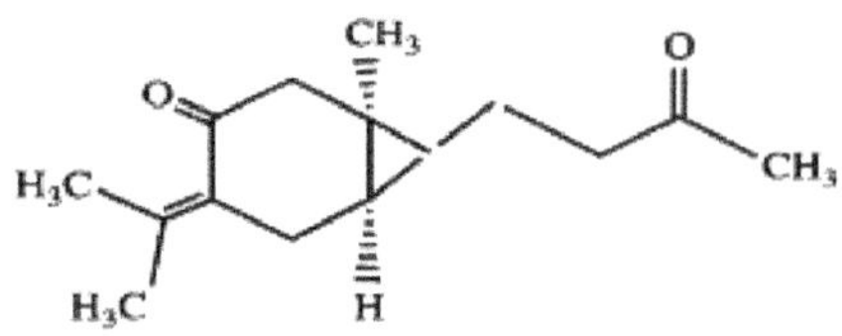

Curcumenone

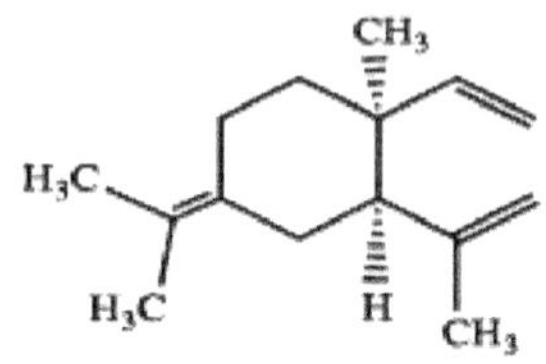

beta-Elemene

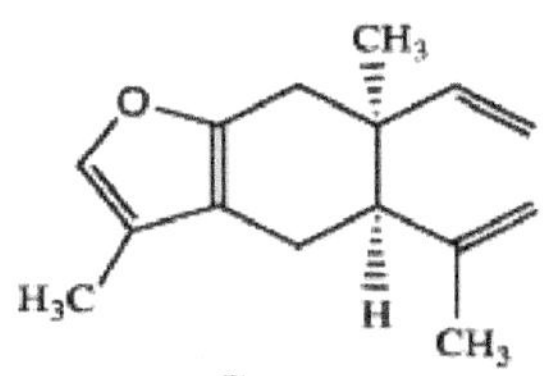

Curzerene

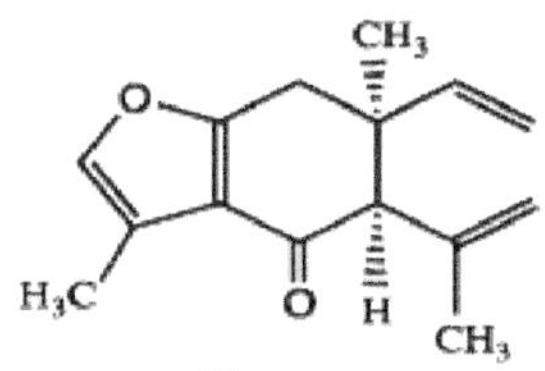

Curzerone

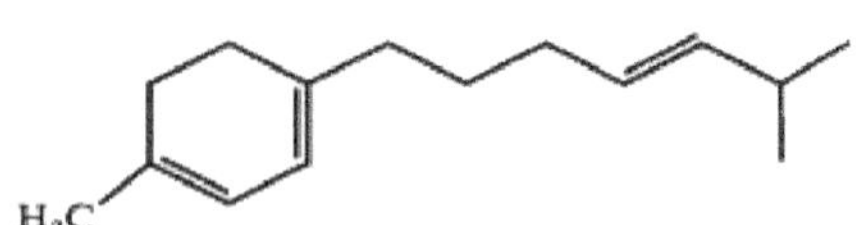

Zingeberene

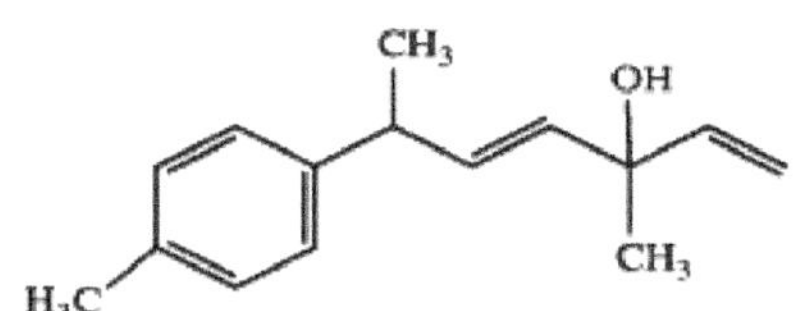

Curcumene

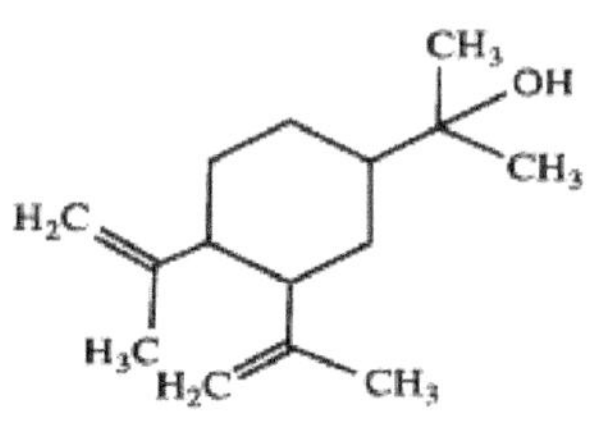

Elemol

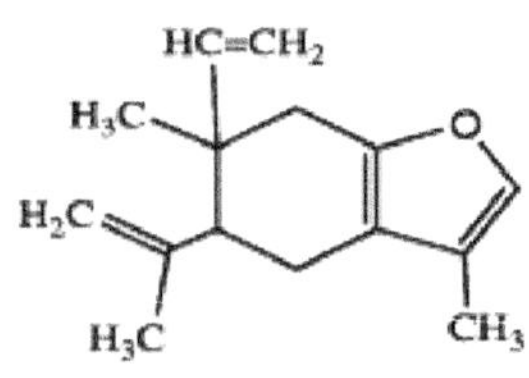

Curzerene

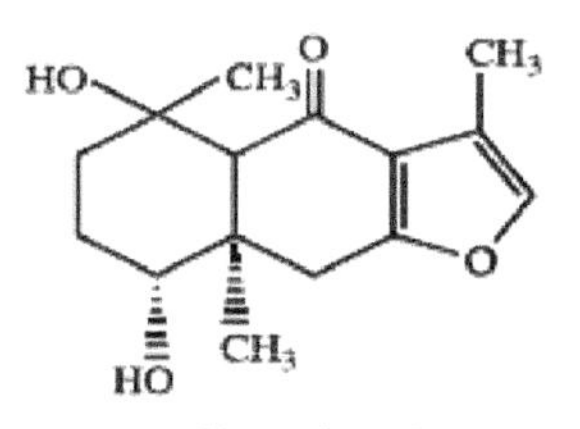

Curcolonol

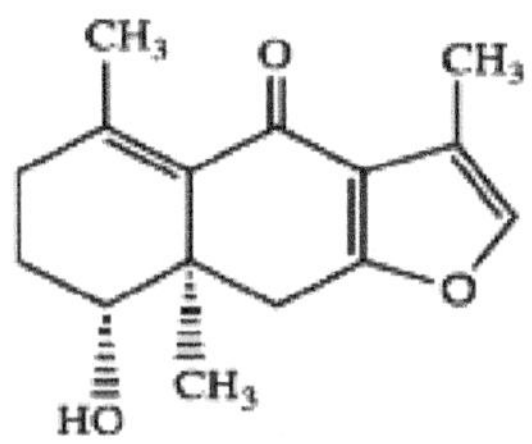

Curcolone

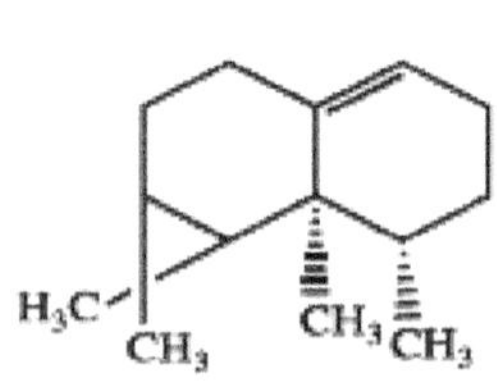

Clarene

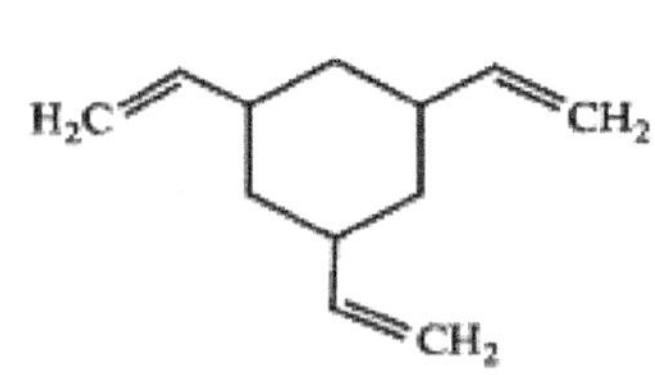

b-Elemene

Figure 5.3 *(Continued)*

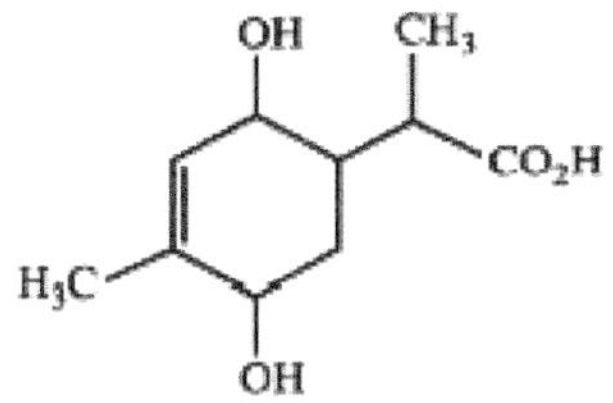
2-(2,5-Dihydroxy-4-methylcyclohex -3-enylpropanoic acid)

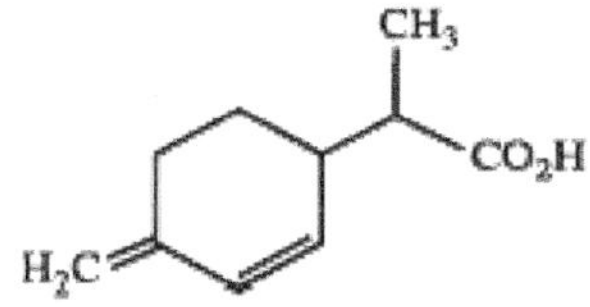
a-Phellandrene

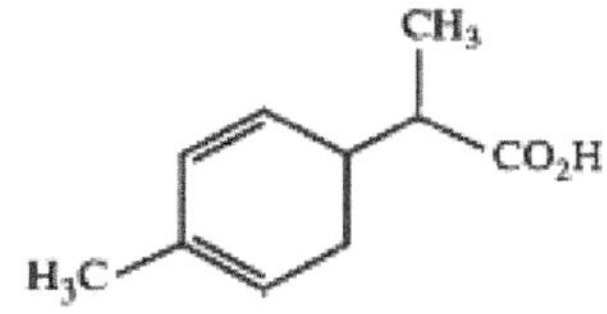
b-phellandrene

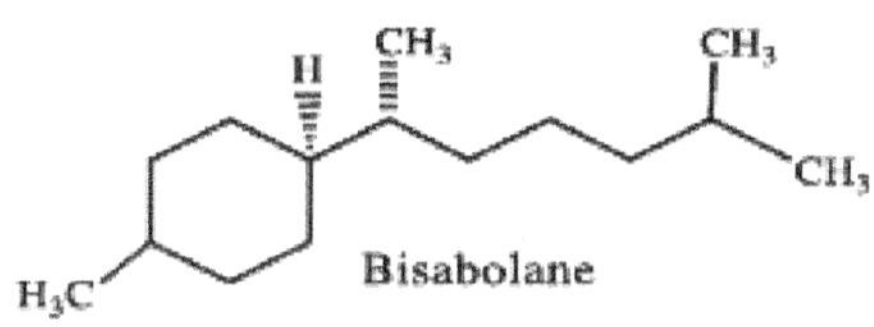
Bisabolane

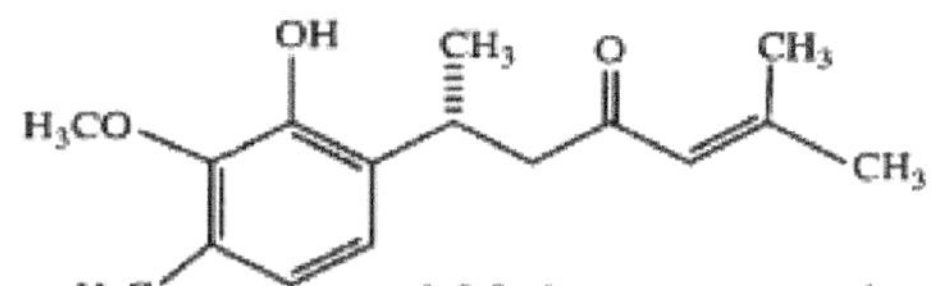
ar-3-Methoxyturmeronol

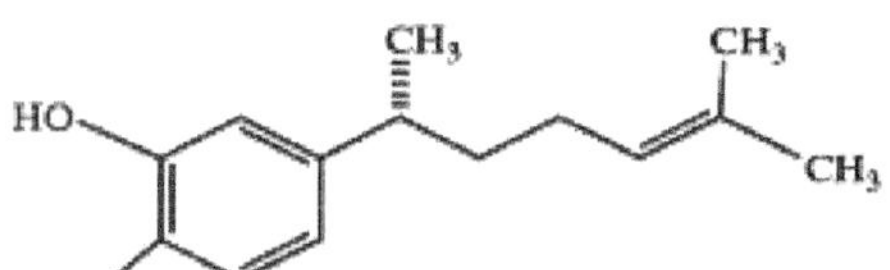
Xanthorrhizol

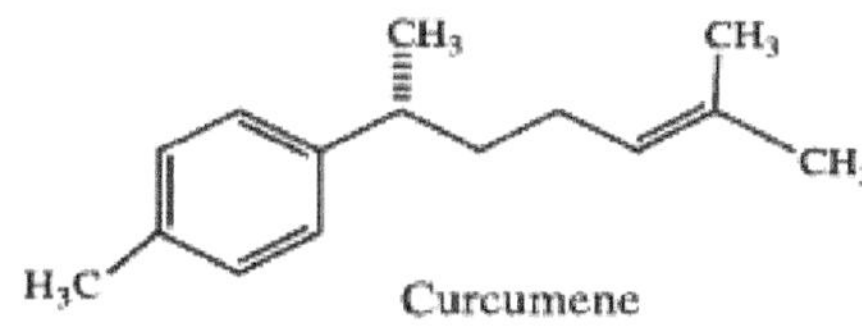
Curcumene

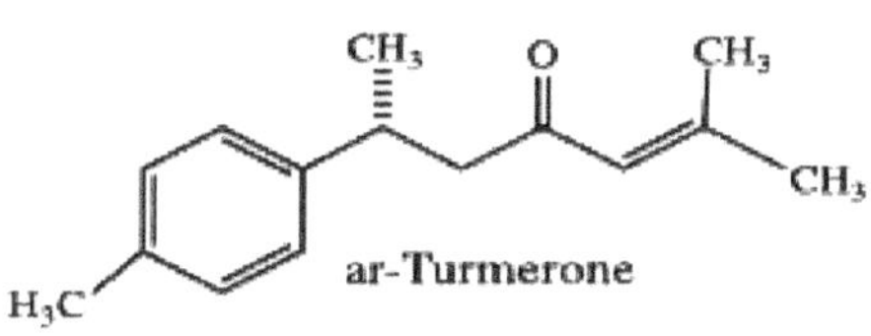
ar-Turmerone

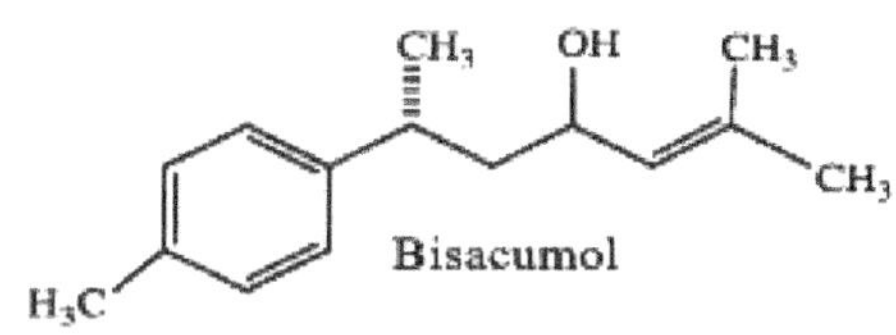
Bisacumol

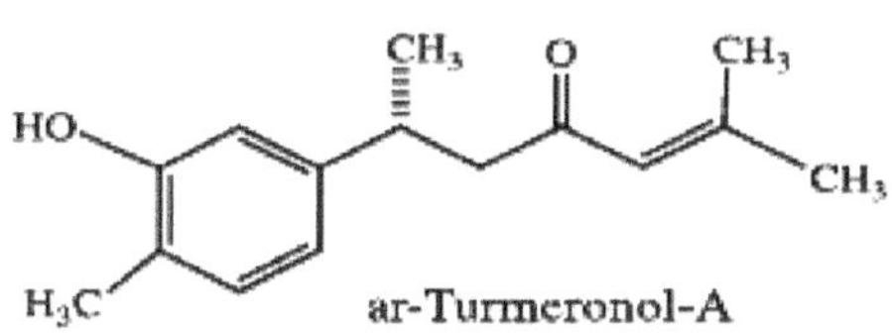
ar-Turmeronol-A

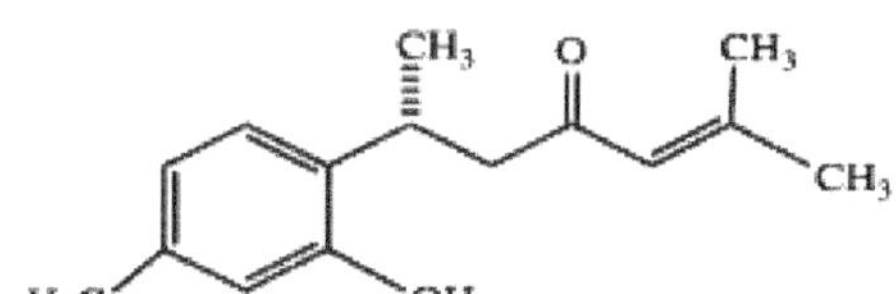
ar-Turmeronol-B

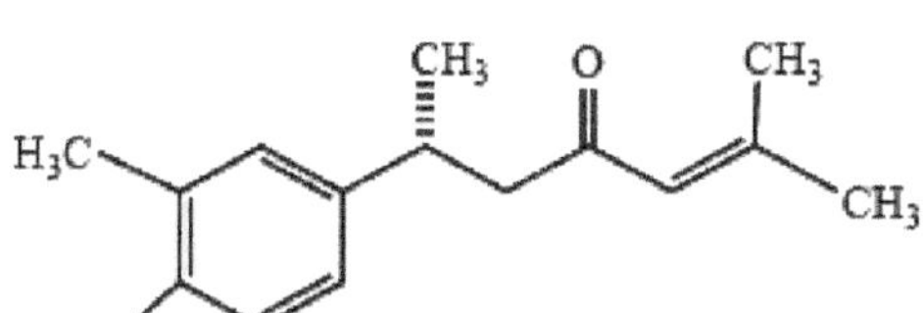
(6S)-2-Methyl-6-(4-formaylphenyl)-2-hepten-4-one

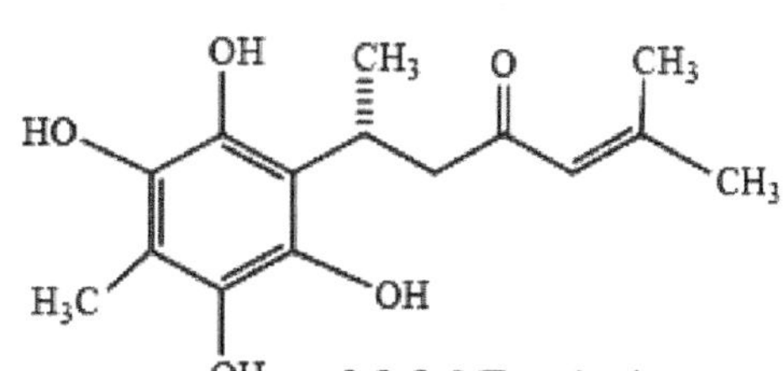
2,3,5,6-Tetrahydroxy-ar-turmerone

Figure 5.3 *(Continued)*

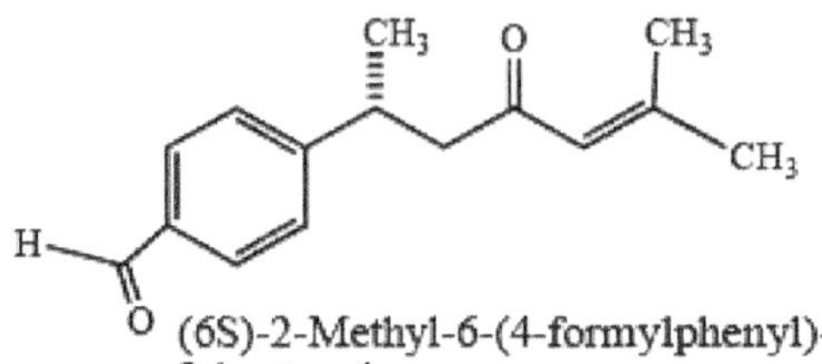

(6S)-2-Methyl-6-(4-formylphenyl)-2-hepten-4-one

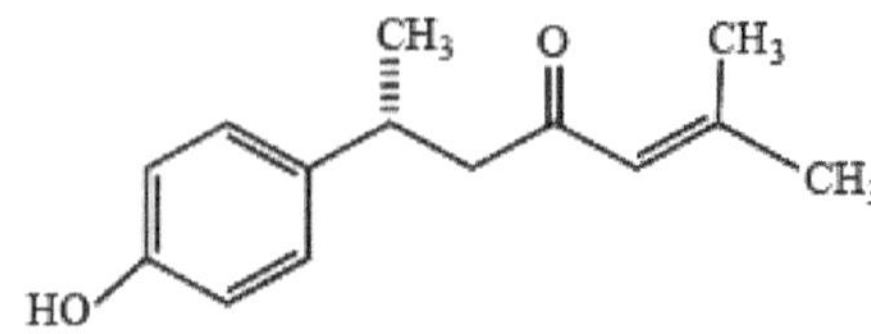

(6S)-2-Methyl-6-(4-hydroxyphenyl)-2-hepten-4-one

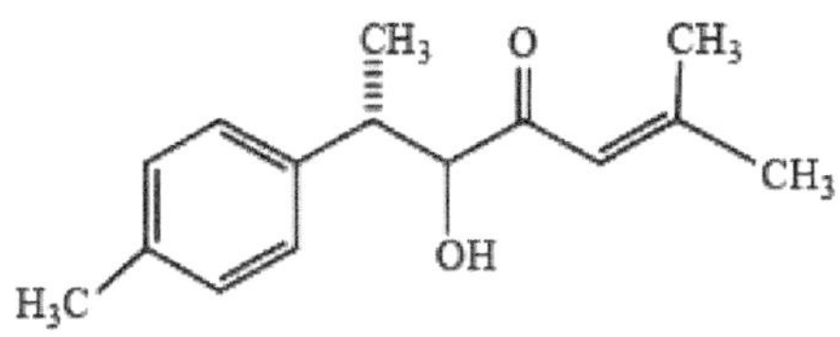

5-Hydroxy-ar-turmerone

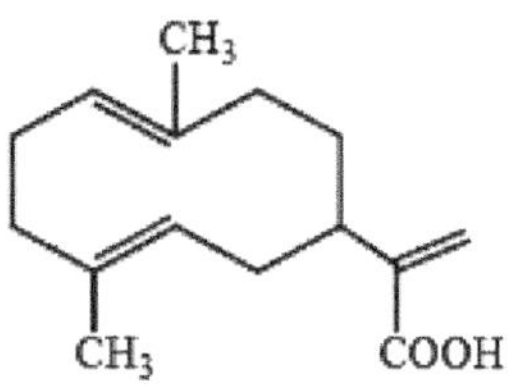

Germacerene A acid

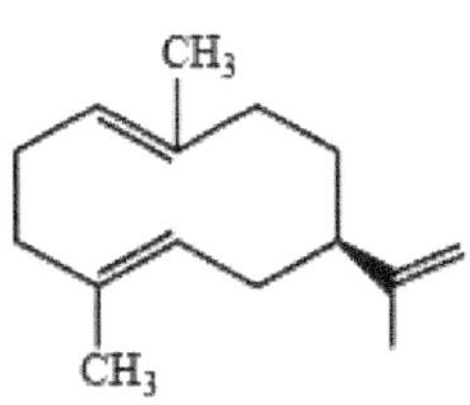

Germacene A

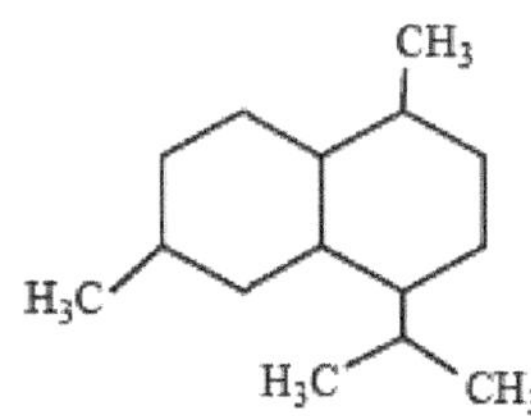

Cadinane

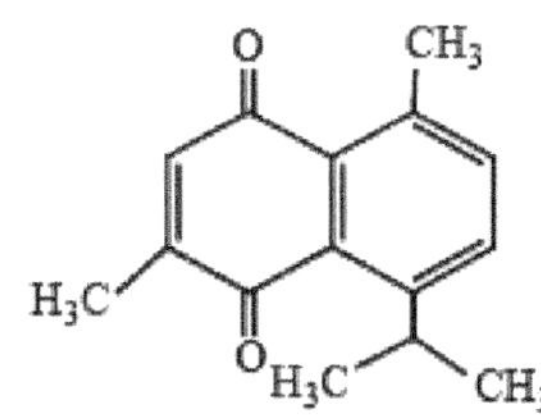

Cadalenequinone

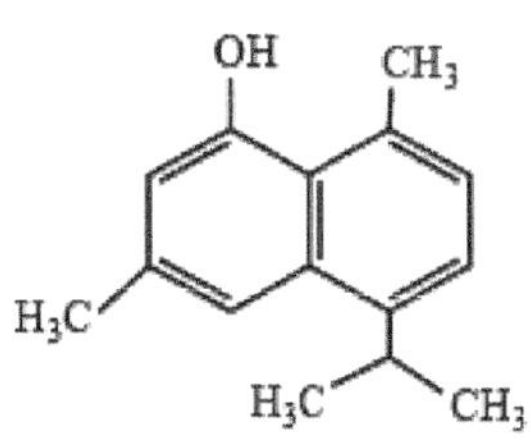

8-Hydroxycadalene

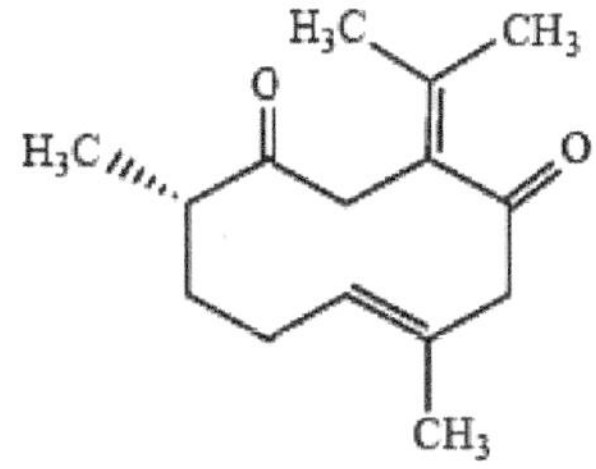

Dehydrocurdione

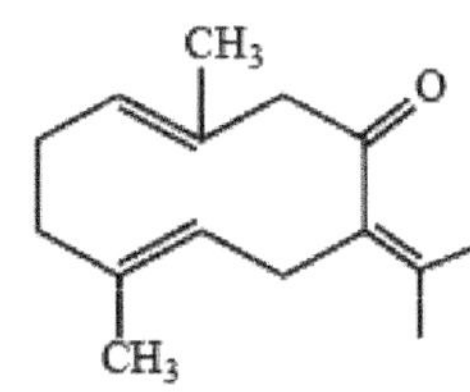

Germacrone

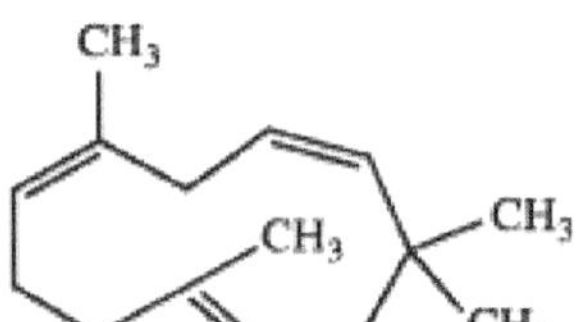

a-Humulene

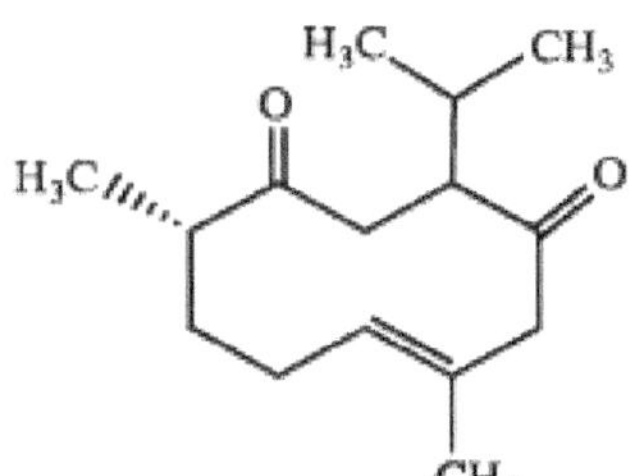

Dihydrocurdione

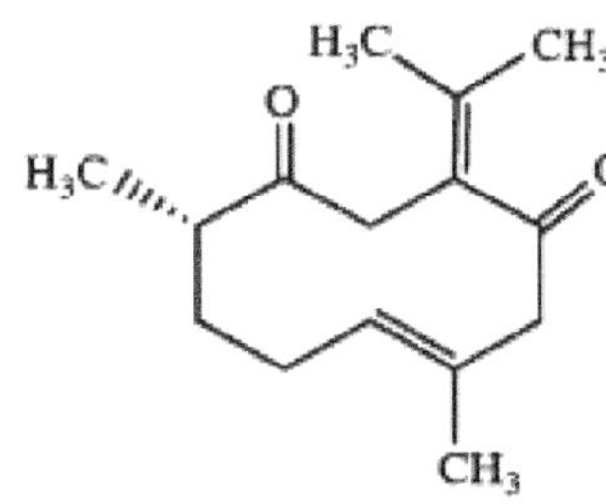

Curdione

Figure 5.3 (*Continued*)

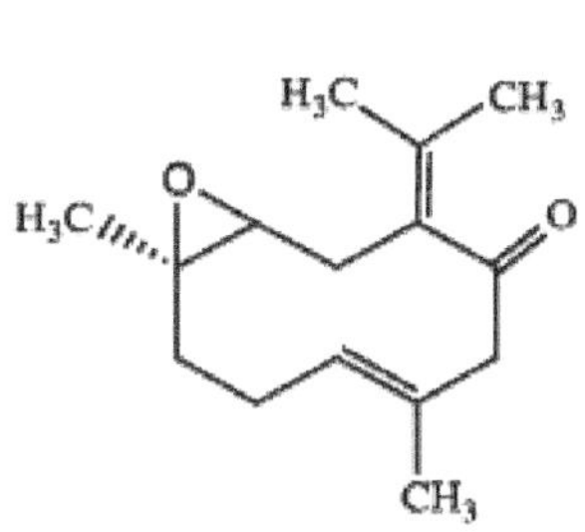
Germacrone-4.5-epoxide

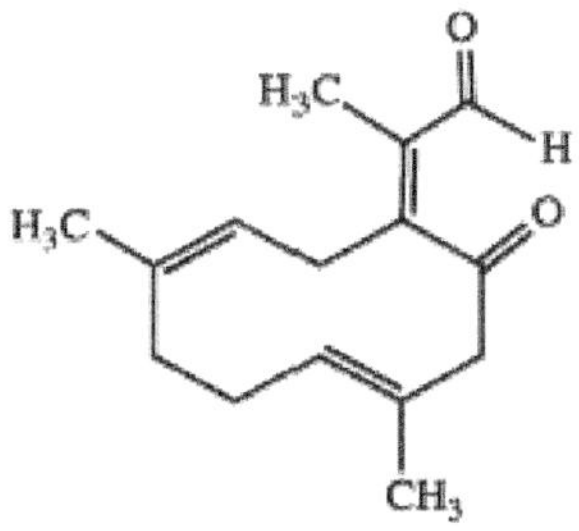
Germacrone-13-al

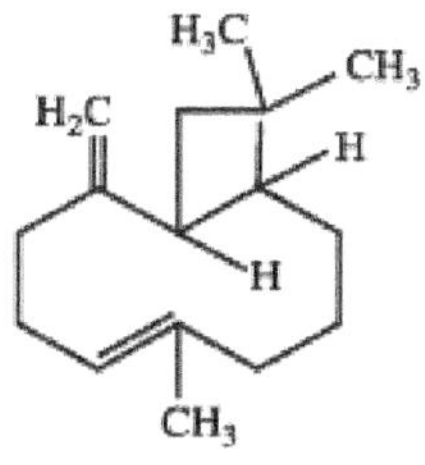
b-Cryophyllene

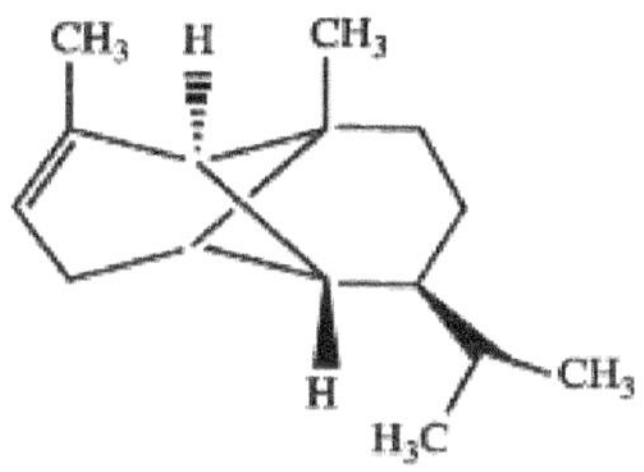
a-Copaene

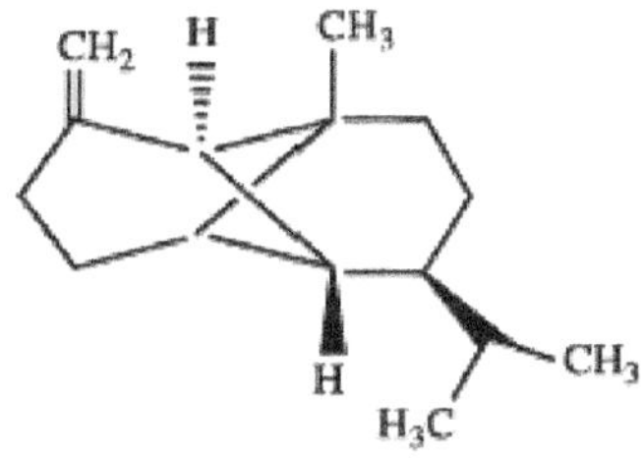
b-Copaene

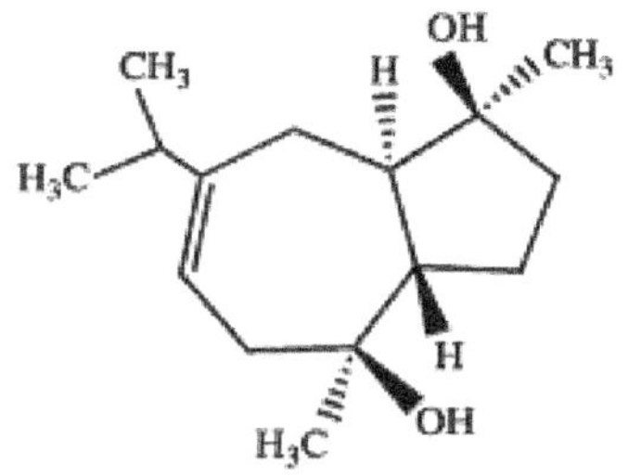
Curcumadiol

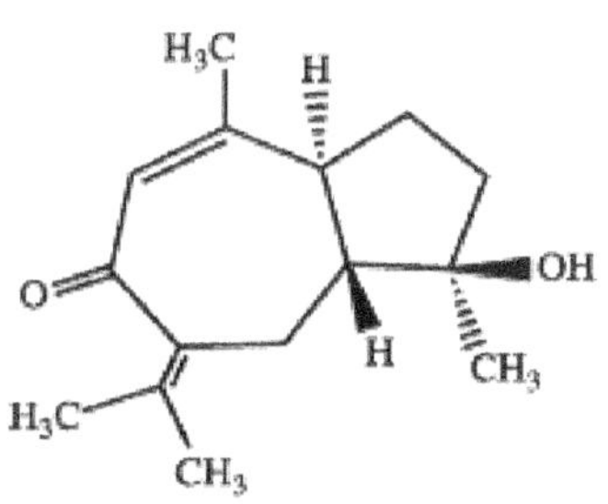
Epicurcumenol

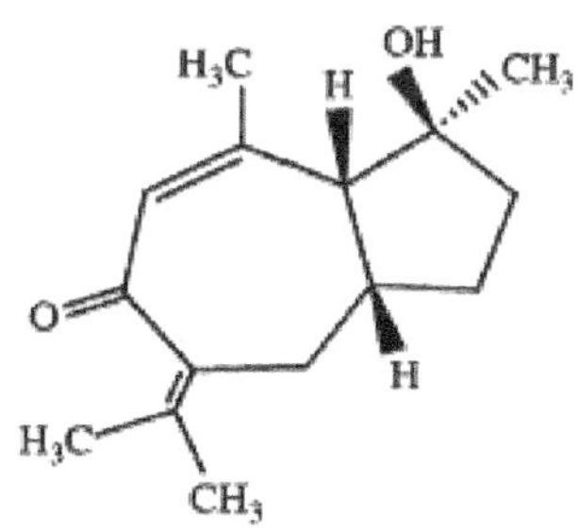
Epiprocurcumenol

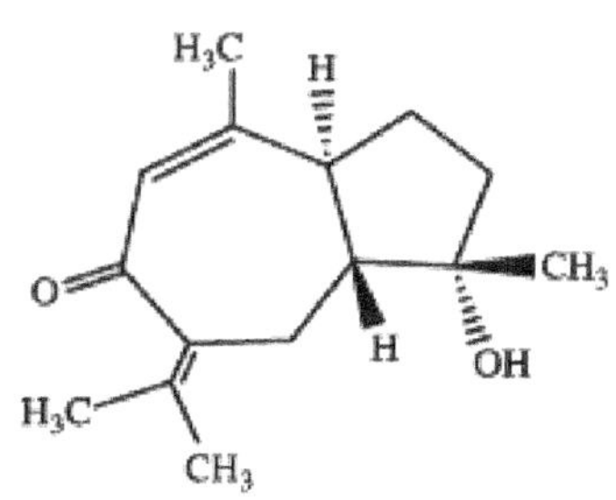
Procucumenol

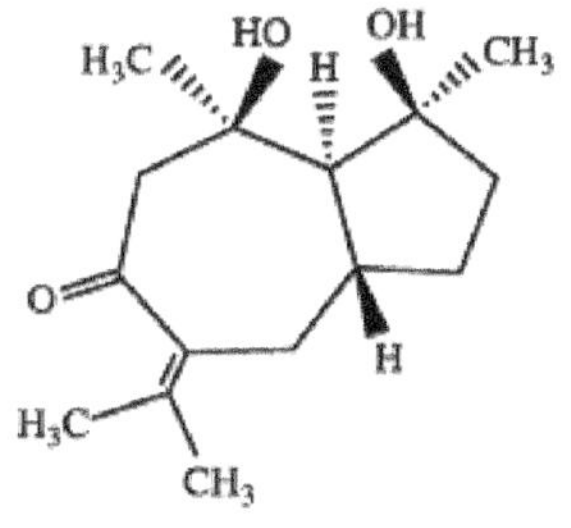
Isozedoaronediol

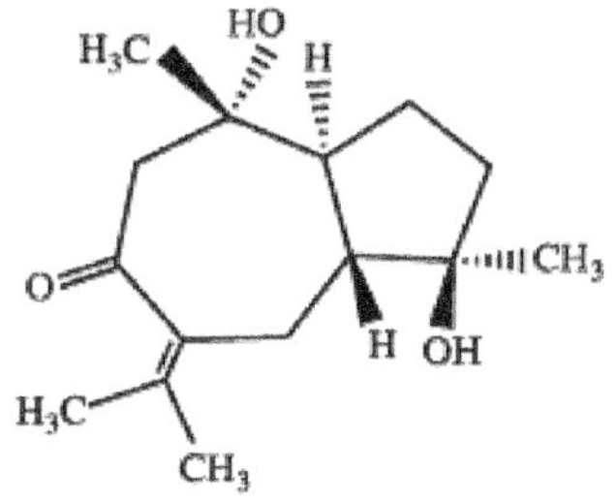
Zedoaronediol

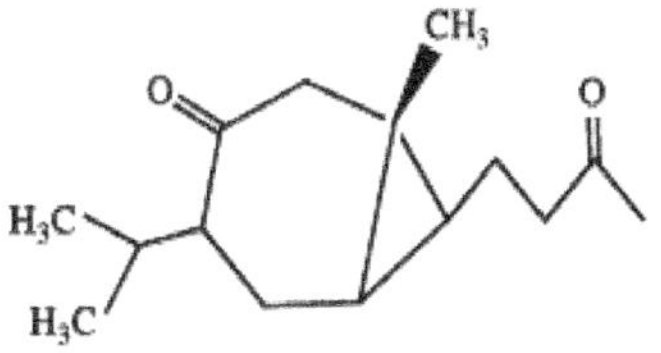
Curcuminone

Figure 5.3 (*Continued*)

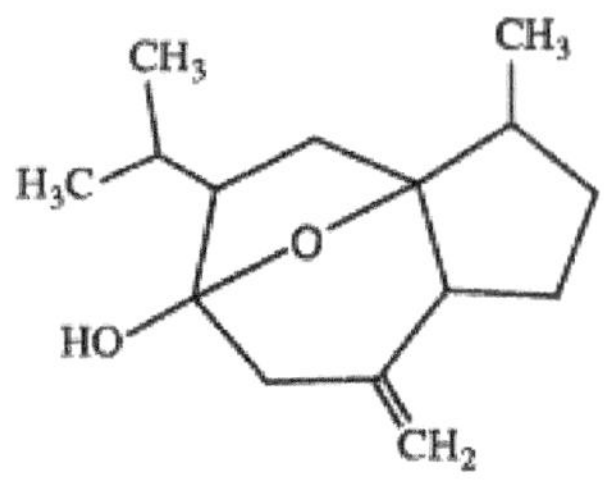

Curcumol

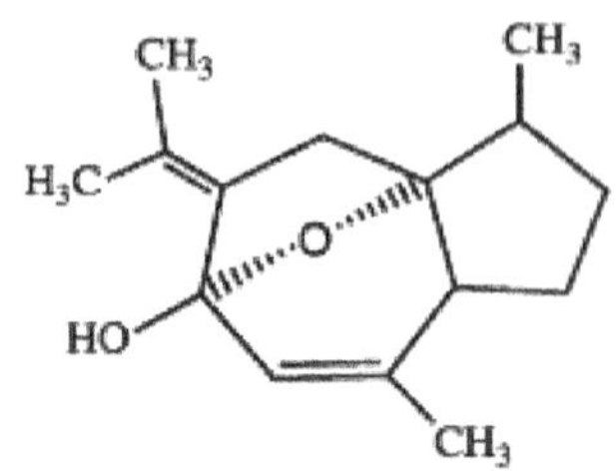

Curcumenol

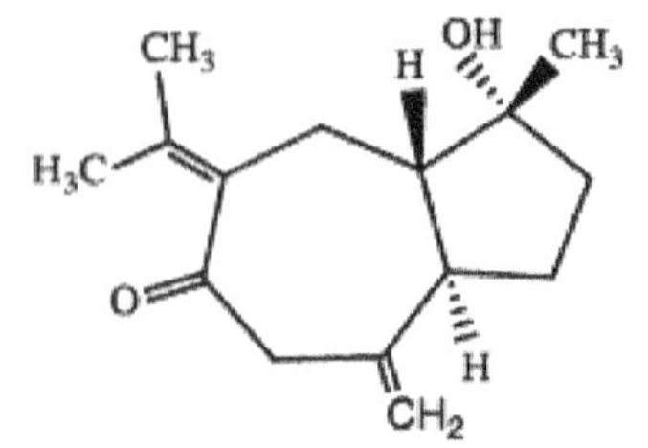

Isocurcumenol

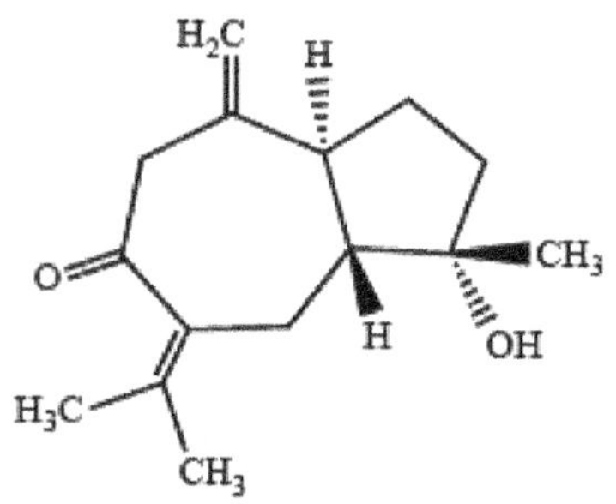

Isoprocurcumenol

Figure 5.3 *(Continued)*

which, curcumin, demethoxycurcumin and bisdemethoxycurcumin were also identified by comparing their chromatographic and spectral data with those of authentic standard compounds.

Variability in volatile oil constituents: There is wide variability in volatile oil constituents of turmeric. There are so many factors which can interfere with the chemical variation of essential oils in plants, such as temperature, humidity, luminosity, altitude, pluviometry, ultraviolet radiation, soil and nutrient conditions, seasonality, circadian cycle, method of collection, drying and part of the plant (Manzan, Toniolo, Bredow and Povh 2003; Sanagi, Ahmad and Smith 1993; Gopalan, Goto, Kodama and Hirose 2000). Due to these factors, the relative composition of turmeric varies considerably with the geographical origin (Garg, Bansal, Gupta, and Kumar 1999; Janaki and Bose 1967) and different agroclimatic zones (Sastry 1970). Maturity also plays an important role in the variation of composition of essential oil. As the plant ages, the concentrations of sesquiterpenes increase, whereas, those of monoterpenes decline in the rhizomes. The maximum curcumin content was found at the age of approximately 9 months after planting, after which there was a decline in the total curcumin content (Zhang and Yang 1988). Furthermore, the essential oil content positively correlates with the levels of phosphorous and potassium in the soil (Verghese 1993). Regarding the effect of solar radiation on the essential oil yield, *C. longa* cultivated under full sunlight treatment showed a greater yield (3.0%) than that under treatment with shading (50% of sunlight) (1.90%) (Verghese and Joy 1989). Major compounds identified are: *p*-cymene (25.4%), 1, 8-cineole (18%), *cis*-sabinol (7.4%) and β-pinene (6.3%). Rhizome oil analysed through GC-MS, gave 17 chemical constituents, of which turmerone (24%), *ar*-turmerone (18%) and germacrone (11%) as major components in turmeric from Chinese origin. Rhizome oil of fresh turmeric from French Polynesia identified 20 components, with zingiberene (16.7%), *ar*-turmerone (15.5%) and α-phellandrene (10.6%) being the major ones. GC and GC-MS of the rhizome oils of turmeric from Bhutan exhibited *ar*-turmerone (16.7–25.7%), α-turmerone (30.1–32.0%) and β-turmerone (14.7–18.4%) as major constituents. Leaf oil from turmeric has also been studied for the chemical constituents and the turmeric grown in North India shows 20 compounds by GC-MS, dominated by monoterpenes, sesquiterpene hydrocarbons and oxygenated sesquiterpenes (57.1%, 3.3% and 2.1%, respectively).

(Leela et al. 2002) detected fifty four components in leaf oil, out of which 30 compounds, which constituted 91% of the oil, could be identified. The major constituents found in the oil from leaves were α-phellandrene (32.6%), terpinolene (26.0%), 1, 8-cineole (6.5%) and *p*-cymene (5.9%).

Monoterpenes from green leaves and fresh rhizomes of *C. longa* L. grown in India by GC and GC-MS showed the absence of β-pinene from the rhizome oil, but it had all the other components of the leaf oil in different proportions. Car-3-ene, α-terpinene, γ-terpinene and terpinolene were also included in the rhizome volatiles. Variations in the volatle oil composition were also reported from the rhizomes collected from the sub-Himalayan region of the Tarai in India. Table 5.1 shows the composition of essential oils from leaf, flower, root and rhizome of *C. longa* L.

Two major ketonic sesquiterpenes-*ar*-turmerone and turmerone ($C_{15}H_{20}O$ and $C_{15}H_{22}O$) are responsible for the aroma of turmeric. In addition, *p*-cymene, β-sesquiphellandrene and sesquiterpene

Table 5.1 Composition of Essential Oils from Leaf, Flower, Root and Rhizome of *Curcuma Longa* L.

	Concentration (%)			
Component	**Leaf**	**Flower**	**Root**	**Rhizome**
α-Pinene	2.1	0.4	0.1	0.1
β-Pinene	2.8	0.1	0.1	tr
Myrcene	2.3	0.2	tr	0.1
α-Phellandrene	32.6	–	0.1	0.1
δ-Carene	1.1	0.6	–	–
α-Terpinene	1.3	0.1	–	–
p-Cymene	5.9	1.6	303	3.0
β- Phellandrene	3.2	tr	–	tr
1,8-Cineole	6.5	4.1	0.7	2.4
Z-β-Ocimene	0.2	–	–	–
E-β-Ocimene	0.4	–	–	–
γ-Terpinene	1.5	–	–	–
Terpinolene	26.0	7.4	0.1	0.3
Linalool	0.7	1.1	0.1	–
1,3,8-Paramenthatriene	0.2	0.3	–	–
p-Methylacetophenone	0.1	0.3	tr	tr
p-Cymen-8-ol	0.8	26.0	1.5	0.3
α-Terpineol	0.4	1.1	0.1	0.2
Thymol	0.3	–	0.1	–
Carvacrol	0.1	–	0.3	0.1
γ-Curcumene	0.1	tr	0.4	0.1
ar-Curcumene	0.2	1.9	7.0	6.3
α-Zingiberene	0.5	0.8	tr	tr
β-Bisabolene	–	0.9	2.3	1.3
β-Sesquiphellandrene	0.3	1.1	tr	2.6
E-Nerolidal	0.1	1.1	–	–
Dehydrocurcumene	–	–	4.3	2.2
ar-turmerone	0.1	1.2	46.8	31.5
Turmerone	0.9	1.0	–	10.0
Curlone	0.2	0.3	0.6	10.6
Curcuphenol	tr	tr	0.6	0.5
6S,7R Bisabolene	0.1	0.4	1.2	0.9
Other	9.0	48.0	30.3	27.8

alcohols are also present in turmeric rhizomes and the ratio of turmerone to *ar*-turmerone is 60:40, while analysis of the volatile oils from commercial oleoresin has a ratio of 80:20. Maturity of turmeric also affects the major components of rhizome oil as turmeric grown in Sri Lanka has indicated *ar*-turmerone (24.7–48.9%) and turmerone (20–39%) as the major compounds and the structure of the two turmerones can be further described as 2-methyl-6-(4-methylcyclohexa-2,4-dien-1-yl) hept-2-en-4-one (5, α-turmerone) and 2-methyl-6-(4-methylenecyclohex-2-en-1-yl)hept-2-en-4-one (2, β-turmerone). GC-MS studies of rhizome oils have revealed thirty-five components out of which turmerone (49%); *ar*-curcumene (15%) and *ar*-turmerone (6.4%) are the major compounds. Analyses of five different *Curcuma* species exhibit variations even in the major components, like *ar*-turmerone (2.6–70.3%), α-turmerone (trace–46.2%) and zingiberene (trace–36.8%). *ar*-Turmerone is also known as a potential antivenom agent.

Various extraction techniques have been reported to maximize the yield of essential oil and pigments, especially *ar*-turmerone (α- and β-), turmerone and the curcuminoids. (Manzan, Toniolo, Bredow and Povh 2003). Higher yields can be obtained by varying the autoclave pressure and distillation time. Therefore, extraction with volatile solvents, maximum yield (5.49 wt%) is obtained using 0.175, 0.124 and 0.088 mm particles at 40°C and 6 hours. Supercritical fluid extraction (SFE), mainly by supercritical carbon dioxide ($SCCO_2$), can be used to extract volatile oils from natural products and does not produce substantial thermal degradation or organic solvent contamination. (Sanagi, Ahmad and Smith 1993) have reported a process for the direct analysis of turmeric using online coupling of supercritical fluid extraction (SFE) with supercritical fluid chromatography (SFC) and the extraction rate of turmeric oil in SC-CO_2 can be measured as a function of pressure, temperature and flow rate at constant extraction time. The total oil yield decreases with temperature at constant flow rate, but increases with flow rate at constant pressure and temperature. The optimum pressure for the extraction yield is found to be 22.5 MPa. GC-MS analysis of the hydro-distilled oil indicates β-turmerone (11–36%), α-turmerone (19–24%) and *ar*-turmerone (4–14%). The biochemical constituents of the turmeric oil extracted using SC-CO_2 has been compared with that of steam-distilled oil by GC-MS (Gopalan, Goto, Kodama and Hirose 2000). *ar*-turmerone and turmerone constituted about 60% of the total oil out of the 21 components identified. Analysis of the cyclohexane extract of turmeric by GC-MS coupled with Pseudo Sadtler etention indices reveals a series of saturated and unsaturated fatty acids, along with sesquiterpenes. The fatty acids reported are: tetradecanoic acid, *cis*-9-hexadecenoic acid, hexadecanoic acid, *cis-cis*-9, 12-octadecenoic acid, *cis-trans*-9-octadecenoic acid and octadecenoic acid. Comparative studies on the constituents of turmeric oil from leaves, flowers, rhizomes and roots showed that the oils from rhizomes and roots had more similarities as compared to the oil from leaves and flowers, thus indicating the presence of biogenetically linked characters (Leela et al. 2002). The oils extracted from flowers and leaves were dominated by monoterpenes, whereas the major part of the oil from roots and rhizomes contained sesquiterpenes. Essential oil from the leaf of turmeric also showed α-phellandrene (47.7%) and terpenoline (28.9%) as the major constituents in another study.

5.4 ISOLATION OF CURCUMINOIDS

Curcuminoids *viz.* curcumin, demethoxycurcumin and bisdemethoxycurcumin are really tedious to isolate from turmeric. The purified extract of these three forms is available in the market with curcumin (75–81%), demethoxycurcumin (1.5–1.9%) and bisdemethoxycurcumin (2.2–6.6%). Isolation methods of curcuminoids are summarized in Table 5.2.

1. Rhizomes, extracted with hexane, remove fatty components followed through benzene extraction to obtain yield of 1.1% which is ahead purified by way of crystallization from ethanol to yield orange-yellow needles (Janaki and Bose 1967).

Table 5.2 Physico-Chemical Properties of Curcuminoids

Chemical Name	Di-Cinnamoyl Methane	4-Hydroxy Cinnamoy (Feruloyl) Methane	Bis-4-Hydroxy Cinnamoyl Methane
Common name	Curcumin	Demethoxycurcumi n	Bisdemethoxycurcumin
Molecular weight	368.4	338.0	308.1
Melting point (°C)	184–186	172.5–174.5	224
		Solubility	
Hexane or ether	Insoluble	Insoluble	Insoluble
Alcohol or acetone	Soluble	Soluble	Soluble
		Relative Rf	
Chloroform: ethanol (25:1)	1.00	0.66	0.41

2. Similarly, (Sastry 1970) reported isolation of curcumin and related demethoxy compounds by extraction with organic solvents, but the yield was very poor (1.5–2.0%), whereas hot and cold percolation methods gave higher yield.
3. Some other investigations involving organic solvents for extraction purpose are the hit choices adopted for the isolation of curcuminoids (Zhang and Yang 1988; Verghese and Joy 1989; Verghese, 1993; Xianchun, Xie, Shan and Qi 1993).
4. SC-CO_2 extraction modified by 10% ethanol is more efficient (Baumann, Rodrigues and Viana 2000), but the high operating pressure creates problems. Microwave-assisted extraction (MAE) technique for selective and rapid extraction of curcuminoids is also possible.

Turmeric powder irradiated for 2 and 4 minutes with microwave showed marginally higher extraction of curcuminoids in 60 min by acetone (Dandekar and Gaikar 2002). The composition and yield of curcuminoids extract using the existing techniques, *viz.* soxhlet, low-pressure solvent extraction, hydrodistillation and SC-CO_2 extraction were also compared (Braga, Leal, Carvalho and Meireles 2003). The lowest (2.1%) yield was obtained with hydrodistillation, whereas highest was obtained by soxhlet extraction with ethanol. A mixture of ethanol and isopropanol is the best combination for SC-CO_2 extraction.

Analysis: The most common method used for the analysis is direct spectrophotometric absorption between 420 and 430 nm (BSI 1983), however, other compounds absorbing in this region influence the values. Therefore, only pure standards are to be used. Direct flourimetric method for analysis of curcumin also has problems due to the relative flourescence intensities of the three curcuminoids, which are in the ratio of 1:2.2:10.4 at equimolar concentrations (Tonnesen and Karlsen 1983). Other methods, based on the molar absorptivity by developing coloured complexes with alkalis, strong acids and boric acid, are not consistent (Janssen and Gole 1984) due to unstable nature of the complex. IR, NMR and MS can also be used for the identification and characterization of curcuminoids (Roughley and Whiting 1973; Govindarajan 1980; Tonnesen and Karlsen 1986). Commercial produce of curcumin contains the three curcuminoids that can be estimated using HPLC (Tonnesen and Karlsen 1983), but separation is strongly dependent on the chromatographic conditions. C18 columns are preferred for HPLC analysis due to the labile nature of the curcuminoids (Khurana and Ho 1988). Using aternary mobile phase gradients in chromatography on polar adsorbents can be used as an alternative to the incremental gradient elution, with the aim of eliminating sudden changes in composition of the adsorbed solvent layer on the surface of adsorbent because of steep changes in polarity of the mobile phase during gradient elution and their negative influence on chromatographic separations). Mobile phase, e.g. methanol, 2% acetic acid, acetonitrile; the three curcuminoids can be separated and quantified. The advantage of this method lies in the estimation of individual curcuminoids from varieties of turmeric rhizomes (Jayaprakasha, Negi, Anandha Ramakrishanan and Sakariah 2001).

Structure-activity relationships: The anti-inflammatory activity of curcumin and its derivatives is associated with the hydroxyl and phenol groups in the molecule, which are also essential for the inhibition of prostaglandins, PG synthetase and leucotriene synthesis (LT) (Kiuchi, Shibuya and Sankawa 1982); Iwakami et al. 1986). (Claeson et al. 1993; Claeson et al. 1996) suggested that the anti-inflammatory action and the antiparasitic activity were associated with the β-dicarbonylic system with conjugated double bonds (dienes) (Araujo et al. 1999). The better skin penetration and lipophylicity is attributed to the presence of a diene ketone system. Calebin-A, a novel curcuminoid isolated from turmeric, protects neuronal cells from β-amyloid insult. The hydroxyl group at para-position of this compound is most critical for the expression of biological activity (Kim, Park and Kim 2001).

5.5 OLEORESIN

Turmeric oleoresin functions chiefly as a food colourant and secondarily imparts characteristic mild spicy aroma which is compatible with mustard, pickles, relish formulae, etc. The product from industrial practice using good, clean turmeric, with a curcuminoid content of 4.5–5.0%, is a highly viscous, deep brownish-orange product, obtained in a yield of about 12%. This analyses 30–40% as curcumin, 15–20% volatile oil and has the characteristic fresh, clean, mildly pungent, woody-pungent, woody-spicy aroma of turmeric. Alcohol and acetone are very good extractants and the yields can also be expected to be high. Ethylene dichloride has the advantage of having relatively selective extraction of the flavour constituents and it is water immiscible, non-flammable and has sufficiently low boiling point, therefore requiring no refrigeration. However, impurities in the solvent, such as traces of high-boiling fractions accumulating in the product and creating off-flavour problems, and possible corrosivity to the equipment in the presence of water, are serious problems to contend with. Many manufacturers prefer acetone as their choice of solvent for extraction of good-quality turmeric oleoresin (Govindarajan 1980). Soxhlet extraction of turmeric powder with acetone gives in 4–5 hours, a yield of about 5.0%, containing 42% curcuminoids; prolonged extraction up to 24 hours gave only a fractionally higher yield of curcuminoids. Extraction carried out by batch and counter-current cold percolation in columns resulted in slightly more yield of 5.7–5.9% with 42.2–46.0% curcuminoids and extraction efficiencies varying from 72.9–81.7%. Acetone as solvent was slightly superior to alcohol and ethylenedichloride. Dilution with an appropriate vegetable oil, polysorbates or propylene glycol, the oleoresin gives a bright yellow liquid with the typical turmeric aroma and a slightly bitter and pungent taste (Govindarajan 1980). The oleoresin may also be spray-dried as powder on to the sugar matrix, such as maltodextrin, and can be used as a colourant in dry cereals or beverages. Spray -dried turmeric oleoresin has advantage over ground turmeric powder as it is devoid of starch, the predominant component in dried rhizome, proteins and other fibres (Buescher and Yang 2000). The aroma of turmeric is, however, due to high-boiling components and is rather difficult to remove unless curcuminoids are isolated by crystallization (Perotti, 1975).

5.6 HEALTH BENEFITS

Several pharmacological and medicinal benefits of turmeric are widely known (Eigner and Scholz 1999). The rhizome extracts of turmeric, apart from the ethanolic and methanolic ones, have been examined for their biological activities and have been in use for centuries.

i. **Anti-inflammatory activity:** Curcuminoids *(because constituents of Curcumin are curcuminoids which in one form or the other play an important role in various health* *benefitting effects of Turmeric to humans)* and other constituents of turmeric are well known for their anti-inflammatory activity. Turmeric extract, volatile oils from turmeric and curcuminoids were reported to possess this property in different experimental models of inflammation in mice, rats, rabbits and pigeons (Ghatak and Basu 1972). Thus, curcuminoids are effective against carrageen an induced oedema in rats (Srivastava, Dikshit, Srimal and Dhawan 1985) and mice (Srimal and Dhawan 1985). Administration of curcuminoids to patients who have under gone surgery or suffered from trauma, benefitted through reduced inflammation to a comparable level with phenylbutazone (Satoskar, Shah and Shenoy 1986). Oral administration of curcumin at a dose of 3 mg/kg was also found to be effective in reducing inflammation associated with various forms of arthritis (Srimal and Dhawan 1983). The antirheumatic properties of curcuminoids were also tested successfully in patients with diagnosed rheumatoidarthritis. Curcumin also enhances wound healing in diabetic rats and mice (Sidhu 1999) and in H_2O_2-induced damage in human keratinocytes and fibroplasts (Phan, See, Lee and Chan 2001). The wound healing effect is expressed by inhibiting the activation of NF-κB transcription factor through the prevention of the p65 unit of the factor.

ii. **Antioxidant effect:** Curcuminoids are natural phenolic compounds with potent antioxidant properties, which were reported as early as 1976 (Sharma 1976). Both turmeric and curcumin inhibit generation of superoxide and hydroxyl free radicals (Subramonian, Sreejayan, Devasagayam and Singh 1994; Ruby et al. 1995). The antioxidant properties of curcumin in the prevention of lipid peroxidation are also well recognized (Sreejayan and Rao 1994). The three forms of the pigment have dual prolonged antioxidant activity, e.g. preventing the formation of free radicals, as well as intervening in their propagation. In fact, the antioxidant activity has been attributed to its unique conjugated structure, which includes two methoxy phenols and an enol form of β-diketone, with the typical radical trapping ability as a chain-breaking antioxidant (Masuda et al. 2002). Curcumin also has the potential to prevent oxidative damage to the arterial wall. Thus, administration of 500 mg of curcuminoids daily to healthy humans for 7 days reduced lipid peroxides by 33% and blood cholesterol by 29%, indicating a possible role of curcumin in reducing cardiovascular diseases (Soni and Kuttan 1992). An *in vitro* study on the comparison of the antioxidant activity of curcuminoids and tetrahydro-curcumin using rabbit erythrocyte membrane showed highest activity with tetrahydro-curcumin (Osawa, Sugiyama, Inayoshi and Kawakishi 1995).

iii. **Antimutagenic and anticancerous property:** The anticancer action of curcumin has been studied in a standard model of radiation induced tumour in rat mammary gland (Inano et al. 2000). In animal studies, curcuminoids inhibited capsaicin-induced mutagenic changes in mouse bone marrow. Additionally, mice maintained on turmeric or curcuminoid enriched diets, when challenged with carcinogens, excreted low levels of mutagenic metabolites, as well as carcinogens (Usha et al. 1994). Curcumin is reported to prevent DNA damage, even in individuals who may be genetically susceptible to the toxic effects of xenobiotic exposures, and is also able to exert antimutagenic/anticarcinogenic properties at levels as low as 0.1–0.5% in the diet (Polasa, Naidu, Ravindranath and Krishnaswamy 2004).

iv. **Chemopreventive and bioprotectant property:** Numerous studies have been published on the positive effects of turmeric, both in the prevention of cancer and in the recovery from chemotherapy and radiation treatment (Jung et al. 2005). In addition to its capacity to intervene in the initiation and growth of cancer cells and tumours, and to prevent their subsequent spread throughout the body by metastasis, curcumin increases sensitivity of cancer cells to certain drugs commonly used to combat cancer, rendering chemotherapy more effective. Curcuminoids can also act as photochemoprotective agents that provide protection against UVB radiation-induced oxidative stress. This inhibition of UVB radiation induced damage can reduce the incidence of skin cancer (Afaq, Adhami and Ahmad 2002).

v. **Antiviral, antimicrobial and antiparasitic activity:**

a. **Antiviral:** An interesting property of curcuminoids is their anti-HIV effect, which has been demonstrated during *in vitro* and *in vivo* experiments, including a limited number of human studies. HIV infection is characterized by a complex command system, the structural part of which is called 'long terminal repeat' (LTR), which results in virus activation or inactivation. Drugs that interfere with LTR may be of potential therapeutic value in delaying active HIV infection and the progression of AIDS. Curcumin has been found to inhibit activation of the LTR and to decrease HIV replication effectively (Li, Zhang, Dezube and Crumpacker 1993).

b. **Antimicrobial:** Curcuminoids have also been shown to exhibit antimicrobial properties. The antibacterial effects of alcoholic extract of turmeric, curcumin and oil from turmeric have been studied by Bhavani Shankar and Srinivasamurthy (1979). Extracts from turmeric, as well as the active principles, the curcuminoids, were found to inhibit the growth of numerous Gram-positive and Gram-negative bacteria, fungi and the intestinal parasite, *Entamoeba histolytica.* The ethanol extract of turmeric has been reported to have anti-amoebic activity against *E. histolytica in vitro* (Dhar et al. 1968). Curcumin at concentrations of 2.5–50.0 mg/100 ml inhibited *in vitro* growth of *Staphylococcus aureus* (Bhavani Shankar and Srinivasamurthy 1979). Interestingly, the antibacterial and antiviral activities of curcumin were enhanced significantly by illumination with visible light (Pervaiz, 1990). Curcumin also inhibits *in vitro* production of aflatoxins/toxins produced by the mould *Aspergillus parasiticus*, which may grow and contaminate poorly preserved foods and is a potent biological agent causing injury to the liver, often resulting in liver cancer (Jayaprakasha, Negi, Anandha Ramakrishanan and Sakariah 2001).

c. **Antiparasitic:** Curcumin possesses leishmanicidal effects *in vitro* and is more potent than the standard leishmaniasis drug, pentamidine. LD50 for leishmanicidal activity *in vitro* is found to be 37.6 + 3.5 µM (Koide et al. 2002).

d. **Antidiabetic property:** The efficacy of turmeric and curcumin on the blood sugar and polyol pathway in diabetic albino rats showed significant reduction in blood sugar and glycosylated haemoglobin levels. This could be due to a decreased influx of glucose into the polyol pathway, leading to an increased NADPH/NADP ratio and elevated activity of the antioxidant enzyme, glutathione peroxidase. The activity of sorbitol dehydrogenase, an enzyme that catalyses the conversion of sorbitol to fructose, is also lowered significantly on treatment with turmeric or curcumin. Dietary curcumin can alleviate dangerous secondary complications induced by diabetes. The beneficial effects of dietary curcuminon diabetic nephropathy are probably mediated through the hypolipidaemic effects of curcumin. Curcumin has been proved to be an effective hypolipidaemic agent. One study validated the role of dietary curcumin in maintaining healthy serum cholesterol levels in diabetic rats. Employing a high-cholesterol diet for the diabetic rats, curcumin exhibited a lowering of cholesterol and phospholipidin treated animals as compared with curcumin-free controls. Liver cholesterol, triglycerides and phospholipid elevated under diabetic conditions were lowered by dietary curcumin. Curcumin induces a higher rate of cholesterol catabolism, which is evidenced by the higher activity of liver cholesterol-7α-hydroxylase.

vi. **Other pharmacological properties:**

a. **Antiangiogenic effect:** Studies on the effect of curcumin on the growth of Ehrlich ascites tumour cells and endothelial cells *in vitro* prove curcumin to be a potent angio-inhibitory compound, as demonstrated by inhibition of angiogenesis in two angiogenesis assay systems *in vivo*, e.g. peritoneal angiogenesis and chorioallantoic membrane assay. The angio-inhibitory effect of curcumin *in vivo* is corroborated by the results on down-regulation of the expression of proangiogenic genes by curcumin (Gururaj et al. 2002).

b. **Antithrombotic effect:** In a comparative study on the protective effects of curcumin and aspirin in mice subjected to induced thrombotic challenge, curcumin exhibited a dose-related antithrombotic effect.

c. **Hepatoprotective effect:** Curcumin and turmeric protect the liver against several toxicants both *in vitro* and *in vivo*. Oral administration of curcumin (30 mg/kg body weight) for 10 days lowered the liver and serum lipid peroxide levels, serumalanine aminotransferase (ALAT), aspartate amino transferase (ASAT) and lactate dehydrogenase (LDH), enhanced by i.p. injection of iron in rats.

d. **Alzheimer's disease:** Accumulation of β-amyloid proteins in the brain is one of the hallmarks of Alzheimer's disease. Dietary curcumin at low dose (160 ppm) and high dose (5000 ppm) significantly lowered oxidized proteins and interleukin1-β, a pro-inflammatory cytokine elevated in the brains of mice. With low dose, the astrocytic marker, GFAP, was reduced and insoluble β-amyloid (Abeta), soluble Abeta and plaque burden were reduced significantly by 43–50% (Lim et al 2001).

vii. **Coronavirus disease 2019 (COVID-19):** Since December 2019, COVID-19 caused by severe acute respiratory syndrome coronavirus 2 (SARS-CoV-2) virus, has spread rapidly throughout the planet and also the World Health Organization (WHO) declared it a global pandemic on 11 March 2020. As of May 2020, more than 212 countries across the globe had reported over 4 million cases and over 280,000 deaths. Till date, no particular medicine is available for its cure or prevention, although some antiretroviral drugs and hydroxychloroquine are recommended in a few centres. From the studies it is evident that numerous curcumin derivatives have antiviral properties. Studies using neuraminidase activation assay showed that five active curcumin derivatives decreased H1N1-induced neuraminidase activation in H1N1-infected lung epithelial cells. Tetramethylcurcumin and curcumin even down regulates nucleoprotein expression (Lai, Yan and Liao 2020). Turmeric is additionally beneficial in the management of influenza virus infections by various researchers. Richart, Li and Mizushina (2018) observed that monoacetylcurcumin and curcumin both inhibited influenza virus infection, but via different pathways. Significant antiviral activity of turmeric against H5N1 (highly pathogenic avian influenza) virus in Madin–Darby canine kidney (MDCK) cells *in vitro* by interfering with viral haemagglutination (HA) activity has also been observed. The effects of anti-H5N1 virus activity by turmeric extracts were demonstrated by upregulation in the tested MDCK cells of the mRNA expression of the genes for tumour necrosis factor-α and interferon-β, which are potent antiviral agents. Curcumin is also beneficial in other viral disorders such as AIDS (Prasad and Tyagi 2015) due to its inhibitory activity against HIV protease and integrase along with its synergistic action on other therapeutic drugs (Ciavarella, Motta and Valente 2020). It also inhibits other viruses such as hepatitis B, hepatitis C, zika, chikungunya and dengue. Respiratory distress syndrome with fulminant hypercytokinaemia and multiorgan failure is the leading cause of mortality with COVID-19. Curcumin has been found to attenuate influenza, a virus-induced lung tissue injury by blocking nuclear factor κB signalling and inhibiting the production of inflammatory cytokines. Curcumin is a natural ligand of peroxisome proliferator-activated receptor-γ, which represses the inflammatory process by reducing cytokine production; therefore, it might play a similar role to protect against lung injury associated with COVID. Turmeric has been used for centuries with a good safety profile. It has shown promising efficacy against influenza and viral infections by regulating the immune response to prevent injury to pulmonary tissue. Well-defined randomized studies should be performed to evaluate the efficacy of turmeric derivatives against SARS-CoV-2 and assess its value as a possible treatment for this deadly virus.

5.7 CONCLUSION

Turmeric has been used in Ayurveda, Unani and Siddha systems of medicines from ancient times. Literature survey shows wide spectrum of pharmacological activities of turmeric either in the form of powder, extracts or in its isolated compounds with minimum side effects. The methoxy group on the phenyl ring, phenolic, and 1, 3-diketone systems in the curcumin play important role for various pharmacological activities. Several products have been launched in National and International markets for various diseases fortified with curcumin or turmeric. Extensive work has been reported on this plant, but still it requires exploring more work for drug development. Curcumin is safe, non-toxic and strong natural antioxidant in comparison with other phyto-antioxidants.

REFERENCES

Afaq, F., Adhami, V.M. and Ahmad, N. 2002 Botanical antioxidants for chemoprevention of photocarcinogenesis. *Plant Biosciences* 7(7): 84–92.

Araujo, C.A.C., Alegrio, L.V., Castro, D., Lima, M.E.F., Gomescardoso, L. and Leon, L.L. 1999. Studies on the effectiveness of diarylheptanoid derivatives against *Leishmania amazonensis*. *Memórias do Instituto Oswaldo Cruz* 94: 791–794.

Asai, A. and Miyazawa, T. 2000. Occurrence of orally administered curcuminoid as glucuronide and glucuronide/sulfate conjugates in rat plasma. *Life Sciences* 67: 2785–2793.

ASTA. 2002. *A Concise Guide to Spices, Herbs, Seeds and Extractives*. American Spice Trade Association, Washington DC, pp. 48–50.

Baumann, W., Rodrigues, S.V. and Viana, L.M. 2000 Pigments and their solubility in and extractability by supercritical CO2. Part 1. The case of curcumin. *Brazilian Journal of Chemical Engineering* 17, 323–328.

Bhavani Shankar, T.N. and Srinivasamurthy, V. 1979. Effect of turmeric fractions on the growth of some intestinal and pathogenic bacteria in vitro. *Indian Journal of Experimental Biology* 17: 1363–1366.

Braga, M.E.M., Leal, P.F., Carvalho, J.O.E. and Meireles, M.A.A. 2003. Comparison of yield, composition, and antioxidant activity of turmeric (*Curcuma longa* L.) extracts obtained using various techniques. *Journal of Agricultural and Food Chemistry* 51: 6604–6611.

BSI. 1983. *British Standard Methods of Test For Spices and Condiments, Part 13*. BSI, London.

Buescher, R. and Yang, L. 2000. Turmeric. In: Lauro, G.L. and Fancis, F.J. (eds) *Natural Food Colorants. Science and Technology*. Marcel Dekker, New York, pp. 205–226.

China, E. C. O. T. P. O. P. S. R. O. 2005. *The Pharmacopoeia of People's Republic of China*. Chemical Industry Press, Beijing.

Ciavarella C, Motta I, Valente S. 2020. Pharmacological (or synthetic) and nutritional agonists of PPAR-γ as candidates for cytokine storm modulation in COVID-19 disease. *Molecules* 2020; 25: 2076.

Claeson, P., Panthong, A., Tuchinda, P., Reutrakul, V., Kanjanapothi, D. and Taylor, W.C. 1993. Three non-phenolic diarylheptanoids with anti-inflammatory activity from *Curcuma xanthorrhiza*. *Planta Medica* 59: 451–454.

Claeson, P., Pongprayoon, U., Sematong, T., Tuchinda, P., Reutrakul, V. and Soontornsaratune, P. 1996. Non-phenolic linear diarylheptanoids from *Curcuma xanthorrhiza:* a novel type of topical anti-inflammatory agents: structure activity relationship. *Planta Medica* 62: 236–240.

Dandekar, D.V. and Gaikar, V.G. 2002. Microwave assisted extraction of curcuminoids from *Curcuma longa*. *Separation Science and Technology* 37: 2669–2690.

Dhar, M.L., Dhar, M.M., Dhawan, B.M., Mehrotra, B.N. and Ray, C. 1968. Screening of Indian plants for biological activity. *Indian Journal of Experimental Biology* 6: 232.

Eigner, D. and Scholz, D. 1999. *Ferulaasafoetidaand Curcuma longa* in traditional medicinal treatment and diet in Nepal. *Journal of Ethnopharmacology* 67: 1–6.

Garg, S.N. Bansal, R.P. Gupta, M.M. and Kumar, S. 1999. Variation in the rhizome essential oil and curcumin contents and oil quality in the land races tumeric *Curcuma longa* of North Indian plains. *Flavour and Fragrance Journal* 14: 315–318.

Ghatak, N. and Basu, N. 1972. Sodium curcuminate as an effective anti-inflammatory agent. *Indian Journal of Experimental Biology* 10: 235–236.

Gopalan, B., Goto, M., Kodama, A. and Hirose, T. 2000. Supercritical carbon dioxide extraction of turmeric (*Curcuma longa*). *Journal of Agricultural and Food Chemistry* 48: 2189–2192.

Govindarajan, V.S. 1980. Turmeric – chemistry, technology and quality. *CRC Critical Reviews in Food Science and Nutrition* 12(3): 200–213.

Gururaj, A., Belakavadi, M., Venkatesh, D., Marme, D. and Salimath, B. 2002. Molecular mechanisms of antiangiogenic effect of curcumin. *Biochemical and Biophysical Research Communications* 297(4): 934–942.

Inano, H., Onoda, M., Inafuku, N. et al. 2000. Potent preventive action of curcumin on radiation-induced initiation of mammary tumorigenesis in rats. *Carcinogenesis* 12: 1835–1841.

Iwakami, S., Shibuya, M., Tseng, C.F., Hanaoka, F. and Sankawa, U. 1986. Inhibition of arachidonate 5-lipoxygenase by phenolic compounds. *Chemical and Pharmaceutical Bulletin* 34: 3960–3963.

Janaki, N. and Bose, J.L. 1967. An improved method for the isolation of curcumin from turmeric, *Curcuma longa* L. *Journal of the Indian Chemical Society* 44: 985–989.

Janssen, A. and Gole, T. 1984. Thin-layer chromatographic determination of curcumine (turmeric) in spices. *Chromatographia* 18: 546–549.

Jayaprakasha, G.K. Rao, L.J.M. and Sakariah, K.K. 2002. Improved HPLC method for determination of curcumin, demethoxycurcumin, and bisdemethoxycurcumin. *Journal of Agriculture and Food Chemistry* 50: 3668–3672.

Jayaprakasha, G.K., Negi, P.S., Anandha Ramakrishanan, C. and Sakariah, K.K. 2001. Chemical composition of turmeric oil – a by-product from turmeric oleoresin industry and its inhibitory activity against different fungi. *Zeitschrift fur Naturforschung* 56c: 40–44.

Jung, E.M., Lim, J.H., Lee, T.J., Park, J., Choi, K.S. and Kwon, T.K. 2005. Curcumin sensitizes tumor necrosis factor-related apoptosis-inducing ligand (TRAIL)-induced apoptosis through reactive oxygen species mediated up-regulation of death receptor 5 (DR5). *Cancer Biology* 26(11): 1905–1913.

Kapoor, L.D. 1990. *Handbook of Ayurvedic Medicinal Plants.* CRC Press, Boca Raton, Florida, p. 185.

Khurana, A. and Ho, C.T. 1988. High-performance liquid chromatographic analysis of curcuminoids and their photooxidative decomposition compounds in *Curcuma longa* L. *Journal of Liquid Chromatography* 11: 2295–2304.

Kim, D.S.H.L., Park, S.Y. and Kim, J.Y. 2001. Curcuminoids from *Curcuma longa* L. (Zingiberaceae) that protect PC12 rat pheochromocytoma and normal human umbilical vein endothelial cells from ba(1–42) insult. *Neuroscience Letters* 303: 57–61.

Kiso, Y., Suzuki, Y., Oshima, Y. and Hikino, H. 1983. Stereostructure of curlone asesquiterpenoid of Curcuma longa Rhizomes. *Phytochemistry* 22: 596–597.

Kiuchi, F., Shibuya, M. and Sankawa, U. 1982. Inhibitors of prostaglandin biosynthesis from *Alpinia officinarum*. *Chemical and Pharmaceutical Bulletin* 30: 2279–2282.

Koide, T., Nose, M., Ogihara, Y., Yabu, Y. and Ohta, N. 2002. Leishmanicidal effect of curcumin in vitro. *Biological and Pharmaceutical Bulletin* 25: 131–133.

Krishnamurthy, N., Mathew, A.G., Nambudiri, E.S., Shivasankar, S., Lewis, Y.S. and Natarajan, C.P. 1976. Oil and oleoresin of turmeric. *Tropical Science* 18(1): 37–45.

Lai Y., Yan Y. and Liao S. 2020. 3D-quantitative structure–activity relationship and antiviral effects of curcumin derivatives as potent inhibitors of influenza H1N1 neuraminidase. *Archives of Pharmacal Research* 43: 489– 502.

Leela, N.K., Aldo, T., Shaji, P.M., Sinu, P.J. and Chempakam, B. 2002. Composition of essential oils from turmeric (*Curcuma longa* L.). *Acta Pharmaceutica* 52: 137–141.

Li, C.J., Zhang, L.J., Dezube, B.J. and Crumpacker, C.S. 1993. Three inhibitors of human type 1 immunodeficiency virus long terminal repeat directed gene expression and virus replication. *Proceedings of the National Academy of Science (USA)* 90: 1839–1842.

Lim, G.P., Chu, T., Yang, F., Beech, W., Frantschy, S.A. and Cole, G.M. 2001. The curry spice curcumin reduces oxidative damage of amyloid pathology in an Alzheimer transgenic mouse. *Journal of Neurosciences* 21: 8370–8377.

Manzan, A.C.C.M., Toniolo, F.S., Bredow, E. and Povh, N.P. 2003. Extraction of essential oil and pigments from *Curcuma longa* (L.) by steam distillation and extraction with volatile solvents. *Journal of Agricultural and Food Chemistry* 51: 6802–6807.

Masuda, T., Toi, Y., Bando, H., Maekawa, T., Takeda, Y. and Yamaguchi, H. 2002. Structural identification of new curcumin dimers and their contribution to the antioxidant mechanism of curcumin. *Journal of Agricultural and Food Chemistry* 50: 2524–2530.

Ohshiro, M., Kuroyanagi, M. and Ueno, A. 1990. Structures of sesquiterpenes from Curcumalonga. *Phytochemistry* 29: 2201–2205.

Osawa, T., Sugiyama. Y., Inayoshi, M. and Kawakishi, S. 1995. Antioxidative activity of tetrahydro curcuminoids. *Bioscience, Biotechnology and Biochemistry* 59(9): 1609–1612.

Perotti, A.G. 1975. Curcumin – a little known but useful vegetable colour. *Industrial Aliment for Products and Vegetables* 14(6): 66–68.

Pervaiz, S. 1990. Antitumor and antiviral activity of curcumin and light. *Proceedings of the Annual Meeting of the American Association of Cancer Research Abstract* A 2325.

Phan, T.T., See, P., Lee, S.T. and Chan, S.Y. 2001. Protective effects of curcumin against oxidative damage on skin cell *in vitro*: its implication for wound healing. *Journal of Trauma* 51: 927–931.

Polasa, K., Naidu, A.N., Ravindranath, I. and Krishnaswamy, K. 2004. Inhibition of B(a)P induced strand breaks in presence of curcumin. *Mutation Research* 557: 203–213.

Prasad, S. and Tyagi, A.K. 2015. Curcumin and its analogues: a potential natural compound against HIV infection and AIDS. *Food & Function* 6: 3412–3419.

Radha, K.M., Anoop, K.S., Gaddipati, J.P. and Richkab, C.S. 2006. Multiple biological activities of curcumin: a short review. *Life Sciences* 78(18): 2081–2087.

Remadevi, R. Surendran, E. and Kimura, T. 2007. Turmeric in Traditional Medicine. In: Ravindran, P.N. Nirmal Babu, K. and Sivaraman, K. (eds) *Turmeric: The Genus Curcuma*. CRC Press, Boca Raton, London, New York, pp. 409–436.

Richart S.M., Li Y.L., Mizushina Y. 2018. Synergic effect of curcumin and its structural analogue (Monoacetylcurcumin) on anti-influenza virus infection. *Journal of Food Drug Analysis* 26: 1015–23.

Roth G.N., Chandra A. and Nair, M.G. 1998. Noval bioactivities of Curcuma longa constituents. *Journal of Natural Products* 61: 542–545.

Roughley, P.J. and Whiting, D.A. 1973. Experiments in the biosynthesis of curcumin. *Journal of the Chemical Society – Perkin Transactions-I* 20: 2379–2388.

Ruby, A.J., Kuttan, G., Dinesh Babu, K., Rajasekharan, K.N. and Kuttan, R. 1995. Antitumor and antioxidant activity of natural curcuminoids. *Cancer Letters* 94: 79–83.

Sanagi, M.M., Ahmad, U.K. and Smith, R.M. 1993. Application of supercritical fluid extraction and chromatography to the analysis of tumeric. *Journal of Chromatographic Science* 31: 20–25.

Sasikumar, B. 2005. Genetic resources of *Curcuma*: diversity, characterization and utilization. *Plant Genetics Resources* 3: 230–251.

Sastry, B.S. 1970. Curcumin content of turmeric. *Research and Industry* 15: 258–260.

Satoskar, R.R., Shah, S.J. and Shenoy, S.G. 1986. Evaluation of anti-inflammatory property of curcumin (diferuloyl methane) in patients with postoperative inflammation. *International Journal of Clinical Pharmacology and Theoretical Toxicology* 24: 651–654.

Sharma, O.P. 1976. Antioxidant activity of curcumin and related compounds. *Biochemical Pharmacology* 25: 1811–1812.

Sidhu, G.S. *et al.* 1999. Curcumin enhances wound healing in streptozotocin induced diabetic rats and genetically diabetic mice. *Wound Repair and Regeneration* 7: 362–374.

Soni, K.B. and Kuttan, R. 1992. Effect of oral curcumin administration on serum peroxides and cholesterol levels in human volunteers. *Indian Journal of Physiology and Pharmacology* 36(4): 273–275 and 239–243.

Sreejayan, N. and Rao, M.N.A. 1994. Curcuminoids as potent inhibitors of lipid peroxidation. *The Journal of Pharmacy and Pharmacology* 46: 1013–1016.

Srimal, R.C. and Dhawan, B.N. 1985. Pharmacological and clinical studies on *Curcuma longa. Monograph, Hamdard National Foundation*, New Delhi, section 313(ii), 131–138.

Srimal, R.C. and Dhawan, N. 1983. Pharmacology of diferuloyl methane (curcumin), a non-steroidal anti-inflammatory agent. *The Journal of Pharmacy and Pharmacology* 25: 447–452.

Srivastava, R., Dikshit, M., Srimal, R.C. and Dhawan, B.N. 1985. Antithrombotic effect of curcumin. *Thrombosis Research* 40: 413–417.

Subramonian, M., Sreejayan Rao, M.N.A., Devasagayam, T.P.A. and Singh, B.B. 1994. Diminuition of singlet oxygen induced D.N.A. damage by curcumin and related antioxidants. *Mutation Research* 311: 249–255.

Suh, J., Jo, Y., Kim, N.D., Bae, S.J., Jung, J.H. and Im, K.S. 2002. Cytotoxic constituents of the leaves of *Ixerissonchifolia. Archives of Pharmacal Research* 25: 289–292.

Tainter, D.R. and Grenis, A.T. 2001. *Spices and Seasonings. A Food Technology Handbook.* 2nd edn. Wiley-VCH, New York.

Tanaka, K. Kuba, Y. Sasaki, T. Hiwatashi, F. and Komatsu, K. 2008. Quantitation of curcuminoids in *Curcuma* rhizome by near-infaredspectroscopic analysis. *Journal of Agriculture and Food Chemistry* 56: 8787–8792.

Tonnesen, H.H. and Karlsen, J. 1983. High-performance liquid chromatography of curcumin and related compounds. *Journal of Chromatography* 259: 367–371.

Tonnesen, H.H. and Karlsen, J. 1986. Studies on curcumin and curcuminoids. VII. Chromatographic separation and quantitative analysis of curcumin and related compounds. *Zeitschrift fur Lebensmittel-Untersuchung und Forschung* 182: 215–218.

Usha, K., Nagabhushan, M., Nair S.C., Rao P.U. and Rahiman, M.A. 1994. The possible mode of action of cancer chemopreventive spice, turmeric. *Journal of the American College of Nutrition* 13: 519.

Verghese, J. 1993. Isolation of curcumin from *Curcuma longa* L. rhizome. *Flavour and Fragrance Journal* 8(6): 315–319.

Verghese, J. and Joy, M.T. 1989. Isolation of the colouring matter from dried turmeric (*Curcuma longa* L.) with ethyl acetate. *Flavour and Fragrance Journal* 4: 31–32.

Zhang, L. and Yang, Z. 1988. Extraction of curcumine from rhozoma *Curcumas longae. Faming Zhuanli Shenqing Gongkai Shuongshu CN* 87, 101, 355 (Cl. C09B61100), Oct 12, 10.

CHAPTER 6

Bioactive Compounds in Ginger

Jessica Pandohee and Lisa F. M. Lee Nen That

CONTENTS

6.1 INTRODUCTION

Ginger is a plant root or rhizome that is used as a spice across the world. The ginger plant (*Zingiber officinale* Roscoe) is known to be from South-East Asian countries and as part of the spice trade from Asia to Europe, a century ago, ginger was transported to other countries and is now widespread across the globe. Nowadays, ginger is grown in tropical countries in Asia (Bangladesh, China, India, Indonesia, Japan, Malaysia, Taiwan, the Philippines, Sri Lanka, Thailand), Africa (Ghana, Mauritius, Nigeria, Sierra Leone), America (Brazil, Costa Rica, Hawaii, Jamaica, Trinidad and Tobago) and Oceania (Australia, Fiji, Pacific Ocean islands) (Nair 2019).

The importance of ginger goes beyond its use as a spice in the food industry. In China and India, ginger is called a medicinal spice and is used in Traditional Chinese Medicine and Ayurveda to treat common ailments such as nausea and vomiting. The tales of the practices and effectiveness in the use of ginger and its attribution to health benefits have led to an increased

DOI: 10.1201/9781003215387-6

demand in the characterization of ginger, its contents and the compounds that are providing all the desired properties. In the last decades, ginger has been shown to have anticancerous, anti-inflammatory and antioxidant properties, amongst others. Over 400 compounds have been detected in ginger and its extracts; amongst those, gingerols, shogaols and zingerones were found to be unique to ginger.

A search in Scopus (June 2020) with the keywords "ginger" resulted in 11,328 publications. The bioactive compounds from ginger with the highest number of publications were "6-gingerol," "6-shogaol," "zingerone," "10-gingerol," "8-gingerol" and "6-paradol" with a total of 1092, 506, 502, 231, 200 and 97, respectively. Figure 6.1 shows the search results of the number of publications including keywords for "ginger" and its bioactive compounds and how they increased with time. As from the year 2000, the increase in publications of the bioactive ginger components showed an increase in interest in their function and benefits. Moreover, there was a marked increased as from 2010 with 6-gingerol and 6-shogaol being published significantly.

Considering the increased interest in bioactive compounds present in ginger, there is no doubt that an updated review of their chemistry, characterization and health benefits will shed the light on the new uses and applications of the bioactive compounds in ginger. With a focus on the bioactive compounds in ginger that have been most studied in the last 10 years, this chapter provides a discussion on the novel findings and their functions in improving human health.

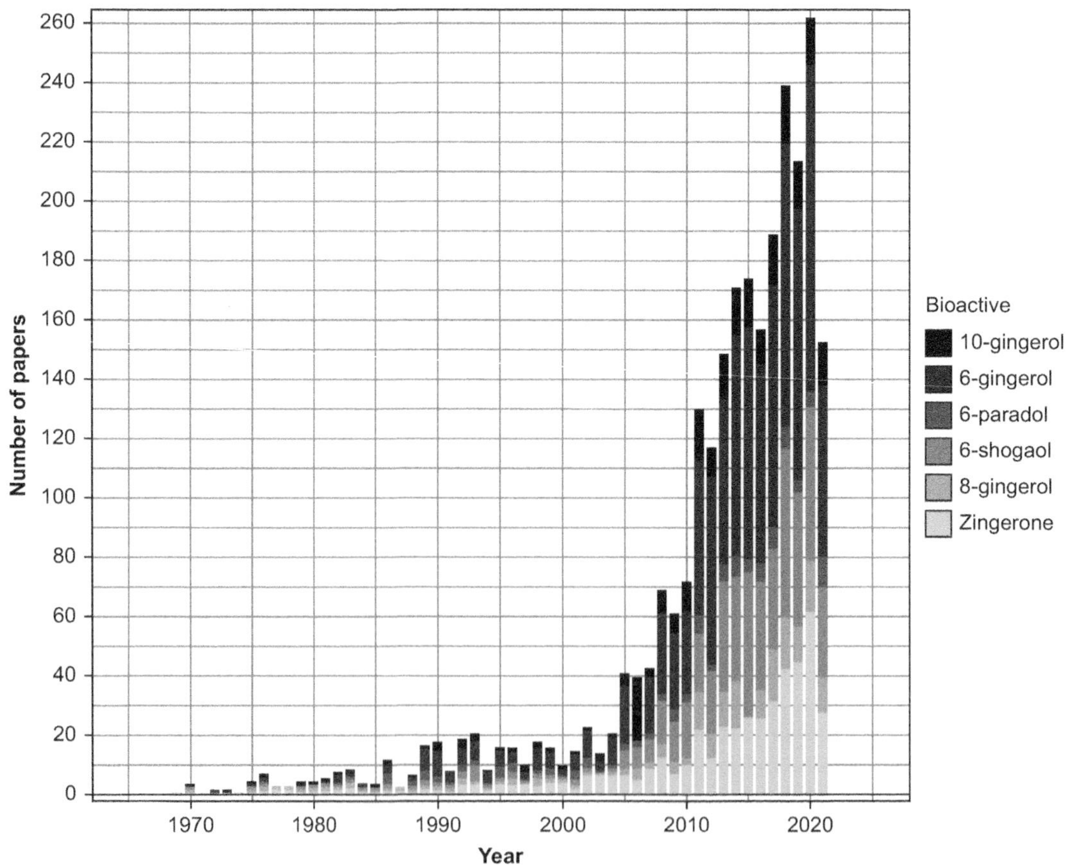

Figure 6.1 Number of publications featuring bioactive compounds in ginger since 1970.

6.2 CHEMISTRY, PROPERTIES, AND CHARACTERIZATION OF BIOACTIVE COMPOUNDS IN GINGER

6.2.1 Extraction and Isolation

Ginger is used in many different forms; raw, dried, powdered, oleoresin or essential oil. Different bioactive compounds are present in the various forms of ginger products, and the extent and concentration of those compounds depend primarily on the extraction methods and the type of solvents used. Dried and powdered ginger are primary products of ginger; they are obtained by allowing raw ginger to be air-dried in the sun (traditional method) or through instrumentation including freeze-drying, infra-red or microwave (An et al. 2016). The bioactive compound profiles of dried and powdered ginger differ slightly from raw ginger, with volatile bioactive compounds being lost during the drying process. A comparison of air drying to modern drying processes showed that air drying of ginger and the use of infra-red were effective in retaining volatile bioactive compounds (An et al. 2016). Moreover, the formation of 6-, 8-, and 10-shogaol from gingerols observed in hot air drying was shown to possess higher antimicrobial and antioxidant activities (Ghasemzadeh et al. 2018).

Oleoresin and essential oil are secondary products of ginger. Ginger oleoresin is extracted from powdered ginger; optimized methods include supercritical carbon dioxide extraction, active carbon stratum separation, macro-pore resin adsorption, enzyme-assisted partitioning and solvent extraction (Cheng-Lun, Derong, and Le 2008; Varakumar, Umesh, and Singhal 2017). Solvent extraction is by far the most common method used to produce ginger oleoresin, with carbon dioxide or with solvents such as acetone, ethanol, ethyl ether and hexane being used, with or without ultrasound (Supardan et al. 2012; Trung Dung, Duc, and Thanh 2018; Zancan et al. 2001). The use of ultrasound has been shown to not affect the content of major bioactive compounds in the oleoresins and has the advantage of increasing the yield in a shorter time (Supardan et al. 2012). Ginger oleoresin contains the majority of the flavour- and aroma-attributing compounds such as 6-, 8-gingerols and 8-shogaol together with hundreds of other bioactive metabolites while the use of enzyme-assisted extraction is preferred for higher extraction yields of the bioactive compounds (Varakumar, Umesh, and Singhal 2017).

Ginger essential oil on the other hand uses less harsh extraction methods such as hydro-distillation or steam distillation (dos Santos Reis et al. 2020; Argo et al. 2020). The light yellow to yellow oil contains a range of volatile compounds such as sesquiterpene hydrocarbons, monoterpene hydrocarbons, and oxygenated compounds (Akshitha et al. 2020). Munda et al. (2018) reported high amounts of oxygenated volatile compounds such as zingiberene, ar-curcumene, citral, β-bisabolene, geranial, camphene in ginger essential oil that conferred high therapeutic values such as analgesic and antioxidant.

6.2.2 Characterization and Identification

Ginger has been found to contain over 400 compounds; of which some are phytochemicals of different chemical classes with various physicochemical properties (Figure 6.2) and existing in a range of concentration (Mao et al. 2019; Mengmeng Zhang et al. 2021). Liquid chromatography mass spectrometry (LC-MS) is a technique that is useful in the profiling of non-volatile bioactive compounds and allows the simultaneous analysis of several bioactive compounds in a ginger extract including 6-, 8- and 10-gingerols, zingerone and 8- and 10-shogaols (Wang et al. 2009). In ginger samples, the application of LC-MS in ginger characterization and analysis can also be used for authentication of ginger products due to its high accuracy in identifying chemical compounds (Jiang et al. 2006). Using liquid chromatography, Li et al. (2016) also showed that the levels of zingerone, 6-, 8- and 10-gingerols, together with their phenolic content, vary in different ginger products (fresh, dried, stir-fried and carbonized).

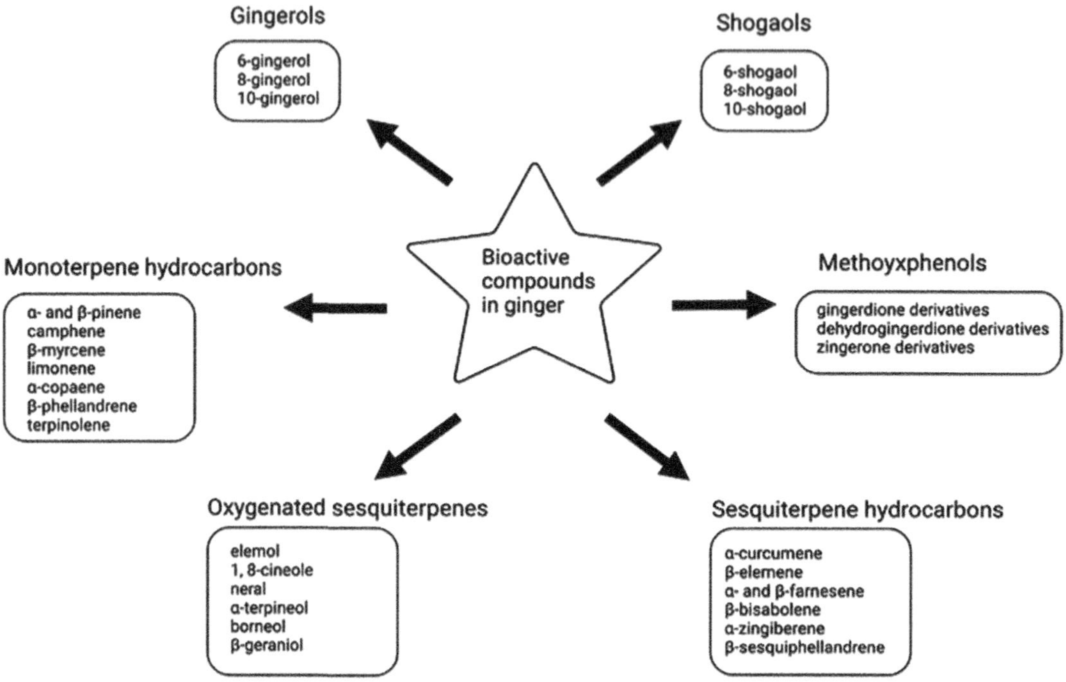

Figure 6.2 Classification chart showing the different classes and bioactive compounds identified in ginger.

The volatile bioactive compounds of ginger are studied to a lesser extent because they also exist in other natural products. Ginger essential oil has been shown to contain monoterpene hydrocarbons, sesquiterpene hydrocarbons and oxygenated hydrocarbons (Al-Dhahli et al. 2020). Gas chromatography mass spectrometry (GC-MS) is the method of choice for the analysis of the bioactive compounds that are volatile by nature, for example, curcumene, bisabolene, farnesene and zingiberene, terpineol, geraniol, myrcene and limonene, amongst others (Ashraf et al. 2017). Using GC-MS, Abdullahi et al. (2020) showed that the volatile bioactive compounds (with camphene, bisacurone epoxide, eucalyptol, β-phellandrene and α-zingiberene as major volatile compounds) present in ginger essential oil provided antimicrobial properties which inhibited phytopathogens. Moreover, the use of solid-phase microextraction and two dimensional gas chromatography allowed the identification of 36 volatile compounds in the headspace of ginger oil; the authors also depicted the complexity and variety of the stereochemical isomers in the volatile extract (Shao et al. 2003).

6.2.3 Encapsulation

The effectiveness of the role and benefits that the bioactive compounds present in ginger will deliver depends on their stability, bioavailability, delivery to target proteins or cells and their absorption (Noore et al. 2021). Encapsulation involves the incorporation of analytes of interest, such as the bioactive compounds in ginger, in a capsule. With time, advances in technologies and repeated optimization in the preparation of different types of encapsulation, the method of encapsulation nowadays also includes attachment of molecules to nanoparticles that provide similar functions to the traditional encapsulation methods (micro-encapsulation) with the added benefits of being sensitive, targeted and stable (Fathi, Mozafari, and Mohebbi 2012; Assadpour and Mahdi Jafari 2019). Figure 6.3 shows the most common encapsulation methods that have been applied with the bioactive compounds in ginger.

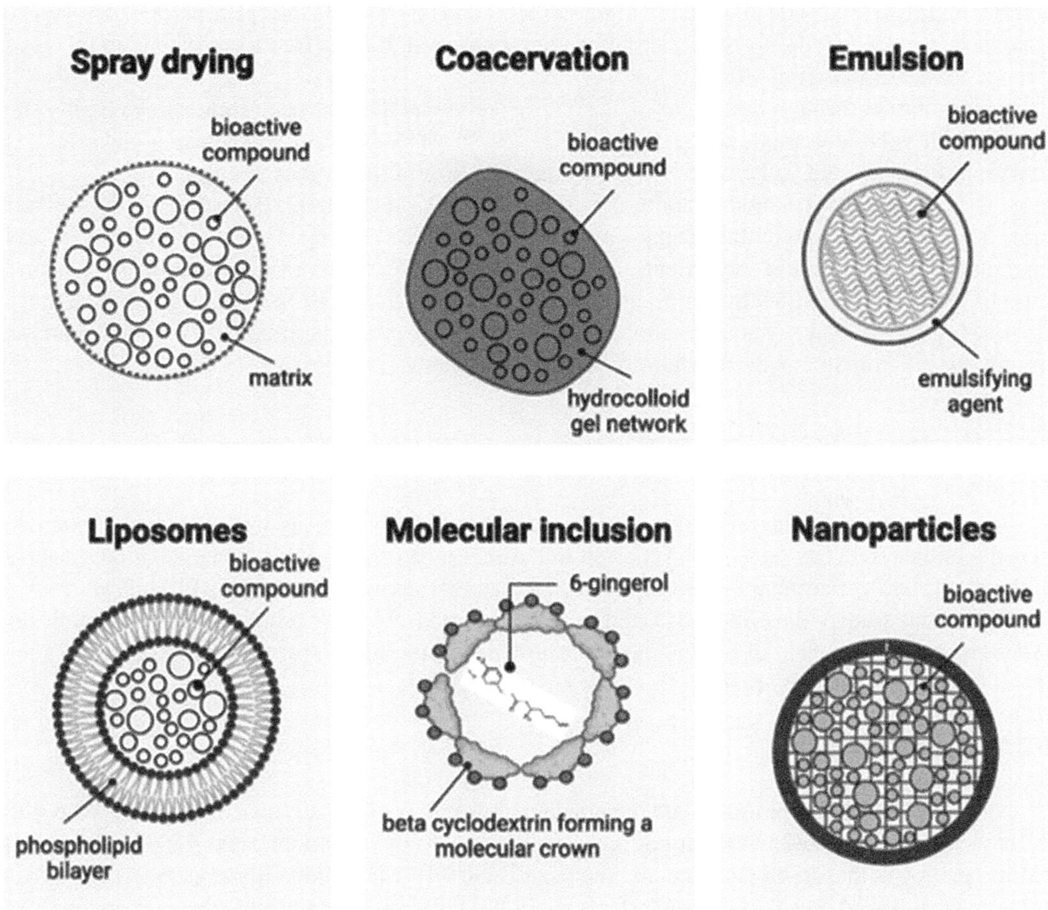

Figure 6.3 Types of encapsulation methods used on bioactive compounds extracted from ginger.

The most common and easiest method of encapsulation is spray-drying; in a nutshell, spray-drying involves homogenizing the core material (bioactive compounds in ginger) with the wall material (very often saccharides) to form an emulsion that is then dried using hot air resulting in droplets with diameter ranging from 1 to 1000 μm (Simon-Brown et al. 2016). Highly efficient approaches for spray-drying include ginger oleoresin encapsulated in maltodextrin and skim milk and ginger essential oil encapsulated in maltodextrin/whey protein isolate (Setyaningsih, Kurniawan, and Muna 2020; Touré et al. 2011). One of the disadvantages of spray-drying is that it provides low resistance to oxidation (Mohammed et al. 2020). Coacervation encapsulation, which takes advantage of oppositely charged gels/polymers, has been shown to reduce the degradation of ginger essential oil (Tavares and Noreña 2020; Ferreira and Nicoletti 2021).

Encapsulation through emulsions offers improved stability and bioavailability of bioactive compounds, therefore ensuring that the right amount of active ingredient is being delivered to the target organ (Lu, Kelly, and Miao 2016). This is particularly important in drug delivery systems (Xu et al. 2016). Cao et al. (2021) reported higher availability, release, and anti-tumor properties of zingerone-self-microemulsion drug delivery system compared to its free version. Liposomes, which involve encapsulating bioactive compounds within lipid-based vesicles, offer improved physical stability, solubility and oxidation resistance (Ajeeshkumar et al. 2021). Furthermore, higher antioxidant activity of ginger ethanolic extract and higher *in vitro* and *in vivo* anticancer properties of 6-shogaol have been reported in the form of nanoliposomes compared to their free form (Ganji and Sayyed-Alangi 2017; Kemkar et al. 2018).

Encapsulation through molecular inclusion is a technique that has been shown to be highly successful with volatile compounds (Perinelli et al. 2020). While molecular inclusion encapsulation has been demonstrated to be efficient with gingerols, little research has been performed with the monoterpenes and oxygenated sesquiterpenes in ginger (Perinelli et al. 2020). Nanoparticles is the most recent development in the encapsulation industry and has displayed great potential in increasing bioavailability of active ingredients, absorption at target receptors in cells, and being resistance time in the body. Initial testing in mice demonstrated early promises in the non-toxicity and efficacy of 6-gingerol and 6-shogaol nanoparticles in reducing pro-inflammatory cytokines therefore improving inflammatory bowel syndrome (Mingzhen Zhang et al. 2016).

6.3 EFFECT OF GINGER BIOACTIVE ON HEALTH DISORDERS

The nutraceutical value of ginger can be attributed to its bioactives which have had extensive benefits on health. They have been reported to have a range of properties as they exhibit powerful antioxidant, anti-inflammatory and anticancer activities (Semwal et al. 2015; El Halawany et al. 2017). Several studies have demonstrated that they carried out their protective role in health disorders by restoring levels of compounds and proteins that could disrupt signalling pathways and impair function in all the organs.

6.3.1 Effect on Cancer

Numerous studies have looked at the anticancer properties of ginger bioactivities and these studies have all demonstrated the ability of these compounds to induce apoptosis through different signalling pathways in a dose dependent manner (see Table 6.1) (Mahomoodally et al. 2021). Moreover, it has been reported that they had no effect on normal cells and certain bioactive compounds such as 10-gingerol have been considered as adjuvant with chemotherapeutic drugs (Martin et al. 2020).

6.3.2 Neurological Disorders

Ginger bioactives have demonstrated their protective role in neurological disorders including Alzheimer's disease, epilepsy and cerebral ischemia due to their ability to passively diffuse in the blood brain barrier (Simon et al. 2020). Inflammation and oxidative stress have contributed to neurological disorders such as Alzheimer's disease (AD) and Parkinson's disease (Adetuyi and Farombi 2021). However, studies by Adetuyi and Farombi (2021) have demonstrated how 6-gingerol (20, 50 mg/kg) has protected lipopolysaccharide (LPS)-induced rats suffering from oxidative stress, inflammation, cognitive impairments and amyloidogenesis. The ginger bioactive improved cognitive deficiency such as impaired locomotor and motor activity and showed improvement in escape latency. Moreover, it restored the antioxidant enzyme activity such as catalase and decreased levels of proinflammatory cytokines, IL-1β and iNOS expression. After suffering from weight loss caused by LPS, rats had gained weight and 6-gingerol protected both the cerebral cortex and hippocampus as there were reduced abnormal deposits, haemorrhagic lesions and impaired neurons in the brain. Recently, Berntsson et al. (2021) aimed to demonstrate the molecular mechanisms by which 6-gingerol had a positive effect on the pathology of AD using biophysical experiments. Characteristic features of AD included aggregation of amyloid-Aβ peptide and 6-gingerol was shown to interact with the amyloid plaques but not the monomers. This suggested the role of 6-gingerol as a potential drug against Alzheimer's disease.

Table 6.1 Selected Studies Looking at the Effect of Bioactive Compounds Against Cancer Cell Lines (GI: Growth Inhibition, IC_{50}: Concentration of Drug Required for 50% Inhibition)

Bioactive Compound	Cancer Type	Cell Line Used	Findings	References
6-gingerol 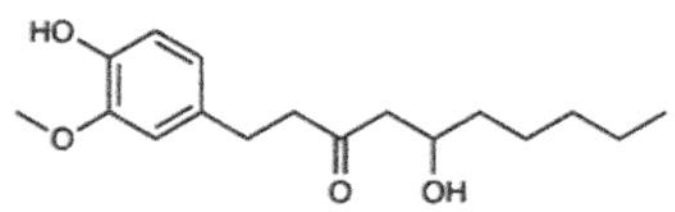	cervical	HeLa, CaSki, SiHa	GI (50 μM) (24h) - HeLa (20 %), CaSki (23 %), SiHa (28 %). Induced apoptosis by reactivation of p53 proteasome in cancer cells. Damage in DNA and cell cycle arrest was at G2/M phase.	Rastogi et al. (2015)
	gastric	HGC-27	Decrease in viability (50%) - 2 μg/ml, cell cycle arrest was at G1 phase. Cell proliferation and invasion were suppressed through regulation of the PI3K/AKT pathway.	Luo et al. (2019)
	kidney	786-O, ACHN	Inhibition of tumour migration induced by phosphorylation of YAP protein at Ser^{127} which is involved in Hippo-YAP pathway. This led to inhibition of YAP nuclear localization.	Xu et al. (2020)
8-gingerol 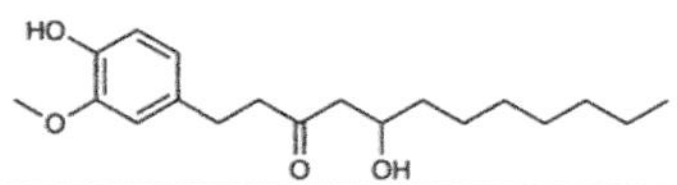	colorectal	HCT116, DLD1	IC_{50} for HCT116- 118.2±7.37 μM (24h), 77.4±4.70 μM (48h), 61.8±3.57 μM (72h), IC50 for DLD1- 100.3±6.32 μM (24h), 53.7±2.24 μM (48h), 34.5±2.33 μM (72h), G0/G1 cell cycle arrest. Apoptosis was mediated through the EGRF/STAT/ERK pathway.	Hu et al. (2020)
10-gingerol 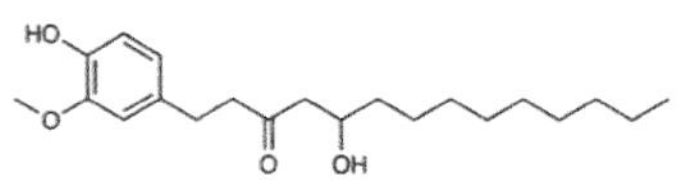	colon	HCT116	IC_{50} - 30 μM (24h), Increasing doses of 10-gingerol reduced spread of colon cancer cells through MAPK pathway leading to activation of caspase 9, caspase 3 and poly-ADB-ribose polymerase (PARP).	Ryu and Chung (2015)
	ovarian	HEY, OVCAR3, SKOV-3	GI (200 μM) (72h) - HEY (71±14 %), OVCAR3 (33±5 %), SKOV-3 (38±7 %). Cell cycle arrest was at G2 phase.	Rasmussen, Murphy, and Hoskin (2019)
	breast	MDA-MB-231/ IR	IC_{50} - 121.2 μM (24h), 101.4 μM (48h), 10-gingerol targeted lipid rafts in cell membranes involved in the PI3K/Akt pathway. Apoptosis was induced by activation of caspase-7, PARP and caspase-3.	Ediriweera et al. (2020)
6-shogaol 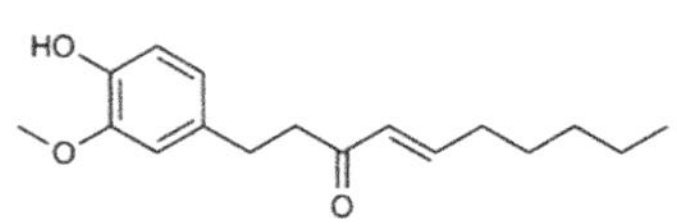	cervical	HeLa cells	IC_{50} - 14.75±0.94 μM (24h). Proteomic analysis revealed that 14-3-3 signalling pathway could be involved in the apoptosis and cell cycle arrest at G2/M phase.	Liu et al. (2012)
	breast	MCF-7, T47D	IC_{50} (72h) - 23.3 μM (MCF-7), 0.5±0.1 μM (T47D), cell cycle arrest was observed at G2/M phase and apoptosis was mediated through the Notch signalling pathway.	Bawadood et al. (2020)
	endometrial	Ishikawa	IC_{50} - 24.91 μM (24h), cell cycle was interrupted at G2/M phase, leakage of Ca2+ from ER and change in mitochondrial membrane potential led to apoptosis.	Ma et al. (2020)

(Continued)

Table 6.1 Selected Studies Looking at the Effect of Bioactive Compounds Against Cancer Cell Lines (GI: Growth Inhibition, IC_{50}: Concentration of Drug Required for 50% Inhibition) (*Continued*)

Bioactive Compound	Cancer Type	Cell Line Used	Findings	References
6-paradol 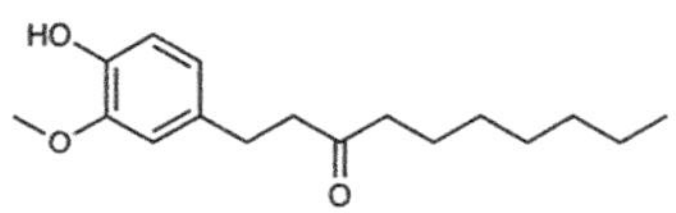	pancreas	MIA PaCa-2, SW1990	Molecular docking analysis revealed interaction between 6-paradol and EGFR to inhibit metastasis of cancer cells and could involve degradation by proteasome through the P13K/AKT signalling pathway. Suppression of tumour growth by 6-paradol was also observed in rats.	Jiang et al. (2021)
zingerone 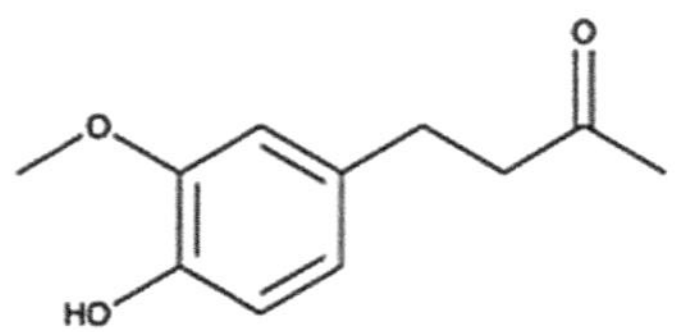	prostate	PC-3	IC_{50} - 20 μM (24h), Induced apoptosis through the PI3K/Akt/mTOR signalling pathway and suppressed metastasis of cancerous cells by inhibition of cell adhesion and migration.	Qian et al. (2020)
zingiberene 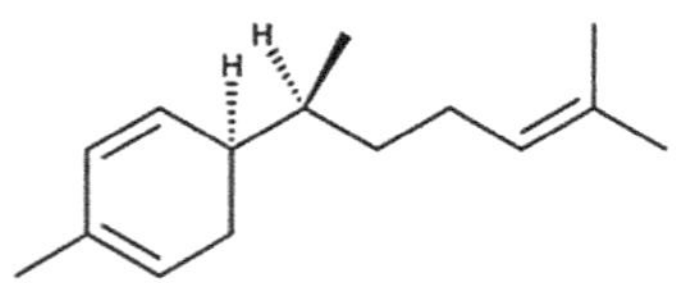	liver	HepG2	G0/G1 was at cell cycle arrest. Regulation of cell migration and invasion were mediated through the miR-16/cyclin B axis.	Peng et al. (2020)

Zingerone is another bioactive compound that had a beneficial effect on neurological disorders such as status epilepticus (SE) (Rashid et al. 2021). Neuroinflammation is the prerequisite for SE and zingerone has been known for anti-inflammatory properties. Indeed, zingerone allowed decrease of acetylcholine esterase to promote production of acetylcholine as it could inhibit the innate immune response which in turn suppressed inflammation. The ginger bioactive had also suppressed proinflammatory cytokines and proteins such as NF-κB, TNF-α, IL-6, IL-1β. It also protected from oxidative stress as antioxidant enzymes (SOD, CAT) were increased. An increase in latency to seizure, and a decrease in convulsion and escape latency in Morris water maze were reported in rats treated with zingerone. Moreover, several physiological changes were observed upon treatment with the bioactive namely alleviation of necrosis and nuclear degeneration.

6.3.3 Renal Disorders

Occurrence of kidney dysfunction has been the consequence of other diseases such as diabetes mellitus, cancer, bacterial infections, and the ginger bioactives have alleviated these renal pathological changes. Several studies have looked at the effect of 6-gingerol on diabetic nephropathy (Song, Dang, and Kumar 2019; Almatroodi et al. 2021). The compound attenuated oxidative stress and inflammation in the kidney by regulating key factors involved and levels of biomarkers of renal dysfunction (creatinine, blood urea) were also suppressed upon treatment of 6-gingerol. Moreover, Almatroodi et al. (2021) reported that 6-gingerol improved the renal damage such as congestion, lymphocyte infiltration, enlarged glomerular basement and reduction in collagen deposition. According to Song, Dang, and Kumar (2019), diabetic rats had elevated levels of hepatic parameters which were attenuated by 6-gingerol. The bioactive compound also decreased the elevated levels of inflammatory mediators (8-OHdG, COX-2, TGF-β, NF-κB) in a dose-dependent manner.

Side effects associated with the chemotherapeutic drug Adriamycin led to nephrotoxicity (Elshopakey et al. 2021) and the administration of zingerone (25 mg/kg b.w.) in rats protected them from oxidative stress and inflammation. Level of biomarkers for renal damage namely serum urea, kidney injury molecule and creatinine increased, and histopathological findings included infiltration of inflammatory cells, degeneration of renal cells, congestion in renal corpuscles in rats suffering from renal damage. However, treatment with zingerone attenuated these cytotoxic effects. Rats suffering from side effects of gentamicin, an antibiotic used for infections by gram-negative bacteria had similar health problems (Hosseinzadeh et al. 2020). When they were treated with zingerone (10 mg/kg per day), the fluctuations in levels of oxidative stress, inflammation, biomarkers of renal damage were also alleviated. Moreover, histopathological modifications including necrosis in proximal tubule, infiltration of inflammatory cells, congestion of red blood cells were reduced.

6.3.4 Cardiovascular Disorders

Bioactive compounds have demonstrated cardioprotective activities by attenuating the pathological changes caused by both external and internal factors (C. Liu et al. 2019). Indeed, C. Liu et al. (2019) reported that zingerone alleviated modifications caused by cardiac remodelling in rats that could affect its normal function and eventually lead to clinical heart failure. Addition of zingerone (50, 250 μmol/L) downregulated the mRNA expression levels of proteins (atrial natriuretic peptide and β-myosin heavy chain) involved in promoting cardiac hypertrophy and it elevated levels of markers involved in fibrosis (transforming growth factor β1, connective tissue growth factor, and collagen I and collagen III). Furthermore, the metabolite attenuated cardiac dysfunction, oxidative stress and inflammation and decreased diameter and wall thickness of left ventricle. The mechanism of action of zingerone could be through the activation of eNOS/Nrf2 pathway by modulation of the proteins heme oxygenase-1, Kelch-like ECH-associated protein and nucleus-Nrf2.

6-gingerol has exhibited a positive effect on rats suffering from myocardial fibrosis (Han et al. 2020). Indeed, the electrocardiogram demonstrated that 6-gingerol (10, 20 mg/kg per day) decreased elevated levels of J-point and heart rate in the rats. Moreover, pathological and morphological changes including enlarged heart, swelling in myocardial cell, inflammatory cells infiltration, hypertrophy, collagen deposition were attenuated with 6-gingerol. Levels of cardiac markers including creatine kinase (CK) and lactate dehydrogenase (LDH) and calcium concentration, which were also significantly increased were reversed by 6-gingerol indicating its role in improving cardiac damage. 6-gingerol restored levels of oxidative markers (MDA, GSSG, SOD, CAT, GSH) and proinflammatory cytokines (TNF-α, IL-6, c-fos and c-jun. A decrease in apoptosis at both low (51.51%) and high dose (58.81%) of 6-gingerol was observed by regulating protein expression of Bax, Caspase-3 and ratio of Bax/Bcl-2. 6-gingerol carries out its protective role by modulating the TLR4/MAPKs/NF-κB signalling pathway.

6.3.5 Hepatic Disorders

Ginger metabolites have had a beneficial effect in patients suffering from liver disorders. Indeed, side effects caused by the antibiotic drug, vancomycin against gram-positive bacteria was attenuated by zingerone (25 mg/kg b.w., 50 mg/kg b.w.) (Kucukler et al. 2020). Markers of liver function including ALP, ALT and AST were elevated upon addition of vancomycin but decreased significantly after treatment with zingerone. Necrotic changes such as hydropic degeneration and coagulative necrosis were observed in rats suffering from the side effects of vancomycin. The administration of the bioactive compound alleviated these pathological changes. Zingerone also reduced oxidative stress by increasing levels of antioxidant enzymes (SOD, CAT and GPx) and suppressed level of proinflammatory cytokines (TNF-α, IL-1β) and transcription factor, NF-κB to reduce inflammation. Apoptosis was also inhibited by zingerone by regulating protein levels involved in apoptotic pathways (Bax, cytochrome *c*, Bcl-2, p53, NF-κB, caspase-8, procaspase-3).

6-gingerol is another bioactive that has demonstrated a protective role in the liver. M. Hong et al. (2020) reported that 6-gingerol (40 mg/kg) decreased elevated levels of liver biomarkers (ALT, AST) in mice suffering from polymicrobial sepsis. Moreover, 6-gingerol has an antioxidative effect as it alleviated levels of serum ROS. The metabolite also downregulated the expression levels of inflammatory cytokines TNF-α, IL-6, Cxcl2, Ccl1, Ccl2. Increasing expression levels of proteins (NLRP3, caspase-1) and cytokine IL-1β, involved in pyroptosis also reduced indicating the role of the bioactive in suppressing inflammation and pyroptosis through the Nrf2/HO-1 pathway. The above findings were further confirmed as pyroptosis was also inhibited by 6-gingerol in raw 264.7 macrophage cells. Findings from Alsahli et al. (2021) also showed that 6-gingerol demonstrated both antioxidant and anti-inflammatory activities in diethylnitrosamine-induced liver injury in rats by regulating key markers involved in the two processes. 6-gingerol reduced elevated levels of liver biomarkers and serum lipids. Severe pathological damage (congestion, inflammatory cells infiltration, blood vessel dilation and oedema) were reduced in rats treated with 6-gingerol. Furthermore, 6-gingerol also decreased the collagen fibre deposition in liver tissue.

6.3.6 Respiratory Disorders

Disorders in the respiratory system have been alleviated by a number of bioactives. A study by Gungor et al. (2020) demonstrated that zingerone attenuated the side effects caused by bleomycin. It is a cytotoxic antibiotic drug for treating cancer and has caused idiopathic pulmonary fibrosis in rats. Zingerone restored levels of MDA, antioxidant enzymes (SOD, GPx) and pro-inflammatory cytokines (TNF-α, IL-1β). The bioactive compound also reduced inflammatory changes in the bronchi and thickening of the interalveolar septum in the rats. Wang et al. (2016) reported that 6-shogaol (10, 20, 40 mg/kg) contributed to the decrease of elevated levels of myeloperoxidase in

LPS-induced lung injury in a dose dependent manner by regulation of the NF-κB signalling pathway. 6-shogaol also attenuated histopathological features observed in lungs including inflammation by high number of neutrophils, congestion in lungs and thickening of alveolar wall. Molecular docking studies also showed that 6-shogaol bound deep into the active site of NF-κB protein model flanked by Tyr57, Val58, Cys59, Tyr143, Lys145 and Lys146.

According to W. Hong et al. (2021), 6-gingerol provided protection to ventilator-induced lung injury. Dose-dependent pre-treatment of 6-gingerol (5mg/kg, 10mg/kg, 20mg/kg) mitigated effects of lung oedema, alveolar collapse, inflammation and oxidative stress caused by mechanical ventilation. Protective effect of 6-gingerol was possible through the modulation of NF-κB pathway by upregulating expression level of PPARγ. This in turn reduced levels of proinflammatory cytokines and biomarkers for oxidative stress (MDA and SOD). 6-gingerol has also demonstrated its beneficial effect in sepsis in the liver (See section 6.3.5). Yang et al. (2021) reported here that 6-gingerol (5, 25, 50 mg/kg) also protected against sepsis in the lung and reduced oxidative stress and inflammation in a dose-dependent manner. It also alleviated the characteristic features observed in septic mouse model which included congestion, infiltration by inflammatory cells, alveolar collapse. Beneficial role of 6-gingerol occurred by inhibiting the NF-κB pathway and promoting the Nrf2/HO-1 pathway.

6.3.7 Immunological Disorders

The beneficial effect of ginger bioactive compounds has been reported on autoimmune diseases. Ali et al. (2021) tested the effect of 6-, 8- and 10-gingerols on lupus and showed that NETosis was suppressed by different concentrations (0.1, 1, 10, 100 μM) of gingerols. NETosis is the activation of neutrophil extracellular traps (NETs) that consist of chromatin, antimicrobial proteins released by neutrophils forming an intertwined network (Brinkmann et al. 2004). They reduced NETosis elicited by several stimuli namely *Escherichia coli* LPS, phorbol 12-myristate 13-acetate, lupus and antiphospholipid syndrome autoantibodies. This was possible as gingerols are powerful antioxidant and inhibit production of H_2O_2 necessary for NETosis. Further results revealed that 6-gingerol could regulate proteins (cAMP specific phosphodiesterase, PKA) involved in the cAMP pathway. In rats, 6-gingerol also decreased elevated levels of antiphospholipid which plays a direct role in thrombosis formation.

Zingerone (25 mg/kg b.w.) exhibited anti-inflammatory activity in rats suffering from Freund's complete adjuvant (FCA)-induced rheumatoid arthritis (Bashir et al. 2021). Elevated levels of pro-inflammatory proteins and cytokines (NF-κB, TNF-α, TGF-β, IL-1β, IL-6, Hs-CRP) were reduced after administration of zingerone. Zingerone also increased levels of anti-inflammatory cytokine IL-10 in rats suffering from rheumatoid arthritis. It also reduced oxidative stress by reducing MDA levels and increasing low levels of SOD, GPx and catalase in both hepatic and joint tissues. The administration of 2% ginger reduced allergic symptoms such as sneezing in rats suffering from allergic rhinitis (Kawamoto et al. 2016). Levels of both IgE and mast cells involved in allergic reactions were also lowered and suppression of gene expression of IL-4, IL-10 and IL-13 in a dose-dependent manner after 7 days was observed. 6-gingerol had an effect on T cell receptor-mediated signal transduction indicating the suppression of both Th2 and Th1 cytokine and proliferation of T cells. This could be due to the downregulation of the phosphorylation of MAPKs and calcium mediated IL-2 expression. Moreover, transcription factors involved in cytokine expression namely NF-κB and c-fos were suppressed after pre-treatment with 6-gingerol.

6.3.8 Metabolic Disorders

Ginger bioactives have been tested for their efficacy in metabolic disorders including diabetes and obesity. Ahmad et al. (2018) reported that zingerone (50, 100 mg/kg b.w.) has demonstrated a hypoglycaemic role in diabetic rats. Variations in serum glucose levels, serum insulin and serum amylase were attenuated upon administration with zingerone which could indicate that zingerone had a regenerative effect in the pancreas. Furthermore, zingerone caused a significant decrease in lipid markers (TC, TG, LDL-C) and an increase in HDL-C levels in a dose dependent manner. As zingerone has antioxidant properties, markers for oxidative stress (SOD, CAT, GPx, GSH) were significantly increased in diabetic rats treated with zingerone. NF-κB plays a role in diabetes and indeed the level of NF-κBp65 was increased in diabetic rats but was attenuated with zingerone. Inflammation was also recorded in diabetic rats. However, the addition of zingerone reduced levels of proinflammatory cytokines (IL1-β, IL-6, IL-2 and TNF-α). Studies have demonstrated the role of 6-shogaol in reducing elevated blood glucose level in diabetic rats (Yi et al. 2019). Similar results were observed for these diabetic rats where inflammation and oxidative stress were reduced. Findings further revealed the increase in the expression level of renal Nrf-2 by 6-gingerol against oxidative stress. Moreover, pathological changes such as abnormal morphology in islets structure and β-cell observed in pancreas were reduced with treatment of 6-shogaol. Diabetes is usually associated with disorders in other organs such as diabetic nephropathy (As described in section 6.3.3) which were alleviated by 6-shogaol in this study.

Ginger bioactives induce browning in adipose tissue which could promote energy expenditure for heat production over energy storage to combat obesity (Wang, Li, et al. 2019; Sampath et al. 2021). The role of 6-gingerol was to upregulate expression levels of transcription markers (PGC-1α, PRDM16, UCP1, C/EBPβ) involved in browning in a dose dependent manner (Wang, Zhang, et al. 2019). Mitochondrial biogenesis was also induced by 6-gingerol as gene expression (*Tfam* and *Nrf1*) involved were significantly upregulated. 6-gingerol also promoted mitochondrial respiration and energy metabolism as ratio of p-AMPK to AMPK, which represents AMPK activity was increased. Lipolysis was also increased by upregulating the expression levels of HSL and PLIN. Further studies by Sampath et al. (2021) demonstrated that 6-gingerol had a stronger efficacy than 6-shogaol in browning. 6-gingerol was more effective than 6-shogaol in upregulation of expression level of PPARγ which is involved in adipocyte differentiation. High oxygen consumption was reported in both cells by both bioactives with 6-gingerol having higher oxygen consumption rate. Molecular docking studies revealed that both bioactives had high binding affinity with β3-AR found on adipose tissue and 6-gingerol had a higher binding score demonstrating that it was a better agonist than 6-shogaol in the process of browning of white adipose tissue.

6.3.9 Antibacterial, Antimicrobial, Antiviral, Antifungal, Anthelmintic Activity

Several studies have looked at the effect of ginger bioactive compounds against bacteria, viruses, fungi and helminths (Table 6.2). These compounds have all had varying results in inhibition concentration, cytotoxic concentration and lethal dose.

6.3.10 Anti-Emetic Effect

The occurrence of both vomiting and nausea associated with chemotherapy and post-surgery and its possible treatment by ginger bioactives have been investigated. Several studies in rats and minks have revealed that these bioactives had an anti-emetic effect because they acted on neurotransmitters namely substance P, dopamine and 5-HT that caused emesis (Tian et al. 2020; Cheng et al. 2020; W. Qian et al. 2016). According to W. Qian et al. (2016), gingerol suppressed activation of dopamine in cisplatin-induced rats. These rodents do not suffer from vomiting and pica behaviour is induced which is considered an index for nausea and vomiting. Gingerol reduced the elevated levels of elements involved in release of dopamine namely dopamine D2 receptor, tyrosine hydroxylase. It also increased the dopamine transporter which induces reuptake of dopamine and inhibit transmission of the neurotransmitter. The effect of 6-gingerol on

Table 6.2 Antibacterial, Antiviral, Antifungal, Anthelmintic Activity of Ginger Bioactives (CC_{50}: Concentration Required to Reduce 50% of Cell Viability, IC_{50}: Concentration of Drug Required for 50% Inhibition, LD_{50}: Dose Required to Kill 50% of the Population, MBC: Minimum Bactericidal Concentration, MIC: Minimum Inhibition Concentration)

Bioactive Compound	Activity	Organisms Tested	MIC	Findings	References
6-gingerol	antibacterial	*Mycobacterium tuberculosis*	strain H37Rv – 12 μg/ml, strain MDR – 3 μg/ml, strain XDR – 50 μg/ml	Reduction of bacillary load after 30 days of treatment, adjuvant role of 6-gingerol upon co-treatment with drugs, promotion of pro-inflammatory cytokines and targeted immune responses against *M. tuberculosis.*	Bhaskar et al. (2020)
	antibacterial	*Bacillus subtilis*	No effect	No antimicrobial activity	Chiaramonte et al. (2021)
	antibacterial	*Micrococcus roseus*	No effect	No antimicrobial activity	Chiaramonte et al. (2021)
	antifungal	*Candida albicans*	1000 μg/ml	Biofilm formation suppressed at concentrations 10, 50, 100 μg/ml, inhibition of hyphal growth and protective role of 6-gingerol in infected nematode *Caenorhabditis elegans.*	Lee et al. (2018)
	antiviral	Dengue virus	IC_{50} – 14.70 μM (DENV-1), 14.17 μM (DENV-2), 78.76 μM (DENV-3), 112.84 μM (DENV-4)		Halim et al. (2021)
	antiviral	Chikungunya virus	IC_{50} – 0.038 mM (post treatment of 6-gingerol), IC_{50} – 0.031 mM (full treatment)	CC_{50} – 0.21 mM (48 h), inhibition of viral replication.	Hayati et al. (2021)
	anthelmintic	*Angiostrongylus cantonensis*	–	Larvicidal effect: – log LD_{50} (μM): 3.557 ± 0.023 (24 h), 4.105 ± 0.033 (48 h), 4.877 ± 0.023 (72 h)	Lin et al. (2010)
8-gingerol	antifungal	*C. albicans*	–	Biofilm formation suppressed at concentrations 10, 50, 100 μg/ml.	Lee et al. (2018)
10-gingerol	antibacterial	*Porphyromonas gingivalis*	6 μg/ml	MBC – 6 μg/ml	Park, Bae, and Lee (2008)
	antibacterial	*Porphyromonas endodontalis*	8 μg/ml	MBC – 4 μg/ml	Park, Bae, and Lee (2008)
	antibacterial	*Prevotella intermedia*	14 μg/ml	MBC – 14 μg/ml	Park, Bae, and Lee (2008)
	antifungal	*C. albicans*	–	no effect	Lee et al. (2018)
	anthelmintic	*A. cantonensis*	–	Larvicidal effect: – log LD_{50} (μM): 4.609 ± 0.031 (24 h), 4.413 ± 0.034 (48 h), 4.429 ± 0.031 (72 h), 100 % loss of spontaneous movement: 10 μM (72 h)	Lin et al. (2010)

(*Continued*)

Table 6.2 Antibacterial, Antiviral, Antifungal, Anthelmintic Activity of Ginger Bioactives (CC_{50}: Concentration Required to Reduce 50% of Cell Viability, IC_{50}: Concentration of Drug Required for 50% Inhibition, LD_{50}: Dose Required to Kill 50% of the Population, MBC: Minimum Bactericidal Concentration, MIC: Minimum Inhibition Concentration) (*Continued*)

Bioactive Compound	Activity	Organisms Tested	MIC	Findings	References
12-gingerol	antibacterial	*P. gingivalis*	15 μg/ml	MBC – 10 μg/ml	Park, Bae, and Lee (2008)
	antibacterial	*P. endodontalis*	15 μg/ml	MBC – 10 μg/ml	Park, Bae, and Lee (2008)
	antibacterial	*P. intermedia*	30 μg/ml	MBC – 20 μg/ml	Park, Bae, and Lee (2008)
6-shogaol	antibacterial	*Staphylococcus aureus*	0.625 – 1.25 mg/ml	MBC – 1.25 mg/ml	Rigane, Mnif, and Salem (2018)
	antibacterial	*Pseudomonas savastanoa*	0.625 – 1.25 mg/ml	MBC – 1.25 mg/ml	Rigane, Mnif, and Salem (2018)
	antibacterial	*Salmonella enterica*	1.25 – 2.5 mg/ml	MBC – 2.5 mg/ml	Rigane, Mnif, and Salem (2018)
	antibacterial	*Listeria monocytogenes*	0.625 – 1.25 mg/ml	MBC – 2.5 mg/ml	Rigane, Mnif, and Salem (2018)
	antibacterial	*Escherichia coli*	0.625 – 1.25 mg/ml	MBC – 2.5 mg/ml	Rigane, Mnif, and Salem (2018)
	antibacterial	*Agrobacterium tumefaciens*	0.156 – 0.3125 mg/ml	MBC – 0.3125 mg/ml	Rigane, Mnif, and Salem (2018)
	antifungal	*C. albicans*	>2000 μg/ml	Biofilm formation suppressed at concentrations 10, 50, 100 μg/ml, inhibition of hyphal growth and protective role of 6-shogaol in infected nematode *C. elegans.*	Lee et al. (2018)
	anthelmintic	*A. cantonensis*	–	Larvicidal effect: – log LD_{50} (μM): 3.620 ± 0.026 (24 h), 4.095 ± 0.105 (48 h), 4.782 ± 0.018 (72 h)	Lin et al. (2010)
8-shogaol	antifungal	*C. albicans*	–	no effect	Lee et al. (2018)
10-shogaol	antifungal	*C. albicans*	–	no effect	Lee et al. (2018)
	anthelmintic	*A. cantonensis*	–	Larvicidal effect: – log LD_{50} (μM): 3.057 ± 0.023 (24 h), 3.180 ± 0.029 (48 h), 3.180 ± 0.019 (72 h)	Lin et al. (2010)
hexahydrocurcumin	anthelmintic	*A. cantonensis*	–	Larvicidal effect: – log LD_{50} (μM): 3.584 ± 0.016 (24 h), 4.287 ± 0.022 (48 h), 4.351 ± 0.039 (72 h), 100 % loss of spontaneous movement: 10 μM (72 h)	Lin et al. (2010)
Zingerone	antibacterial	*Pseudomonas aeruginosa*	–	Suppression of quorum sensing activity led to inhibition of biofilm formation, reduction in motility and production of virulence factors.	Kumar et al. (2015)

5-HT_3 system to inhibit emesis was also investigated in rats (Cheng et al. 2020). Findings revealed that 6-gingerol (50, 100 mg/kg) improved pica behaviour in a dose-dependent manner. To reduce emesis, 6-gingerol downregulated levels of 5-HT, its metabolite 5-HIAA, 5-HT_3 receptor and enzyme tyrosine hydroxylase (TH) which promote production of 5-HT in the medulla oblongata and ileum. Monoamine oxidase-A (MAO-A) and serotonin transporter (SERT) were increased to induce degradation of 5-HT after administration with 6-gingerol. Tian et al. (2020) confirmed findings reported by W. Qian et al. (2016) and Cheng et al. (2020). They also studied the effect of gingerol on substance P in rats and in minks. Emesis induced histopathological modifications in ileum including infiltration of inflammatory cells, irregular arrangement of villi and decrease in content of Nissl granules. However, gingerol alleviated these symptoms and also regulated emesis by reducing levels of substance P, receptor neurokinin-1 and precursor of substance P (preprotachykinin). The degradation of substance P is caused by neutral endopeptidase and was increased upon administration of gingerol.

6.4 CONCLUSION

Ginger has been considered a valuable nutraceutical and here, an overview of the recent research has provided further insight into how bioactives of ginger have had beneficial effects on health disorders. These bioactives are unique to ginger and their wide range of health properties have enabled them to perform multiple functions aimed at attenuating the debilitating effects associated with these disorders. However, it has also been important to acknowledge that there are factors that play a role in promoting their efficacy. Moreover, the evaluation of the beneficial effect of ginger bioactives in experimental studies has not been consistent across different research studies (Sang et al. 2020). Some of these factors include ginger compound used, concentration and stability of bioactive and method of delivery to target proteins in the body. An approach aimed at standardizing these factors in studies would give an optimized and maximized efficacy of metabolites. Furthermore, as new knowledge is uncovered, there are more questions to elucidate. With all this information, further research is necessary to validate the therapeutic value of ginger.

REFERENCES

Abdullahi, A., A. Khairulmazmi, S. Yasmeen, I. S. Ismail, A. Norhayu, M. R. Sulaiman, O. H. Ahmed, and M. R. Ismail. 2020. "Phytochemical profiling and antimicrobial activity of ginger (*Zingiber officinale*) essential oils against important phytopathogens." *Arabian Journal of Chemistry* 13 (11): 8012–8025. https://doi.org/10.1016/j.arabjc.2020.09.031.

Adetuyi, B. O., and E. O. Farombi. 2021. "6-Gingerol, an active constituent of ginger, attenuates lipopolysaccaride-induced oxidation, inflammation, cognitive deficits, neuroplasticity, and amyloidogenesis in rat." *Journal of Food Biochemistry* 45: e13660. https://doi.org/10.1111/jfbc.13660.

Ahmad, B., M. U. Rehman, I. Amin, M. R. Mir, S. B. Ahmad, A. Farooq, S. Muzamil, I. Hussain, M. Masoodi, and B. Fatima. 2018. "Zingerone (4-(4-hydroxy-3-methylphenyl) butan-2-one) protects against alloxan-induced diabetes via alleviation of oxidative stress and inflammation: Probable role of NF-κB activation." *Saudi Pharmaceutical Journal* 26: 1137–1145. https://doi.org/10.1016/j.jsps.2018.07.001.

Ajeeshkumar, K. K., P. A. Aneesh, N. Raju, M. Suseela, C. N. Ravishankar, and S. Benjakul. 2021. "Advancements in liposome technology: Preparation techniques and applications in food, functional foods, and bioactive delivery: A review." *Comprehensive Reviews in Food Science and Food Safety* 20: 1280–1306. https://doi.org/10.1111/1541-4337.12725.

Akshitha, H. J., K. Umesha, N. K. Leela, M. S. Shivakumar, and D. Prasath. 2020. "Quality attributes and essential oil profiling of ginger (*Zingiber officinale* Rosc.) genotypes from India." *Journal of Essential Oil Research* 32 (5): 456–463. https://doi.org/10.1080/10412905.2020.1789000.

Al-Dhahli, A. S., F. A. Al-Hassani, K. M. Alarjani, H. M. Yehia, W. M. Al Lawati, S. N. H. Azmi, and S. A. Khan. 2020. "Essential oil from the rhizomes of the Saudi and Chinese *Zingiber officinale* cultivars: Comparison of chemical composition, antibacterial and molecular docking studies." *Journal of King Saud University - Science* 32 (8): 3343–3350. https://doi.org/10.1016/j.jksus.2020.09.020.

Ali, R. A., A. A. Gandhi, L. Dai, J. Weiner, S. K. Estes, S. Yalavarthi, K. Gockman, D. Sun, and J. S. Knight. 2021. "Antineutrophil properties of natural gingerols in models of lupus." *JCI Insight* 6 (3): e138385. https://doi.org/10.1172/jci.insight.138385.

Almatroodi, S. A., A. M. Alnuqaydan, A. Y. Babiker, M. A. Almogbel, A. A. Khan, and A. H. Rahmani. 2021. "6-Gingerol, a bioactive compound of ginger attenuates renal damage in streptozotocin-induced diabetic rats by regulating the oxidative stress and inflammation." *Pharmaceutics* 13: 317. https://doi.org/10.3390/pharmaceutics13030317.

Alsahli, M. A., S. A. Almatroodi, A. Almatroudi, A. A. Khan, S. Anwar, A. G. Almutary, F. Alrumaihi, and A. H. Rahmani. 2021. "6-Gingerol, a major ingredient of ginger attenuates diethylnitrosamine-induced liver injury in rats through the modulation of oxidative stress and anti-inflammatory activity." *Mediators of Inflammation* 2021: 6661937. https://doi.org/10.1155/2021/6661937.

An, K., D. Zhao, Z. Wang, J. Wu, Y. Xu, and G. Xiao. 2016. "Comparison of different drying methods on Chinese ginger (*Zingiber officinale* Roscoe): Changes in volatiles, chemical profile, antioxidant properties, and microstructure." *Food Chemistry* 197 (April): 1292–1300. https://doi.org/10.1016/j.foodchem.2015.11.033.

Argo, B. D., M. B. Hermanto, D. W. Andriani, and J. S. Rosadhani. 2020. "The effect of ginger oil extraction using Microwave Assisted Hydro-Distillation (MAHD) method on zingiberene content." *IOP Conference Series: Earth and Environmental Science* 542. Institute of Physics Publishing. https://doi.org/10.1088/1755-1315/542/1/012002.

Ashraf, S. A., E. Al-Shammari, T. Hussain, S. Tajuddin, and B. P. Panda. 2017. "In-vitro antimicrobial activity and identification of bioactive components using GC–MS of commercially available essential oils in Saudi Arabia." *Journal of Food Science and Technology* 54 (12): 3948–3958. https://doi.org/10.1007/s13197-017-2859-2.

Assadpour, E., and S. M. Jafari. 2019. "A systematic review on nanoencapsulation of food bioactive ingredients and nutraceuticals by various nanocarriers." *Critical Reviews in Food Science and Nutrition* 59 (19): 3129–3151. https://doi.org/10.1080/10408398.2018.1484687.

Bashir, N., S. B. Ahmad, M. U. Rehman, S. Muzamil, R. R. Bhat, M. R. Mir, G. A. Shazly, M. A. Ibrahim, G. M. Elossaily, A. Y. Sherif, and M. Kazi. 2021. "Zingerone (4-(four-hydroxy-3-methylphenyl) butane-two-1) modulates adjuvant-induced rheumatoid arthritis by regulating inflammatory cytokines and antioxidants." *Redox Report* 26 (1): 62–70. https://doi.org/10.1080/13510002.2021.1907518.

Bawadood, A. S., F. A. Al-Abbasi, F. Anwar, A. M. El-halawany, and A. M. Al-Abd. 2020. "6-Shogaol suppresses the growth of breast cancer cells by inducing apoptosis and suppressing autophagy via targeting notch signaling pathway." *Biomedicine & Pharmacotherapy* 128: 110302. https://doi.org/10.1016/j.biopha.2020.110302.

Berntsson, E., S. Paul, S. B. Sholts, J. Jarvet, A. Barth, A. Gräslund, and S. K. T. S. Wärmländer. 2021. "6-Gingerol interferes with amyloid-beta (Aβ) peptide aggregation." *BioRxiv*. https://doi.org/10.1101/2021.01.03.425159

Bhaskar, A., A. Kumari, M. Singh, S. Kumar, S. Kumar, A. Dabla, S. Chaturvedi, V. Yadav, D. Chattopadhyay, and V. P. Dwivedi. 2020. "6-gingerol exhibits potent anti-mycobacterial and immunomodulatory activity against tuberculosis." *International Immunopharmacology* 87: 106809. https://doi.org/10.1016/j.intimp.2020.106809.

Brinkmann, V., U. Reichard, C. Goosmann, B. Fauler, Y. Uhlemann, D. S. Weiss, Y. Weinrauch, and A. Zychlinsky. 2004. "Neutrophil extracellular traps kill bacteria." *Science* 303 (5663): 1532–1535. https://doi.org/10.1126/science.1092385.

Cao, X., Q. Zhu, Q. Wang, M. Adu-Frimpong, C. Wei, W. Weng, R. Bao, Y. Wang, J. Yu, and X. M. Xu. 2021. "Improvement of oral bioavailability and anti-tumor effect of zingerone self-microemulsion drug delivery system." *Journal of Pharmaceutical Sciences* 110 (7): 2718–2727. https://doi.org/10.1016/j.xphs.2021.01.037.

Cheng, Q., X. Feng, Q. Meng, Y. Li, S. Chen, G. Wang, and K. Nie. 2020. "[6]-Gingerol ameliorates cisplatin-induced pica by regulating the TPH/MAO-A/SERT/5-HT/5-HT3 receptor system in rats." *Drug Design, Development and Therapy* 14: 4085–4099. https://doi.org/10.2147/DDDT.S270185.

Cheng-Lun, L., T. Derong, and L. Le. 2008. "Research on the extracting and anti-oxidation dynamic characteristics of ginger oleoresin." *International Journal of Food Science and Technology* 43 (3): 517–525. https://doi.org/10.1111/j.1365-2621.2006.01486.x.

Chiaramonte, M., R. Bonaventura, C. Costa, F. Zito, and R. Russo. 2021. "[6]-Gingerol dose-dependent toxicity, its role against lipopolysaccharide insult in sea urchin (*Paracentrotus lividus* Lamarck), and antimicrobial activity." *Food Bioscience* 39: 100833. https://doi.org/10.1016/j.fbio.2020.100833.

Ediriweera, M. K., J. Y. Moon, Y. T. Nguyen, and S. K. Cho. 2020. "10-Gingerol targets lipid rafts associated P13K/Akt signaling in radio-resistant triple negative breast cancer cells." *Molecules* 25: 3164. https://doi.org/10.3390/molecules25143164.

El Halawany, A. M., N. S. El Sayed, H. M. Abdallah, and R. S. El Dine. 2017. "Protective effects of gingerol on streptozotocin-induced sporadic Alzheimer's disease: Emphasis on inhibition of β-amyloid, COX-2, alpha-, beta - secretases and APH1a." *Scientific Reports* 7: 2902. https://doi.org/10.1038/s41598-017-02961-0.

Elshopakey, G. E., R. Almeer, S. Alfaraj, G. Albasher, M. E. Abdelgawad, A. E. A. Moneim, and E. A. Essawy. 2021. "Zingerone mitigates inflammation, apoptosis and oxidative injuries associated with renal impairment in adriamycin-intoxicated mice." *Toxin Reviews*. https://doi.org/10.1080/15569543.2021.1923528.

Fathi, M., M. R. Mozafari, and M. Mohebbi. 2012. "Nanoencapsulation of food ingredients using lipid based delivery systems." *Trends in Food Science and Technology* 23 (1): 13–27. https://doi.org/10.1016/j.tifs.2011.08.003.

Ferreira, S., and V. R. Nicoletti. 2021. "Microencapsulation of ginger oil by complex coacervation using atomization: Effects of polymer ratio and wall material concentration." *Journal of Food Engineering* 291 (February). https://doi.org/10.1016/j.jfoodeng.2020.110214.

Ganji, S., and S. Z. Sayyed-Alangi. 2017. "Encapsulation of ginger ethanolic extract in nanoliposome and evaluation of its antioxidant activity on sunflower oil." *Chemical Papers* 71 (9): 1781–1789. https://doi.org/10.1007/s11696-017-0164-1.

Ghasemzadeh, A., H. Z. E. Jaafar, A. Baghdadi, and A. Tayebi-Meigooni. 2018. "Formation of 6-, 8- and 10-shogaol in ginger through application of different drying methods: Altered antioxidant and antimicrobial activity." *Molecules* 23 (7). https://doi.org/10.3390/molecules23071646.

Gungor, H., M. Ekici, M. O. Karayigit, N. H. Turgut, H. Kara, and E. Arslanbas. 2020. "Zingerone ameliorates oxidative stress and inflammation in bleomycin-induced pulmonary fibrosis: Modulation of the expression of TGF-β1 and iNOS." *Naunyn-Schmiedeberg's Archives of Pharmacology* 393: 1659–1670. https://doi.org/10.1007/s00210-020-01881-7.

Halim, J. A. N., S. N. Halim, D. Denis, S. Haryanto, E. Dharmana, R. Hapsari, R. T. Sasmono, and B. Yohan. 2021. "Antiviral activities of curcumin and 6-gingerol against infection of four dengue virus serotypes in A549 human cell line in vitro." *Indonesian Journal of Biotechnology* 26 (1): 41–47. https://doi.org/10.22146/ijbiotech.60174.

Han, X., P. Liu, M. Liu, Z. Wei, S. Fan, X. Wang, S. Sun, and L. Chu. 2020. "[6]-Gingerol ameliorates ISO-induced myocardial fibrosis by reducing oxidative stress, inflammation, and apoptosis through inhibition of TLR4/MAPKs/NF-κB pathway." *Molecular Nutrition and Food Research* 64: 2000003. https://doi.org/10.1002/mnfr.202000003.

Hayati, R. F., C. D. Better, D. Denis, A. G. Komarudin, A. Bowolaksono, B. Yohan, and R. T. Sasmono. 2021. "[6]-gingerol inhibits Chikungunya virus infection by suppressing viral replication." *BioMed Research International* 2021: 6623400. https://doi.org/10.1155/2021/6623400.

Hong, M., L. Hu, Y. Zhang, Y. Xu, X. Liu, P. He, and Y. Jia. 2020. "6-gingerol ameliorates sepsis-induced liver injury through the Nrf2 pathway." *International Immunopharmacology* 80: 106196. https://doi.org/10.1016/j.intimp.2020.106196.

Hong, W., F. X. Zhi, T. H. Kun, F. J. Hua, L. H. Ling, F. Fang, C. Wen, W. Jie, and L. C. Yang. 2021. "6-gingerol attenuates ventilator-induced lung injury via anti-inflammation and antioxidative stress by modulating the PPARγ/NF-κB signalling pathway in rats." *International Immunopharmacology* 92: 107367. https://doi.org/10.1016/j.intimp.2021.107367.

Hosseinzadeh, A., M. Goudarzi, M. Y. Karimi, L. Khorsandi, S. Mehrzadi, and M. A. Mombeini. 2020. "Zingerone ameliorates gentamicin-induced nephrotoxicity in rats." *Comparative Clinical Pathology* 29: 971–980. https://doi.org/10.1007/s00580-020-03129-5.

Hu, S., X. Yao, Y. Hao, A. Pan, and X. Zhou. 2020. "8-Gingerol regulates colorectal cancer cell proliferation and migration through the EGRF/STAT/ERK pathway." *International Journal of Oncology* 56: 390–397. https://doi.org/10.3892/ijo.2019.4934.

Jiang, H., Z. Xie, H. J. Koo, S. P. McLaughlin, B. N. Timmermann, and D. R. Gang. 2006. "Metabolic profiling and phylogenetic analysis of medicinal *Zingiber* species: Tools for authentication of ginger (*Zingiber officinale* Rosc.)" *Phytochemistry* 67 (15): 1673–1685. https://doi.org/10.1016/j.phytochem.2005.08.001.

Jiang, X., J. Wang, P. Chen, Z. He, J. Xu, Y. Chen, X. Liu, and J. Jiang. 2021. "[6]-Paradol suppresses proliferation and metastases of pancreatic cancer by decreasing EGFR and inactivating PI3K/AKT signaling." *Research Square*. https://doi.org/10.21203/rs.3.rs-506985/v1.

Kawamoto, Y., Y. Ueno, E. Nakahashi, M. Obayashi, K. Sugihara, S. Qiao, M. Iida, M. Y. Kumasaki, I. Yajima, Y. Goto, N. Ohgami, M. Kato, and K. Takeda. 2016. "Prevention of allergic rhinitis by ginger and the molecular basis of immunosuppression by 6-gingerol through T cell inactivation." *Journal of Nutritional Biochemistry* 27: 112–122. https://doi.org/10.1016/j.jnutbio.2015.08.025.

Kemkar, K., L. Sathiyanarayanan, A. Sathiyanarayanan, and K. Mahadik. 2018. "6-Shogaol from ginger oleoresin loaded liposomes using DMPG-Na as a carrier enhances the *in-vitro* and *in-vivo* anticancer activity." *Journal of Applied Pharmaceutical Science* 8 (2): 1–10. https://doi.org/10.7324/JAPS.2018.8201.

Kucukler, S., E. Darendelioğlu, C. Caglayan, A. Ayna, S. Yıldırım, and F. M. Kandemir. 2020. "Zingerone attenuates vancomycin-induced hepatoxicity in rats through regulation of oxidative stress, inflammation and apoptosis." *Life Sciences* 259: 118382. https://doi.org/10.1016/j.lfs.2020.118382.

Kumar, L., S. Chhibber, R. Kumar, M. Kumar, and K. Harjai. 2015. "Zingerone silences quorum sensing and attenuates virulence of *Pseudomonas aeruginosa*." *Fitoterapia* 102: 84–95. https://doi.org/10.1016/j.fitote.2015.02.002.

Lee, J., Y. Kim, P. Choi, J. Ham, J. G. Park, and J. Lee. 2018. "Antibiofilm and antivirulence activities of 6-gingerol and 6-shogaol against *Candida albicans* due to hyphal inhibition." *Frontiers in Cellular and Infection Microbiology* 8: 299. https://doi.org/10.3389/fcimb.2018.00299.

Li, Y., Y. Hong, Y. Han, Y. Wang, and L. Xia. 2016. "Chemical characterization and antioxidant activities comparison in fresh, dried, stir-frying and carbonized ginger." *Journal of Chromatography B: Analytical Technologies in the Biomedical and Life Sciences* 1011 (February): 223–232. https://doi.org/10.1016/j.jchromb.2016.01.009.

Lin, R., C. Chen, L. Chung, and C. Yen. 2010. "Larvicidal activities of ginger (*Zingiber officinale*) against *Angiostrongylus cantonensis*." *Acta Tropica* 115: 69–76. https://doi.org/10.1016/j.actatropica.2009.12.007.

Liu, C., Q. Wu, Z. Cai, S. Xie, M. Duan, Q. Xie, Y. Yuan, W. Deng, and Q. Tang. 2019. "Zingerone attenuates aortic banding-induced cardiac remodelling via activating the eNOS/Nrf2 pathway." *Journal of Cellular and Molecular Medicine* 23: 6466–6478. https://doi.org/10.1111/jcmm.14540.

Liu, Q., Y. Peng, L. Qi, X. Cheng, X. Xu, L. Liu, E. Liu, and P. Li. 2012. "The cytotoxicity mechanism of 6-shogaol-treated HeLa human cervical cancer cells revealed by label-free shotgun proteomics and bioinformatics analysis." *Evidence-Based Complementary and Alternative Medicine* 2012. https://doi.org/10.1155/2012/278652.

Lu, W., A. L. Kelly, and S. Miao. 2016. "Emulsion-based encapsulation and delivery systems for polyphenols." *Trends in Food Science and Technology* 47: 1–9. https://doi.org/10.1016/j.tifs.2015.10.015.

Luo, Y., L. Zha, L. Luo, X. Chen, Q. Zhang, C. Gao, X. Zhuang, S. Yuan, and T. Qiao. 2019. "[6]-Gingerol enhances the cisplatin sensitivity of gastric cancer cells through inhibition of proliferation and invasion via PI3K/AKT signaling pathway." *Phytotherapy Research* 33: 1353–1362. https://doi.org/10.1002/ptr.6325.

Ma, R., Z. Ni, F. Zhang, Y. Zhang, M. Liu, K. Thakur, J. Zhang, S. Wang, and Z. Wei. 2020. "6-Shogaol mediated ROS production and apoptosis via endoplasmic reticulum and mitochondrial pathways in human endometrial carcinoma Ishikawa cells." *Journal of Functional Foods* 74: 104178. https://doi.org/10.1016/j.jff.2020.104178.

Mahomoodally, M. F., M. Z. Aumeeruddy, K. R. R. Rengasamy, S. Roshan, S. Hammad, J. Pandohee, X. Hu, and G. Zengin. 2021. "Ginger and its active compounds in cancer therapy: From folk uses to nano-therapeutic applications." *Seminars in Cancer Biology* 69: 140–149. https://doi.org/10.1016/j.semcancer.2019.08.009.

Mao, Q., X. Xu, S. Cao, R. Gan, H. Corke, T. Beta, and H. Li. 2019. "Bioactive compounds and bioactivities of ginger (*Zingiber officinale* Roscoe)." *Foods* 8 (6): 185. https://doi.org/10.3390/foods8060185.

Martin, A. C. B. M., R. Tomasin, L. Luna-Dulcey, A. E. Graminha, M. A. Naves, R. H. G. Teles, V. D. da Silva, J. A. da Silva, P. C. Vieira, B. Annabi, and M. R. Cominetti. 2020. "10-gingerol improves doxorubicin anticancer activity and decreases its side effects in triple negative breast cancer models." *Cellular Oncology* 43: 915–929. https://doi.org/https://doi.org/10.1007/s13402-020-00539-z.

Mohammed, N. K., C. P. Tan, Y. A. Manap, B. J. Muhialdin, and A. S. M. Hussin. 2020. "Spray drying for the encapsulation of oils—A review." *Molecules* 25 (17): 3873. https://doi.org/10.3390/molecules25173873.

Munda, S., S. Dutta, S. Haldar, and M. Lal. 2018. "Chemical analysis and therapeutic uses of ginger (*Zingiber officinale* Rosc.) essential oil: A review." *Journal of Essential Oil-Bearing Plants* 21 (4): 994–1002. https://doi.org/10.1080/0972060X.2018.1524794.

Nair, K. P. 2019. Ginger as a spice and flavorant. In: *Turmeric (Curcuma longa L.) and Ginger (Zingiber officinale Rosc.) - World's Invaluable Medicinal Spices*. Springer, Cham. https://doi.org/10.1007/978-3-030-29189-1_26

Noore, S., N. K. Rastogi, C. O'Donnell, and B. Tiwari. 2021. "Novel bioactive extraction and nano-encapsulation." *Encyclopedia* 1 (3): 632–664. https://doi.org/10.3390/encyclopedia1030052.

Park, M., J. Bae, and D. Lee. 2008. "Antibacterial activity of [10]-gingerol and [12]-gingerol isolated from ginger rhizome against periodontal bacteria." *Phytotherapy Research* 22: 1446–1449. https://doi.org/10.1002/ptr.2473.

Peng, X., R. Luo, J. Li, A. He, X. Wang, H. Wan, Y. Cai, W. Dong, and J. Lin. 2020. "Zingiberene targets the miR-16/cyclin-B1 axis to regulate the growth, migration and invasion of human liver cancer cells." *JBUON* 25 (4): 1904–1910.

Perinelli, D. R., G. F. Palmieri, M. Cespi, and G. Bonacucina. 2020. "Encapsulation of flavours and fragrances into polymeric capsules and cyclodextrins inclusion complexes: An update." *Molecules* 25 (4): 5878. https://doi.org/10.3390/molecules25245878.

Qian, S., H. Fang, L. Zheng, and M. Liu. 2020. "Zingerone suppresses cell proliferation via inducing cellular apoptosis and inhibition of the PI3K/AKT/mTOR signaling pathway in human prostate cancer PC-3 cells." *Journal of Biochemical and Molecular Toxicology* 35: e22611. https://doi.org/10.1002/jbt.22611.

Qian, W., X. Cai, Y. Wang, X. Zhang, H. Zhao, Q. Qian, Z. Yang, Z. Liu, and J. Hasegawa. 2016. "Effect of gingerol on cisplatin-induced pica analogous to emesis via modulating expressions of dopamine 2 receptor, dopamine transporter and tyrosine hydroxylase in the vomiting model of rats." *Yonago Acta Medica* 59: 100–110.

Rashid, S., A. F. Wali, S. M. Rashid, R. M. Alsaffar, A. Ahmad, B. L. Jan, B. A. Paray, S. M. A. Alqahtani, A. Arafah, and M. U. Rehman. 2021. "Zingerone targets status epilepticus by blocking hippocampal neurodegeneration via regulation of redox imbalance, inflammation and apoptosis." *Pharmaceuticals* 14: 146. https://doi.org/10.3390/ph14020146.

Rasmussen, A., K. Murphy, and D. W. Hoskin. 2019. "10-Gingerol inhibits ovarian cancer cell growth by inducing G2 arrest." *Advanced Pharmaceutical Bulletin* 9 (4): 685–689. https://doi.org/10.15171/apb.2019.080.

Rastogi, N., S. Duggal, S. K. Singh, K. Porwal, V. K. Srivastava, R. Maurya, M. L. B. Bhatt, and D. P. Mishra. 2015. "Proteasome inhibition mediates p53 reactivation and anticancer activity of 6-Gingerol in cervical cancer cells." *Oncotarget* 6 (41): 43310–43325. https://doi.org/10.18632/oncotarget.6383.

Rigane, G., S. Mnif, and R. B. Salem. 2018. "One step purification of 6-shogaol from *Zingiber officinale* Rosco, a phenolic compound having a high effectiveness against bacterial strains." *Revue Roumaine de Chimie* 63 (1): 5–10.

Ryu, M. J., and H. S. Chung. 2015. "(10)-Gingerol induces mitochondrial apoptosis through activation of MAPK pathway in HCT116 human colon cancer cells." *In Vitro Cellular & Developmental Biology* 51: 92–101. https://doi.org/10.1007/s11626-014-9806-6.

Sampath, S. J. P., S. Birineni, S. Perugu, N. Kotikalapudi, and V. Venkatesan. 2021. "Therapeutic efficacy of 6-gingerol and 6-shogaol in promoting browning of white adipocytes vis-à-vis enhanced thermogenesis portrayed in high fat milieu." *Food Bioscience* 42: 101211. https://doi.org/10.1016/j.fbio.2021.101211.

Sang, S., H. D. Snook, F. S. Tareq, and Y. Fasina. 2020. "Precision research on ginger: The type of ginger matters." *Journal of Agricultural and Food Chemistry* 68: 8517–8523.

Santos Reis, N. d., N. B. de Santana, I. M. de Carvalho Tavares, O. A. Lessa, L. R. dos Santos, N. E. Pereira, G. A. Soares, R. A. Oliveira, J. R. Oliveira, and M. Franco. 2020. "Enzyme extraction by lab-scale hydrodistillation of ginger essential oil (*Zingiber officinale* Roscoe): Chromatographic and micromorphological analyses." *Industrial Crops and Products* 146 (April). https://doi.org/10.1016/j.indcrop.2020.112210.

Semwal, R. B., D. K. Semwal, S. Combrinck, and A. M. Viljoen. 2015. "Gingerols and shogaols: Important nutraceutical principles from ginger." *Phytochemistry* 117: 554–568. https://doi.org/10.1016/j.phytochem.2015.07.012.

Setyaningsih, D., D. Kurniawan, and N. Muna. 2020. "Encapsulation of ginger oleoresin with a combination of maltodextrin and skim milk powder as wall material." *IOP Conference Series: Earth and Environmental Science* 472. Institute of Physics Publishing. https://doi.org/10.1088/1755-1315/472/1/012016.

Shao, Y., P. Marriott, R. Shellie, and H. Hügel. 2003. "Solid-phase micro-Extraction - Comprehensive two-dimensional gas chromatography of ginger (*Zingiber officinale*) volatiles." *Flavour and Fragrance Journal* 18:5–12. https://doi.org/10.1002/ffj.1133.

Simon, A., A. Darcsi, Á. Kéry, and E. Riethmüller. 2020. "Blood-brain barrier permeability study of ginger constituents." *Journal of Pharmaceutical and Biomedical Analysis* 177: 112820. https://doi.org/10.1016/j.jpba.2019.112820.

Simon-Brown, K., K. M. Solval, A. Chotiko, L. Alfaro, V. Reyes, C. Liu, B. Dzandu, E. Kyereh, A. G. Barnaby, I. Thompson, Z. Xu, and S. Sathivel. 2016. "Microencapsulation of ginger (*Zingiber officinale*) extract by spray drying technology." *LWT - Food Science and Technology* 70 (July): 119–25. https://doi.org/10.1016/j.lwt.2016.02.030.

Song, S., M. Dang, and M. Kumar. 2019. "Anti-inflammatory and renal protective effect of gingerol in high-fat diet/streptozotocin-induced diabetic rats via inflammatory mechanism." *Inflammopharmacology* 27: 1243–1254. https://doi.org/10.1007/s10787-019-00569-6.

Supardan, M. D., A. Fuadi, P. N. Alam, and N. Arpi. 2012. "Solvent extraction of ginger oleoresin using ultrasound." *MAKARA of Science Series* 15 (2): 163–167. https://doi.org/10.7454/mss.v15i2.1066.

Tavares, L., and C. P. Z. Noreña. 2020. "Encapsulation of ginger essential oil using complex coacervation method: Coacervate formation, rheological property, and physicochemical characterization." *Food and Bioprocess Technology* 13 (8): 1405–1420. https://doi.org/10.1007/s11947-020-02480-3.

Tian, L., W. Qian, Q. Qian, W. Zhang, and X. Cai. 2020. "Gingerol inhibits cisplatin-induced acute and delayed emesis in rats and minks by regulating the central and peripheral 5-HT, SP, and DA systems." *Journal of Natural Medicines* 74: 353–370. https://doi.org/10.1007/s11418-019-01372-x.

Touré, A., H. B. Lu, X. Zhang, and X. Xueming. 2011. "Microencapsulation of ginger oil in 18DE maltodextrin/whey protein isolate." *Journal of Herbs, Spices and Medicinal Plants* 17 (2): 183–195. https://doi.org/10.1080/10496475.2011.583137.

Trung Dung, N., T. H. Duc, and N. D. B. Thanh. 2018. "Optimization of ginger oleoresin extraction from fresh ginger by using microwave-assisted energy." *Vietnam Journal of Science and Technology* 56: 229. https://doi.org/10.15625/2525-2518/56/4a/13126.

Varakumar, S., K. V. Umesh, and R. S. Singhal. 2017. "Enhanced extraction of oleoresin from ginger (*Zingiber officinale*) rhizome powder using enzyme-assisted three phase partitioning." *Food Chemistry* 216 (February): 27–36. https://doi.org/10.1016/j.foodchem.2016.07.180.

Wang, W., C. Li, X. Wen, P. Li, and L. Qi. 2009. "Simultaneous determination of 6-gingerol, 8-gingerol, 10-gingerol and 6-shogaol in rat plasma by liquid chromatography-mass spectrometry: Application to pharmacokinetics." *Journal of Chromatography B: Analytical Technologies in the Biomedical and Life Sciences* 877 (8–9): 671–679. https://doi.org/10.1016/j.jchromb.2009.01.021.

Wang, J., D. Li, P. Wang, X. Hu, and F. Chen. 2019. "Ginger prevents obesity through regulation of energy metabolism and activation of browning in high-fat diet-induced obese mice." *Journal of Nutritional Biochemistry* 70: 105–115. https://doi.org/10.1016/j.jnutbio.2019.05.001.

Wang, J., L. Zhang, L. Dong, X. Hu, F. Feng, and F. Chen. 2019. "6-Gingerol, a functional polyphenol of ginger, promotes browning through an AMPK-dependent pathway in 3T3-L1 adipocytes." *Journal of Agricultural and Food Chemistry* 67: 14056–14065. https://doi.org/10.1021/acs.jafc.9b05072.

Wang, J., L. Zhou, H. Zhao, and S. Cai. 2016. "Examination of the protective effect of 6-shogaol against LPS-induced acute lung injury in mice via NF-κB attenuation." *Archives of Biological Sciences* 68 (3): 633–639. https://doi.org/10.2298/ABS151012055W.

Xu, Y., Q. Wang, Y. Feng, C. K. Firempong, Y. Zhu, E. Omari-Siaw, Y. Zheng, Z. Pu, X. Xu, and J. Yu. 2016. "Enhanced oral bioavailability of [6]-Gingerol-SMEDDS: Preparation, *in vitro* and *in vivo* Evaluation." *Journal of Functional Foods* 27 (December): 703–710. https://doi.org/10.1016/j.jff.2016.10.007.

Xu, S., H. Zhang, T. Liu, Z. Wang, W. Yang, T. Hou, X. Wang, D. He, and P. Zheng. 2020. "6-Gingerol suppresses tumor cell metastasis by increasing YAP^{ser127} phosphorylation in renal cell carcinoma." *Journal of Biochemical and Molecular Toxicology* 35: e22609. https://doi.org/10.1002/jbt.22609.

Yang, F., L. Deng, M. Chen, Y. Liu, and J. Zheng. 2021. "6-Gingerol ameliorates sepsis induced acute lung injury by regulating nuclear factor-κB and nuclear-factor erythroid 2-related factor 2/heme oxygenase-1 signaling pathways." *Current Topics in Nutraceutical Research* 19 (3): 255–260. https://doi.org/10.37290/ctnr2641-452X.19:255-260.

Yi, J., Z. Ryoo, J. Ha, D. Oh, M. Kim, and S. Kim. 2019. "Beneficial effects of 6-shogaol on hyperglycemia, islet morphology and apoptosis in some tissues of streptozotocin-induced diabetic mice." *Diabetology & Metabolic Syndrome* 11: 15. https://doi.org/10.1186/s13098-019-0407-0.

Zancan, K. C., M. O. M. Marques, A. J. Petenate, and M. A. A. Meireles. 2001. "Extraction of ginger (*Zingiber officinale* Roscoe) oleoresin with CO_2 and co-solvents: A study of the antioxidant action of the extracts." *Journal of Supercritical Fluids* 24 (1): 57–76. https://doi.org/10.1016/S0896-8446(02)00013-X.

Zhang, M., R. Zhao, D. Wang, L. Wang, Q. Zhang, S. Wei, F. Lu, W. Peng, and C. Wu. 2021. "Ginger (*Zingiber officinale* Rosc.) and its bioactive components are potential resources for health beneficial agents." *Phytotherapy Research* 35: 711–742. https://doi.org/10.1002/ptr.6858.

Zhang, M., E. Viennois, M. Prasad, Y. Zhang, L. Wang, Z. Zhang, M. K. Han, B. Xiao, C. Xu, S. Srinivasan, and D. Merlin. 2016. "Edible ginger-derived nanoparticles: A novel therapeutic approach for the prevention and treatment of inflammatory bowel disease and colitis-associated cancer." *Biomaterials* 101 (September): 321–340. https://doi.org/10.1016/j.biomaterials.2016.06.018.

CHAPTER 7

Bioactive Compounds in Black Pepper

Navneet Kaur, Bhairav Prasad, and Shilpi Ahluwalia

CONTENTS

7.1 INTRODUCTION

Plants have always benefitted mankind in several ways and have been used as food or as traditional medicine as time immemorial. In fact, even today majority of world's population till rely on natural products having medicinal properties for fulfilling their health care needs. Resistance of certain pathogens towards multiple synthetic drugs/antibiotics have also inclined the researchers towards development of plant-based pharmaceuticals. Herbs and spices are chief constituents of our diet and have been used for various purposes such as preservative, seasoning, colouring agents, flavouring agents are rich in many bioactive compounds. Black pepper is one such plant having wide range of

DOI: 10.1201/9781003215387-7

applications which are proven scientifically. *Piper nigrum* L. recognized as black pepper is one of the main spice available in the market (Feng et al. 2020). Fully developed unripe dried berries is the main part of the pepper plant that is being used as a spice from time immemorial (Wilson 2016).

Not only it has commendable ability to be used as seasoning in various dishes, it also provides a characteristic aroma and biting taste. In addition, it is claimed to have antifungal, antibacterial, aphrodisiac and medicinal properties. It has proven to be beneficial in treating cold, cough and digestive problems and is widely used in traditional herbal medicines. Even in modern days, black pepper is being used an antioxidant, GI stimulant, immunomodulator, anti-inflammatory substance etc. (Lee et al. 2020; Joshi et al. 2018). These properties can be attributed due to the presence of piperine and other vast range of polyphenolic compounds found in black pepper (Feng et al. 2020; Lee et al. 2020; Joshi et al. 2018).

Black pepper contains carbohydrates, proteins, starch, alkaloids, amides, flavonoids, phenols and some essential oils (Zorica et al., 2019). The major constituents present in essential oil include α-pinene, β-pinene, β-caryophyllene, phellandrene, cotral, limonene, linalool etc. (Dosoky et al. 2019). Along with this, there are certain other compounds present in black pepper such as β-carotene, lauric acid, palmitic acid, myristic acid and piperine. Piperine, an alkaloid is the major component in *Piper nigrum* contributing for the pungency of black pepper. Piperine, besides having pharmacological effects like anti-cancer, antioxidant, antidepressant, anti-inflammatory etc. many biological activities have also been reported in literature which include increasing the bioavailability of nutrients and certain drugs thereby; increasing the therapeutic potential and efficacy of drugs (Rather and Bhagat, 2018).

Black pepper oil is highly popular because of its aroma not only in food and beverage industry but also in cosmetic industry. Abundant uses of essential oil and various other components of black pepper, the extraction techniques are gaining a lot of attention. Steam distillation method is used for pepper oil extraction from dried black pepper, but this method does not involve extraction of pungent compound (piperine). Piperine is solvent extractable component of black pepper which can be obtained by repeated extractions with organic solvents like acetone, ethanol, dichroloethane, petroleum ether followed by removal of solvent at reduced pressures. Several advanced extraction techniques are also employed nowadays which include supercritical CO_2 extraction, ultrasound assisted extraction, microwave assisted extraction, ionic liquid-based extraction etc.

This chapter gives a comprehensive outlook about the phytochemical constituents of black pepper with a special reference to piperine (bioactive compound), its different traditional as well as modern extractions methods with its pharmacological importance.

7.2 BOTANICAL ASPECTS OF BLACK PEPPER

Black Pepper (*Piper nigrum*) is amongst the oldest and most important spice. It belongs to family *Piperaceae,* with different varieties that vary in yield and other physical characteristics (Shango et al. 2021). It has a biting flavour which makes it distinct from all other spices, called "King of Spices," due to its wide usage and great commercial importance. The characteristic aroma and pungency of black pepper is due to the presence of piperine which is the major bioactive compound of black pepper (Joshi et al. 2018; Srinivasan 2007).

P. nigrum is a woody, perennial, climbing plant, or "vine," that can grow up to 4–6 m and can reach 8–15 m in wild conditions. It has two types of branches: orthotropic branches, which grow tall straight and become woody after climbing a support and plagiotropic branches, which are actually fruiting branches. The plant contain pendant spikes on fruiting branches, which vary in length depending upon the variety of the plant. The high yielding varieties have 70–100% bisexual flowers and due to presence of bisexual flowers, *P. nigrum* is considered as self-pollinating plant. These plants have green unripe fruits, which on ripening turn red, and black on drying (Figure 7.1).

Figure 7.1 Black pepper unripe fruit (green) and dried seeds (black).

Vietnam, Indonesia, India, Brazil and China are major pepper producing countries in world. In India, most of the country's pepper produce is cultivated in Kerala, and some in Karnataka and Tamil Nadu. There are many varieties found and grown in South India, namely, Karimunda, Kottanda, Neelamundi, Balamkotta, Morata, Uddagare, etc. (Ratish and Rou 2019). Till now, many improved varieties have also been prepared and are cultivated widely. Some of those are Sreekara, Subhakara, Panchnami, Pournami, IISR- Thevam, PLD-2, etc. Sreekara is climate friendly and among the highest yielding varieties as it bears 98% bisexual flowers and contains 5.1% piperine which gives a superior quality pepper. Subhakara, Pournami and Panchami give moderately high yield and contains 3.4, 4.1, 4.7% piperine respectively.

Black pepper is the most popular spice owing to its multiple uses and offering a number of health benefits. It can be used in kitchen as an antioxidant or a flavouring agent, and is used traditionally to treat many diseases since ancient times (Joshi et al. 2018; Srinivasan 2007).

7.3 PHYTOCHEMICAL ANALYSIS OF BLACK PEPPER

Black pepper mainly consists of carbohydrates (37.4%), proteins (25.5%), fats (5.3%), moisture content (4.7%) including certain minerals like potassium, calcium, phosphorous and magnesium (Lee et al. 2020). The phytochemical screening of black pepper reveals the presence of various chemical compounds viz. alkaloids, phenols, flavonoids, lignans, terpenes, amides, steroids etc. and many researchers have able to qualitatively analyse and purify these compounds (Zorica et al. 2019). Some of the chief constituents found in essential oil of black pepper which is responsible for odour are α-pinene, β-pinene, sabinene, limonene, citral, linalool, mycrene, phellandrene (Figure 7.2) (Zorica et al. 2019). Antioxidants such as β-carotene, lauric acid, palmitic acid and piperine are also found in black pepper while compounds like 2,3-diethyl-5-methylpyrazine and 2-isopropyl-3-methoxypyrazine are responsible for musky off flavour in black pepper (Lee et al. 2020).

Black pepper is popular for its flavour and pungent taste because of piperine and certain essential oils. Volatile oils are responsible for imparting particular aroma of black pepper and constitute about 0.4–0.7% of black pepper while, piperine constituting 2–9% of black pepper is responsible for giving pungency (Gorgani et al. 2017). Piperine constitutes 98% of total alkaloids responsible for imparting pungency to black pepper and therefore this chapter deals majorly with studying the properties extraction techniques and therapeutic uses of piperine.

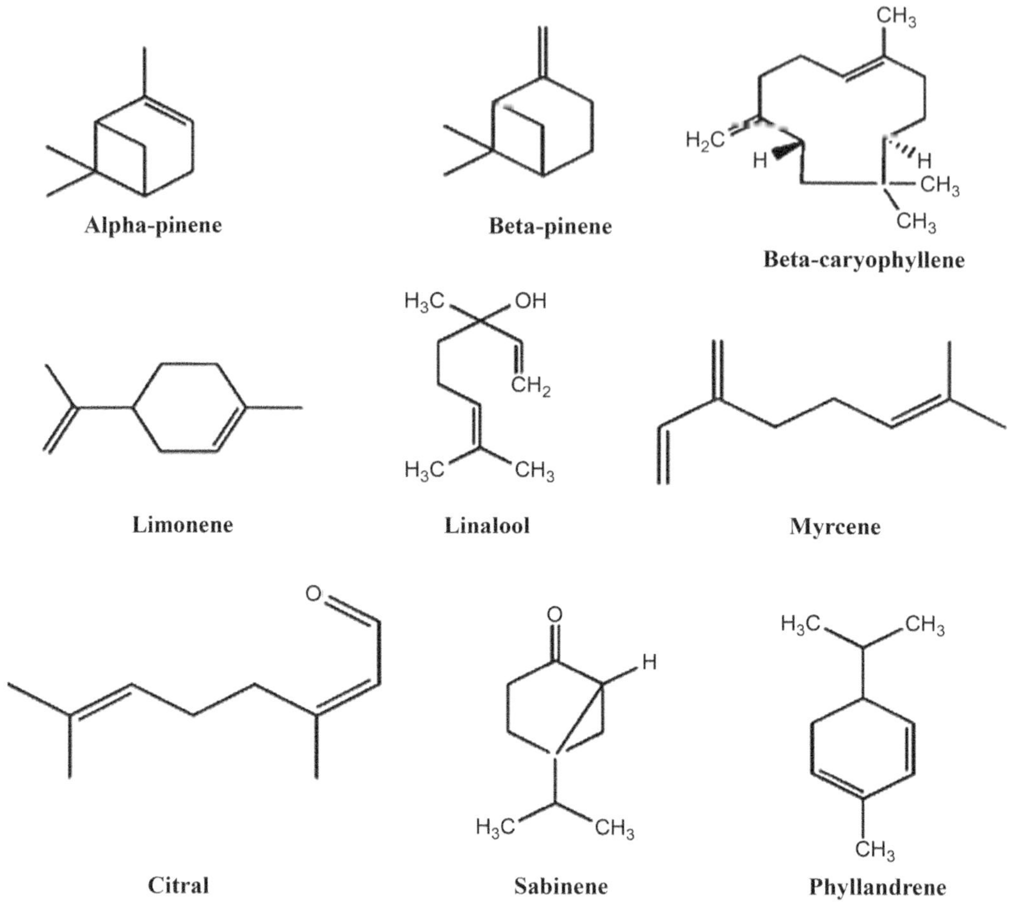

Figure 7.2 Compounds present in black pepper essential oil.

Alkaloid piperine isolated from *Piperaceae* family is the most abundant and a natural pharmacological active compound. It was first isolated in the year 1819 by Hans Christian Oersted, which in pure form represented a yellow coloured crystalline compound having molecular formula of $C_{17}H_{19}NO_3$ (piperoylpiperidine) and melting point of 128–130°C (Shityakov et al. 2019).

Four isomeric forms of piperine (Figure 7.3) are found in *Piper nigrum* L., and these are: (1) trans-trans isomer: Piperine, (2) cis-trans isomer: Isopeperine, (3) cis-cis isomer: Chavicine and (4) and trans-cis isomer: Isochavicine (4). The intriguing structure of the piperine consists of an aliphatic chain connecting an aromatic moiety and an amide moiety (Tiwari et al. 2020). Though researchers have also reported presence of some other alkaloids like piperanine, piperettine, piperidine, piperylin A, piperolein B possessing very small percentage of pungency (Gorgani et al. 2017).

7.4 THERAPEUTIC EFFECTS OF BIOACTIVE COMPOUNDS (PIPERINE)

7.4.1 Anti-Inflammatory Effect

Various herbs and spices because of their anti-inflammatory properties have been used as medicines to treat number of ailments. This anti-inflammatory property of the spices can be used to

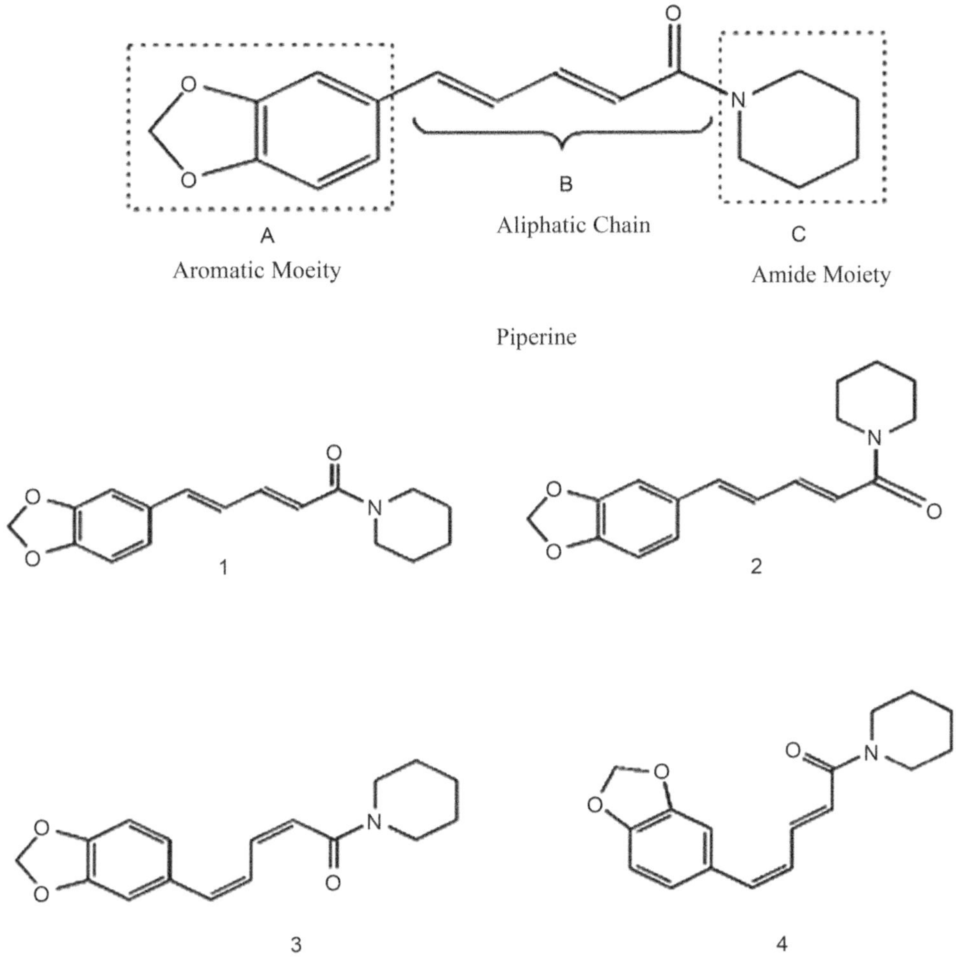

Figure 7.3 Structure of piperine and isomers of piperine: (1) trans-trans, (2) cis-trans, (3) cis-cis and (4) trans-cis (Tiwari et al. 2020).

tackle cytokine storms for management of various highly infectious and fatal diseases like Covid-19 (Kunnumakkara et al. 2021).

According to various *in-vitro* and *in-vivo* studies conducted, the anti- inflammatory role of piperine obtained from *Piper nigrum* has been proved to be effective against bacterial sepsis (Liang et al. 2016; Kunnumakkara et al. 2021), acute pancreatitis (Kunnumakkara et al. 2021) and lupus nephritis (Peng et al. 2018; Kunnumakkara et al. 2021).

Piperine administration helped in significantly reducing ATP induced pyroptosis by suppressing AMPK (AMP activated protein kinase) signalling. Piperine treatment severely affected systemic IL-1 β levels thereby suppressing systemic inflammation and pyroptosis (Liang et al. 2016).

Evaluation of anti-inflammatory effect of piperine on rats in which inflammation was induced by administration of carrageenan showed some favourable results. In the study a test group that were administered piperine (100 mg/kg) showed lesser infiltration of lymphocytes into the ankle joints. In the same study, arthritis and oedema were induced by administration of carrageenan also provided positive results when a dose of 100 mg/kg of piperine was administered and a considerable decrease in paw volume was observed (Bang et al. 2009).

7.4.2 Antioxidant Effect

A free radical is any molecular atom capable of existing independently and containing an unpaired electron in the outer shell making them highly reactive and unstable. They tend to lose electron when acted upon by a reducing agent. There are various reasons that can lead to production of free radicals like exposure to U.V radiations, industrial and environmental pollutants, tissue injury, infections, and autoimmune disorders etc. (Gorgani et al. 2017). In human body free radicals act as intermediates of various physiological reactions and thus play a vital role when present in appropriate concentrations in appropriate circumstances. However, in case of "oxidative stresses" the free radicals can even cause damage to the body. The body has developed a mechanism to deal with such stresses by means of naturally occurring antioxidants that are found in the human body. These antioxidants act as reducing agents. A perfect balance between the antioxidants and free radicals ensures that there is no damage caused by the free radicals to the body (Reena et al. 2018; Alkadi 2020).

Reactive oxygen species (ROS) and reactive nitrogen species (RNS) are the two most important free radicals. Oxidative stresses that result due to ROS, have been known to cause serious diseases including cancer, coronary heart disease (CHD), and osteoporosis (Nimse and Pal 2015).

Antioxidants inhibits these free radical reactions thus preventing the potential cell injuries. Apart from the antioxidants that are naturally present in the human body various food products have also been known to have significant amount of antioxidants. Antioxidants such as vitamin E, ascorbic acid, carotenoids and various polyphenols have proved to have beneficial biological effects. These not only prevents free radical induced cell injuries but are further claimed to be anti-inflammatory, anti-aging, anti-cancerous and anti-atherosclerosis in nature (Xu et al. 2017).

The antioxidant activity of black pepper is because of phenolic compounds, including phenolic acids and flavonoids. In-vitro studies have revealed that black pepper extracts have shown scavenging activity against various free radicals including DPPH (1,1-diphenyl-2-picrylhydrazyl) free radical, hydroxyl free radical, superoxide free radical (Agbor et al. 2006).

Piperine extracted from *Piper nigrum*, has the antioxidant activity that can help to deal with the oxidative stress. Both *in-vitro* and *in-vivo* clinical trials have shown that piperine is effective in preventing lipid peroxidation and thus can prevent the damage of cell membranes and organelles. AAPH [2,2′-Azobis(2-amidinopropane) dihydrochloride] is a free radical inducer and AAPH has the ability to cause oxidative damage in erythrocytes, further resulting in haemolysis. Piperine derivatives were defensive against AAPH-mediated oxidatively damaged erythrocytes and this protective effect may be derived from the intracellular antioxidant enzyme system (Qin et al. 2020). Because of the antioxidant activity of piperine, it is claimed to be beneficial for liver health and combatting other life threatening diseases like cancer. (Gorgani et al. 2017).

7.4.3 Anti-Diabetic Effect

Diabetes mellitus is a disease caused by the disorders in the metabolism of carbohydrates, proteins and lipids due to insufficient insulin production by pancreatic cells (type 1) or because of increased insulin resistance in peripheral tissues (type 2). These metabolic disorders mainly hyperglycemia (high blood glucose) results in establishing many microvascular complications like neuropathy, nephropathy, retinopathy and macro vascular complications like coronary heart disease, peripheral vascular diseases etc. affecting the quality of life and life expectancy of diabetic people. Different treatments of diabetes mellitus are available which include anti-diabetic drugs (metformin, phenoformin), sulfonylureas, amylin analogs, DPP-IV inhibitors, and insulin. But these conventional treatments do have some limitations of high cost, bioavailability and other side effects like weight gain, liver toxicity, gastrointestinal problems and many more (Srinivas et al. 2020) which opens the route to alternative traditional medicines. These medicines are mostly bioactive compounds from plant products and have advantages of easy availability, low cost with no side effects. Piperine from black pepper is one such compound that controls metabolism of glucose and activity of gluconeogenic enzymes of the liver; thereby has been used in managing type 2 diabetes (Srinivas et al. 2020).

Piperine and its derivatives activates the peroxisome proliferator-activated receptor (PPAR γ) showing anti-diabetic effect (Wang et al. 2020; Kharbanda et al. 2016). PPAR are transcription factors belonging to family of nuclear receptors and have a prominent role in regulating glucose and lipid metabolism. PPAR has been categorized into three different isoforms i.e. PPARα, PPARγ and PPARδ. Though all isoforms contribute in glucose regulation and lipid homeostasis they differ in their expression areas like PPARα is expressed in heart, kidney, liver, muscles, PPARδ regulates lipid metabolism in skeletal muscle, adipose tissue, skin and heart while PPARγ is expressed in smooth muscles and endothelial cells. Piperine has also been known of having bioenhancing property improving the bioavailability of other drugs like tetracycline, streptomycin, nimusulide, ciprofloxacin, dapsone, etc. (Peterson et al. 2019; Chinta et al. 2017). Piperine was used with an anti-diabetic drug metaformin and helped in lowering blood glucose and showed better effect when used in combination (Atal et al. 2016). In a recent study, piperine phytosomes formulations exhibited noteworthy improvement in the absorption and bioavailability of domperidone as compared to pure drug suspension under same conditions (Islam et al. 2021).

7.4.4 Anti-Cancer Activity

Cancer is a deadly disease and its development in a body is a very slow process which involves the malignant transformation of genetically altered cell to an abnormal mass of cells. Though cancer is difficult to treat but scientists believe around 30–40% cancer deaths can be prevented by dietary and life style modifications and reducing exposure to carcinogens present in environment. Cancer development is divided into three different stages i.e. initiation, promotion and progression stage and growth of cancerous cells can be disturbed using pharmacological drugs. Piperine is one such pharmacological agent which has cancer and tumour preventive properties. It has been reported in preventing the multiplication and survival of cancer cells through different mechanisms such as altering redox homeostasis, activating autophagy, reducing the self-renewal process of cancer cells, modulating endoplasmic reticulum stress (Rather and Bhagat 2018). Piperine is also shown to amend activity of enzymes and transcription factors thereby assisting in inhibiting invasion, metastatis and angiogenesis of cancer cells. Chemo-preventive drugs induce apoptosis and retard tumour growth, in a similar way piperine activates the apoptotic process by different mechanisms such as formation of ROS, decline in mitochondrial membrane potential and cellular apoptosis (Rather and Bhagat 2018).

Piperine successfully inhibited the growth of HeLa cells with ROS generation and nuclear condensation. Increased ROS production lead to oxidation of biomolecules which results in mitochondrial DNA mutations, aging, and cell death; which implicates that increased intracellular ROS production can lead to cellular damage of HeLa cells (Jafri et al. 2019). Matrix metalloproteinases (MMPs) are zinc binding metalloproteinases which promote tumour growth and metastatic process. Piperine (35–280 μmol/L) significantly reduced the expression of MMP-9 and MMP-3 which inhibited the growth of metastatic 4T1 cells (Lai et al. 2012). Piperine has been exhibited to inhibit the growth and proliferation of prostate cancer cells by inducing autophagy (Zeng and Zeng 2018; Ouyang et al. 2013).

Angiogenesis process involves the formation of new blood vessels and is associated with cancer development allowing the cancer cells to grow and metastasize. Piperine interrupts the angiogenic process by inhibiting proliferation, migration and tube formation by human unbilical vein endothelial cells (HUVECs) (Doucette et al. 2013).

In a latest study, piperine was extracted from black pepper and anti-CD 133 monoclonal antibodies were created containing nanoliposomal complexes (PMC) and their effect on cancer stem cells was studied. The results showed that the PMC's repressed the growth of cancer stem cells by 55.02% which proved the potential of PMC in cancer therapies (Thao et al. 2021).

7.4.5 Anti-Allergic Effect

Allergic rhinitis (AR) is very common health problem around the world affecting people of all age groups. It is associated with sleep disorders leading to poor work performance, decreased learning ability eventually affecting the quality of life. Allergic rhinitis is an inflammatory disease and a common cause for asthma and conjunctivitis. In the initial phase, the inhaled antigens (mostly environmental allergens, dust, pollens, moulds, cockroach residues) pass through nasal mucosal epithelial cells and bind to IgE antibodies at mast cells which are distributed over entire nasal mucosa. In this antigen-antibody reaction, certain chemical mediators like histamine, prostaglandins and leukotrienes are released by mast cells which irritate the sensory nerve terminals and blood vessels of nasal mucosa causing sneezing, swelling, watery rhinorrhoea, nasal congestion etc. While in the later phase i.e. after 8–10 hours of exposure to antigen, in response to cytokines there is migration of mast cells, eosinophils and basophiles into the nasal tissue. Many oral antihistamines and corticosteroids are used as treatment for AR but they may have undesirable adverse effects on our body such as headache, sedation, constipation, blurry vision which could be more serious if higher doses of steroids are taken for long periods.

Piperine has been reported to have anti-allergic properties as it helps in stabilization of mast cells by inhibiting the release of chemical mediators (Tiwari et al. 2020). It significantly reduces sneezing, rubbing and redness due to release of histamine. Piperine was reported as the major component for inhibiting airway inflammation by supressing cytokines such as interleukin (4, 5, 13) and also increased gene expression of transforming Growth Factor –β and CC chemokine receptor-3 (Kim and Lee 2009). In another study, anti-allergic effect of *P. nigrum* fruit extract was studied in ovalbumin induced allergic rhinitis and the oral therapy of extract inhibited the early phase nasal symptoms like sneezing and rubbing. *Piper nigrum* fruit extract worked by supressing the accumulation of inflammatory cells and also prohibited the activation of STAT3 and NFκBp65 signalling which helps in increasing the anti-inflammatory cytokines (Thi et al. 2019). Some of the recent physiological studies of piperine and black pepper are shown in Table 7.1.

7.4.6 Antimicrobial Activity of Black Pepper

Antimicrobial agents have been used since long because of their ability to kill microorganisms or inhibit their growth. Their effectiveness against various microorganisms viz. bacteria, viruses, fungus, parasites, pesticides etc. is helpful in preventing many types of diseases and infections. Many synthetic chemicals have been used as antimicrobial agents but resistance of microorganism against plant antimicrobials are comparatively less due to the presence of different natural constituents in one single plant. Researchers have studied the inhibitory effects of black pepper and its extracts against many gram positive and negative bacteria such as *Pusedomonas aeruginosa, Bacillus cereus, Listeria monocytogenes, Staphylococcus aureus, Salmonella typhii, E. coli* etc. Gram positive bacteria are more susceptible to black pepper extract as compared to gram negative due to difference in their cell wall and cell membrane composition (Kaho et al. 2019).

Black pepper petroleum extract (BPPE) shows inhibitory effect against *Listeria monocytogenes* and *Salmonella typhimurium*, because of its ability to impair cell wall permeability which results in considerable cell contents loss and separation of macromolecular elements from intracellular substances. The active components of BPPE are able to enter the cell and interact with enzymes of glycolytic pathway deterring and disturbing metabolism of cell. Furthermore, BPPE also disrupted

Table 7.1 Some Recent Physiological Effects of Piperine and Black Pepper

Effects Reported	Reference
Black pepper and piperine prevented anaemia of inflammation by downregulating BMP-6 and IL-6 induced hepcidin expression and STAT3 proteins	Banerjee et al. (2021)
Effective treatment strategy of piperine against cancer stem cells of hepatocarcinogenesis	Tiwari et al. (2021)
Antioxidant activity is piperine similar to other well-known antioxidants like ascorbic acid, tyrosine etc.	Carp et al. (2021)
Antifilarial effectiveness of novel piperine derivatives	Joardar et al. (2021)
Effect of aqueous extracts of black pepper, turmeric and ajwa date on experimentally induced diabetes on albino rats	Sarfraz et al. (2021)
Therapeutic effect of piperine against allergic encephalomyelitis by inhibiting dihydroorotate dehydrogenase	Liu et al. (2020)
Protective effect of piperine in renal function due to its anti-inflammatory and anti-oxidative properties	Mohammadi et al. (2020)
Piperine as effective therapeutic agent for treating UV irradiation induced skin inflammation	Jaisin et al. (2020)
Protective effect of piperine against LPS induced mastitis by activating PPARγ	Shan et al. (2020)
Protective effects of piperine on lipid metabolism in obese mice	Yafang et al. (2020)
Application of piperine rich herbal mixture for anti-inflammatory activity	Rathapon et al. (2020)
Piperine for the treatment for spatial memory impairment and myelin repair in experimental demyelination model	Roshanbakhsh et al. (2020)

the TCA pathway which cause abnormal cellular metabolism which eventually lead to cell death (Chen et al. 2019). Swathy et al. (2018) prepared nano-emulsion of black pepper oil by high pressure homogenization technique and the minimum inhibitory concentration of black pepper nano emulsion (0.234 μl/ml) inhibited the formation of biofilm by *Pseudomonas aeruginosa* (Swathy et al. 2018).

The inhibitory action of black pepper is not limited to bacteria only; many researchers have also reported the antimicrobial action of piperine against fungal strains. Piperine was able to inhibit the formation of biofilm and hyphal morphogenesis of *Candida albicans* (Priya and Pandian 2020). In a recent study, piperine is effective against biofilm and hyphal growth of *Candida albicans (ATCC10231)* at less than 15 μg/ml and 5 μg/ml concentration respectively and proved as fungicidal at higher concentrations (Thakre et al. 2021). Piperine showed its anti-fungal activity by affecting the membrane integrity which lead to oxidative stress and apoptosis of *Candida albicans*. The solvent used for extraction of antimicrobial components from black pepper also affects the efficacy of the extract. According to Abdallah and Abdalla (2018), maximum inhibition of 36 mm and 38 mm was achieved for gram negative and gram positive bacteria respectively, when ethanol was used as solvent for extracting antimicrobial components from black pepper. Apart from this, piperine has been reported of enhancing the antimicrobial action of synthetic antibiotics. Piperine and piperlongumine were able to improve the effectiveness of two antibiotics i.e. rifampicin and tetracycline against *Staphylococcus aureus* and also helped in minimizing the adverse effects associated with them (Mgbeahuruike et al. 2019). Maitra and Shilpi (2017), demonstrated that con combining piperine and ciprofloxacin lead to reduction in drug dosage with enhanced inhibition efficacy against *E. coli* and *B. subtillus*.

7.5 ENHANCEMENT OF BIOAVAILABILITY BY PIPERINE

Bioavailability or bioavailability enhancers are the substances which when given in combination with a drug, improves the activity of that drug in several ways which includes increasing the absorption of drug, acting as a receptor for drug, creating target cells more responsive to drugs and

Table 7.2 Bio-Enhancing Activity of Piperine (Dudhatra et al. 2012)

Nutrients	Examples
Amino acids	Lysine, methionine, valine, tryptophan, leucine, isoleucine, phenylalanine, threonine
H_2O soluble vitamins	Vitamin C, Vitamin B (B_1, B_2, B_6, B_{12}), Folic acid, Niacinamide
Fat soluble vitamins	Vitamin A, D, E, K
Minerals	Calcium, Iodine, Iron, Zinc, Magnesium, Potassium, Copper, Manganese, Selenium
Antimicrobial agents	Amoxicillin, ofloxacin, penicillin, erythromycin, ciprofloxacin, norfloxacin, azithromycin, cefixime, ampicillin, fluconazole etc.
Non-steroidal anti-inflammatory drugs (NSARDs)	Diclofenac, ibuprofen, naproxen, ketorolac, indomethacin
Corticosteroids	Dexamethasone, prednisolone, betamethasone

amplifying the drug target interactions. Improvement in bioavailability of the drug helps in lowering the drug dose and frequency which reduces its adverse side effects and can make an expensive drug more accessible. Many evidences have shown the bioavailability character of piperine. It is known that piperine when given in combination enhances the bioavailability of various drugs or xenobiotic agents. Research suggests that there are two different mechanisms which explains the role of piperine as bioenhancer (Dudhatra et al. 2012): (1) It helps in rapid absorption of nutrients and drugs by increasing the amount of blood supply to gastrointestinal tract, preventing the breakdown of drugs by decreasing HCl secretions and increasing emulsification in the gut and certain enzymes which play a role in transport of nutrients to intestinal cells and (2) by inactivating and inhibiting enzymes which participate in the biotransformation of drugs. Action of piperine includes inhibition of p-glycoprotein and cytochrome P450 3A4 which play an important role in metabolism of many drugs. It also helps in absorption by stimulating the transport of amino acids and prevents the production of glucuronic acid hence permitting an active form of drug to enter body. Many studies have shown the bio-enhancing activity of piperine in which it has enhanced the absorption of many nutrients and drugs (Table 7.2).

There are many drugs which have benefitted from the bio-enhancement property of piperine which include cardiovascular drugs, antiprotozoal drugs, antiulcer drugs, anticancer drugs along with the drugs which are used in the treatment of respiratory disorders, central nervous system, allergies, acid reflux etc. The basic aim of using this technology is to enhance the bio efficacy of those drugs which are expensive, toxic and have poor bioavailability.

7.6 TECHNIQUES FOR THE EXTRACTION OF BIOACTIVE COMPOUNDS FROM BLACK PEPPER

Pepper is a naturally occurring herbs and spice plant and contain major bio-active component "piperine." Although, piperine exerts numerous medicinal properties and biological applications, it still has no wide applicability due to pitiable solubility in liquids. To overcome the problem of poor solubility and enhance the bioavailability of piperine in aqueous medium advanced extraction methods are required (Chopra et al. 2016). There are several piperine extraction methods employed till now, few of them are discussed below (Figure 7.4).

7.6.1 Soxhlet and Double Modified Soxhlet Extraction

This is the oldest and most common method for extraction. Initially it was used for the extraction of lipid from various biological sources, but over the time it was adopted for extraction of vital

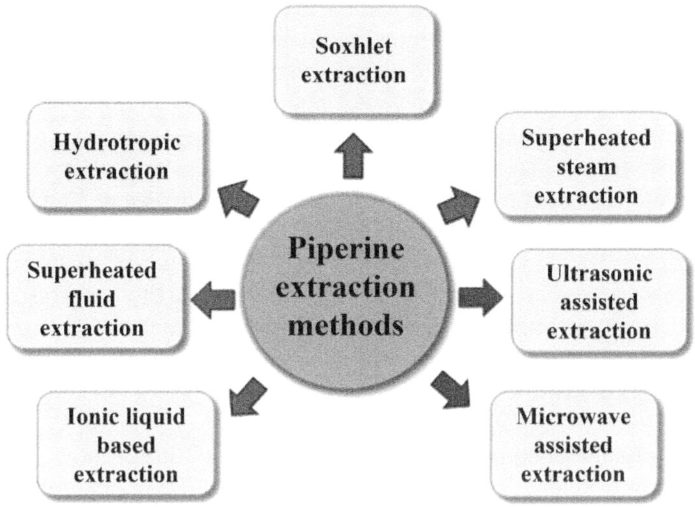

Figure 7.4 Different methods of piperine extraction.

bio-compounds from different sources. Soxhlet extraction method involves keeping the sample in a thimble which is placed in main soxhlet extractor. The extraction solvent is heated to generate vapour and the vapour travel through the extractor which extracts the volatile components from the sample and takes them to the condenser. The process continues and repeats the same cycle while some amount of extracted compound remains in the flask. This cycle run up to several hours to days as to get the desired solute (Gorgani et al. 2017; Azmir et al. 2013).

DBSA (Double –Bypass Soxhlet Apparatus) is a tailored Soxhlet apparatus which has binary side arm instead of single side arm and it is also suggested that DBSA considerably reduces time od extraction and thereby enhancing the extraction cycle. The DBSA was used for the extraction of piperine from *P. nigrum* and a comparison was made with conventional soxhlet method and results showed that DBSA was able to extract 3.9% piperine with cycle time 8 min in 12 hrs, while in the conventional soxhlet method the piperine yield was 3.8% with cycle time 16 min in 22 hrs (Subramanian et al. 2016). Similarly, 9.56 ± 0.83 mg/g of piperine was obtained from the root of *Piper nigrum* using methanol as a solvent in 8 hrs (Rajopadhye et al. 2012). Yamaguchi et al. (2011), extracted piperine from black pepper corns and a production yield of 54.9, 52.7, and 56.6 mg/g of piperine was reported for solvent extraction, soxhlet extraction and supercritical CO_2 extraction respectively DBSA was used for the recovery of piperine in a similar way where 40 gm of raw plant material and 250 ml methanol as solvent was used and the extraction cycle ran up to 24 hrs. The extract was further concentrated in boiling water bath and final yield was estimated (Subramanian et al. 2016).

7.6.2 Superheated Steam Extraction

It is also called steam distillation or hydro-distillation and the most ancient method for extraction of bioactive compounds from its native sources. The extraction mechanism of this method includes mixing of parent material with water in an apparatus, followed by boiling above its boiling temperature to generate steam, which passes through condenser and extract is collected. Steam is designated as superheated when it has been heated above its saturation temperature.

Hydro distillation technique was used for extracting piperine from Vietnam's black pepper (*Piper nigrum*). The extraction parameters were optimized using Response Surface Methodology that included extraction temperature 150°C, water- to material ratio was kept 1:21 gm/ml, with an extraction period of 5.2 hrs gave corresponding yield of 2.45%. GCMS (Gas Chromatography-Mass Spectroscopy) was performed with the crude extract to identify main constituents. A total of 26 major compounds of vital oils were found and the major components included 3-Carene (29.21%), followed by D-limonene (20.94%), β-caryophyllene (15.05%), β-pinene (9.77%) and α-pinene (4.69%) respectively (Tran et al. 2019). Similarly, under optimal hydro distillation conditions, compounds were extracted from Vietnam green pepper seeds and a total 19 key ingredients of crucial oil were identified with GCMS and the main constituents of the essential oil were found to be maximum 35.21% (3-Carene), followed by 21.54% (D-limonene), 10.05% (β-Caryophyllene), 9.17% (β-Pinene), 7.37% (Sabinene), 4.45% (α-Pinene) and 4.09% (Elemene) respectively (Dao et al. 2020). Hydro distillation is a simple, economical and a practical method for large scale production.

7.6.3 Hydrotropic Extraction

Hydrotrophs are organic compounds that increase the solubility of phyto-constituents such as aldehydes, hydrocarbons, ketones, esters, acids and alcohols. These hydrotrophs are extensively used in phytochemical extraction due to their low cost and surface permeability. The hydrotrops are amphipathic compound like surfactants possess both hydrophilic and hydrophobic moiety. The most common hydrotrops such as sodium butyl monoglycol sulphate and sodium alkyl benzene sulfonate are widely used for extracting water insoluble phyto-constituents (Dongre et al. 2011). Hydrotrops interact with cellulose and phospholipids layer of the plant cell wall and penetrate inside the cell interacting with water insoluble constituents, breaking the laminar crystalline structure and allowing them free from the cell structure and hence diffusing them to the external environment. The hydrotrops have gained importance from last few decades for the mining of expensive biological compounds from its aqueous solution (Raman and Gaikar 2002a). A good quality hydrotrope is one that has both aqueous solubility and hydrophobicity and it is the equilibrium between these two opposite attributes that determine the solubilization capacity of hydrotropes (Lee et al. 2003). Hydrotropic solutions are amphiphilic in nature, they interact with the complex biomatrix of plant cell by disrupting plant cell structures and extract hydrophobic components. Hydrotrope solutions interact with the outer pericarp of the pepper and make it permeable and enhance the extraction of piperine (Raman and Gaikar 2002a). Hydrotropes such as sodium butyl monoglycol sulphate and sodium alkayl benzene sulponate were used for the selective isolation of piperine from black pepper (*Piper nigurm*) with 90% purity and free from oleoresin. Raman and Gaikar (2002a) reported that aqueous hydrotopes caused permeabilization of the cell membrane in *Piper nigurm* fruit. The hydrotropes first adsorb to the pericarp of the fruit followed by penetration and disorganizing the cellular architect of the fruit and making the cell permeabilization thus enhancing the extraction of piperine. Subbarao et al. (2012) reported that temperature plays a vital role in hydrotropic extraction. Increase in temperature of extraction lead to more permeability of hydrotropic solution because of more lysis of the plant cells and their crystalline structure, therefore enhancing the extraction of desired products (However, the purity of extract decreases at higher temperatures. The rise in temperature facilitates more cell lysis which leads to decrease in selectivity of the desired product due to undesired oleoresin extraction and other impurities (Mishra and Gaikar 2009). The optimum temperature of hydrotropic extraction of piperine is 30°C, as at this temperature the hydrotrops only weaken the liquid lamellar structure of the cell membrane without affecting cellulose polymer (Raman and Gaikar 2002a).

7.6.4 Supercritical Fluids (Dense CO_2) Extraction

Supercritical Fluids (SCFs) are the substances either liquid or gas when used above their critical temperature and pressure, therefore coexist as liquids and gases. These SCF are now widely used in phyto-extraction to get new and novel plant based vital bio-compounds. This method has gained tremendous interest amongst spice and aromatic crop industries. The Supercritical fluid extraction (SFE) is considered as the rapid, most efficient, selective and unsoiled method for the extraction of plant based organic compounds. The SCFs are excellent medium for reaction and separation; their critical temperature and pressure make them to have liquid like-densities and become excellent career for solutes (Ahluwalia et al. 2013). It was also reported that the piperine solubility increased in dense CO_2 at critical and super critical condition. The enhanced piperine solubility in CO_2 is due to the increase in density at critical and near critical condition (Kumoro et al. 2009). The SCFs are excellent solvents due to their low viscosity, less surface tension and large molecular diffusivity enabling them to allow efficient mass transfer, quick diffusion into sample matrix and selective extraction of compounds of interest. Although, this is an excellent method of extraction in comparison with the conventional methods, but high cost of operation is the only limitation of using this technology (Herrero et al. 2010; Herrero et al. 2006).

The Dense CO_2 is widely used in phyto-extraction of novel products. This method involves in the forcing of supercritical CO_2 (CO_2 under high temperature and pressure) into the ground plant material in the extraction vessel, where it reacts with the plant materials and break down the trichomes which facilitate the dissolution of crystalline matrix of plant materials and extracting essential compounds. The extraction of piperine through supercritical CO_2 was done at optimal setting viz. temperature 1°C, pressure 73.8 bar, extraction time 2–5 hrs, and yield of 81–98% organic compounds recorded (Khaw et al. 2017). Nagavekar and Singhal (2018) extracted pepper oleoresin from *Piper nigrum* using SFE and the optimum conditions were calculated by evaluating the properties of piperine and dense CO_2 using Hilderbrandt solubility equation. Supercritical fluids at 50°C/250 bar/60 min extracted 87.61 and 61.7% of piperine and oleoresin respectively and the extraction was up lifted to 184.7 and 95.9% of piperine and oleoresin respectively by adding 30% ethanol as co-solvent at the similar optimum conditions. Optimization of SCO_2 process was done for piperine extraction from Malabar black pepper using full factorial design (3 level) and optimized conditions achieved were pepper (20 g), mean particle diameter (0.42mm), extraction time (45 min) and flow rate of CO_2 (2 L/min) (Dutta and Bhattacharjee 2016).

7.6.5 Ultrasonic-Assisted Extraction (UAE)

The conventional extraction techniques such as solvent extraction and soxhlet extraction have been widely used in phytochemical extraction since ancient times, but these methods have limitations of high labour, solvent wastage and lengthy extraction times. To conquer these problems, new extraction techniques have been employed and Ultrasonic-Assisted Extraction (UAE) technique is one amongst them. This technique uses ultrasound energy and is based on the principle of thermal effect and acoustic cavitation phenomenon. The UAE generates ultrasonic waves that produce cavitation of small bubbles in the solvent. The sound waves and the collapsing cavitation bubbles induce fragmentation, pore formation, erosion, increased absorption and swelling index in plant's cellular matrix resulting in increased solubilization of bioactive material in solvent due to increase in surface area and mass transfer rate. Hence, cavitational force and thermal effects disintegrates the cell wall and crystalline structure and facilitates the desired compounds to move out to the external environment (Mason et al. 2011). The UAE can generate high pressures equal or greater than 1000 bar and temperature of equal or greater than 5000 K. Combined effect of high pressure and temperature produce high energies and some other Vander wall forces which disintegrate the lamellar structure and helps in better circulation of solvent increasing the efficiency of extraction (Yang and Zhang 2008). UAE has many advantages such higher yield, less extraction time, less laborious, and low solvent wastage and therefore very popular as an extraction technique (Mason et al. 2011). While considering UAE for extraction, certain properties of solvent are very important which include the physico-chemical properties such as surface tension, vapour pressure, viscosity and polarity index (Sun et al. 2011). The piperine yield of 4.35 mg/gm, 4.32 mg/gm and 4.08 mg/gm was extracted from *Piper nigrum* using three different solvents viz. acetone, ethanol and hexane respectively by UAE. The results showed that acetone and ethanol gave very close and similar yield due their close polarity index (Rathod and Rathod 2014). In an advanced version of this method, Gorgani et al. (2017) used sequential microwave UAE for the extraction of piperine from black pepper and effect of various factors on extraction yield was studied. A yield of 46.6 mg piperine/gm pepper was achieved when ethanol was used as solvent at 0.15 mm particle size, 20:1 solvent solid ratio, microwave power of 100 W for one minute and temperature of 50°C for 30 minutes. The extraction yield obtained by this method was higher than the yield obtained by Soxhlet, microwave assisted and ultrasound assisted extraction methods.

7.6.6 Ionic Liquid-Based Extraction

Ionic liquid (IL)-based extraction has gained importance in the last two decades (Angell et al 2012). Ionic liquids are the molten salts, made up of cations and anions having low melting point typically below 100°C. ILs has a number of distinguished properties over conventional solvents viz. high polarity, low melting point, low vapour pressure and high electrochemical stability (Aparicio et al. 2010; Wasserscheid and Welton 2008). The configuration and physio-chemical properties of ionic liquids play a significant role in the extraction of products from its native material and therefore greatly used for extracting bioactive compounds (Cao et al 2009). It has been suggested that in ILs extraction a magical force i.e. hydrophobic interaction acts between the solvent and bio-active compound responsible for the extraction of desired products. For the extraction of piperine, dried black pepper powder was mixed with 1-alkyl-3-methylimidazolium ionic solutions and extraction was done using ultrasonic extraction technique (Tiwari et al. 2020). In a study, piperine was isolated from black pepper by 1-alkyl-3-methyl-imidazoliumchlorides with varying alkyl chain lengths and concluded that the extraction efficiency of the process depends on critical micelle concentration of the ionic liquids (Ressmann et al 2013). The same mechanism was proposed by some other researchers while extracting different plant based vital compounds viz. tannin (Chowdhury et al. 2010), rutin and quercetine (Wu et al. 2012) and curcumin (Xu et al. 2015).

7.6.7 Microwave-Assisted Extraction (MAE)

MAE is an extraction method that combines both microwave radiation technology with conventional solvent extraction method. The microwaves are electromagnetic waves possessing dual properties viz. electric and magnetic and both the fields lies perpendicular to each other. Microwave radiations are basically used to internally heat the solvent and plant materials. The microwaves are low wave length (1 m -1 mm) and high frequency (300 MHz to 300 GHz) and lies between infrared and radio frequencies. Currently, microwave radiation is gaining interest due to its exponential applications in various field including extraction of bio-active compounds from plants and their products. Some recent studies have also shown the popularity of MAE for the extraction of bio-active compounds from various types of botanical resources. The wave energy generated by MAE is transformed to heat energy which is absorbed by cell matrix causing evaporation of volatile organic compounds from the matrix following recovery by condensation process (Chan et al. 2011). This is a selective extraction technique which rely on selective heating of bioactive compound based on its dielectric constant. The wave energy converted to heat either by its ionic conduction or dipole rotation. Any rise in temperature and pressure cause absorption of high amount of wave energy by extraction solvent accelerating MAE (Wang et al. 2008). It is reported that the MAE results in enhanced extraction of desired products and the high achievement of yield in MAE is because of heat and mass transfer mechanisms occurring in same direction and same timings within the material (Mason et al. 2011). A distinguished factor of MAE is its volumetric heating; the produced heat is evenly distributed throughout the sample and deposits energy within them. It was also reported that the hydration of sample before MAE results in more extraction of the desired product. In a study it was shown that the piperine extraction was increased by 6 folds when the pepper soaked in water for 4 hrs prior to MAE (Raman and Gaikar 2002b). Similarly, various extraction solvents viz. ethanol, toluene, heptanes, dichloromethane and petroleum ether were evaluated for the extraction of piperine from *P. nigrum* by MAE. They found that the nonpolar solvent viz. petroleum ether, yield highest extraction efficiency with actual piperine content of 94% along with purity of 85%. In contrast, the piperine yield was merely 75% and 80%, with purity of 60% and 72% using non polar solvent ethanol and dichloromethane respectively and it was concluded that enhanced yield can be achieved by using nonpolar solvents for extraction (Raman and Gaiker 2002b). Also, the nonpolar solvents are neutral and do not cause any hindrance, therefore the microwave dissemination is high in these solvents favouring high extraction. While with the polar solvent the case is reverse due to the ionizing potential the microwave dissemination is reduced, hence the yield is low. The two important factors viz. microwave power and extraction temperature play crucial role in MAE. The increase in microwave power results in more solvent loss. It was reported that using high microwave radiation to about 300 and 400 W increased the solvent loss by 16 and 20% respectively. Similarly, the solvent loss is reduced to 8% by just dropping the power level to 150 W (Raman and Gaikar 2002b). Therefore, these two parameters i.e. temperature and microwave power are interlinked variables. It was also seen that rise in microwave power raised the extraction temperature and the increased temperature enhanced the solvent power due to decrease in surface tension and viscosity (Mandal et al. 2007).

In one such study, microwave reflux extraction method was used and effect of four extraction parameters (irradiation time, feed particle size, microwave power level, feed solvent ratio) on piperine yield was studied and reported irradiation time plays a significant role in the extraction process and should be optimized to prevent degradation of bioactive compound of interest (Olalere et al. 2018).

7.7 IDENTIFICATION/DETECTION OF BIOACTIVE COMPOUNDS

Various quantitative and analytical techniques have been used in literature for the detection of piperine after its extraction from black pepper by the methods discussed above. Kjeldahl method is one of the oldest technique used to indirectly detect piperine through the evaluation of total nitrogen in black pepper. As it measures the total nitrogen which include all the nitrogenous compounds such as cavicine, piperittine and some free amino acids, the experimental value of piperine is generally higher than the actual amount. Several other methods are also used for detection of bioactive compounds from black pepper such as high-performance liquid chromatography (HPLC), liquid chromatography mass spectrometry (LCMS), gas chromatography mass spectrometry (GCMS) along with many colorimetric techniques; UV visible spectrometry being the most common approach (Gorgani et al. 2017). Colorimetric method generally involves hydrolysis of methylenedioxy group of piperine by a chromotropic acid which results in the formation of formaldehyde giving a definite colour (Shityakov et al., 2019).

Alyaseen et al. (2018) isolated piperine and identification was done using thin layer chromatography method. Ethanol and hexane (7:3) were used as mobile phase, silica gel G as stationary phase and dragendoff's reagent was used as spray agent. Retention factor (Rf) value of 0.23 was obtained which was close to standard Rf value of piperine of 0.25 given in literature.

HPLC has been the method of choice because of its higher sensitivity compared to UV visible technique. Ajmal (2018) developed a simple and rapid method for the quantitative determination of piperine from *P. nigrum*. The analysis was carried out in RP-HPLC with C18 column using acetonitrile and aqueous acetic acid as mobile phase and quantification was done using UV detector at 343 nm. In another study by Rathapon et al. (2020), piperine content of 85.1 mg/g was obtained using HPLC technique in gradient phase using acetonitrile and water as mobile phase and diode array detector was set at 340 nm. In a research by Rodriguez et al. (2021) HPLC method (reversed phase) was used to quantify piperine with multi wavelength detection at 150 mm × 4.6 mm I.D. C18 column using acetronitrile and acetic acid as mobile phase and this method was able to recover 86.1% of piperine. High performance thin layer chromatography technique which is improved version of TLC was used in another study for the extraction of piperine in herbal formulations (Vyas et al. 2013).

It has been reported that gas chromatography method has an upper edge in separating and identifying alkaloids present in black pepper and also evaluates the degradation of piperine.

In one such study, gas chromatography (GCMS) technique for the detection of piperine (Juan et al. 2020). Helium gas at a flow rate of 1ml/min and quadrupole triple axis detector with electronic impact ionization was used while mass spectra was obtained at 70 eV and identification of compounds was done using mass spectra database and library NIST MS search software. In a recent research, three techniques i.e. GCMS, LCMS and DART MS (Direct Analysis in Real Time Mass Spectrometry) were used for the detection of piperine from black pepper and confirmed DART-MS as a simple and high throughput technique (Jian et al. 2021). Basically DART-MS is the latest mass spectrometry technique that doesn't require extractions and chromatographic separations and able to identify organic compounds in a variety of matrices. It has an atmospheric pressure ion source attached with mass spectrometer which ionizes solids, liquids and gases instantly under ambient conditions. It is rapid method and offers a high speed alternative to LCMS and GCMS techniques.

While there are numerous methods for detecting piperine from black pepper, the chemical characterization of piperine can be achieved by techniques like nuclear magnetic resonance spectroscopy (NMR) and Fourier transform infra-red technique (FTIR).

7.8 CONCLUSION

The history of using natural products specially spices and herbs for culinary and medicinal use is well-established. Among spices, black pepper holds an important position due to its characteristics flavour and pungency. It is an indispensable ingredient in many food preparations and have numerous pharmacological applications. Black pepper consists of 5-9% piperine which is the major component responsible for the pungency of black pepper. Piperine is known to enhance the bioavailability of various drugs such as ibuprofen, diclofenac, naproxen, omeprazole etc. It has been reported to have spectrum of therapeutic benefits and has been used for the treatment of diseases like cancer, inflammation, asthma, depression and many other diseases. Using a combination of piperine with medicinal drugs helped in increasing the bioavailability and reducing the dose thereby decreasing the side effects and toxicity of the drugs. However, the limitation of low solubility of piperine in aqueous environments limits its pharmacological activities which emphasizes the need for the development of new techniques and methods which are sustainable and environment friendly. Supercritical CO_2 extraction, microwave and ultrasound assisted extraction are some of the advanced techniques used nowadays for the effective separation of piperine from black pepper; though all these techniques have their own benefits and limitations. The future perspective should be to study the mechanism by which the bioavailability of other drugs can be enhanced so that new and advanced formulations can be developed with targeted delivery.

REFERENCES

Abdallah, E.M. and Abdalla, W.E. 2018. Black pepper fruit (*Piper nigrum* L.) as antibacterial agent: A mini-review. *Journal of Bacteriology and Mycology* 6(2): 141–145. doi: 10.15406/jbmoa.2018.06.00192

Agbor, G.A., Vinsonb, J.A., Obenc, J.E. and Ngogangd, J.Y. 2006. Comparative analysis of the in vitro antioxidant activity of white and black pepper. *Nutrition Research* 26(12): 659–663. doi: 10.1016/j.nutres.2006.09.019

Ahluwalia, S., Shivhare, U.S. and Basu, S. 2013. Supercritical CO2 extraction of compounds with antioxidant activity from fruits and vegetables waste—A Review. *Focusing on Modern Food Industry* 2: 43–62.

Ajmal, H. 2018. Isolation, identification and quantitative analysis of piperine from *Piper nigrum* linn. of various regions of Kerala by RP-HPLC method. *World Journal of Pharmacy and Pharmaceutical Sciences* 7(6): 1023–1049.

Alkadi, H. 2020. A Review on free radicals and antioxidants. *Infectious disorders - Drug targets* 20(1): 16–26. doi: 10.2174/1871526518666180628124323

Alyaseen, F., Hassan, B., Abdulhussein, H. and Alsafee, B. 2018. Extraction, isolation and chemical identification of piperine alkaloid from black pepper seeds and its antibacterial activity. *Plant Archives* 18(2): 2171–2176.

Angell, C. A., Ansari, Y. and Zhao, Z. 2012. Ionic liquids: past, present, and future. *Faraday Discuss* 154: 9–27.

Aparicio, S., Atilhan, M. and Karadas, F. 2010. Thermophysical properties of pureionic liquids: review of present situation. *Industrial and Engineering Chemistry Research* 49: 9580–9595.

Atal, S., Atal, S., Vyas, S. and Phadnis, P. 2016. Bio-enhancing effect of piperine with metformin on lowering blood glucose level in alloxan induced diabetic mice. *Pharmacognosy Research* 8: 56–60.

Azmir, J., Zaidul, I., Rahman, M. et al. 2013. Techniques for extraction of bioactive compounds from plant materials: a review. *Journal of Food Engineering* 117: 426–436.

Banerjee, S., Katiyar, P., Kumar, L. et al. 2021. Black pepper prevents anemia of inflammation by inhibiting hepcidin over-expression through BMP6-SMAD1/IL6-STAT3 signaling pathway. *Free Radical Biology and Medicine* 168: 189–202. https://doi.org/10.1016/j.freeradbiomed.2021.03.019.

Bang, J.S., Oh, D.H., Choi, H.M. et al. 2009. Anti-inflammatory and antiarthritic effects of piperine in human interleukin 1β-stimulated fibroblast-like synoviocytes and in rat arthritis models, *Arthritis Res Ther* 11(2): R49. Published online 2009 Mar 30. doi: 10.1186/ar2662

Cao, X., Ye, X., Lu, Y. et al. 2009. Ionic liquid-based ultrasonic-assisted extraction of piperine from white pepper. *Analytica Chimica Acta* 640: 47–51.

Carp, O.E., Moraru, A., Pinteala, M. and Arvinte, A. 2021. Electrochemical behaviour of piperine. Comparison with control antioxidants. *Food Chemistry* 339: 128110. https://doi.org/10.1016/j.foodchem.2020.128110.

Chan, C. H., Yusoff, R., Ngoh, G. C. et al. 2011. Microwave-assisted extractions of active ingredients from plants. *Journal of Chromatography A* 1218: 6213–6225.

Chen, W., Tang, H. Jiang, N. et al. 2019. Antibacterial effect of black pepper petroleum ether extract against *Listeria monocytogenes* and *Salmonella typhimurium*. *Journal of Food Quality* 1–10 pages. https://doi.org/10.1155/2019/2356161

Chinta, G., Coumar, M. S. and Periyasamy, L. 2017. Reversible testicular toxicity of Piperine on male albino rats. *Pharmacognosy magazine* 13(Suppl3): 525–532. https://doi.org/10.4103/pm.pm_405_16

Chopra, B., Dhingra, A.K., Kapoor, R.P. and Prasad, D.N. 2016. Piperine and its various physicochemical and biological aspects: A Review. *Open Chemistry Journal* 3: 75–96.

Chowdhury, S.A., Vijayaraghavan, R. and MacFarl, D. 2010. Distillable ionic liquid extraction of tannins from plant materials. *Green Chemistry* 12: 1023–1028.

Dao, T.P., Tran, T. H., Nhan, N.P.T. et al. 2020. Optimization of essential oil yield from Vietnamese green pepper (*Piper nigrum*) using hydro-distillation method. *Materials Science and Engineering* 736: 022039. doi: 10.1088/1757-899X/736/2/022039

Dongre, P., Kannur, D.M., Kosambiya, V. et al. 2011. Significant role of hydrotropes in extraction of phytoconstituents – a review. *International Journal Pharmaceutical Science and Research* 2(4): 730–734. http://dx.doi.org/10.13040/IJPSR.0975-8232.2

Dosoky, N.S., Satyal, P., Barata, L.M., da Silva, J.K.R. and Setzer, W.N. 2019. Volatiles of black pepper fruits (*Piper nigrum* L.). *Molecules* 24(23): 42–44. doi:10.3390/molecules24234244

Doucette, C.D., Hilchie, A.L., Liwski, R. and Hoskin, D.W. 2013. Piperine, A dietary phytochemical, inhibits angiogenesis. *Journal of Nutrition and Biochemistry* 24(1): 231–239. doi: 10.1016/j.jnutbio.2012.05.009 22902327

Dudhatra, G.B., Mody, S.K., Awale, M.M. et al. 2012. A comprehensive review on pharmacotherapeutics of herbal bioenhancers. *Scientific World Journal* 637953. doi: 10.1100/2012/637953.

Dutta, S. and Bhattacharjee, P. 2016. Modeling of supercritical carbon dioxide extraction of piperine from Malabar black pepper. *Materials Today: Proceedings* 3(10): 3238–3252. https://doi.org/10.1016/j.matpr.2016.10.005.

Feng, Y., Dunshea, F.R. and Suleria, H.A.R. 2020. LC-ESI-QTOF/MS characterization of bioactive compounds from black spices and their potential antioxidant activities, *Journal of Food Science and Technology* 57: 4671–4687.

Gorgani, L., Mohammadi, M., Najafpour, G.D. and Nikzad, M. 2017. Piperine-The bioactive compound of black pepper: from isolation to medicinal formulations. *Comprehensive Reviews in Food Science and Food Safety* 16: 124–140.

Gorgani, L., Mohammadi, M., Najafpour, G.D. *et al.* 2017. Sequential microwave-ultrasound-assisted extraction for isolation of Piperine from black pepper (*Piper nigrum* L.). *Food Bioprocess Technology* 10: 2199–2207. https://doi.org/10.1007/s11947-017-1994-0

Herrero, M., Cifuentes, A. and Ibanez, E. 2006. Sub-and supercritical fluid extraction of functional ingredients from different natural sources: plants, food-by-products, algae and microalgae: A review. *Food Chemistry* 98: 136–148.

Herrero, M., Mendiola, J.A., Cifuentes, A. et al. 2010. Supercritical fluid extraction: recent advances and applications. *Journal of Chromatography A* 1217: 2495–2511.

Islam, N., Irfan, M., Hussain, T. et al. 2021. Piperine phytosomes for bioavailability enhancement of domperidone. *Journal of Liposome Research* 4: 1–9. doi: 10.1080/08982104.2021.1918153

Jafri, A., Siddiqui, S., Rais, J. et al. 2019. Induction of apoptosis by piperine in human cervical adenocarcinoma via ROS mediated mitochondrial pathway and caspase-3 activation. *EXCLI Journal* 18: 154–164. https://doi.org/10.17179/excli2018-1928

Jaisin, Y., Ratanachamnong, P., Wongsawatkul, O., Watthammawut, A., Malaniyom, K. and Natewong, S. 2020. Antioxidant and anti-inflammatory effects of piperine on UV-B-irradiated human HaCaT keratinocyte cells. *Life Sciences* 263: 118607. https://doi.org/10.1016/j.lfs.2020.118607.

Jian, L., Jianghao, S., Pei, C., Jared, F., Virginia, B., Mengliang, Z. 2021. Chemical analysis and classification of black pepper (*Piper nigrum* L.) based on their country of origin using mass spectrometric methods and chemometrics. *Food Research International* 140: 109877. https://doi.org/10.1016/j.foodres.2020.109877.

Joardar, N., Shit, P., Halder, S. et al. 2021. Disruption of redox homeostasis with synchronized activation of apoptosis highlights the antifilarial efficacy of novel piperine derivatives: An in vitro mechanistic approach. *Free Radical Biology and Medicine* 169: 343–360. https://doi.org/10.1016/j.freeradbiomed.2021.04.026.

Joshi, D.R., Shrestha, A. and Adhikari, N. 2018. A review on diversified use of the king of spices: *Piper nigrum* (black pepper). *International Journal of Pharmaceutical Sciences and Research* 9(10): 4089–4101.

Juan, V.M., Genesis, B.C., Juan, G.V., Julia, G.G., Erique, R.C. and Jorge, M.T. 2020. Bioautography and GC-MS based identification of piperine and trichostachine as the active quorum quenching compounds in black pepper. *Heliyon* 6(1): e03137. doi: 10.1016/j.heliyon.2019.e03137

Kaho, Z.M., Kadum, A. and Hadi, A.A. 2019. Evaluation of antibacterial activity of *Piper nigrum* extract against *Streptococcus mutans* and *Escherichia coli*. *Journal of Pharmaceutical Sciences and Research* 11(2): 367–370.

Kharbanda, C., Alam, M.S., Hamid, H. et al. 2016. Novel piperine derivatives with antidiabetic effect as PPAR-γ Agonists. *Chemical Biology and Drug Design* 88: 354–362. https://doi.org/10.1111/cbdd.12760

Khaw, K.Y., Parat, M. O., Shaw, P.N. and Falconer, J.R. 2017. Solvent supercritical fluid technologies to extract bioactive compounds from natural sources: A review. *Molecules* 22: 1186. https://doi.org/10.3390/molecules22071186.

Kim, S.H. and Lee, Y.C. 2009. Piperine inhibits eosinophil infiltration and airway hyper responsiveness by suppressing T cell activity and Th2 cytokine production in the ovalbumin-induced asthma model. *Journal of Pharmacy and Pharmacology* 61(3): 353–359. doi: 10.1211/jpp/61.03.0010

Kumoro, A., Singh, H. and Hasan, M. 2009. Solubility of piperine in supercritical and near critical carbon dioxide. *Chines Journal of Chemical Engineering* 17: 1014–1020.

Kunnumakkara, J.B., Rana, V., Parama, D. et al. 2021. COVID-19, cytokines, inflammation, and spices: How are they related?. *Life Sciences* 119201. doi: 10.1016/j.lfs.2021.119201

Lai, L.H., Fu, Q.H., Liu, Y. et al. 2012. Piperine suppresses tumor growth and metastasis in vitro and in vivo in a 4T1 murine breast cancer model. *Acta Pharmaceutica Sinica* 33(4): 523–30. doi: 10.1038/aps.2011.209

Lee, J., Lee, S.C., Acharya, G. et al. 2003. Hydrotropic solubilization of paclitaxel: analysis of chemical structures for hydrotropicproperty. *Pharmaceutical Research* 20: 1022–1030.

Lee, J.G., Chae, Y., Shin, Y. and Kim, Y.J. 2020. Chemical composition and antioxidant capacity of black pepper pericarp. *The Korean Society of Applied Biological Chemistry* 63: 35. doi: https://doi.org/10.1186/s13765-020-00521-1

Liang, Y.D., Bai, W.J., Li, C.G. et al. 2016. Piperine suppresses pyroptosis and interleukin-1β release upon ATP triggering and bacterial infection. *Frontiers in Pharmacology* 7: 390. doi: 10.3389/fphar.2016.00390

Liu, Z., Hu, Q., Wang, W. et al. 2020. Natural product piperine alleviates experimental allergic encephalomyelitis in mice by targeting dihydroorotate dehydrogenase. *Biochemical Pharmacology* 177: 114000. https://doi.org/10.1016/j.bcp.2020.114000.

Maitra, J. and Shilpi 2017. Synergistic effect of Piperine, extracted from *Piper nigrum* with Ciprofloxacin on *Escherichia coli*, *Bacillus subtilis*. *Der Pharmacia Sinica* 8(3): 29–34.

Mandal, V. Mohan, Y. and Hemalatha, S. 2007. Microwave assisted extraction—aninnovative and promising extraction tool for medicinal plant research. *Pharmacogonosy Review* 1: 7–18.

Mason, T. Chemat, F. and Vinatoru, M. 2011. The extraction of natural products using ultrasound or microwaves. *Current Organic Chemistry* 15: 237–247.

Mgbeahuruike, E.E., Stalnacke, M., Vuorela, H. and Holm, Y. 2019. Antimicrobial and synergistic effects of commercial Piperine and Piperlongumine in combination with conventional antimicrobials. *Antibiotics (Basel)* 4; 8(2): 55. doi: 10.3390/antibiotics8020055

Mishra, S.P. and Gaikar, V.G. 2009. Hydrotropic extraction process for recovery of forskolin from Coleus forskohlii roots. *Industrial and Engineering Chemistry Research* 48: 8083–8090.

Mohammadi, M., Najafi, H., Yarijani, Z.M., Vaezi, G. and Hojati, V. 2020. Protective effect of piperine in ischemia-reperfusion induced acute kidney injury through inhibition of inflammation and oxidative stress. *Journal of Traditional and Complementary Medicine* 10(6): 570–576. https://doi.org/10.1016/j.jtcme.2019.07.002.

Nagavekar, N. and Singhal, R.S. 2018. Enhanced extraction of oleoresin from *Piper nigrum* by supercritical carbon dioxide using ethanol as a co-solvent and its bioactivity profile. *Journal of Food Process Engineering* 41(5): e12670. doi: https://doi.org/10.1111/jfpe.12670

Nimse, S.B. and Pal, D.K. 2015. Free radicals, natural antioxidants, and their reaction mechanisms, *RSC Advances* 5: 27986–28006.

Olalere, O. A, Abdurahman, N. H. Yunus, Rosli, M. O. et al. 2018. Evaluation of optimization parameters in microwave reflux extraction of piperine-oleoresin from black pepper (Piper nigrum). *Journal of Basic and Applied Sciences* 7(4): 626–631. https://doi.org/10.1016/j.bjbas.2018.07.006.

Ouyang, D.Y., Zeng, L.H., Pan, H. et al. 2013. Piperine inhibits the proliferation of human prostate cancer cells via induction of cell cycle arrest and autophagy. *Food and Chemical Toxicology* 60: 424–430.

Peng, X., Yang, T., Liu, G., Liu, H., Peng, Y. and He, L. 2018. Piperine ameliorated lupus nephritis by targeting AMPK-mediated activation of NLRP3 inflammasome. *International Journal of Immunopharmacology* 65: 448–457.

Peterson, B., Weyers, M., Steenekamp, J.H., Steyn, J.D., Gouws, C. and Hamman, J.H. 2019. Drug bioavailability enhancing agents of natural origin (Bioenhancers) that modulate drug membrane permeation and pre-systemic metabolism. *Pharmaceutics* 11(1): 33. doi: 10.3390/pharmaceutics11010033

Priya, A. and Pandian, S.K. 2020. Piperine impedes biofilm formation and hyphal morphogenesis of *Candida albicans*. *Frontiers in Microbiology* 13(11): 756. doi: 10.3389/fmicb.2020.00756

Qin, B., Yang, K. and Cao, K. 2020. Synthesis and anti-oxidative activity of piperine derivatives containing phenolic hydroxyl. *Journal of Chemistry* 2020: 1–9. https://doi.org/10.1155/2020/2786359

Rajopadhye, A.A, Namjoshi, T.P. and Upadhye, A.S. 2012. Rapid validated HPTLC method for estimation of piperine and pipelongumine in root of *Piper longum* extract and its commercial formulation. *Brazilian Journal of Pharmacognosy* 22: 1355–1361.

Raman, G. and Gaikar, V.G. 2002a. Extraction of piperine from *Piper nigrum* (black pepper) by hydrotropic solubilization. *Industrial and Engineering Chemistry Research* 41: 2966–2976.

Raman, G. and Gaikar, V.G. 2002b. Microwave-assisted extraction of piperine from *Piper nigrum*. *Industrial Engineering Chemistry Research* 41: 2521–2528.

Rathapon, A., Papawee, S., Sukitaya, V., Asira, F. and Arunporn, I. 2020. Application of film-forming solution as a transdermal delivery system of piperine-rich herbal mixture extract for anti-inflammation. *Heliyon* 6(6): e04139. https://doi.org/10.1016/j.heliyon.2020.e04139.

Rather, R. A. and Bhagat, M. 2018. Cancer chemoprevention and piperine: Molecular mechanisms and therapeutic opportunities. *Frontiers in Cell and Developmental Biology* 6: 10. https://doi.org/10.3389/fcell.2018.00010

Rathod, S.S. and Rathod, V.K. 2014. Extraction of piperine from *Piper longum* using ultrasound. *Industrial Crops and Products* 58: 259–264.

Ratish, M.A.V. and Roy, S. 2019. Pepper cultivation in Kerala. *International Journal of Advanced Research in Commerce, Management & Social Science* 02(2): 35–42.

Reena, S.R., Sundar, A.S.W., Sandhya, S.M., Dhanya, S. and Gopal, L. 2018. A review on free radicals and antioxidants. *Indo American Journal of Pharmaceuticals sciences* 5(11): 11541–11548.

Ressmann, A.K., Zirbs, R., Pressler, M. et al. 2013. Surface-active ionic liquids for micellar extraction of Piperine from black pepper. *Z. Naturforsch* 68b: 1129–1137. doi: 10.5560/ZNB.2013-3196

Rodriguez, E.L., Zhang, C., Woolfork, A.G. et al. 2021. Analysis of curcumin and piperine in biological samples by reversed-phase liquid chromatography with multi-wavelength detection. *Journal of Chromatography B Analytical Technologies in Biomedical and Life Sciences* 1: 1162: 122487. doi: 10.1016/j.jchromb.2020.122487

Roshanbakhsh, H., Salmani, M.E., Dehghan, S., Nazari, A., Javan, M. and Pourabdolhossein, F. 2020. Piperine ameliorated memory impairment and myelin damage in lysolecethin induced hippocampal demyelination. *Life Sciences* 253: 117671. https://doi.org/10.1016/j.lfs.2020.117671.

Sarfraz, M., Khaliq, T., Hafizur, R.M., Raza, S.A. and Ullah, H. 2021. Effect of black pepper, turmeric and ajwa date on the endocrine pancreas of the experimentally induced diabetes in wister albino rats: A histological and immunohistochemical study. *Endocrine and Metabolic Science* 4: 100098. https://doi.org/10.1016/j.endmts.2021.100098.

Shan, Y., Xueshibojie, L., Duo, Y., Changyong, E. and Jinghui, Y. 2020. Piperine protects LPS-induced mastitis by inhibiting inflammatory response. *International Immunopharmacology* 87: 106804. https://doi.org/10.1016/j.intimp.2020.106804.

Shango, A.J., Majubwa, R.O. and Maerere, A.P. 2021. Morphological characterization and yield of pepper (*Piper nigrum* L.) types grown in Morogoro District, Tanzania. *CABI Agriculture and Bioscience* 2: 6. doi.org/10.1186/s43170-021-00028-9

Shityakov, S., Bigdelian, E., Hussein, A. A. et al. 2019. Phytochemical and pharmacological attributes of piperine: A bioactive ingredient of black pepper. *European Journal of Medicinal Chemistry* 176: 149–161. 10.1016/j.ejmech.2019.04.002.

Srinivas, B.K., Selvaraj, J., Priya, V.V. and Gayathri, R., 2020. Study on the effect of piperine on carbohydrate metabolic enzymes on experimental diabetic rats. *Drug Invetion Today* 14(3): 76–79.

Srinivasan, K. 2007. Black pepper and its pungent principle- piperine: A Review of Diverse Physiological Effects. *Critical Reviews in Food Science and Nutrition* 47(8): 735–748.

Subbarao, C.V., Chakravarthy, I.K. and Bharadwaj, A. 2012. Functions of hydrotropes in solutions. *Chemical Engineering and Technology* 35: 225–237.

Subramanian, R., Subbramaniyan, P. and Ameen, J.N. et al. 2016. Double bypasses soxhlet apparatus for extraction of piperine from *Piper nigrum. Arabian Journal of Chemistry* 9: S537– S540. https://doi.org/10.1016/j.arabjc.2011.06.022.

Sun, Y., Liu, D., Chen, J. et al. 2011. Effects of different factors of ultrasound treatment on the extraction yield of the all-trans-β-carotene from citrus peels. *Ultrasonic Sonochemistry* 18: 243–249.

Swathy, J.S., Prabhakar, M., John, T., Amitava, M. and Natarajan, C. 2018. Antimicrobial potency of high-energy emulsified black pepper oil nanoemulsion against aquaculture pathogen. *Aquaculture* 491: 210–220. https://doi.org/10.1016/j.aquaculture.2018.03.045.

Thakre, A., Jadhav, V., Kazi, R. et al. 2021. Oxidative stress induced by piperine leads to apoptosis in *Candida albicans. Medical Mycology* 59(4): 366–378. https://doi.org/10.1093/mmy/myaa058

Thao, D.T., Minh, L.N., Anh, T.T.M., Thi, N.N., Hue, P.T.K. and Van, K.P. 2021. The improved anticancer activities of piperine nanoliposome conjugated CD133 monoclonal antibody against NTERA-2 cancer stem cells. *Natural Product Communications*. doi: 10.1177/1934578X21998184

Thi, T.B., Chun, H.P., Eunjin, H. et al. 2019. The protective role of *Piper nigrum* fruit extract in an ovalbumin-induced allergic rhinitis by targeting of NFκBp65 and STAT3 signalings. *Biomedicine & Pharmacotherapy* 109: 1915–1923. https://doi.org/10.1016/j.biopha.2018.11.073.

Tiwari, A., Modi, S.J., Gabhe, S.Y. and Kulkarni, V.M. 2021. Evaluation of piperine against cancer stem cells (CSCs) of hepatocellular carcinoma: Insights into epithelial-mesenchymal transition (EMT). *Bioorganic Chemistry* 110: 104776. https://doi.org/10.1016/j.bioorg.2021.104776.

Tiwari, A., Mahadik, K.R. and Gabhe, S.Y. 2020. Piperine: A comprehensive review of methods of isolation, purification, and biological properties. *Medicine in Drug Discovery* 7: 100027. https://doi.org/10.1016/j.medidd.2020.100027.

Tran, T.H., Ha, L.K., Nguyen, D.C. et al. 2019. The study on extraction process and analysis of components in essential oils of black pepper (*Piper nigrum* L.) seeds harvested in Gia Lai Province, Vietnam. *Processes* 7: 56. doi: 10.3390/pr7020056

Vyas, J., Itankar, P., Tauqeer, M., Kelkar, A., Agrawal, M. 2013. Development of HPTLC method for estimation of piperine, guggulsterone E and Z in polyherbal formulation. *Pharmacognosy Journal* 5: 259–264.

Wang, Y., Yao, Y., Liu, J. et al. 2020. Synthesis and biological activity of Piperine derivatives as potential *PPARγ* agonists. *Drug design, development and Therapy* 14: 2069–2078. https://doi.org/10.2147/DDDT.S238245

Wang, Y., You, J., Yu, Y. et al. 2008. Analysis of ginsenosides in Panax ginseng in high pressure microwave-assisted extraction. *Food Chemistry* 110: 161–167.

Wasserscheid, P. and Welton, T. 2008. *Ionic Liquids in Synthesis.* Wiley Online Library.

Wilson, L. 2016. Spices and Flavoring Crops: Fruits and Seeds. *Encyclopaedia of Food and Health*, 73–83. doi: 10.1016/b978-0-12-384947-2.00647-4

Wu, H., Chen, M. and Fan, Y. 2012. Determination of rutin and quercetin in Chinese herbal medicine by ionic liquid-based pressurized liquid extraction–liquid chromatography–chemiluminescence detection. *Talanta* 88: 222–229.

Xu, D. P., Li, Y., Meng, X. et al. 2017. Natural antioxidants in foods and medicinal plants: Extraction, assessment and resources. *International Journal of Molecular Sciences* 18(1): 96. https://doi.org/10.3390/ijms18010096

Xu, J., Wang, W., Liang, H. et al. 2015. Optimization of ionic liquid-based ultrasonic-assisted extraction of antioxidant compounds from *Curcuma longa* L. using Response Surface Methodology. *Industrial Crops and Products* 76: 487–493.

Yafang, D., Yuzhong, C., Xiaoya, F. et al. 2020. Effects of piperine on lipid metabolism in high-fat diet induced obese mice. *Journal of Functional Foods* 71: 104011. https://doi.org/10.1016/j.jff.2020.104011.

Yamaguchi, T. Kamezawa, K., Iwaya, K. et al. 2011. Analysis of piperine in peppers using on-line SFE-UHPLC with photodiode array detection. *American Laboratory* 43: 29–31.

Yang, Y. and Zhang, F. 2008. Ultrasound-assisted extraction of rutin and quercetin from *Euonymus alatus* (Thunb.) Sieb. *Ultrasonic Sonochemistry* 15: 308–313.

Zeng, Y. and Zeng, Y. 2018. Piperine depresses the migration progression via downregulating the Akt/mTOR/MMP-9 signaling pathway in DU145 cells. *Molecular Medicine Reports* 17: 6363–6370. https://doi.org/10.3892/mmr.2018.8653

Zorica, S.R., Milica, P., Marina, D. et al. 2019. Piperine-A major principle of black pepper: A review of its bioactivity and studies. *Applied Sciences* 9: 4270. doi: 10.3390/app9204270

CHAPTER 8

Bioactive Compounds in Clove

Poornima, Aziza Begum, Mukul Sain, and Amritpal Singh

CONTENTS

DOI: 10.1201/9781003215387-8

8.1 INTRODUCTION

Clove (*Syzygium aromaticum*) is a spice that is famous for its distinct aromatic scent and health benefits. The word Clove is derived from the Latin word "*clavus*," which means nail because the Clove bud's shaft and head resemble a nail. Clove resides in the family of "Myrtaceae," which is inherent to the eastern part of Indonesia, in particular to volcanic islands of the North Moluccas, also known as the spice island. It was first used in 200 BCE as a breath freshener by the Chinese courtier (Han dynasty), and later in the early 4th century CE, the Clove trade was spread to the Middle East, and then it reached Europe in the Middle East Ages. Later, in the 18th century, it was first brought to Mauritius by Pierre Poivre (French administrator) from Indonesia and then by an Arab trader to Zanzibar, which was further widespread to different countries like Madagascar, Zanzibar, India, Sri Lanka, Malaysia, and Mediterranean parts like Turkey (Worrall 2012, Gosavi et al. 2020 and Dey and Mukherjee 2021). The world's largest Clove producer is Indonesia (produced around 123,399 metric tons in 2018 with a cultivation area of 561,200 hectares), contributing approximately 73.7% to the global market, followed by Madagascar, Zanzibar, India, and Sri Lanka (FAOSTAT 2020). Furthermore, in Indonesia, out 90% of the total harvest is purchased and used to manufacture Kreteks by Indonesia Tobacco Industry. At present, the Clove market is predicted to reach 3.1% CAGR during 2021–2026, which can be impacted by COVID-19 trade restrictions (Mordor Intelligence 2020).

The Clove tree can be grown ideally in a tropical environment (200 m above sea level) with high humidity and sunlight exposure at a temperature ranging between 18 and 27°C in rich and loamy soil and starts flowering only after 6–7 years of plantation. This tree of Clove is evergreen, medium-sized, and can grow up to a height of 20 m with an average life span of up to 100 years. The leaves are generally pink in colour and turn to dark green when matured; they are very aromatic, single, opposite with oval, lanceo, or oval-lanceolate shape, which varies according to the variety of trees (Nurdjannah and Bermawie 2012). The number of flowers usually ranges 15–50 with inflorescences pattern, has a hermaphrodite shape with fleshy hypanthium surrounded by sepal, and grows at the terminal end of branches. The unmatured flower buds are green in colour, and when mature, they turn into flushed pink-bright red colour, and this indicates that they can be harvested. The Clove flower consists of four petals and several pistils. At the end stage, the Clove head is formed where the stamen remains intact and covered inside by petals which in a dried state appears as Clove bud (head) (Nurdjannah and Bermawie 2012). In addition, flowers sprout into small and elongated dark red coloured berries as a fruit. So, unopened buds in clusters entailing long calyx with sepals and unopened petals are usually harvested (size 1.5–2.0 cm length) and are dried to form a complete Clove bud (central ball). The detailed illustration of Clove is depicted in Figure 8.1 with three different varieties of Clove are found, i.e., Penang, Zanzibar, and the Ripe fruit of the Clove plant.

Further, the taste of Clove can be described as a sharp, pungent, and aromatic taste with mild bitterness. Over 2000 years, Cloves are used by Asians in the preparation of different meals like soups, food meals like flavoured rice, curries, tea, and herbal medicine formulation, especially for their intense and strong flavour and aroma. When Clove is used in cooking, it tastes similar to nutmeg giving the food a mild and complementary odour and taste. Also, World Health Organization (WHO) established standards for Clove consumption, i.e., 2.5 mg/kg of Clove per day is acceptable in humans.

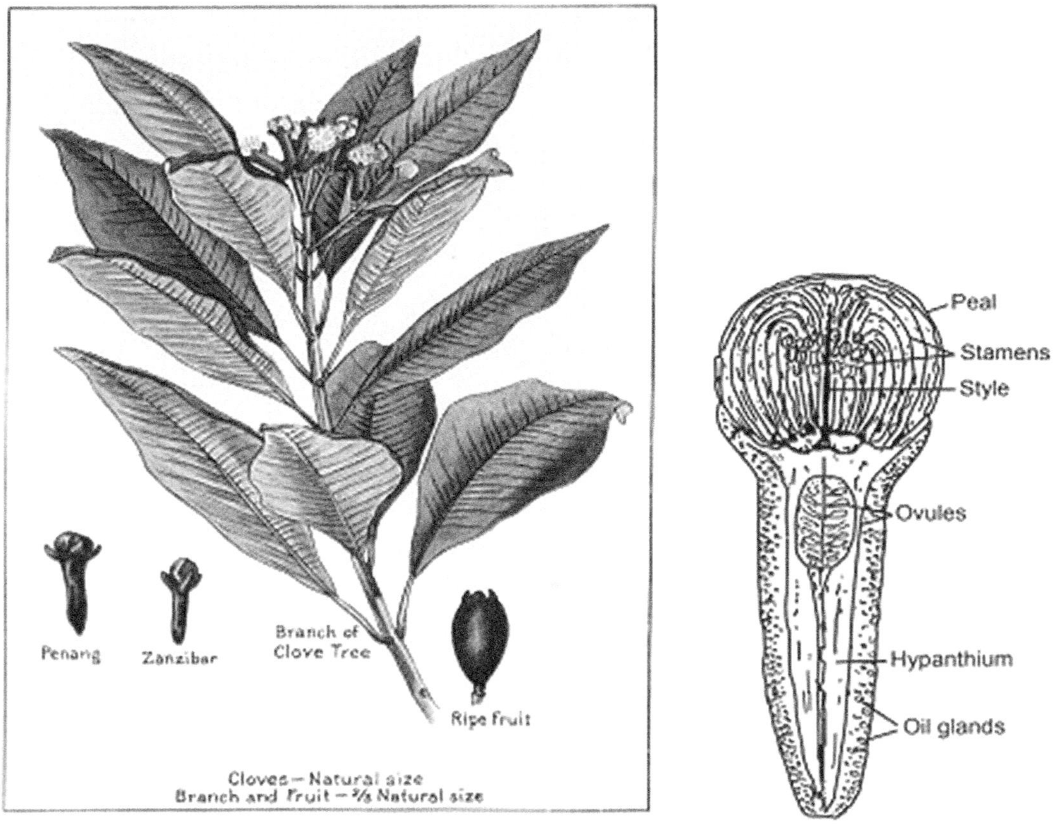

Figure 8.1 Pictorial representation of Clove, its varieties, and individual Clove bud (*Source:* Ayushi et al. 2020).

8.1.1 Taxonomical Classification of Clove

The taxonomical classification of Clove (Daniel et al. 2009; Cortés-Rojas et al. 2014) is given below:

Kingdom – Plantae
Sub kingdom – Tracheobionta
Super division – Spermatophyta
Division – Magnoliphyta
Class – Magnoliopsida
Subclass – Rosidae
Order – Myrtales
Family – Myrtaceae
Genus – *Syzygium*
Species – *aromaticum* (L.)

8.1.2 Common Names of Clove

Clove is vastly cultivated and consumed worldwide from generation to generation, and different names are used in different parts of the globe due to linguistic diversity. Clove is called by unique names in India and internationally, as mentioned in Table 8.1.

Table 8.1 Unique Names of Clove in India and Internationally

Language	Name
India	
Sanskrit	Devkusuma, Devapuspa, Lavanga, Sriprasum, Chandanpushpaks
English	Clove, Clovos, Caryophyllus, Carophyllus
Botanical	*Eugenia caryophyllus, Syzygium aromaticum*
Hindi	Lavang, Laung
Marathi	Lavang
Malayalam	Grampus, Karayampu
Kannada	Krambu, Daevakusuma, Lavanga
Tamil	Kirampu, Kiraambu, Grambu
Bengali	Lavanga
Gujrati	Lavang
Oriya	Labanga
Urdu	Laung
International	
Arabic	Kabshqarunfil, Kabshqaranful
Bulgarian	Karamfil
Chinese	Ding xiang
Dutch	Kruindnagel
Danish	Nellike
French	Giroflier
German	Nelke
Greek	Garifalo
Georgian	Georgian Mixaki
Hungarian	Szegfu
Indonesian	Cengke
Italian	Chiodo di garofano
Japanese	Girofla, Choji, Kurobu
Korean	Jeonghyang
Latvian	Krustnaglinas
Nepalese	Lwaang
Swedish	Kryddnejlika, Kryddnejlikor
Spanish	Clavo, Clavo de olor

Source: Milind and Deepa (2011), Ayushi et al. (2020), and Yadav et al. (2020).

8.2 CHEMICAL COMPOSITION OF CLOVE

The Clove buds are usually harvested when the buds reach their full size before the corolla opens, and the hypanthium turns to reddish colour. These harvested buds are separated either by hand or thresher machine from the stems, and finally, they are either sun-dried or artificially dried using a suitable method (Hidayat and Nurdjannah 1992). The Clove is available in three major forms, which include dried Clove bud, grounded Clove, and Clove oil. Firstly, dried Clove bud's chemical composition varies according to its harvesting time and processing conditions (Nurdjannah et al. 1997). The chemical characteristics at different maturity stages of Clove bud are shown in Table 8.2. Generally, in Clove, volatile oil content is around 14–21%, along with eugenol, gallotannic acid, eugenol acetate, caryophyllenes (in α and β form), methyl furfural, resin, gum, fibre, and small quantities of ketones, esters, alcohol (Jagapriya and Saranya 2020). However, more than twenty

Table 8.2 Chemical Characteristics of Clove Bud at Different Maturity Stage

	Maturity Stage of Bud			
Component (in %)	**Fallen Flower**	**Undeveloped Bud**	**Fully Developed Bud**	**Over-Ripe Bud**
		Clove Bud		
Water	12.3	5.0	12.8	6.5
Oil	13.9	14.9	16.4	16.1
Ash	4.7	3.8	4.7	6.11
Fibre	11.8	10.8	8.5	13.3
Si	0.15	0.11	0.11	0.10
		Clove Bud Oil		
Total Eugenol	91.0	90.0	94.0	93.0
Eugenol	54.2	55.6	68.5	72.2
Eugenol acetate	34.4	36.5	22.0	9.4

Source: Nurdjannah et al. (1997).

volatile compounds were found, and the quality of Clove mainly depends on its essential oil content, which gives Cloves its characteristics aroma, and the aroma profile is affected by the ratio of eugenol and eugenol acetate. Also, a higher amount of eugenol and eugenol acetate gives a harsher and smoother aroma, respectively (Nassar et al. 2007). The chemical characteristics of Clove buds are available in three major forms, i.e. in the form of extracted oil from Clove bud, Clove stem, and Clove leaf, as shown in Tables 8.2 and 8.3.

Secondly, the ground Clove is obtained either through milling or grinding, which is usually processed at 25–35°C. Since this involves the application of heat, various techniques are developed and used to minimize the heat loss of volatile compounds such as cryogenic grinding and/or cooling, chilling or refrigeration of conventional grinding chamber, etc. Finally, Clove oil is known for its characteristic composition of eugenol, caryophyllene (β form), and eugenol acetate content which varies according to agroecological parameters, pre-and post-harvesting methods,

Table 8.3 Volatile Constituent of Oil from Different Parts of Clove

Compound (%)	Clove Bud Oil (CBO)	Clove Stem Oil	Clove Leaf Oil
Eugenol	80–90	76.4–84.8	38.3–95.2
Eugenol acetate	15	1.5–8.0	51.2–1.5
β-Caryophyllene	5–12	7.3–12.4	6.3–0.2
α-Caryophyllene	1.34	0.07	0.22
Α-Humulin	10–12	1.0–1.4	1–1.44
Vanillin	1.82	0.89	1.15
δ-Cadinene	1.18	0.31	0.44
Benzyl alcohol	1.10	0.59	0.57
α-Terpineol	0.92	0.93	0.96
β-Terpineol	0.81	0.01	0.02
Benzaldehyde	0.69	–	0.01
2-Nonanol	0.64	–	0.02
α-Terpinene t-limonene	0.53	0.14	–
2-Heptanone + 1.8-cineol	0.50	0.11	0.10
2-Hexanone	0.48	0.13	0.09
β-Pinene	0.44	0.16	0.09

Source: Raina et al. (2001) and Gosavi et al. (2020).

Table 8.4 Physical and Chemical Attributes of Clove Oil Derived from Bud, Stem, and Leaf

Compound	Bud (%)	Stem (%)	Leaf (%)
Colour	Colourless or Yellow	Pale Yellow	Dark Brown
Essential oil	15–20	5–10	1–4
Specific gravity	1004–1057	1048–1056	1030–1060
Refractive index	1528–1538	1534–1538	1520–1540
Optical activity	0–(−1.58)	0–(−1.50)	−1.58
Total eugenol	85–93	89–85	78–93
Eugenol (True)	~61.62	~80.0	~77.1
Eugenol acetate	~18.7	~2.1	Trace
β-Caryophyllene	~15.27	~12.70	~17.02

Source: Nurdjannah (2004).

processing conditions, and extraction techniques (Purseglove et al. 1981 and Singh and Goswami 2000). Clove oil is either colourless or pale yellow extracted from a bud, leaf, or stem, and its characteristics constituents are shown in Table 8.4. The chemical structure of eugenol, eugenol acetate, and β-caryophyllene are shown in Figure 8.2 (Guenther 1950). Hence, each Clove essential oil differs from one another in the chemical composition, colour, and flavour (aroma), and presence of secondary metabolites are influenced by the nature of the soil, climate, genetics, cultivation, harvesting, and processing techniques. The non-volatile component of Clove includes tannins (10–13% in Clove buds), tri-terpenes (~2%), sterols (stigma-sterol, sitosterol, and campesterol in traces), and flavonoid (iso-biflorin, C-glucoside chromone) (Yadav et al. 2020).

8.3 IMPORTANT BIOACTIVE COMPONENTS IN CLOVE OR CLOVE OIL

Clove contains more than twenty bioactive components as identifies using various chromatographic techniques. The bioactive components of both food and medicinal use found in Clove majorly include polyphenols such as flavonoids, hydroxycinnamic acids, hydroxybenzoic acids, and hydroxyphenyl propenols. Among different polyphenolic compounds, the main compound is eugenol; it is contained in fresh plant material ranges between 9.38–14.65 g/100 g, phenolic acid-like gallic acid is also present in higher concentrations, i.e., around 783.50 mg/100 g fresh weight, hydrolyzable tannins constitutes around 2.38 g/100 g which is determined using Gas Chromatography-Mass

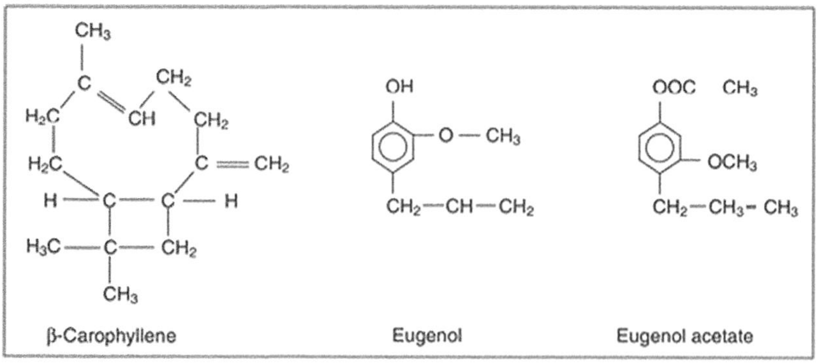

Figure 8.2 Chemical structure of important constituent of Clove (*Source:* Guenther 1950).

Table 8.5 Major Bioactive Compounds of Clove in Different Solvent Extract

Clove Extract	Content
Essential oil	Eugenol, β-caryophyllene, α-humulene, and Eugenol Acetate
Ethanol extract	Eugenol, β-caryophyllene, Eugenol Acetate, Flavonoids, Tannins, and Alkaloids
Hexane extract	Eugenol, Eugenol Acetate, β-caryophyllene, and Flavonoids
Water extract	Anthraquinones, Eugenol, Eugenol derivates, Flavonoids, Saponins, Tannins, and Trans-caryophyllene

Source: Kaur and Kaushal (2019) and Hiwandika et al. (2021).

Spectrometry (GC-MS) analysis (Hussain et al. 2017). Other minor compounds found are ferulic acid, ellagic acid, caffeic acid, salicylic acid, and flavonoids like quercetin, kaempferol, limonene, β-pinene, farnesol, 2-heptanone, benzaldehyde, ethyl hexanoate, etc. (Table 8.3). However, the concentration of various bioactive compounds in the extracted oil varies depending upon the method of extraction from the Clove, as shown in Table 8.5. These bioactive compounds are shown to be associated with many functional and medicinal health benefits, which will be discussed in further sections.

8.3.1 Eugenol

Among ample plants available, eugenol content in herbs and spices like Clove pepper (from fruit), holy basil (from leaves), betel pepper (from leaves), cinnamon (from bark), bay (from leaves), turmeric (from leaves and essential oil), nutmeg (from seeds) and thyme (from the shoot) contain approximately 36, 17.85, 3.52, 4.2–4.9, 1.34, 2.1, 0.32 and 0.021 mg per g respectively but the highest concentration of eugenol is found in Clove (flowers and buds), i.e., 180 mg per g (Raja et al. 2015). It is a colourless or pale yellow, completely soluble in organic solvents, slightly less soluble in water, has 164.2 g/mol molecular mass, and can be easily extracted from buds, stem, or leaves using conventional distillation method using sodium hydroxide (4–7% concentration) which yield around 75% of eugenol and to get 98% purity, hexane can be used as washing solution (Nurdjannah et al. 1997) or common method used for extraction and isolation through steam distillation. Eugenol has many pharmacological activities like anaesthetic, antioxidant, antimicrobial, anti-inflammatory, neuroprotective, antidiabetic, anticarcinogenic, and hypolipidemic efficiency. World Health Organization (WHO) has given Eugenol a GRAS (Generally Recognized as Safe) status, which is regarded as non-mutagenic and safe when consumed less than 1.5 g/kg body weight/day. According to WHO, the daily quantity acceptable per day can be around 2.5 mg/kg of human body weight (Gülçin et al. 2012; Cortés-Rojas et al. 2014). Hence, its application is found in aromatherapy and as a food flavouring, and also as an essential ingredient in perfumes, soaps, detergents, soaps, beverages, and food.

Similarly, iso-eugenol (includes 4-hydroxy-3-methoxy-1-propenylbenzene, 1-hydroxy-2-methoxy-4-propenylbenzene, 4-propenylguaiacol 2-methoxy-4-propenylphenol), which is popular for acting as a commercial precursor for vanillin, can be produced by isomerizing eugenol extracted from Clove leaf oil. It is also utilized in the making of non-alcoholic beverages, cigarettes, chewing gums, ice-cream, etc. (Hariono 2009 and Khalil et al. 2017).

8.3.2 β-Caryophyllene

It is a natural bicyclic sesquiterpene; it is one of the main components found in Cloves (approximately 9%). Similar to eugenol, β-Caryophyllene has many beneficial properties to humans, including antioxidant, anti-inflammatory, antidepressant, antitumor, neuroprotective, analgesic, and anti-allergic activity (Hiwandika et al. 2021).

8.4 METHODS OF CLOVE OIL EXTRACTION

Clove oil is often extracted from different parts of the Clove plant, which includes flowers, buds, leaves, and stems. Most essential or aromatic oils from herbs or spices are extracted using various techniques, particularly steam distillation, hydrodistillation, and solvent extraction (traditional method); supercritical CO_2 extraction, ultrasound-assisted, and microwave-assisted (latest technology) (Guan et al. 2007). The concentration and yield of various volatiles compounds in extracted oil vary depending on the methods employed. However, the traditional methods were widely used because of cost-effective or economic processes, but these methods result in the loss of certain oil characteristics. The negative aspect of this conventional method is that it results in hydrolysis, thermal degradation, water solubilization, and loss of very volatile constituents or heat sensitive aroma compounds, which affects the final product quality and its storage (Frohlich et al. 2019). The chief compounds of Clove oil include eugenol, iso-eugenol, eugenol acetate, and caryophyllene, and their concentration in the oil varies with extraction techniques. The methods of extraction are given in the subsection below.

8.4.1 Hydro and Steam Distillation

The procedure followed for extraction of oil from Clove includes subjecting dried and grounded Cloves (100 g) for 4 to 6 and 8 to10 hours of hydro and steam distillation, respectively. Then the collection of volatiles distillates is done till no oil drops out and it is saturated with sodium chloride (NaCl) and a smaller amount of organic solvent like ether. After mixing in the funnel, two separate layers are formed, i.e., ether layer and hydro layer, and are separated. Also, dehydrate (dehydrating) the hydro-layer remains can be done using anhydrous sodium sulphate. To concentrate the oil, the ether layer is further heated in a pre-set water bath maintained at 60°C and after concentrating it as per the requirements. Finally, it is weighed and stored at refrigeration conditions until used for further analysis. The average yield using hydrodistillation is approximately 11.5%, containing 50.5–53.5% eugenol (Ratri et al. 2020). Figure 8.3 shows the extraction of essential oil using steam distillation method.

Steam distillation procedure: The equipment used are distillation vessel, condenser, boiler, steam pipe, separator, funnel, and oil collecting tank. The steam distillation process works on the fact that certain organic compounds with low vapour pressure are immiscible with water. When we mix organic compounds with water, it will distil when the vapour pressure reaches atmospheric pressure, and this should occur below the boiling point of water. In the extraction of Clove oil, attach a steam generator to the apparatus; allow it to boil for the steam generation with an adequate water supply. The supply of heat or steam can be controlled using a safety valve. After steaming for sufficient time (20–30 min), collect the distillate, which shows a pale milky appearance. Then with the help of a dropping funnel which acts as the receiver, it can be used to collect the distillate for further processing; for 100 g of Clove, it yields 50–60 cm^3 distillates. Then, remove the funnel from the rest of the apparatus, add around 8 cm^3 cyclohexane, stopper it and shake the mixture till two separate layers are formed. Separate the bottom aqueous layer by slowly running off from the funnel, filter it to remove any solids present, and finally, the ethereal extract layer is dehydrated and concentrated (Rao and Pandey 2007 and Ratri et al. 2020), as mention earlier, leaving behind the Clove bud oil.

8.4.2 Solvent Extraction Method

It is the most extensively and commonly method to extract most of the essential oils from herbs and spices. Food grade solvents *viz.,* ethanol, hexane, methanol, and petroleum ether are used in this method. The advantage of this method is that it can be used to extract the essential oil of the material, which is delicate towards pressure and temperature, substances that yield a lower amount of oil and the extracted oil has a finer aroma. The most important disadvantage of this particular solvent extraction method is the presence of undesirable soluble residues such as waxes and pigments,

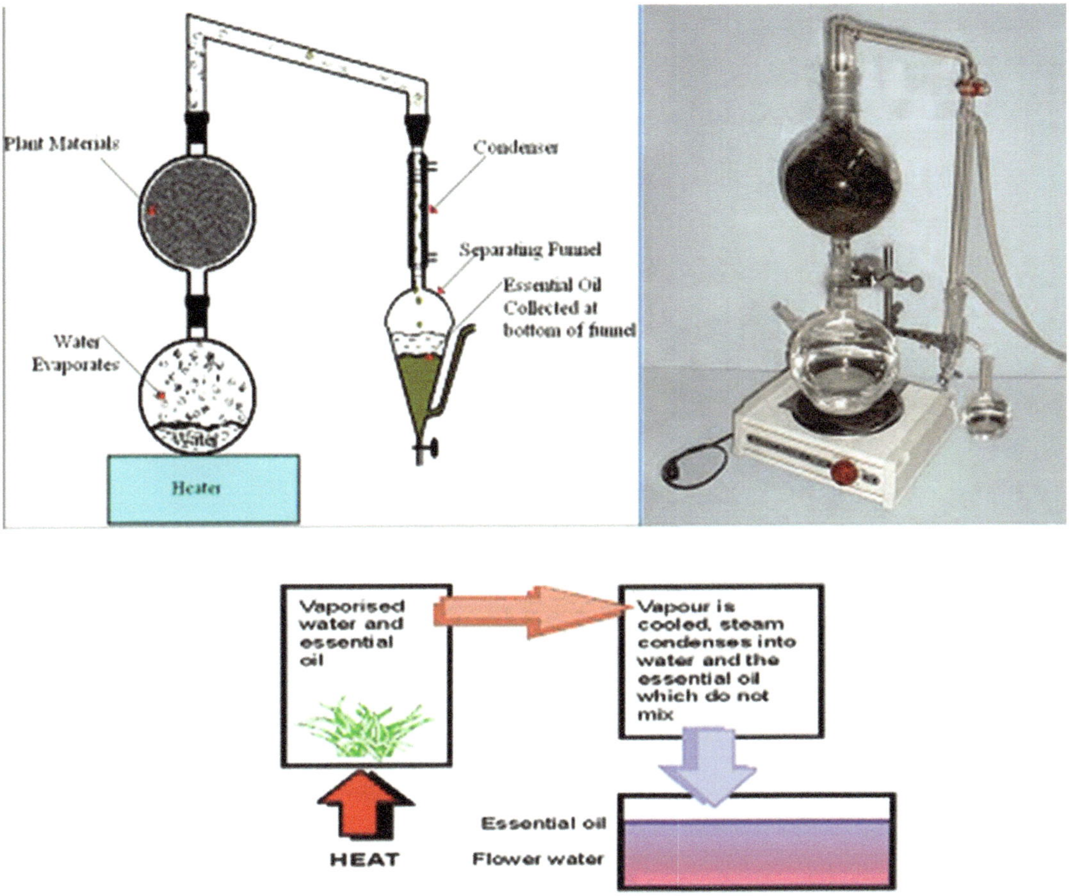

Figure 8.3 Extraction of Essential oil using Steam Distillation Method (top left) using steam distiller (top right) with flow diagram (bottom) (*Source:* Rao and and Pandey 2007).

which results in an undesirable aroma (Dasari and Goud 2014). However, this method is widely used to extract eugenol from the Clove plant material, and the extraction apparatus is shown in Figure 8.4. In this method, the Clove is treated with solvent, and it forms "concrete." This concrete is a waxy aromatic compound that chemically reacts when mixed with solvents used (as mentioned earlier), and oil components are released. Typically, 30 g of Cloves buds are grounded and wrapped in filter papers. Then this warp is transferred to the thimble section and inserted into a 500 ml capacity reflux flask. Using suitable organic solvent (250 ml ethanol), the extraction is further carried out in the Soxhlet apparatus for around 6 h. Finally, the extracted oil is concentrated at 50°C in a rotary vacuum evaporator. This conventional method can be improved in terms of efficiency and efficacy by using a batch process where it uses a reactor equipped with a four-bladed agitator and a motor with 1200 rpm capacity (Garkal et al. 2012).

8.4.3 Supercritical Carbon Dioxide Extraction

It is one of the most efficient methods used for essential oil extraction where instead of steam or heated water, carbon dioxide (CO_2) is utilized as solvent at the end of the extraction stage. Supercritical carbon dioxide (SC-CO_2) extraction operates at low temperatures, i.e., 35–38°C, in contrast to steam distillation (50–100°C). The advantage of this method is that the extracted oil is

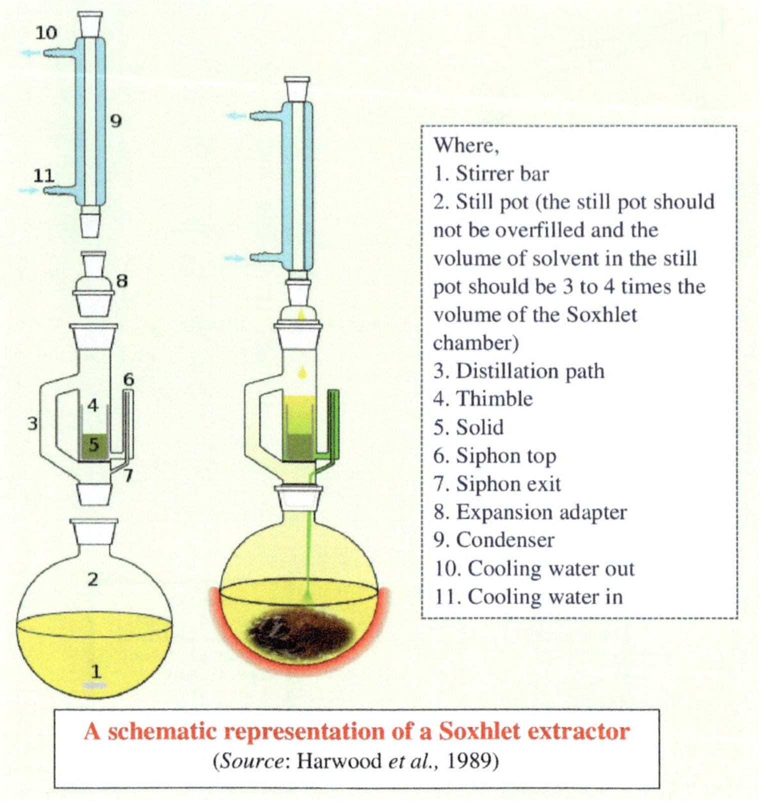

Figure 8.4 Schematic representation of Soxhlet apparatus used for essential oil extraction (*Source:* Dasari and Goud 2014).

thicker in consistency and preserves a greater number of volatile compounds compared to Clove oil obtained by steam distillation (Wenqiang et al. 2007). The efficiency in extracting oil from Clove is higher in the SC-CO_2 extraction method in contrast to steam distillation (Gopalakrishnan et al. 1990, Gopalakrishnan 1994 and Reverchon and Marrone 1997). In addition, the chemical composition of SC-CO_2 extracted Clove oil was found to be very similar to that of traditional methods.

In this method, the Speed superfluid extraction (SFE) instrument is used to pressurize liquid CO_2, and at desired pressure, this pressurized CO_2 is charged into the extraction column (specifications: 195 mm length, 32 ml with 14.40 inner diameter column), which consists of powdered Clove and glass beads. The column is heated to the required temperature in an oven, and temperature is controlled with the help of a thermocouple. Then, SC-CO_2 with dissolved Clove volatile components is passed through the pre-heated micrometre valve and reaches the end of this valve to ambient pressure (for 15 g Clove bud powder, generally CO_2 is passed at the flow rate of 2 l/min). Finally, at the collect vial of the instrument, the extract is precipitated at ambient temperature and pressure (Khajeh et al. 2004). A schematic diagram to set up a supercritical fluid extraction unit is illustrated in Figure 8.5.

8.4.4 Microwave-Assisted Extraction

To overcome the drawbacks of conventional extraction methods of Clove oil manufacturing resulted in discovering new extraction procedures, and one such method is microwave-assisted

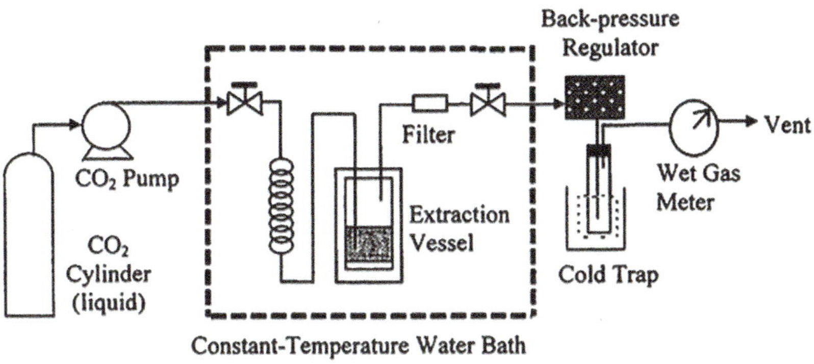

Figure 8.5 Schematic diagram of supercritical fluid extraction (*Source:* Markom et al. 2001).

extraction (MWAE). This method is recognized as a green extraction approach, and the oil obtained has similar characteristics, rapid extraction, and better-quality Clove oil in contrast to the traditional extraction method. Various techniques like microwave steam distillation (MWSD), coaxial microwave-assisted hydrodistillation (coaxial MWHD), microwave-assisted hydrodistillation (MWHD), and microwave assisted hydro-diffusion and gravity (MWHG) is developed. Among all these mentioned techniques, coaxial MWHD is found to be more efficient in terms of safety, energy & time saving (30% and 400%, respectively), cost-effective, and easy to scale up the configuration (Khalil et al. 2017). A study directed by González-Rivera et al. (2016) using headspace (HS) GC-MS and liquid chromatography consisting Ultra Violet/visible diode array/fluorescence detector (HPLC-DAD-FD) revealed that eugenol and other essential oil extracted via coaxial MWHD method had better thermal stability compared to that conventional hydrodistillation extracted Clove oil (Figure 8.6) where respective abbreviations are HS-GCMS = Head Space Gas Chromatography

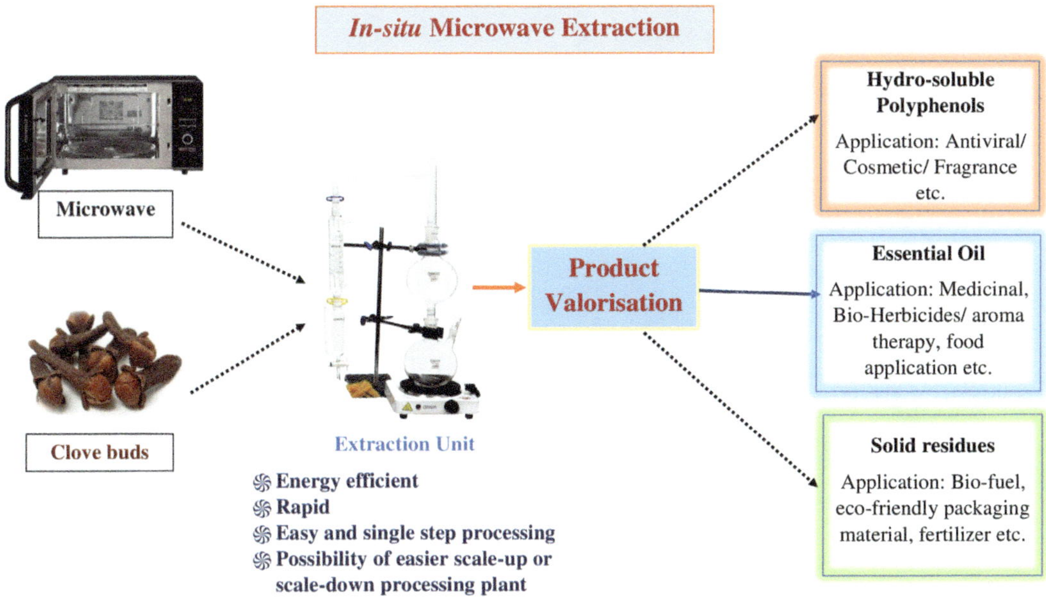

Figure 8.6 Schematic representation of Clove bud extraction by in-situ microwave assisted extraction and its application (*Source:* Gonzalez-Rivera et al. 2021).

Mass Spectrometry, HPLC-DAD-FD = High performance liquid chromatography with a fluorescence detector and diode array detector, TG-FTIR = Thermogravimetric analysis coupled with Fourier transform infrared spectroscopy. In addition, the time required for extraction was relatively less than the conventional extraction process, and the concentration of polyphenols was also higher. Conclusively, it can be stated that MWAE can be used as a replacement for conventional techniques at an industrial scale.

8.4.5 Ultrasound-Assisted Extraction

Another greener approach for extraction using ultrasound (US) has been developed and is has significantly enhanced the extraction process and decreased the energy demand. The advantages of this method are eco-friendly, improved oil quality, lower thermal degradation of constituents, no residues, rapid execution, higher yield, and easily extract handling. Lavinia et al. (2013) showed that among three different treatments (viz., 450, 900, and 1350 ml/min) for Clove oil extraction, 1350 mL/min had the highest extraction rate. This finding was supported by Tekin et al. (2015) and they concluded that at fixed US frequency (54 kHz), Clove oil yield was very much dependent on the temperature employed. The schematic representation of the USAE apparatus is shown in Figure 8.7. As a result, USAE can become an effective tool for commercial Clove oil production. Hence, the number of components and the quality of the Clove oil extracts differ from its source, harvesting, method of handling, extraction methods, processing, and storage, etc. various literature reports related to yield and characterization of Clove bud extracts using the aforementioned methods is shown in Table 8.6.

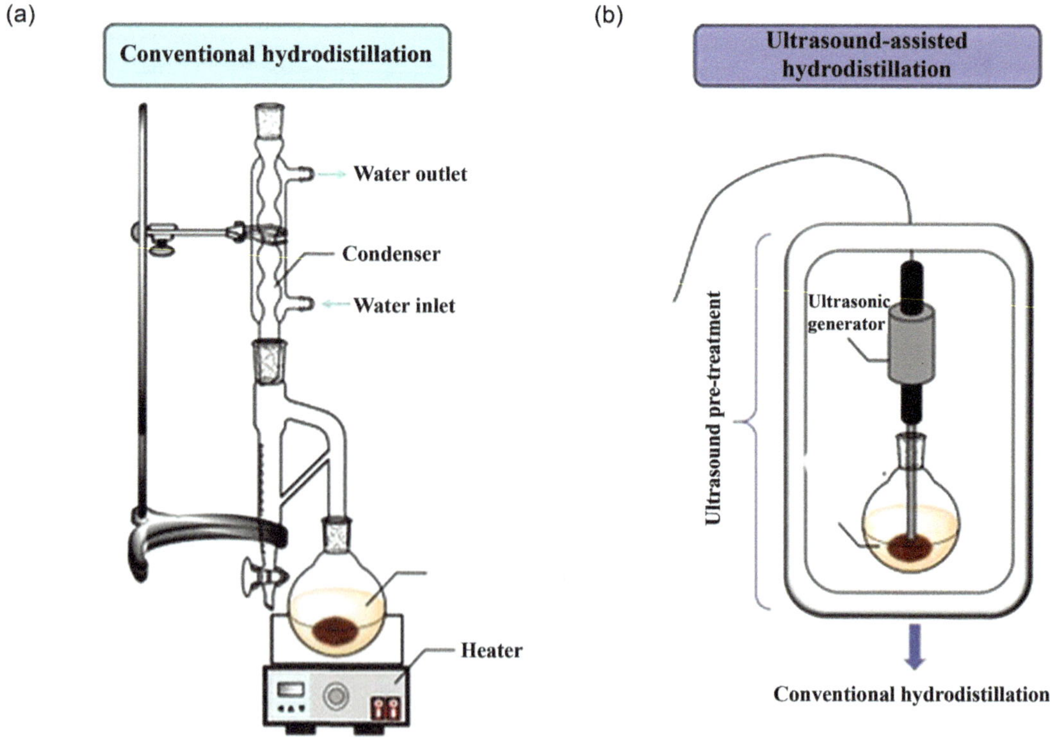

Figure 8.7 Schematic representation of ultrasound-assisted extraction of essential oil (*Source:* Chen et al. 2021).

Table 8.6 Nutritional Composition of Clove

Composition (mg)	USDA (Ground Form)
Water (g)	5.40–6.86
Food energy in Kcal	323
Protein (g)	5.98
Fat (g)	20.06
Carbohydrate (g)	61.22
Ash (g)	5.88
Calcium (g)	0.646
Phosphorus	105
Sodium	243
Potassium	1102
Iron	8.68
Thiamin	0.115
Riboflavin	0.267
Niacin	1.458
Ascorbic acid	80.81
Vitamin A (RE)	53.0

Source: Tainter and Grenis (1993).

8.5 METHODS OF CHARACTERIZATION OF BIOACTIVE COMPOUNDS IN CLOVE OIL USING GAS CHROMATOGRAPHY USING MASS SPECTROMETRY DETECTION

Many studies have proved that Clove or Clove oil contains more than 20 identifiable bioactive compounds, and its concentration varies depending on different parameters, as mentioned in section 2. The currently available extraction method as such the traditional methods, SC-CO_2 extraction, MWAE, and/or USAE methods. Among different methods available to isolate, identify and characterize bioactive components, GC-MS is most widely used in Clove or Clove oil analysis which will not only identify and quantify but also reveal the pharmacological property of Clove components (Park et al. 2021). A detailed analysis was carried out by Amelia et al. (2017) to determine the chemical composition of Clove oil from Java and Manado variants using GC-MS analysis. In this experiment, dried Clove bud (moisture content around ~15%) were cut to a size of approximately 0.5 mm, and isolation was done by conducting a steam distillation method. The dried Clove oil was subjected to GC-MS analysis using the Agilent 7890 gas chromatograph coupled with Mass Spectrometry (Agilent 5975C) detector where, HP-5MS UI capillary column having film of 0.25 μm thickness with 30 m × 0.25 mm dimensions was used with methyl phenyl silox, which was cross-linked 5%, carrier gas used was helium (streaming at 1 mL/min), and the temperature in oven was increased at a rate from 6°C per min to 200°C from 40°C and further continued at 30°C per min up to 200–280°C. Finally, post-run was done at 280°C for 10 min. GC-MS analysis was conducted by injecting 1 μL of dried Clove oil sample in absolute methanol at 0.1% and operated by scan mode and splitless. Here, as an external standard, quinaldine was used. Using Wiley mass spectra library, the chemical elements found were determined by comparing the results of the chromatogram with the reference retention time. The results obtained showed that among different constituents, the highest retention time was 24.85 min by benzyl benzoate (0.065 & 0.016%)[a] and the lowest retention time was 9.22 min by 2-Heptanol acetate (~0.013), eugenol (55.606 &74.647%)[a], Caryophyllene (14.846 & 12.799%)[a], α-Humulene (2.756 & 1.535%)[a] and Eugenyl acetate (20.545 & 8.708%)[a] had retention time of 16.96, 18.23, 18.86 and 20.41 min, respectively (Note: [a] denotes Java and Manado Clove extracted oil respectively).

8.6 NUTRITIONAL COMPOSITION OF CLOVE

Apart from being a natural source of the essence, Cloves do contain other significant nutrients, such as carbohydrates, proteins, fat, vitamins, minerals, etc. Based on a different method of chemical analysis as discussed in the earlier section of the chapter, the nutrients present in Clove per 100 g are represented in Table 8.5. From the nutritional composition, it can be concluded that Clove contains a higher amount of fat and carbohydrate and has relatively more calories.

8.7 FUNCTIONAL PROPERTIES OF CLOVE

Currently, with the increase in consumer awareness, concern towards the safety of food preservation, the addition of food additives and food preservatives is increasing. Cohort studies have been carried out to explore, identify and develop safe food additives and preservatives so that it has not had any threat to the well-being of a human. The natural sources of various beneficial active ingredients like antioxidants, antimicrobial agents, etc., are obtained from plant-based oil, which has been documented and is being used for many decades till date in foods (Oiye and Muroki 2002). For ages, Clove and Clove bud oil are used in food as an aroma and taste enhancer, preservatives, and medicinal condiment, etc. With the advancement in research and technological development, Clove has shown many functional properties like antifungal, antibacterial, and antiseptic agents. (Sulieman et al. 2007). In the earlier days, it was used as a preserving agent to cover up decomposition in sweetmeats, sauces, meat, and syrups (Tajkarimi et al. 2010). Despite having potential application as a food preservative, it remains mostly as a food condiment. However, further *in vitro* and *in vivo* analyses are required to certify the safety and toxic effect of Clove oil for application in the food industry. Different functional properties of Clove are discussed below:

8.7.1 Antifungal Property

Many studies have shown that Clove oil exhibits antifungal properties against various fungi like Penicillium spp., Aspergillus spp., Eurotium spp., etc., and against their secondary metabolites like aflatoxins. Clove oil has the strongest inhibition towards the growth of Penicillium spp., especially *Penicillium digitalis* and *Penicillium italicum* mycelia when there is direct contact with the oil or its vapour. This may be due to its high phenolic content, which gets vaporized easily (Plaza et al. 2004). Aflatoxin produced by *Aspergillus parasiticus* was inhibited when 0.6 mg/ml Clove oil was administrated (Kalemba and Kunicka 2003). Then bioactive compounds in Clove or Clove oil, namely, eugenol, are found to be a potent antifungal agent. Also, in 2004, an experiment was conducted by Chami et al. (2004) to identify the potential of antifungal effect of eugenol, which proved that the Indian CBO (~72% eugenol) showed a poor antifungal property in contrast to Madagascar Clove leaf oil (~90%) which was mainly due to the concentration of eugenol present. Furthermore, studies proving the antifungal properties of Clove or Clove oil are shown in Table 8.7. Hence, it can be stated that using Clove or its oil in food products can act as a natural antifungal agent (preservative), which in turn improves the product's shelf-life.

8.7.2 Antibacterial Property

Clove oil has shown antibacterial properties for many years for its medicinal use against dental problems. Hence, this proves the potential of components present in Cloves as an antibacterial agent against bacteria and moulds (Nychas (1995). Many studies have proved the antibacterial activity of Clove against both pathogenic and non-pathogenic food microbes, which includes *Listeria monocytogenes*, *Vibrio parahaemolyticus*, *Escherichia coli*, *Salmonella typhimurium*, *Campylobacter jejuni*,

Table 8.7 Clove Bud Extracts Isolated by Using Different Approaches

	Conditions for Extraction			Attributes of Extracts	
Method	**Time (min)**	**S/L Ratio (w/w)**	**Solvent**	**Yield % (wt.)**	**Main Compounds**
SHWE	100	–	Superheated water (150°C)	18.3	83.6% Eugenol 16.1% Eugenyl Acetate 1% β-Caryophyllene
SFE	330	–	SC-CO_2 (200 bar at 55°C)	21.5	76. 2% Eugenol 15.7% Eugenyl Acetate 8.1% β-Caryophyllene
MAE	30	1:6.7	H_2O	13.9	90.1% Eugenol
CH	180	1:25	H_2O	12.5	88.1% Eugenol 7.5% Eugenyl Acetate
SE	360	1:20	Hexane	17.5	34.1% Eugenol 10.5% Eugenyl Acetate 9.1% β-Caryophyllene 32% α-Humulene
CH	240	1:10	H_2O	13	87.3% Eugenol 10.4% Eugenyl Acetate 1.4% β-Caryophyllene
SD	240	1:10	Steam	11.5	82.7% Eugenol 15.6% Eugenyl Acetate 0.9% β-Caryophyllene
MWHD	40	1:1	H_2O	3.2	65.2% Eugenol 12.1% β-Caryophyllene
CH	360	1:5	H_2O	9.7	87.1% Eugenol 6.4% Eugenyl Acetate 5.1% β-Caryophyllene
SD	240	1:20	Steam	18.2	81.1% Eugenol 18.1% Eugenyl Acetate 0.7% α-Caryophyllene
SHWE	100	–	Superheated water (150°C)	18.3	83.6% Eugenol 16.1% Eugenyl Acetate 1% β-Caryophyllene
SFE	330	–	SC-CO_2 (200 bar-55°C)	21.5	76.2% Eugenol 15.7% Eugenyl Acetate 8.1% β-Caryophyllene
MAE	30	1:6.7	H_2O	13.1	90.1% Eugenol
CH	120	1:40	H_2O	7.8	49.2% Eugenol 43.5% β-Caryophyllene 3.8% α-Caryophyllene
UAE	45	1:20	Ethanol: H_2O (1:1)	41	9% Eugenol 8% Eugenyl Acetate 26% α-Humulene
UAE	45	1:20	Ethanol: H_2O (1:1)	41	8% Eugenol 5% Eugenyl Acetate 45% α-Humulene
UAE	45	1:20	Ethanol: H_2O (1:1)	39	5% β-Caryophyllene
UAE	45	1:20	Ethanol	28.2	64% Eugenol

Source: González-Rivera et al. (2016).

Where, CH = Hydrodistillation, MAE = Microwave Assisted Extraction, MWHD = Microwave Assisted Hydrodistillation, S/L = Solid/Liquid, SD = Steam Distillation, SE = Solvent Extraction, SFE = Supercritical Fluid Extraction, SHWE = Superheated Water Extraction, UAE = Ultrasound Assisted Extraction.

Table 8.8 Uses of Clove, Clove Oil, Eugenol, and Its Derivative Molecules Against Pharmaceutical and Medicinal Use and Their Mechanism

Extract	Activity	Activity Against	Mode of Action
Eugenol	Antibacterial	*Escherichia coli*	Inhibits bacterial migration, adhesion, expression of virulence factors and fimbriae formation
		Staphylococcus aureus	Hinders biofilm formation and the expression of virulence factors
		Listeria monocytogenes	Prevents bacterial adhesion and biofilm formation
Clove oil		*Pseudomonas aeruginosa*	Prevents expression of virulence factor, migration, formation of biofilm and quorum sensing
Methyl eugenol and Eugenol	Antifungal	*Candida, Penicillium, Fusarium, Aspergillus, Colletotrichum, Botrytis*	It alters fluidity and membrane permeability and hinders biofilm formation
Eugenol	Anticancer	Cancer cells and Tumours	Induce growth inhibition and apoptosis of cancer cells in synergy with anticancer drugs such as gemcitabine effective against melanomas, skin tumours, gastric cancers, leukaemia Would limit angiogenesis of malignant tumours
Eugenol	Antiparasitic	Genus *Leishmania* and *Plasmodium*	Would act by altering membranes of the genus Leishmania (transmission by phlebotomas) and may have an action against Plasmodium Shows anti-malaria action
Clove essence, eugenol, acetyl eugenol, and eugenol derivatives	Anti-inflammatory and Analgesic		Inhibits cyclooxygenases Inhibits production of pro-inflammatory cytokines
Clove and its extract	Antiaesthetic	–	Sequestration of free radicals Chelation of metallic ions Sequestration of superoxide ions.
Clove essence	Repellent and insecticidal	Mosquitos and larvae	Strong repellent capacity against mosquitoes (genus *Aedes, Culex,* and *Anopheles*) Larvicide against mosquito larva action against mites (genus *Tineola*) eugenol and derivatives also develop insecticide properties against lice or weevils (Detailed mechanism still unknown)

Source: Khalil et al. (2017), Hu et al. (2018), and Danthu et al. (2020).

Salmonella enteritidis, *Staphylococcus aureus*, *Bacillus cereus*, *Bacillus subtili*, Pseudomonas spp. (*P. aeroginosa*, *P. putida*), *Aeromonas hydrophila* and *Alcaligenes faecalis,* etc. (refer Table 8.8 for details) (Hoque et al. 2008 and Sulieman et al. 2007). Furthermore, Clove oil has a bacteriostatic effect against foodborne pathogens at a concentration of 0.075% (Smith-Palmer et al. 1998) and inhibitory effects against more than 20 food pathogens, in particular, *S. enteritidis, C. jejuni, S. aureus, E. coli, L. monocytogenes, B. cereus*, etc. which proves that it can be used in food products as an antibacterial agent. In addition, processing at 100°C for about 30 minutes did not affect the antibacterial property, which shows that higher temperature does not change the antibacterial action of Clove oil (Singh et al. 2003 and Sulieman et al. 2007). With the proven evidence from

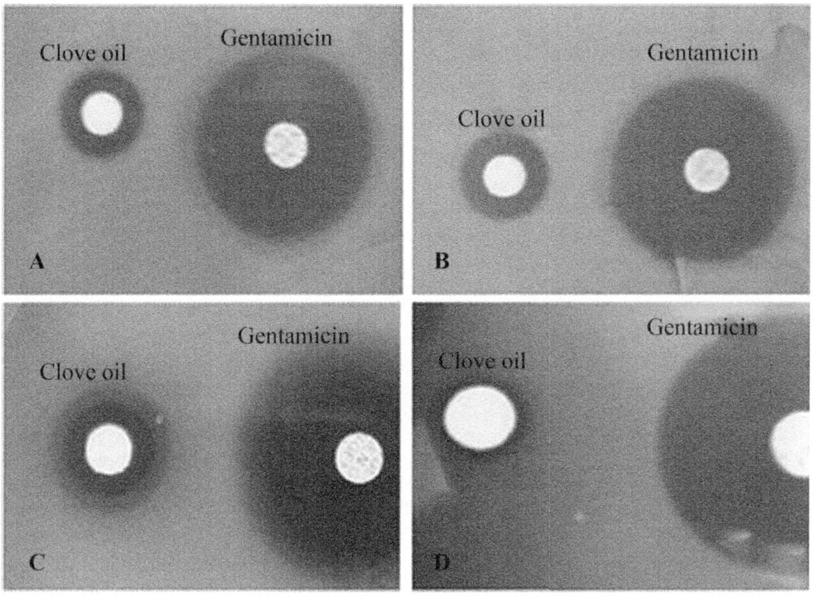

Figure 8.8 The antibacterial activity of Clove oil (*Source:* Chaieb et al. 2007).

Note: 500 μg of Gentamicin and 10 mg of Clove oil or were added onto the seeded agar of paper disks. Disks without antibiotics or clove oil were considered as negative controls. Incubation of culture was done at 37°C for 24 h, and then, the diameter for the inhibition zone was determined. (A) *S. epidermidis* CIP106510; (B) *S. aureus* ATCC 25923; (C) *E. coli* ATCC 35218; (D) *P. aeruginosa* ATCC 27853.

cohort studies, it shall be clearly stated that Clove oil can be regarded as a promising approach as an antibacterial agent with other preservation methods like refrigeration, low oxygen, or high pressure. Figure 8.8 shows the antibacterial activity of Clove oil.

8.7.3 Antioxidant Property

Foods rich in fats or lipids are prone to rancidity due to lipid oxidation. Intensive studies are carried to discover natural antioxidants. Among the different natural antioxidants available, Clove has shown to possess a strong antioxidant activity which varies with the kind of Clove product; for example, ground Clove shows around 1.8% activity, whereas the soluble fraction of petroleum ether and alcohol shows 1.4 and 1.7% antioxidant content, respectively. Studies have also shown that Clove or Clove oil has higher antioxidant activity than butylated hydroxytoluene (BHT), which may be because of eugenol, eugenol acetate, and gallic acid present in clove oil (Kramer 1985 and Lee and Shibamoto 2001). This antioxidant effect of the different bioactive compounds in Clove was identical with that of natural antioxidants like α-tocopherol (vitamin E). Also, comparing eugenol, iso-eugenol, and caryophyllene, iso-eugenol (rarely found) has the highest antioxidant efficiency. A study conducted to determine the effectiveness of various plant origin antioxidants in meat lipids revealed that ground Cloves had higher activity than oregano, sage, ginger, and thyme (Shahidi et al. 1995). Hence, the presence of a variety of aroma chemicals in Clove and other plant materials may explain the improvement in food stability, and this factor can be used in improving the shelf-life of commercial products.

In Clove, the presence of various bioactive compounds has shown their beneficial properties when added to food or consumed; in particular, eugenol, which is present in the highest concentration, has better functional and medicinal properties, and the mechanism of action is shown in

Table 8.9 Mechanism of Eugenol's Action in Functional and Medicinal Properties

Bioactivity	Mechanism of Action
Antioxidant	Inhibitory action on peroxidation of lipids and prevents ROS & RNS formations.
Antimicrobial	Induces cell lysis in both gram negative and positive bacteria Prevents signalling pathways of IKK/NF-kB, ERK, and p38MAPK.
Anti-inflammatory	Inactivates TNFα factors, suppress the activity of COX_{-2}, inhibits prostaglandin synthesis and NF-kappaβ pathways
Anticancer	Trigger apoptosis of cells, targets the pathways of E_2E_1/surviving mechanism, Apoptosis of cancer cells (MCF-7) in the human breast, Suppressing COX_{-2} gene in human colon cells (HT-29), prevents the activity of matrix metalloproteinase (MMP-9) and production of prostaglandin (E_2), decreases oxidation of DNA and inhibits the activity of ERK proteins or related pathways.
Neuro-protective & anti-stress	Impedes peroxidation of lipids and 5-lipoxygenase activity, Inhibits lowering of dopamine concentration, decreases the concentration of MDA, NO, and ROS, in the region of cerebellum and cortex, the augmentation of glutathione takes place, in the region of the brain; attenuation of cytosolic calcium and acetylcholinesterase activity takes place, controls serotonergic system in the region of amygdala, reduces RS-induced upsurge in the levels of plasma corticosterone, influences serotonin (5-HT) levels, lowers norepinephrine levels in the regions of brain and diminishes ulcer index
Antidiabetic	Inhibits the activity of α-glucosidases activity, production of advanced glycated end products (AGE), and attachment of serum albumin with the amygdala, Upregulates antioxidative enzyme levels.
Hypocholesterolemic perspectives	Free radicals are scavenged, antioxidant status is maintained in the body, diminishes damage due to cellular oxidation and serum MDA, rises the SOD enzymes concentration, inhibitory action on the activity of HMG-CoA reductase.

Source: Khalil et al. (2017) and Danthu et al. (2020).

Table 8.9. Hence, the addition of Clove or Clove oil to food products gives the benefit of being a nutraceutical property.

8.8 MEDICINAL PROPERTIES OF CLOVE

Since ancient times, as old as 2000 years ago, traditional cultures in South East Asia have been using Cloves or Clove oil in the preparation of numerous herbal remedies. "Who has the pain that roars in his head as if he were deaf, should often eat the Cloves …" Saint Hildegard quoted in her work Physica about Cloves. Traditional folk healers and modern dentists and pharmacists have been using Clove in easing the symptoms associated with dental decay and toothache. Clove was also used in the preparation of herbal tea by boiling the dried Clove buds in water to help in better digestion, stomach disorders, treat nausea and also as a pain reliever. Cloves are recommended for toothaches due to their bactericidal activity, as it was used in the Middle Ages during the prevailing diseases. It was believed that Cloves prevent contagion during epidemics of plague. Clove oil is an important antibacterial drug is used in dentistry, pharmaceuticals, and aromatherapy. It is used as an analgesic, antiseptic, warming, disinfectant, and antibacterial as it prevents the growth or kills most pathogens, such as *Aspergillus niger, Bacillus substilis, Escherichia coli, Mycobacterium phlei, Penicillium chrysogenum* and *Streptococcus aureus* (Gora et al. 2005).

Few of these therapeutic properties were shown to be associated with eugenol, which plays a crucial role. Clove oil is also used efficaciously for treating pharyngeal mucus, inflamed oral parts, and as topical anaesthesia in dentistry. Balms containing Clove oil acts as an active component has been used to soothe pain associated with rheumatism (Nurdjannah et al. 1997). Hence, bioactive compounds, namely eugenol, have shown to be a potent ingredient promoting numerous health beneficial functions. A vast number of studies have been documented to show the efficiency of eugenol for its medicinal properties, and the mechanism of action is depicted in Tables 8.7 and 8.8. Various lifestyle-related health issues have been vindicated by eugenol is vindicated such as microbial infections, inflammatory actions, elevated blood glucose levels, blood cholesterol irregularity, digestive complications, reproductive derangements, hypertension, nervous disorders, and carcinogenesis (Table 8.7). The documents prove about the effect of eugenol on certain health ailments are listed below.

8.8.1 Anti-Inflammatory Properties of Eugenol

Eugenol (4-ally-2-methoxyphenol) has anti-inflammatory potential as it has the ability to suppress the signalling of tumour necrosis factors and expression of cyclooxygenase $(COX)_{-2}$, and in macrophages, the oligomers of eugenol show inflammatory action against the expression of cytokine and activation of nuclear factor- β kappa. This potential of anti-inflammatory activity is mainly because it inhibits the synthesis of prostaglandin and chemotaxis of associated macrophages or neutrophils (Kim et al. 2003). Also, Huang et al. (2015) studied the anti-inflammation effect of eugenol in mice where acute lung injury was induced due to lipopolysaccharide. They observed that administrating 5 and 10 mg/kg/body weight of eugenol has anti-inflammation action because of the formation of pro-inflammatory cytokines. Koh et al. (2013) further supported that at 50–500 mM concentration of eugenol, an increase in the formation of interleukins-8 was observed, which again exhibited that at 5 mM weak stimulus was seen, and at 500 mM almost inhibition was seen.

The activation of nuclear factor-kβ is prevented by the polyphenols present in Clove oil which are formed in lipopolysaccharide initiated macrophages, and they are induced by inactivation of the activity of tumour necrosis factor (TNF α) and cyclooxygenase (COX_{-2}) in the *in-vitro* method (Kim *et al.* 2003). This study was also supported by Mahapatra et al. (2009), wherein mouse, upon ingestion of 160 mg per kg of bodyweight of eugenol, showed a significant decrease in neutrophils and TNF-α. It was also observed that upon ingestion of eugenol, it provided protection against dysfunctional macrophages, which were chemically induced and balancing the pro-inflammatory mediators.

8.8.2 Anticancer Properties of Eugenol

(S)-6,60-dibromo-dehydrodieugenol is a eugenol associated biphenyl that induces a particular antiproliferative action on the cancer cells of neuroectodermal via partly triggering the mechanism of apoptosis (Pisano *et al.* 2007). Eugenol, in its Epoxide form, is identified as an anticancer drug against breast cancer cells (MCF-7) because it causes cell apoptosis. Further, it suppresses the growth of MCF-7 cells by initiating the death of a cell (Vidhya and Devaraj 2011).

The methyl eugenol content in essential oil obtained from hydrodistillation of *Uvariodendron angustifolium* from barks and roots showed 68.3 and 85.3%, respectively. The methyl eugenol showed cytotoxic properties against MCF-7 cells, and it is probably because of the interaction of the cell membrane with the lipophilic radical. The cell cytotoxicity property was greater in dehydrodiisoeugenol, followed by di-isoeugenol, isoeugenol, eugenol, and lowest by biseugenol (Noudogbessi et al. 2014).

COX_{-2} inhibitors are regarded as anti-inflammatory agents and cancer chemopreventive. The utilization of eugenol as a possible chemotherapeutic drug has been introduced as part of new cancer prevention efforts. The synergistic action of eugenol with chemopreventive medicines are found to lower drug toxicity in natural cells while increasing the cytotoxicity of synthetic compounds administered (Hussain et al. 2012).

Further, Pisano et al. (2007) did an *in-vitro* experiment in order to estimate the antiproliferative property towards melanoma cells by eugenol and its derivatives. They proposed that eugenol showed an inhibitory action through inducing apoptosis and cell cycle arrest. In a B16 Xenograft model, the antiproliferative activity of both eugenol and iso-eugenol was examined, which showed a 40% reduction in tumour size because of the production of ROS which has an inhibitory effect on DNA synthesis (Kim et al. 2006).

8.8.3 Triple Negative Breast Cancer Activity

Many scientific studies have documented the potential of Clove in Triple negative breast cancer. This form of cancer is mainly identified among 10–20% of women, which is mainly caused because of cell growth which is not dependent on human hormones like estrogen, human epidermal growth factor receptor-2, and progesterone (Dent et al. 2007) and treating this type of cancer is more difficult than normal breast cancer. So, scientists have diverted themselves into finding a natural approach that gives a promising result. Since many studies and research have documented the anticancer activity of Clove and the bioactive components present in it, and for this reason, scientists are intensely discovering its potential towards triple negative breast cancer. Besides all these enlisted health and therapeutic effects, further health benefits of Clove are yet to be discovered and classified (Rani and Jena 2021).

8.8.4 Anti-Stress and Neuro-Protective Related Approach of Eugenol

Certain spices like Clove, nutmeg, basil, etc., contain eugenol, which demonstrates anti-stress and neuroprotective properties. It acts as a neuro-protector towards ischemia, amyloid-β peptide, and excitotoxicity through prevention of sciatic nerve transmission and thereby improving neural complications (Anuj et al. 2010). The eugenol found in Clove has shown to possess protecting properties as demonstrated in rats towards restraint stress irritable bowel syndrome through the modification of hypothalamic-pituitary adrenal cortex and brain mono-aminergic pathways. It was demonstrated that when orally administrating 50 mg/kg of eugenol lowered 80% of restraint stress caused due to increase in faecal pellets in comparison with control medication (ondansetron). In addition, eugenol depressed the RS-induced alterations caused in the brain, which occurs due to change in the levels of norepinephrine content and inflated antioxidants (Garabadu et al. 2011).

Eugenol asserts steady functioning of the brain by controlling neurotransmitters levels. In this study, a group of rats were subjected to restrain stress in the model for up to 4 hours. This resulted in increased ulcer index, corticosterone and plasma norepinephrine levels in rats. However, when eugenol was provided for 7 days, it had a noteworthy influence on the hypothalamic pituitary adrenal axis through lowering the stress-induced elevated plasma corticosterone and ulcer index levels. Eugenol has also affected the stress-induced changes in levels of serotonin (5-HT) in all parts of the brain, whereas the reduction in levels of norepinephrine was noticed in all parts except in the hippocampus of the brain. They concluded that the ability of eugenol to contain anti-stress activity is because it has the ability to modulate hypothalamic pituitary adrenal and monoaminergic brain systems (Garabadu et al. 2015).

8.8.5 Antitumour Activity

In-vitro and *in-vivo* studies for antitumour nature have been carried out widely, and it was significantly proved that Clove has the ability to can suppress tumours. The bioactive components found in Clove extract, in particular, ethyl acetate, has both *in vivo* antitumor activity and *in vitro* antiproliferative potential. This ethyl acetate contains a component known as oleanolic acid, which is important for inducing apoptosis of the cell cycle. Hence, cohort studies on the antitumour activity of Clove extract components on human cell lines and lung cancer lines have been carried out, and it has demonstrated that Clove extracts bioactive components exhibit antitumour activity on tumour cells (Liu et al. 2014).

8.8.6 Larvicidal Action

With the change in climatic conditions, new diseases start predominating. Unhygienic conditions upsurge the number of mosquitos, increasing vector-borne related diseases, which include malaria, dengue fever, chikungunya, and yellow fever. These increased numbers of mosquitos act as vehicles or vectors for disease transmission. Dengue fever is found to be very common in India and other parts of the globe. Currently, mosquitos have shown resistance against numerous insecticides, and this paves the way to divert the attention from chemicals mosquito control towards the natural resources. An experiment was conducted by Medeiros-Sousa et al. (2015) on larvae of The *Anopheles darlingi* and *Aedes aegypti*. The results showed that Clove extract that is aqueous eugenol and eugenol extract exhibited a very significant mortality rate of about 76% on larvae after 48 hours of the bioassay. Thus, this result shows that humans can be naturally protected against mosquitoes using eugenol obtained from Clove extract as an optimal larvicidal agent (Rani and Jeena 2021).

8.8.7 Antinociceptive Activity and Anti-Stress Activity

Cohort studies have proven that since the 13th century, Clove's eugenol is responsible for the analgesic activity for joint pains, toothaches, and as an antispasmodic. The mechanism of action is through the activating channels of calcium and chloride molecules in cells of the ganglion. In addition, it acts as an analgesic towards a capsaicin agonist (Ohkubo and Shibata 1997). Further, extracts from the Clove was found to reduce the development of gastric ulcers due to cold restraint, and it also helped in preventing and lowering the biochemical deviations triggered from sound stress like elevated plasma levels of glucose, alanine aminotransferase, cholesterol aspartate-aminotransferase, and corticosterone (Singh et al. 2009)

8.9 OTHER MEDICINAL PROPERTIES AND APPLICATION OF CLOVE

Clove Oil is recommended as an inhaler in the treatment of sore throat, colds, and inflammation of the mucous membranes present in the mouth. It also helps deal with any breathing problems, general weakness, and neuralgia. Clove oil has antioxidant effects. A study on an aqueous and alcoholic extract of Turkish Clove has been carried out. Aqueous extracts oil at different concentrations that is 20, 40, and 60 g/mol exhibited inhibition of linoleic acid lipid peroxidation in an emulsion within the given limits of 93.3–97.9%, while alcoholic extracts of the same concentrations show inhibition between 94.9 and 98.2%. The results obtained make it possible to use Cloves as a convenient and easily available source of natural antioxidants in food supplements and pharmaceutical preparations (Podlewski et al. 2010).

Clove oil has anticonvulsant effects; the Cloves were used as a cure for epilepsy. The experiment was done to determine the anti-stress effect of eugenol, the main component of Clove oil. It was performed on mice. Specified for oil LD50 = 0.64 (0.39–1.22) cm^3/kg. It's also used as a base for the manufacture of dental analgesics. To fill cavities in teeth, it is used with zinc oxide. Clove oil is included in mouthwash and gum solutions, toothpaste, and hand disinfectant treatments. Eugenol is also widely used in dentistry. Zinc-eugenol paste is a good zinc-eugenol paste for the filling of dental canals in treating periodontitis. Furthermore, the paste acts antiseptic, local anaesthetic, and is resistant to moisture (Podlewski et al. 2010).

Clove oil is widely used in aromatherapy. Massages and baths with clove oil were given to relieve pains caused by muscular stress and rheumatism, help with digestion problems, nausea, and flatulence. However, Clove oil can cause skin irritation and allergic reactions. It cannot be applied directly to the skin, neither dissolved in vegetable oil nor massage oil. It is also an important component of perfume products, especially those with an oriental flavour. It is also used for perfuming soaps (Gora et al. 2005). In aromatherapy, there is a risk of undesirable side effects, such as irritation, contact allergy, and phytotoxic reactions. The permissible concentration of the components (expressed in per cent) of Clove oil included in the products, which are applied topically both in contact with human skin and removed from it, are as follows: isoeugenol –0.02 and eugenol –0.50 (Janeczko et al. 2007).

Clove oil boosts concentration and efficiency of thinking. In addition, it revitalizes, energizes, exhibits analgesics, and acts as an "aphrodisiac." It is also used as a good air freshener. It is recommended to mix it with essential oils of basil, citrus, and spices. Clove oil is also used for the production of deodorizing and preparations acting as a deterrent for mosquitoes. Another area in which essential oils were applied (including Clove) is painting. They are used here to dilute the paint as a binder in oil painting. (Podlewski et al. 2010). Clove oil with stalks is used as a fuel additive for diesel engines was interesting. Oil decreased the ignition temperature of the fuel, which allows for its safe storage and transport, but such modifications of fuel, due to the price of oil, are not profitable (Makame Mbarawa 2008).

8.10 APPLICATIONS OF CLOVES IN THE FOOD INDUSTRY

Both Cloves and Clove oil are commonly utilized in the food industry. For centuries, Cloves were used in the kitchen, and unsurprisingly it is one of the oldest spices in the world. The descriptions of the annual gala dinner the professors from the Jagiellonian University used to have in the sixteenth-century report that various meats were richly seasoned with Cloves among others available species (Nowak et al. 2012).

Domestically, Cloves are either used in their whole form or in the ground form for culinary applications and in the food industry as a potent flavouring agent. Whole Cloves are used in the preparation of hams, baked apples, Indian curries, and pickles, etc. In the food industry, the bioactive components present in grounded Clove, Clove oil or oleoresin are used in a very lower concentration for their strong flavour, which gives the product better thermal stability, food stability, and shelf life. Oleoresin obtained from Clove is considered over other forms because it contains both volatile essential oil and non-volatile resinous material, lower risk of bacterial contamination, which mimics the flavour of original ground spice. The Cloves are also used during the preparation of sauces, ketchup, sausages, curry or curry powders, baked foods, confectionery, alcoholic beverages, and also as a seasoning, etc. However, Clover leaf oil is used to a lesser extent as it gives a harsher sensory note and lacks Clover flavour. Hence, it is mainly used to manufacture eugenol and caryophyllene (Weiss 1997).

Clove and cinnamon oil act as natural preservatives. These are harmless constituents of food products. The mixture of these oils (1:1) inhibits the growth of most microorganisms such as mould, yeast, and other microbial cells. This inhibitory effect may occur because of the compounds present in them, such as cinnamic aldehyde and eugenol, which acts by damaging the bacterial cell wall. Thus, these properties can be employed to enhance the storage shelf life of food products. Clove oil can also be used as an antioxidant and antimicrobial addition to oils, such as cotton oil. The addition of oil does not affect the colour and appearance of cottonseed oil, in concentrations of 50–1200 ppm, does not change its smell either (Park et al. 2009). Antifungal actions of Clove oil and eugenol were also used in the food industry. Products may be contaminated with microorganisms during harvesting, technological process, and storage of food. Since ancient times, to inhibit the growth of fungi gaped and spices were used, they also including Cloves (Omidbeyg et al. 2007).

8.10.1 Application in Dairy Products

Ahmed et al. (2012) used a free radical scavenging assay to investigate the antioxidant properties of particular spices, namely black cumin, black pepper, and Clove. Among all, the highest antioxidant potential was displayed by clove powder followed by black cumin and then black pepper, and it was also found that the antioxidant potent was unaffected by storage temperatures. Then clove powder was selected, and Mudaffara cheese was prepared where the clove powder was added to cheese curd (0.2% weight). In addition, spicy Mudaffara cheeses, when kept for 4–6 weeks at room temperature, tasted great, depending on the spices utilized. Clove oil, when administrated in cheese at 0.5 and 1% dose, greatly slowed the growth of *Listeria monocytogenes* when stored at 30°C and 7°C. Clove oil in high amounts has harmed food's sensory characteristics. When the bacterial load is low in food, little amounts of clove oil is sufficient to ensure its safety (Menon and Garg 2001).

8.10.2 Recent Technologies Used to Include Clove Components in Food Products

In the food sector, encapsulation is the most frequent method for incorporating different nutrients into solid, liquid, and gaseous matrices. In the food industry, nanoencapsulation and microencapsulation are used to encapsulate essential oils. Methods like spray and/or freeze drying, liposomal entrapment, inclusion complex, emulsions, and biopolymer-based microparticles are all used to load oil molecules into microparticles (>l m) in microencapsulation. Clove essential oil (CEO) was loaded inside Nanoparticles made of Chitosan, which has a good retention rate of CEO. The capsule wall and essential oil were compatible after CEO encapsulation. CEO-loaded chitosan NPs showed powerful antibacterial action and DPPH radical scavenging properties. Hence, CEO could be encapsulated in Nanoparticles of chitosan, and this strategy possibly is a novel approach in various applications like in foods, active packaging, and other no food uses (Hadidi *et al.* 2020). To order to enhance the antifungal action of CEO, chitosan nanoparticles (ChNPs) encapsulation utilizing an emulsion ionic gelation process. While conducting the experiment, a controlled release of volatile oils from chitosan nanoparticles was followed, and this resulted in improved inhibitory actions. However, in *in-vitro* conditions, encapsulated EO performed better inhibitory action towards *A. niger* isolate than free EO. Using CEO-ChNPs as a natural fungicide was considered improve the storage life of freshly cut fruits and vegetables and advocated, owing to their outstanding antifungal action (Hasheminejad et al. 2019).

Ultrasound (US) was used to synthesis nano-emulsion of hazelnut protein with the inclusion of CEO, and it was successfully prepared. It had average particle size of around 307 nm and lower and this size was varied with the periods and amplitudes of US treatment, allowing the emulsions to be classified as nano-emulsions. The inclusion of CEO in the films of hazelnut protein resulted in antibacterial and antioxidant characteristics, which were strengthened by ultrasonication. Enrichment with CEO improved the mechanical characteristics of films which were due to the improved interaction between protein and oil rather than between the proteins while using the US enhanced tensile strength and elongation at the break because of better homogenous dispersion of oil droplets inside the protein matrix (Gul et al. 2018).

8.11 CONCLUSION

With the increase in knowledge, the current generation is a switch towards organic, herbal, and natural products over the conventional products that are commercially available in the market due to their functional properties alongside nutrition with low side effects. Among well-known spices, Clove is a multipurpose spice that is widely used for since centuries as a mouth freshener, flavouring curries, meat, fruits, cookies, chewing gums, spiced fruits, pickles, chocolates, soft drinks, puddings, sandwiches, pastries, and candies and in addition to an aromatic agent in hot beverages. In addition to the culinary application, Clove has shown many proven functional and beneficial pharmaceutical activities such as antibacterial, antiviral, antioxidant, insecticide, anticancer, antitumour, etc. It is also used in ample of products *viz.* mouthwashes, soaps, antiseptic creams, dental creams, throat sprays, and toothpaste to kill pathogens and relieve sore gums. This medicinal and culinary aspect of Clove is mainly due to distinct chemical composition. Clove in its varied form is rich in eugenol, isoeugenol, eugenol acetate, caryophyllene, anthraquinones, saponins, flavonoids, tannins, oleoresin, and eugenol derivates, etc. also, Clove is used in aromatherapy, alleviating arthritis and rheumatism difficulty, muscular cramps, digestive disorders including loose motion, flatulence, nausea, and dyspepsia. Clove oil inhalation also soothes certain respiratory conditions such as asthma, cold, cough, sinusitis, and bronchitis. Other miscellaneous application of Clove oil includes as a mosquito repellent, herbicide, biopesticides, in the making of eco-friendly packaging, etc. Hence, these advantaged posed by Clove has increased its demand among older adults and middle-aged women which is the main factor to drive its growth in the market.

REFERENCES

Ahmed, M. Bakheit, and M.I. Foda. 2012. "Sensory evaluation and antioxidant activity of new Mudaffara cheese with spices under different storage temperatures." *Journal of Applied Sciences Research* 8: 3143–3150.

Amelia, B., Saepudin, E., Cahyana, A.H., Rahayu, D.U., Sulistyoningrum, A.S., and Haib, J. 2017. "GC-MS analysis of clove (*Syzygium aromaticum*) bud essential oil from Java and Manado." In *AIP Conference Proceedings*, vol. 1862, no. 1, p. 030082. AIP Publishing LLC.

Ayushi, Khan, U.A., Danish, S.M., and Mohammad, Parveen, U. 2020. "A review on biological and therapeutic uses of Syzygium aromaticum Linn. (Clove): Based on phyto-chemistry and pharmacological evidences." *International Journal of Botany Studies* 5: 33–39.

Chaieb, K., Hajlaoui, H., Zmantar, T., Kahla-Nakbi, A.B., Rouabhia, M., Mahdouani, K., and Bakhrouf, A. 2007. "The chemical composition and biological activity of clove essential oil, Eugenia caryophyllata (Syzigium aromaticum L. Myrtaceae): a short review." *Phytotherapy Research: An International Journal Devoted to Pharmacological and Toxicological Evaluation of Natural Product Derivatives* 21(6): 501–506.

Chami, F., Chami, N., Bennis, S., Trouillas, J., and Remmal, A. 2004. "Evaluation of carvacrol and eugenol as prophylaxis and treatment of vaginal candidiasis in an immunosuppressed rat model." *Journal of Antimicrobial Chemotherapy* 54: 909–914.

Chen, G., Sun, F., Wang, S., Wang, W., Dong, J., and Gao, F. 2021. "Enhanced extraction of essential oil from Cinnamomum cassia bark by ultrasound assisted hydrodistillation." *Chinese Journal of Chemical Engineering* 36: 38–46.

Cortés-Rojas, D.F., de Souza C.R.F., and Oliveira, W.P. 2014. "Clove (Syzygium aromaticum): a precious spice." *Asian Pacific Journal of Tropical Biomedicine* 4: 90–96.

Daniel, A.N., Sartoretto, S.M., Schmidt, G. et al. 2009. "Anti-inflammatory and antinociceptive activities A of eugenol essential oil in experimental animal models." *Revista Brasileira de Farmacognosia* 19: 212–217.

Danthu, P., Simanjuntak, R., Fawbush, F. et al. 2020. "The clove tree and its products (clove bud, clove oil, eugenol): prosperous today but what of tomorrow's restrictions?." *World* 161(58,172): 52–915.

Dasari, S.R., and Goud, V.V. 2014. "Effect of pre-treatment on solvents extraction and physico-chemical properties of castor seed oil." *Journal of Renewable and Sustainable Energy* 6(6): 063108.

Dent, R., Trudeau, M., Pritchard, K.I. et al. 2007. "Triple-negative breast cancer: clinical features and patterns of recurrence." *Clinical Cancer Research* 13(15): 4429–4434.

Dey, B.K., and Mukherjee, S.S. 2021. "Potential of clove and its nutritional benefits in physiological perspective: A review.

Frohlich, P.C., Santos, K.A., Palu, F. et al. 2019. "Evaluation of the effects of temperature and pressure on the extraction of eugenol from clove (*Syzygium aromaticum*) leaves using supercritical CO2." *The Journal of Supercritical Fluids* 143: 313–320.

Garabadu, D., Shah, A., Ahmad, A. et al. 2011. "Eugenol as an anti-stress agent: modulation of hypothalamic–pituitary–adrenal axis and brain monoaminergic systems in a rat model of stress." *Stress* 14(2): 145–155.

Garabadu, D., Shah, A., Singh, S., and Krishnamurthy, S., 2015. "Protective effect of eugenol against restraint stress-induced gastrointestinal dysfunction: Potential use in irritable bowel syndrome." *Pharmaceutical Biology* 53: 968–974.

Garkal, D.J., Taralkar, S.V., Kulkarni, P., Jagtap, S., and Nagawade, A., 2012. "Kinetic model for extraction of eugenol from leaves of Ocimum sanctum Linn (Tulsi)." *International Journal of Pharmaceutical Applications* 3(1): 267–270.

Gonzalez-Rivera, J., Duce, C., Campanella, B. et al. 2021. "In situ microwave assisted extraction of clove buds to isolate essential oil, polyphenols, and lignocellulosic compounds." *Industrial Crops and Products* 161: 113203.

González-Rivera, J., Duce, C., Falconieri, D. et al. 2016. "Coaxial microwave assisted hydrodistillation of essential oils from five different herbs (lavender, rosemary, sage, fennel seeds and clove buds): Chemical composition and thermal analysis." *Innovative Food Science & Emerging Technologies* 33: 308–318.

Gopalakrishnan, N. 1994. "Studies on the storage quality of carbon dioxide-extracted cardamom and clove bud oils." *Journal of Agricultural and Food Chemistry* 42(3): 796–798.

Gopalakrishnan, N., Shanti, P.P.V., and Narayanan, C.S., 1990. "Composition of clove (Syzygium aromaticum) bud oil extracted using carbon dioxide." *Journal of the Science of Food and Agriculture* 50(1): 111–117.

Gora, J. "Lis A. 2005. *Najcenniejsze olejki eteryczne. Toruń: Wydawnictwo Uniwersytetu Mikołaja Kopernika.*

Gosavi, N.S., Koli, S.S., Jire, D.S., and Shaikh, A.Z. 2020. Clove (*Syzygium Aromaticum*): A Miraculous Spice.

Guan, W., Li, S., Yan, R., Tang, S., and Quan, C. 2007. "Comparison of essential oils of clove buds extracted with supercritical carbon dioxide and other three traditional extraction methods." *Food Chemistry* 101(4): 1558–1564.

Guanghui, C., Sun, F., Wang, S., et al. 2021. "Enhanced extraction of essential oil from Cinnamomum cassia bark by ultrasound assisted hydrodistillation." *Chinese Journal of Chemical Engineering* 36: 38–46.

Guenther, E., 1950. "The essential oils. Vol. 4. Oil of citronella." *D. Van Nostrand* 65, p. 131.

Gul, O., Saricaoglu, F.T., Besir, A., Atalar, I., and Yazici, F., 2018. "Effect of ultrasound treatment on the properties of nano-emulsion films obtained from hazelnut meal protein and clove essential oil." *Ultrasonics Sonochemistry* 41: 466–474.

Gülçin, İ., Elmastaş, M., and Aboul-Enein, H.Y., 2012. "Antioxidant activity of clove oil–A powerful antioxidant source." *Arabian Journal of Chemistry* 5(4): 489–499.

Hadidi, M., Pouramin, S., Adinepour, F., Haghani, S., and Jafari, S.M., 2020. "Chitosan nanoparticles loaded with clove essential oil: characterization, antioxidant and antibacterial activities." *Carbohydrate Polymers* 236: 116075.

Hariono, A., 2009. "Isoeugenol–local based ingredient–fulfill the next up-trend global demand of natural vanillin thru bioconversion. PT Djasula, Wangi, Jakarta.

Hasheminejad, N., Khodaiyan, F., and Safari, M., 2019. "Improving the antifungal activity of clove essential oil encapsulated by chitosan nanoparticles." *Food Chemistry* 275: 113–122.

Hidayat, T., and Nurdjanah, N., 1992. "Rancangan dan Pengujian Prototipe Alat Perontok Bunga Cengkeh Tipe Aksial." *Buletin Penelitian Tanaman Rempah dan Obat* 7(1): 27–33.

Hiwandika, N., Sudrajat, S.E., and Rahayu, I., 2021. "Antibacterial and antifungal activity of clove extract (Syzygium Aromaticum)." *Eureka Herba Indonesia* 2(2): 93–103.

Hoque, M.M., Inatsu, M., Juneja, V., and Kawamoto, S., 2008. "Antimicrobial activity of cloves and cinnamon extracts against food borne pathogens and spoilage bacteria and inactivation of Listeria monocytogenes in ground chicken meat with their essential oils." *Rep Nat'l Food Res Inst* 72: 9–21.

Hu, Q., Zhou, M., and Wei, S., 2018. "Progress on the antimicrobial activity research of clove oil and eugenol in the food antisepsis field." *Journal of Food Science* 83(6): 1476–1483.

Huang, X., Liu, Y., Lu, Y., and Ma, C., 2015. "Anti-inflammatory effects of eugenol on lipopolysaccharide-induced inflammatory reaction in acute lung injury via regulating inflammation and redox status." *International Immunopharmacology* 26(1): 265–271.

Hussain, A., Priyani, A., Sadrieh, L., Brahmbhatt, K., Ahmed, M., and Sharma, C., 2012. "Concurrent sulforaphane and eugenol induces differential effects on human cervical cancer cells." *Integrative Cancer Therapies* 11(2): 154–165.

Hussain, S., Rahman, R., Mushtaq, A., and Zerey-Belaskri, A.E., 2017. "Clove: A review of a precious species with multiple uses. *IJCBS* 11: 129–133.

Jagapriya, L., and Saranya, S. 2020. "Identification of bioactive compounds and pharmacognostic evaluation of syzygium cumini (l.) And syzygium aromaticum (l.) By gas chromatography-mass spectrometry analysis." *World Journal of Pharmacy and Pharmaceutical Sciences* 9(7): 1628–1638.

Kalemba, D.A.A.K., and Kunicka, A. 2003. "Antibacterial and antifungal properties of essential oils." *Current Medicinal Chemistry* 10(10): 813–829.

Kamel, C., Hajlaoui, H.T., et al. 2007. "The chemical composition and biological activity of clove essential oil, Eugenia caryophyllata (Syzigium aromaticum L. Myrtaceae): a short review." *Phytotherapy Research: An International Journal Devoted to Pharmacological and Toxicological Evaluation of Natural Product Derivatives* 21(6): 501–506.

Kaur, K., and Kaushal, S. 2019. "Phytochemistry and pharmacological aspects of Syzygium aromaticum: A review." *Journal of Pharmacognosy and Phytochemistry* 8: 398–406.

Khajeh, M., Yamini, Y., Sefidkon, F., and Bahramifar, N., 2004. "Comparison of essential oil composition of Carum copticum obtained by supercritical carbon dioxide extraction and hydrodistillation methods." *Food Chemistry* 86(4): 587–591.

Khalil, A.A., ur Rahman, U., Khan, M.R., Sahar, A., Mehmood, T., and Khan, M., 2017. "Essential oil eugenol: sources, extraction techniques and nutraceutical perspectives." *RSC Advances* 7(52): 32669–32681.

Kim, G.C., Choi, D.S., Lim, J.S., et al. 2006. "Caspases-dependent apoptosis in human melanoma cell by eugenol." *The Korean Journal of Anatomy* 39(3): 245–253.

Kim, S.S., Oh, O.J., Min, H.Y., et al. 2003. "Eugenol suppresses cyclooxygenase-2 expression in lipopolysaccharide-stimulated mouse macrophage RAW264. 7 cells." *Life Sciences* 73(3): 337–348.

Koh, T., Murakami, Y., Tanaka, S., Machino, M., and Sakagami, H., 2013. "Re-evaluation of anti-inflammatory potential of eugenol in IL-1β-stimulated gingival fibroblast and pulp cells." *in vivo* 27(2): 269–273.

Kramer, R.E. 1985. "Antioxidants in clove." *Journal of the American Oil Chemists' Society* 62: 111–113.

Lavinia, A., Cravotto, G., Giordana, L., Binello, A., and Chemat, F. 2013. "Ultrasound-assisted extraction of clove buds using batch-and flow-reactors: A comparative study on a pilot scale." *Innovative Food Science & Emerging Technologies* 20: 167–172.

Lee, K.G., and Shibamoto, T. 2001. "Antioxidant property of aroma extract isolated from clove buds [Syzygium aromaticum (L.) Merr. et Perry]." *Food Chemistry* 74(4): 443–448.

Liu, H., Schmitz, J.C., Wei, J., et al. 2014. "Clove extract inhibits tumor growth and promotes cell cycle arrest and apoptosis." *Oncology Research Featuring Preclinical and Clinical Cancer Therapeutics* 21(5): 247–259.

Mahapatra, S.K., Chakraborty, S.P., Majumdar, S., Bag, B.G., and Roy, S., 2009. "Eugenol protects nicotine-induced superoxide mediated oxidative damage in murine peritoneal macrophages in vitro." *European Journal of Pharmacology* 623(1–3): 132–140.

Markom, M., Singh, H., and Hasan, M., 2001. "Supercritical CO2 fractionation of crude palm oil." *The Journal of Supercritical Fluids* 20(1): 45–53.

Mbarawa, M., 2008. "Performance, emission and economic assessment of clove stem oil–diesel blended fuels as alternative fuels for diesel engines." *Renewable Energy* 33(5): 871–882.

Medeiros-Sousa, A.R., Ceretti-Júnior, W., de Carvalho, G.C., Nardi, M.S., Araujo, A.B., Vendrami, D.P., and Marrelli, M.T., 2015. "Diversity and abundance of mosquitoes (Diptera: Culicidae) in an urban park: larval habitats and temporal variation." *Acta Tropica* 150: 200–209.

Menon, K.V., and Garg, S.R., 2001. "Inhibitory effect of clove oil on Listeria monocytogenes in meat and cheese." *Food Microbiology* 18(6): 647–650.

Milind, P., and Deepa, K. 2011. "Clove: a champion spice." *International Journal of Research in Ayurveda and Pharmacy* 2: 47–54.

Nassar, M.I., Gaara, A.H., El-Ghorab, A.H., Farrag, A., Shen, H., Huq, E., and Mabry, T.J., 2007. "Chemical constituents of clove (Syzygium aromaticum, Fam. Myrtaceae) and their antioxidant activity." *Revista Latinoamericana de Química* 35(3): 47.

Noudogbessi, J.P., Gary-Bobo, M., Adomou, A. et al. 2014. "Comparative chemical study and cytotoxic activity of Uvariodendron angustifolium essential oils from Benin." *Natural Product Communications* 9(2): 1934578X1400900232.

Nowak, K., Ogonowski, J., Jaworska, M., and Grzesik, K., 2012. "Clove oil—Properties and applications." *Chemik* 66(2): 145–152.

Nurdjannah, N. 2004. "Diversifikasi penggunaan cengkeh." *Perspektif* 3(2): 61–70.

Nurdjannah, N., and Bermawie, N. 2012. Cloves. In *Handbook of herbs and spices* (pp. 197–215). Woodhead Publishing.
Nurdjannah, N., Yuliani, S., and Yanti, L. 1997. "Pengolahan dan diversifikasi hasil cengkeh." *Monograf Tanaman Cengkeh. Balai Penelitian Tanaman Rempah dan Obat. Badan Penelitian dan Pengembangan Pertanian* 118–30.
Nychas, G.J.E., 1995. Natural antimicrobials from plants. In *New methods of food preservation* (pp. 58–89). Springer, Boston, MA.
Ohkubo, T., and Shibata, M., 1997. "The selective capsaicin antagonist capsazepine abolishes the antinociceptive action of eugenol and guaiacol." *Journal of Dental Research* 76(4): 848–851.
Oiye, S.O., and Muroki, N.M., 2002. "Use of spices in foods." *Journal of food Technology in Africa* 7(2): 39–44.
Omidbeygi, M., Barzegar, M., Hamidi, Z., and Naghdibadi, H., 2007. "Antifungal activity of thyme, summer savory and clove essential oils against Aspergillus flavus in liquid medium and tomato paste." *Food Control* 18(12): 1518–1523.
Park, M.H., Kim, C.J., Lee, J.Y., Kim, I.S., and Kim, S.K., 2021. *Development and validation of a gas chromatography method for the determination of β-caryophyllene in clove extract and its application.* doi: https://doi.org/10.21203/rs.3.rs-402589/v1
Park, M.J., Gwak, K.S., Yang, I., Kim, K.W., Jeung, E.B., Chang, J.W., and Choi, I.G., 2009. "Effect of citral, eugenol, nerolidol and α-terpineol on the ultrastructural changes of Trichophyton mentagrophytes." *Fitoterapia* 80(5): 290–296.
Pisano, M., Pagnan, G., Loi, M. et al. 2007. "Antiproliferative and pro-apoptotic activity of eugenol-related biphenyls on malignant melanoma cells." *Molecular Cancer* 6(1): 1–12.
Plaza, P., Torres, R., Usall, J., Lamarca, N., and Vinas, I., 2004. "Evaluation of the potential of commercial post-harvest application of essential oils to control citrus decay." *The Journal of Horticultural Science and Biotechnology* 79(6): 935–940.
Podlewski, J.K. 2010. Chalibogowska-Podlewska A. Leki współczesnej terapii. Medical Tribune Polska, Warszawa.
Purseglove, J.W., Brown, E.G., Green, C.L., and Robbins, S.R.J., 1981. *Spices Vol. 2* (pp. 447–813). Longman Group Ltd.
Raina, V.K., Srivastava, S.K., Aggarwal, K.K., Syamasundar, K.V., and Kumar, S., 2001. "Essential oil composition of Syzygium aromaticum leaf from Little Andaman, India." *Flavour and Fragrance Journal* 16(5): 334–336.
Raja, M.C., Srinivasan, V., Selvaraj, S., and Mahapatra, S.K., 2015. "Versatile and synergistic potential of eugenol: a review." *Pharmaceutica Analytica Acta* 6(5): 367.
Rani, R., and Jena, M.K., 2021. "Clove (*syzygium aromaticum*): beneficial effects on human health: a review." *Plant Archives* 21(1): 1967–1972.
Rao, V.P.S., and Pandey, D. 2007. Extraction of essential oil and its applications. A Project Report submitted in partial fulfillment of the requirements of Bachelor of Technology (Chemical Engineering). National Institute of Technology Rourkela-769008, Orissa.
Ratri, P.J., Ayurini, M., Khumaini, K., and Rohbiya, A., 2020. "Clove Oil Extraction by Steam Distillation and Utilization of Clove Buds Waste as Potential Candidate for Eco-Friendly Packaging." *Jurnal Bahan Alam Terbarukan* 9(1): 47–54.
Reverchon, E., and Marrone, C. 1997. "Supercritical extraction of clove bud essential oil: isolation and mathematical modeling." *Chemical Engineering Science* 52(20): 3421–3428.
Shahidi, F., Pegg, R.B., and Saleemi, Z.O., 1995. "Stabilization of meat lipids with ground spices." *Journal of Food Lipids* 2(3): 145–153.
Singh, A., Singh, R.K., Bhunia, A.K., and Singh, N., 2003. "Efficacy of plant essential oils as antimicrobial agents against Listeria monocytogenes in hotdogs." *LWT-Food Science and Technology* 36(8): 787–794.
Singh, A.K., Dhamanigi, S.S., and Asad, M., 2009. "Anti-stress activity of hydro-alcoholic extract of Eugenia caryophyllus buds (clove)." *Indian Journal of Pharmacology* 41(1): 28.
Singh, K.K., and Goswami, T.K., 2000. "Cryogenic grinding of cloves." *Journal of Food Processing and Preservation* 24(1): 57–71.
Smith-Palmer, A., Stewart, J., and Fyfe, L., 1998. "Antimicrobial properties of plant essential oils and essences against five important food-borne pathogens." *Letters in Applied Microbiology* 26(2): 118–122.

Sulieman, A.M.E., El Boshra, I.M.O., and El Khalifa, E.A., 2007. "Nutritive value of clove (Syzygium aromaticum) and detection of antimicrobial effect of its bud oil." *Research Journal of Microbiology* 2: 266–271.

Tainter, D.R., and Grenis, A.T., 1993. *Spices and seasonings: a food technology handbook* (No. 633.83 T3).

Tajkarimi, M.M., Ibrahim, S.A., and Cliver, D.O., 2010. "Antimicrobial herb and spice compounds in food." *Food Control* 21(9): 1199–1218.

Tekin, K., Akalın, M.K., and Şeker, M.G. 2015. "Ultrasound bath-assisted extraction of essential oils from clove using central composite design." *Industrial Crops and Products* 77: 954–960.

Vidhya, N., and Devaraj, S.N., 2011. "Induction of apoptosis by eugenol in human breast cancer cells." 49(11): 871–878. PMID: 22126019.

Weiss, E.A., 1997. *Essential oil crops*. Cab International.

Wenqiang, G., Li, S., Yan, R., Tang, S., and Quan, C. 2007. "Comparison of essential oils of clove buds extracted with supercritical carbon dioxide and other three traditional extraction methods." *Food Chemistry* 101: 1558–1564.

Worrall, S., 2012. "The world's oldest clove tree." *BBC News Magazine. Retrieved* 24.

Yadav, S., Gupta, S.K., Bharti, D., and Yogi, B., 2020. "Syzygium aromaticum (clove): a review on various phytochemicals and pharmacological activities in medicinal plant." *World Journal of Pharmaceutical Research* 9(11): 349–363.

CHAPTER 9

Bioactive Compounds in Cumin

Abhirup Mitra and Winny Routray

CONTENTS

DOI: 10.1201/9781003215387-9

9.1 INTRODUCTION

Cumin is famous for its distinctive aroma and is a member of parsley family. Dried cumin seed is derived from *Cuminum cyminum* L. *Cuminum* is a Latin name, which has been derived from Greek word "Kuminon", most probably originated from the old Babyleniun name "ka-ma-nu". In many parts of India, cumin is popularly known as "Jeera". Apart from that name, it has other names, which also differ with respect to the region. It can grow up to 30–50 cm and is usually harvested manually. It is an herbaceous and flowering plant with height of the stem in the range of 20–30 cm and diameter of 3–5 cm. It is also an annual herb, which is considered a family member of Apiaceae family and has a chromosome number of $2n = 14$. Cuminaldehyde, cymene, and terpenoids are the crucial volatile compounds present in cumin. The distinctive fragrance of cumin volatile oil, derived from the dried seeds, is accredited due to the existence of cuminaldehyde, 3p-menthen-7-al, γ-terpinene, p-cymene, β-pinene, p-mentha-1,3-dien-7-al, p-mentha-1, and other existing aldehydes present in cumin.

Lots of previous research studies on cumin have proven that corresponding plant shoots a great resource of essential amino acids, fatty acids, phenolic compounds, and many other metabolites, wherein therapeutic potential of cumin can be peerless. Further stress conditions such as salinity stress have been observed to support the metabolic responses and enhance the concentration of these beneficial components. Previous research also demonstrated that *Nigella sativa*, which is commonly known as black cumin (and is different from *C. cyminum*), has great salubrious prospective, apart from the other wide range of health benefits, including anticancer, antidiabetic, immune-modulatory, antimicrobial, anthelmintic, anti-inflammatory, and analgesic, gastroprotective, hepato-protective, diuretic, antihypertensive, and renal protective effects, mainly accredited to the existence of thymoquinone. Cuminaldehyde, which is also a major component of cumin, is a member of the class of benzaldehydes and is an essential component of essential oils derived from cumin, which has great antifungal and antimicrobial effects against *Penicillium chrysogenum* and *Escherichia coli*. Previous studies on cumin have also shown the anticarcinogenic effects, wherein the cumin seeds act as a potential inhibitor of sarcomas and hepatomas, both squamous cells.

Attributed to the various beneficial effects of the major components of cumin, cumin is not only consumed as a part of several cuisines but also as a nutraceutical component and as a part of various traditional remedies for different health ailments in many parts of the world. This chapter has discussed various components present in cumin along with the extraction, isolation, encapsulation, and characterization apart from the beneficial effects of cumin. A chemical structure of two major compounds of cumin is shown in Figures 9.1 and 9.2.

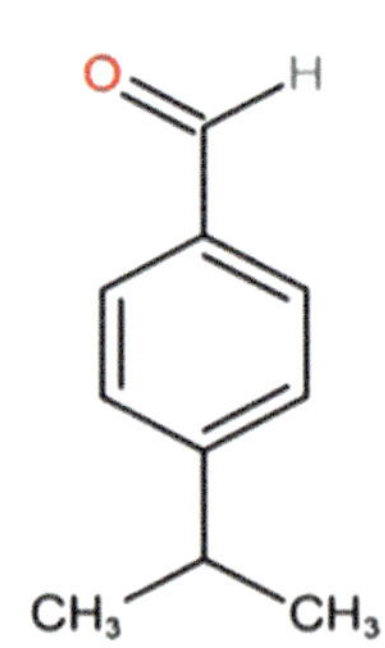

Figure 9.1 Cuminaldehyde.

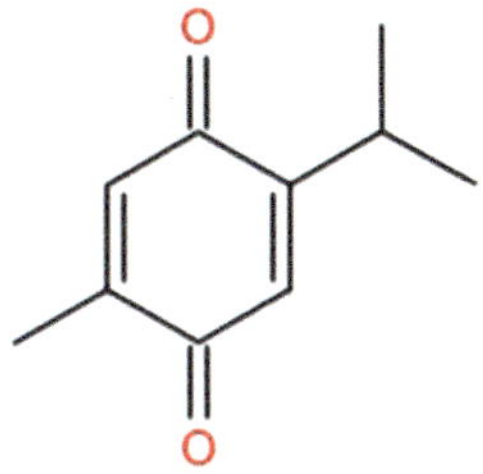

Figure 9.2 Thymoquinone.

9.2 CUMIN VARIETIES

Generally, three major types of cumin are found, which include cumin (*C. cyminum* L.), black cumin (*N. sativa* L.), and bitter cumin (*Cuminum nigrum*). Considering the agro-climatic conditions, different varieties of cumin are cultivated in different regions. In India, in the states of Rajasthan and Gujarat, different varieties of cumin are cultivated such as Gujarat Cumin-1, Gujarat Cumin-2, Gujarat Cumin-3, Gujarat Cumin-4, RZ-19, RZ-209, RZ-223, etc., which also demonstrates the varieties of a spice grown in particular regions. Figure 9.3 shows two different varieties of cumin widely used in Indian cuisine.

9.3 NUTRITIONAL COMPOSITION AND BIOACTIVE COMPONENTS IN CUMIN

Cumin consists of 7% crude fibre, 9.5% total ash, 0.5% acid insoluble ash, 2.3–4.8% volatile oil, 8% moisture, and 25% total ether extract. It is a rich source of monounsaturated fat and vitamins B and E (Srinivasan, 2018). The distinctive aroma of cumin is due to the presence of essential oil consisting of p-cymene (14.7%), p-mentha-1,4-dien-7-al (13.5%), cuminaldehyde (32.6%), and β-pinene (12.7%), which can be easily extracted by steam distillation. According to Shankaracharya (1974), cumin oil is also rich in oxygenated compounds and hydrocarbons. Pande and Goswami (2000) have shown the presence of 15 compounds which constitute 86.4% of the cumin oil, of which only 12 have been identified.

Figure 9.3 Cumin.

Essential oil can be fractionated to enhance the quality. Cuminaldehyde is the major component of essential oil, which exhibits great antifungal property. Agrawal and Gour (1992) have demonstrated the presence of antifungal components in cumin oil and its inhibition capacity against *Fusarium*, the fungus causing cumin wilt. The nutritional composition of cumin seeds is summarized in Table 9.1.

The major components of *N. sativa* seeds include proteins (26.7%), fat (28.5%), carbohydrates (24.9%), total ash content (4.8%), and crude fibre content (8.4%), along with many minerals (Cu, Zn, P, and Fe). Thymoquinone (30–48%), thymohydroquinone, dithymoquinone (nigellone), p-cymene, carvacrol, *t*-anethole, 4-terpineol, sesquiterpene longifolene, α-pinene, and thymol have been identified as the most active components of *N. sativa* (Srinivasan, 2018). Other minor components of *N. sativa* include limonene, carvone, and citronellol, isoquinoline alkaloids (nigellicimine and nigellicimine-*N*-oxide), and pyrazole alkaloids (nigellidine and nigellicine). α-Hederin, which is a water soluble pentacyclic triterpene, has also been identified in *N. sativa* seeds. Unsaturated fatty acids are also found in the black cumin seeds, which include linoleic acid, oleic acid, eicosadienoic acid, and dihomolinoleic acid. Black cumin also contains saturated fatty acids such as palmitic and stearic acids. Among the different sterols present in *N. sativa*, α-sitosterol is the major sterol, followed by stigmasterol (Ali and Blunden, 2003; Cheikh-Rouhou et al. 2008; Mehta et al. 2008; Nickavar et al. 2003; Srinivasan 2018).

There are very few reports regarding the composition and bioactive properties of bitter cumin. Some of the reported components include elements such as Ca, K, Na, Fe, P, Mg, and essential oil (mainly consisting of carvone, limonene, and p-cymene) (Srinivasan 2018).

9.4 NUTRACEUTICAL PROPERTIES OF THE BIOACTIVE COMPONENTS OF CUMIN

As discussed earlier, from ancient times, cumin has been used as food additives to enhance the flavour and taste of food. Apart from that, it has numerous medicinal properties (Figure 9.4) and has been used for the treatment of many health problems. Cumin powder has been used as a part of suppository and poultice; it has been smoked and taken through a pipe orally. Cumin has several health benefits including antioxidant activity, blood platelet aggregation properties, antidiabetic, anticancer, anti-inflammatory, cardioprotective, and pulmonary-protective along with anti-asthmatic properties.

9.4.1 Antioxidant Activity

Several components of cumin have shown antioxidant properties. Antioxidant property of cumin oil is very high due to the presence of flavonoids, monoterpene alcohols, flavours, and other phenolic compounds (De Martino et al. 2009; Najda et al. 2008; Rodov et al. 2010; Ruberto and Baratta 2000; Samojlik et al. 2010). Cumin consists of flavonoids, namely luteolin and apigenin, which have demonstrated high antioxidant potential. In a study by Leung (1980), petroleum ether soluble fraction of cumin seed demonstrated significant antioxidant activity, which could be mainly attributed to cosmosiin and cuminaldehyde, as the petroleum ether extract contains cobalt (2.2 ppm), copaene (30 ppm), copper (9–16 ppm), cosmosiin (1500 ppm), cuminaldehyde (4000–16,000 ppm), cuminic acid (1 ppm) (Lawrence 1976–1980, Leung 1980). In a different study, cuminaldehyde has also shown good scavenging potential against a superoxide anion (Krishnakantha and Lokesh 1993). Supplementation of cumin to diabetic rats has been observed to decrease the free radical-mediated oxidative stress in the cells of the corresponding rats. Keerthiga et al. (2019) and Surya et al. (2005) reported the effectiveness of cumin seed oil-mediated Ag nanoparticles, wherein the antioxidant activity of nano-encapsulated cumin oil has been also further analysed and verified. Further studies in this regard during the past few years are presented in Table 9.2.

Table 9.1 Different Components of Cumin

Types of Nutrients	Name of Nutrient	Applications	Nutrient Value	Sources
Minerals	Iron	Required for red blood cell formation	0.048%	Lal et al. (2014); Singh et al. (2017)
	Copper	Required in the production of red blood cells	0.867 mg	Singh et al. (2017)
	Calcium	–	0.9%	Lal et al. (2014)
	Manganese	Used by the body as a cofactor for the powerful antioxidant enzyme, superoxide-dismutase	3.3 mg	Singh et al. (2017)
	Selenium	–	–	Singh et al. (2017)
	Zinc	Zinc is a cofactor in many enzymes that regulate growth and development, digestion, and nucleic acid synthesis	4.8 mg	Singh et al. (2017)
Electrolytes	Potassium	It is an important component of cell and body fluids that helps controlling heart rate and blood pressure	68 mg	Singh et al. (2017)
	Sodium	–	1788 mg	Singh et al. (2017)
Vitamins	Vitamin A	Antioxidants, essential nutrients	175 IU/100 g	Lal et al. (2014); Singh et al. (2017)
	Riboflavin		0.32 mg	
	Niacin		2.5 mg	
	Folic acid		10 μg	
	Vitamin E	Antioxidants, essential nutrients	3.3 mg	Lal et al. (2014); Singh et al. (2017)
	Pyridoxine		0.435 mg	
	Vitamin C		17.2 mg	
	Vitamin K		5.4 μg	
Phytonutrient	Carotenes	Antioxidants, health beneficial components	762 μg	Singh et al. (2017)
	Zeaxanthin		0 μg	
	Lutein		448 μg	
Others	Energy	Basic nutrients	460 kcal	Lal et al. (2014); Singh et al. (2017)
	Carbohydrate		44.24 g	
	Protein		17.8 g	
	Fat fibres		22.27 g	
	Cholesterol		10.5 g	
Volatile constituents		Part of essential oil (useful against several health ailments)	Relative abundance (%)	Khalil et al. (2018); Wanner et al. (2010)
	α-Thujene			
	α-Pinene		0.47	
	Sabinene		2.44	
	β-Pinene		0.58	
	α-Phellanderene		13.56	
	p-Cymene		0.79	
	Limonene		17.50	
	δ-Carene		0.71	
	1,8-Cineole		0.36	
	γ-Terpinene		0.17	
	Myrtenol		12.15	
	Cuminaldehyde		0.48	
	α-Terpinen-7-al		25.17	
	p-Mentha-1,4-dien-7-al		10.89	
	β-Caryophyllene		7.12	
	cis-Sabinene hydrate		0.11	
	Terpinolene		–	
	trans-Sabinene hydrate		– –	
	trans-Pinocarveol		–	

(Continued)

Table 9.1 Different Components of Cumin (*Continued*)

Types of Nutrients	Name of Nutrient	Applications	Nutrient Value	Sources
	Terpinen-4-ol		–	
	Cymen-8-ol		–	
	α-Terpineol		–	
	p-Isopropylphenol		–	
	p-Menth-3-en-7-al		–	
	p-Mentha-1,3-dien-7-al		–	
	p-Mentha-1,4-dien-7-al		–	
	Cuminic alcohol		–	
	Thymol		–	
	Carvacrol		–	
	Daucene		–	
	Cuminic acid		–	
	Carotol		–	
	β-Caryophyllene		–	
	(*E*)-β-farnesene		–	
	(*E*)-α-bergamotene		–	
	β-Acoradiene		–	

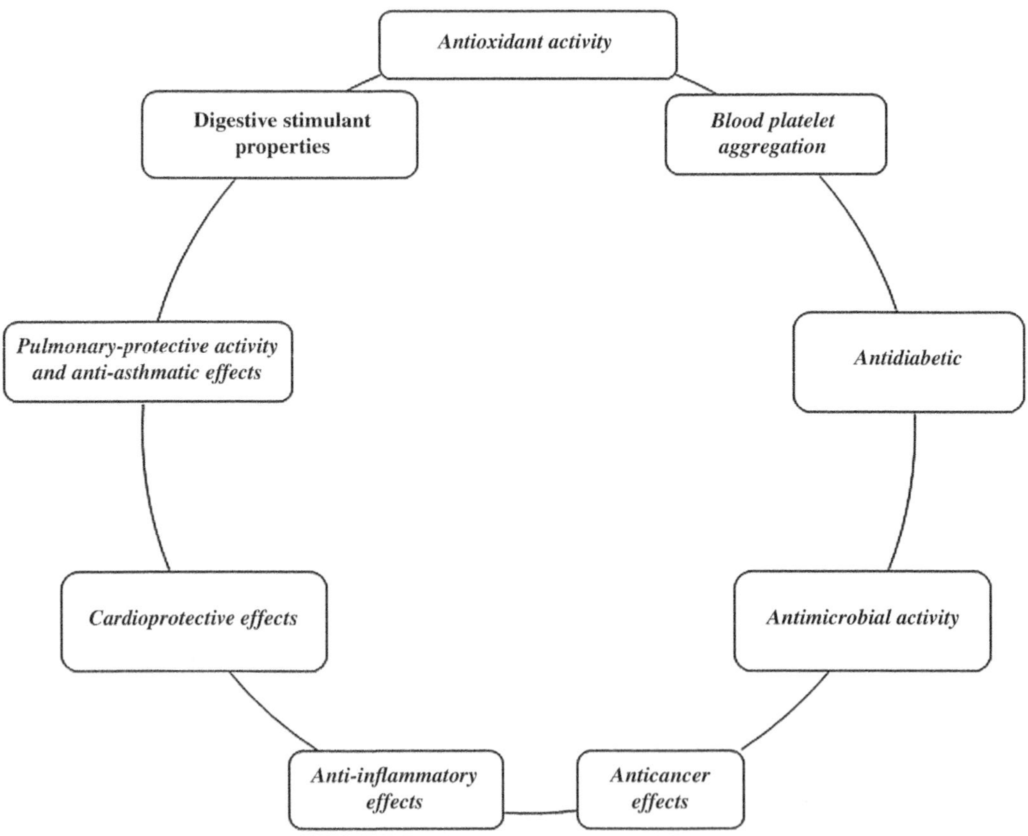

Figure 9.4 Properties of cumin.

Table 9.2 Summary of Certain Compounds and Extracts of Cumin Demonstrating Antioxidant Activities and the Assays Used to Verify the Antioxidant Activities

Sl. No.	Antioxidant Compound	Amount	Sources
1	Total ascorbate	188.67 ± 2.32 μg/g	Dua et al. (2012)
2	Reduced ascorbate	68.99 ± 5.63 μg/g	Dua et al. (2012)
3	Oxidized ascorbate	119.68 ± 5.51 μg/g	Dua et al. (2012)
4	Riboflavin	1.25 ± 0.04 μg/g	Dua et al. (2012)
5	Tocopherol	54.0 ± 2.51 μg/g	Dua et al. (2012)
6	Total phenolic content	10.76 ± 0.15 mg GAE/g	Ali et al. (2021)
7	Total flavonoid content	3.62 ± 0.12 mg QE/g	Ali et al. (2021)
8	Total tannin content	2.04 ± 0.12 mg CE/g	Ali et al. (2021)

9.4.2 Antimicrobial Activity

Alcoholic extract of cumin and cumin essential oil has displayed a significant antimicrobial activity. Cumin oil and alcoholic extract have also showed a successful inhibition capability against *Klebsiella pneumoniae* and other clinical isolates, which improved the cell morphology, helped to decrease urease activity and capsule expression. Cumin has shown great preventive properties against biofilm formation of *Streptococcus pyogenes* and *Streptococcus mutans* (Derakhshan et al. 2010; Shayegh et al. 2008). It also has antifungal property against food, soil, yeasts, mycotoxin, and aflatoxins-producing organisms including pathogens (Boyraz and Özcan 2005; De Martino et al. 2009; Farag et al. 1989a, 1989b; Hajlaoui et al. 2010; Hammad and Youssef, 1995; Mehdi et al. 2009; Razzaghi-Abyaneh et al. 2009; Romagnoli et al. 2010; Skrinjar et al. 2009). Apart from cumin oil, cuminaldehyde has also shown a great antibacterial and larvicidal property. Cumin essential oil and cuminaldehyde were tested against gram-positive and gram-negative bacteria, which were separated from different processed meat sample sources, *Candida albicans* isolates and clinical isolates, wherein inhibitory activity was observed for all microorganisms other than *Pseudomonas* spp. (Wanner et al. 2010). In a different study by Khalil et al. (2018), the antimicrobial activities of several spices obtained from apiaceous fruits were analysed and it was observed that maximum microbial inhibition activities against all basic bacterial strains including *E. coli*, *Bordetella bronchiseptica*, and *Staphylococcus aureus* were observed in the case of cumin, coriander, and caraway oils. The antimicrobial activities were mainly attributed to the monoterpenes (Khalil et al. 2018). Studies with scanning electron microscopy of the mechanism of action causing antimicrobial activity demonstrated that cumin essential oil affects the cell membrane entirety of these microorganisms (Behbahani et al. 2019). This antimicrobial activity can be further enhanced through the encapsulation and nanoparticle production. Cumin oil-mediated Ag nanoparticles showed dose-dependent antimicrobial activity against several oral pathogens including *S. mutans*, *S. aureus*, and *Enterococcus fecalis* (Vignesh et al. 2019). Hence, the alcoholic extract and essential oil have been recommended to be used in the pharmaceutical sector as an antiseptic or a disinfectant (Derakhshan et al. 2007). Biologically active peptides isolated from black cumin seeds, from the thionins family with high cationic properties, have also been reported to demonstrate antimycotic and antiproliferative properties, which can also be further explored and exploited to produce biologically active cumin-based formulations for specific treatments.

9.4.3 Antidiabetic Effects

Cumin helps to oppose metabolic alterations through lowering of a blood urea level and has also been observed to decrease the discharge of creatinine and urea of a diabetic patient (Willatgamuwa et al. 1998). Apart from that, it has also been observed to lower the generally increased plasma urea level of a diabetic patient to 50%. Cumin has been utilized as a herbal antidiabetic drug and has shown a significant effect in human body (Karnic 1991).

Effectiveness of cumin and its extracts have been further supported by the observations during various animal studies. A study on rabbit observed that the oral administration of cumin led to a significant hypoglycemic effect, which resulted in the decrease in the area under the glucose tolerance curve with a hyperglycemic peak. According to Roman-Ramos et al. (1995), the biological activity of cuminaldehyde inhibited α-glucosidase and aldose reductase secluded from a rat. Lee (2005) has demonstrated that in alloxan-induced diabetic rats with hyperlipidemia, when cumin was provided orally, the body weight reduced along with the reduction in plasma and tissue cholesterol, phospholipids, free fatty acids, triglycerides, alkaline phosphatase, aspartate transaminase, and γ-glutamyl transferase (GGT) activities.

During a study (Surya et al. 2005) on diabetic rats (with diabetes induced through 150 mg/kg body weight alloxan monohydrate), cumin supplementation led to elevated levels of catalase, superoxide dismutase, glutathione, glutathione-*S*-transferase, and glutathione peroxidase in the kidney, liver, pancreas, aorta, and intestine of diabetic rats as compared with non-diabetic rats (Surya et al. 2005). In a different study by Surya et al. (2005), the hypolipidemic effect of *C. cyminum* L. was observed in the case of alloxan-induced diabetic rats. Andallu and Ramya (2007) reported the high-density lipoproteins enhancing properties of cumin along with hypolipidemic and cholesterol lowering effects. In certain studies, the effect of cumin extracts has been further enhanced or maintained through combination with controlled concentrations of other components such as glyburide (Kaur et al. 2019).

9.4.4 Cardioprotective Effects and Blood Platelet Aggregation

Consumption of cumin and its extracts has also led to cardioprotective impact occurring through hypotensive and hypolipidemic influence, observed in different types of extracts including the aqueous extract. In addition, cumin consumption significantly affected the inflammation, arterial endothelial nitric oxide synthase expression, and oxidative stresses in hypertensive rats (Kalaivani et al. 2013). During a study, cumin was introduced orally to hypertensive rats with a dose of 200 mg/kg body weight, for nine weeks to ameliorate plasma nitric oxide and decrease systolic blood pressure, wherein the data suggested amelioration of inflammatory and oxidative stress along with increasing endothelial functions in hypertensive rats consuming cumin (Kalaivani et al. 2013). In a study, the extract of cumin in ether inhibited arachidonate-induced platelet agglomeration in human platelets in a dose-depending method, which was demonstrated by Srivastava (1989).

Cardiovascular effects of thymoquinone have been widely investigated, which is a major component of black cumin. This component has been reported to effectively modulate pyroptosis and autophagy and has contributed towards maintenance of structural integrity of pancreatic β cell. The protective characteristics have also been associated with inhibition of DNA damage and increased expression of Nrf2 in the tissues (Sarkar et al. 2021; Xiao et al. 2018).

9.4.5 Digestive Stimulant Properties

C. cyminum is used as a component of remedies or traditional medicines to treat indigestion and hypertension. The effectiveness of cumin in improving digestion has been shown by several research groups; studies mainly focused on its stimulation effects on digestive enzymes. The effects of cumin on digestive enzymes of rat intestinal mucosa and pancreas through single oral administration and/or continuous dietary intake have been studied (Platel and Srinivasan 2000a, 2000b). It was observed that the dietary (1.25%) cumin enhanced the activities of pancreatic amylase, trypsin, and chymotrypsin but minimized the activity of pancreatic lipase (Platel and Srinivasan 2000a), thus leading to enhanced digestion. Furthermore, during the animal-based studies, among the terminal digestive enzymes in small intestine, maltase activity has been reported as significant.

Bile acid secretion has been observed to increase in case of both dietary intake and single dose intake of cumin. A stimulatory effect on the bile flow rate has also been observed, where bile volume has been observed to increase by 25%, attributed to the cumin consumed through dietary intake, whereas single oral dose supplementation did not have a significant effect. Bile acid secretion was also observed to increase up to 70% as compared to control. As bile acids contribute significant to the digestion and absorption of food, mainly during the micelle formation, increase of bile acids attributed to the healthy digestion in individuals (Platel and Srinivasan 2000b; Srinivasan 2018). Furthermore, cumin consumption has also been correlated with the residence of food in gastrointestinal tract, as observed in the case of experimental rats, wherein the food transit time has been reduced by 25%, which also demonstrates the benefits of cumin consumption for the digestive health of the individuals and can be associated with the stimulation of enzyme and bile secretion in the gastrointestinal tract (Platel and Srinivasan 2001; Srinivasan 2018).

9.4.6 Pulmonary-Protective Activity and Anti-Asthmatic Effects

Black cumin has been explored to check its helpful effects on lung injury in rats after pulmonary aspiration (Kanter 2009) and it was observed that it contributes towards hindering the inflammatory pulmonary responses. A black cumin therapy has shown a great reduction in the iNOS activity and increase in surfactant protein D in lung tissues. This study proved that the treatment with cumin might be helpful in lung injury. The ameliorative effect of black cumin oil in rats' lung injury has been noticed (Tayman et al. 2012).

9.4.7 Anticancer Effects

Cumin also has an anticarcinogenic or anticancer property. It has the capacity to hinder the orientation of gastric lamellate cell carcinomas (Gagandeep et al. 2003). A protective effect was also observed in rats against colon cancer after treating with cumin. A research study of Singh et al. (2017) also observed that cumin has the potential to prevent the colon cancer of a cat by inducing a specific carcinogen and also helps in minimizing the activity of mucinase enzyme and β-glucuronidase. It was also observed that the cholesterol, cholesterol/phospholipids ratio, and 3-methylglutaryl COA were reduced in cumin-colon treated rats. During the research by Nalini et al. (1998) and Nalini et al. (2006), the other inhibition activities were also observed, where consumption of cumin-induced benzopyrene for stomach tumorigenesis and similarly 3-methylcholanthrene for uterine cervix tumorigenesis, and 3-methyl-4-dimethyaminoazobenzene for hepatomas.

9.4.8 Immunomodulatory Action

The immunomodulatory properties of black cumin and its extracts have been evaluated, where thymoquinone has been found as the most significant component contributing towards the modulation of cellular and humoral adaptive immune responses. These effects have been more comprehensively covered by Majdalawieh and Fayyad (2015), wherein the application of thymoquinone for regulation of both infectious and non-infectious conditions such as allergies, autoimmunity, and cancer has been emphasized. During the study by Chauhan et al. (2010) with a mice model, it was observed that the oral treatment with cumin restored the T cells (CD4 and CD8) Th1 cytokines' utterance of normal and cyclosporine immune compromised creature. Apart from that, cumin used T lymphocytes and reduced size of adrenal glands along with lowered corticosterone levels and increased thymus and spleen weights in stressed immune compromised mice.

9.4.9 Effect on the Central Nervous System

Cumin oil was observed to minimize the frequency of unbidden activity induced by maximal pentylenetetrazol and electroshock in mice, which was concentration and time dependent and increased with time. According to the research perfomed by Sayyah et al. (2002), cuminaldehyde prevented the oxidation of L-3.5-dihydroxyphenyklalanine (L-DOPA) attributed to its tyrosinase inhibitor property (Kubo and Kinst-Hori 1998).

9.4.10 Estrogenic/Anti-Osteoporotic Activity

The anti-osteoporotic effect was observed in the presence of phytoestrogens during cumin consumption, wherein the urinary calcium excretion was reduced and the calcium content was enhanced leading to the increased strength of bones (Malini and Vanithakumari 1987).

9.4.11 Other Biodynamic Actions

Cumin has shown antitussive effects leading to the relaxant effect by stimulating β-adreno receptors and/or histamine H1 receptors. It has also shown anti-aggregatory activity and inhibition of eicosanoid synthesis through inhibition of arachidonic acid (AA)-induced platelet aggregation, thromboxane B2 production from exogenous AA along with simultaneous increase in the formation of lipoxygenase (Kunzemann and Herrmann 1977; Srivastava 1989).

9.4.12 Bioavailability Enhancer

Cumin has bioenhancer flavonoid glycoside 3′,5-dihydroxyflavone-7-O-β-dgalacturonide-4′-O-β-*d*-glucopyranoside; hence, consumption led to an increase drifampicin levels in rat plasma through the increment of peak concentration, when codosed with this molecule (Sachin et al. 2007, 2009).

There are several other beneficial effects of cumin, which have been reported by various researchers and are being increasingly investigated, including anti-inflammatory effect, analgesic effect, and ameliorative effects on dyslipidemia (Srinivasan 2018). As treatment through the application of natural components is currently being emphasized heavily, research areas are being updated and they include extensive research on the processing methods and their effects on the concentration of the bioactive compounds in the end products along with their retention and bioactivity during the storage.

9.5 EXTRACTION AND ISOLATION OF CUMIN SEED OIL

Traditionally cumin seed oil has been successfully extracted using Soxhlet extraction, steam distillation, and hydrodistillation, which are still prevalent in many parts of the world and are being extensively used in many different research groups. However, the solvent extraction process and the steam distillation process have many demerits such as longer time requirement and require enormous organic solvents (Deng et al. 2006). Apart from that, in these processes, the loss of volatile components can also be high. A lower extraction rate and degradation of unsaturated compounds are also some major issues. Furthermore, sometimes toxic solvent residues can be found in the extract (Gironi and Maschietti 2008; Glišić et al. 2007; Jimenez-Carmona et al. 1999).

Recently, these traditional techniques have been combined with different advanced technologies to increase the efficiency of the process along with the least effect on the quality and quantity of the extract. Mékaoui et al. (2016) combined microwave with steam distillation and hydrodistillation for extraction of essential oils from cryogenically ground cumin seeds, which have been described as accelerated techniques performed at atmospheric pressure. During this process, a high recovery rate of major compounds including cuminaldehyde, p-cymene, β-pinene, and γ-terpinene was observed (Mékaoui et al. 2016). Similar observations were also reported by Benmoussa et al. (2018) regarding rapid extraction of essential oils from Tunisian cumin using microwave hydrodiffusion and gravity. Eikani et al. (2007) compared the volatile oil extracted using superheated water extraction with Soxhlet extraction and hydrodistillation, where it was observed that higher quality oil was obtained through superheated water extraction in terms of oxygenated compounds, which further supported the application of advanced technologies for high-quality volatile cumin oil extraction.

In a study by Kiralan et al. (2014), black cumin oils extracted using solvent-free extraction with cold pressing and solvent-based extraction methods (Soxhlet and microwave-assisted extraction) were evaluated in terms of fatty acids and bioactive compounds (chlorophyll, tocopherols, sterols, carotenoid, and total phenolics) composition and physiochemical properties. It was observed that oils obtained through the cold pressing method had highest concentrations of thymoquinone, benzoic, and p-hydroxy benzoic acids along with high antiradical power (Kiralan et al. 2014). Furthermore, a study of the effects of different solvents on phenolic contents extraction and antioxidant activities of two different varieties of cumin seeds demonstrated that the extracts obtained using 80% acetone had the highest phenolic content and highest antioxidant potential (DPPH-scavenging activity, β-carotene bleaching assay, chelating ability, and reducing power) (Rebey et al. 2012). Deep eutectic solvent-based microwave-assisted extraction also led to high yields of high-quality cumin essential oils in a study by Zhao et al. (2019).

Cumin volatile oil has also been extracted using pretreatment with various enzymes, including cellulase, pectinase, protease, and viscozyme. The oil yield was observed in the range of 3.2–3.3% as compared to 2.7% in a control sample along with no significant change in the cuminaldehyde content. Hence, based on these observations, it was concluded that enzyme-assisted extraction could be an alternative to increase oil yield with high quality (Sowbhagya et al. 2011).

Cumin oils have also been extracted using the supercritical-CO_2 extraction method (Lee 2014). Supercritical extraction is a great alternative to overcome the general problems of long extraction times and degradation of oil, often encountered during the extraction of essential oil. Most common supercritical fluid for supercritical extraction is carbon dioxide (Hawthorne et al. 1993; Jimenez-Carmona et al. 1999; Senorans et al. 2000). In spite of high solubility of the essential oils in supercritical carbon dioxide, the rate of extraction has been comparatively slow with pure carbon dioxide (ca. 80% recovery after 90 min) (Hawthorne et al. 1993). But, the combination of 15 min dynamic extraction with pure carbon dioxide and methylene chloride as a modifier has been observed to give a higher recovery rate. Eikani et al. (1999) compared the cumin essential oil obtained using supercritical extraction with conventional steam distillation, where they identified valuable components present in extract obtained from supercritical extraction, which were generally degraded during steam distillation. In certain cases, the oil obtained using supercritical extraction has been observed to be significantly different from steam distillation methods. In the case of black cumin, a total of 16 compounds were recovered using supercritical extraction, accounting for 99.36% of the extract, wherein α-methyl-benzenemethanol, cuminaldehyde, and γ-terpinene were the major identified compounds. However, the concentration of ρ-cymene and the ratio of γ-terpinene and cuminaldehyde were higher in hydrodistilled essential oil as compared to supercritical CO_2 extract (Pourmortazavi et al. 2005). Hence, the combination of efficient processing methods with extraction methods can lead to the highest retention and recovery of various valuable components of cumin essential oil, which can be further exploited and applied.

9.6 ENCAPSULATION

Encapsulation helps to preserve volatility, aromas, and reactivity of the product and also contributes to greater solidity even under adverse environmental conditions, which is also an important step in case of cumin essential oil applications. Spray drying and freeze drying are the common methods of encapsulation; however, several other methods also have been adapted to produce encapsulated essential oils with highest activities. Spray drying is the oldest and mostly utilized encapsulation process (Balassa et al. 1971). It has several advantages such as less expensive and can be modified depending on the matrix of encapsulation. In addition, it is very much compatible with other equipment (Desai and Jin Park 2005). It can also be used for heat sensitive materials to protect or minimize the damages (Bhandari et al. 2008). The major advantages of freeze drying which makes it appropriate for the encapsulation of cumin essential oils include (i) decrease in heat-based deterioration of the heat sensitive components such as tocopherols, unsaturated fatty acids, etc., (ii) controlled moisture content of the final product, and (iii) easier restoration of oil microcapsules (Oetjen and Haseley 2004). However, it has some major disadvantages such as long processing time, high operation costs, and open porous structure of the final product, which has restricted its industrial application to limited products (Anandharamakrishnan 2014). Coacervation is another technique that has been used for encapsulation and can be used for cumin essential oils.

Apart from the popular encapsulation methods mentioned above, other methods used for oil encapsulation, which could also be applied in case of cumin essential oil application, include single layer emulsions (Kuhn et al. 2014), multilayer emulsions (Fioramonti et al. 2015; Tamnak et al. 2016), molecular inclusion (Li et al. 2011; Petrović et al. 2013; Shrestha et al. 2017) and utilization of particles from gas-saturated solution (Getachew and Chun 2016), encapsulation through utilization of macroscopic beads (Piornos et al. 2017), and a mini-emulsification-solvent evaporation process (Freiberger et al. 2015). In the following subsection, some of the most common materials used for encapsulation and the techniques employed have been briefly mentioned.

9.6.1 Materials Used for Encapsulation of Essential Oils: Specific Considerations and Sample Preparation Methods

Different compounds used for encapsulation of essential oils include starch, gum arabic, alginate, chitosan, and cyclodextrins; however, with increased research, further enhanced components are also being identified and explored.

Gum arabic, modified starch, and maltodextrin were used for the encapsulation of cumin oleoresin (Kanakdande et al. 2007) and the corresponding stability of the different encapsulations was studied under storage conditions for 6 weeks. In all these cases, the highest retention was observed for the volatiles, which was followed by other compounds identified in the oil, including cuminaldehyde, p-cymene, and γ-terpinene. Gum arabic provided the highest protection as compared to modified starch and maltodextrin. In a study by Gholamian et al. (2021), cumin essential oil was filled in calcium alginate hydrogels, wherein the effects of encapsulation and emulsion fabrication methods, hardening time, $CaCl_2$ level, sodium alginate concentration, and the loading capacity of the hydrogels were also studied. Based on the in vitro release of phenolic compounds under simulated conditions, images obtained through scanning electron microscopy and Fourier transform infrared spectrum, calcium alginate hydrogels were concluded as potential carriers for cumin essential oils (Gholamian et al. 2021). Alginate is widely used for encapsulation, wherein multiple emulsions are used with ionic gelation for locking in the essential oils for controlled release (Banerjee et al. 2013; Hosseini et al. 2013a; Noppakundilograt et al. 2015). The spray-drying process has also been used for the encapsulation of essential oils with a combination of other ingredients such as cashew gum and alginate, which has shown a significant result with 45% of essential oil released in 30 h and 95% within 50 h (de Oliveira et al. 2014).

Chitosan has high flexibility and is a good natural product for encapsulation of essential oils, and cumin was successfully encapsulated into chitosan nanoparticles by Amiri et al. (2020), wherein the corresponding nanoparticles displayed a substantial antioxidant trait and a minor anticancer activity. Chitosan can be used for encapsulation through different methods, which can lead to different types of products. In a method demonstrated by Chen et al. (2003), they have shown the formation of a nanogel with strong covalent linkages. Though the essential oil release process is relatively slow as 78% of cumin essential oil in a chitosan–caffeic acid nanogel is released in a week and for complete release it can take almost a month. The formation of nanoparticles with emulsion and ionic gelation is the second most widely used method (Hosseini et al. 2013a, 2013b). In this method, sodium tripolyphosphate is used, which helps to form ionic crosslinking along with divalent cations. A stable and regular nanoparticle is formed, but it has a distinctive release profile as compared with first one. It has an initial burst effect and slower release-related issues (Esmaeili and Asgari 2015; Hosseini et al. 2013a, 2013b; Mohammadi et al. 2015). The release of essential oils is faster under acidic conditions as compared to neutral conditions. Esmaeili and Asgari (2015), Ghaderi-Ghahfarokhi et al. (2017), and Mohammadi et al. (2015) demonstrated this method to create a nano-formulation of liquid additive for food to modify the stability as well as the solubility of essential oil. Another method with chitosan refers to the formation of nanoparticles with the help of nanoprecipitation, which are comparatively smaller than those prepared by the previous method, but have shown a great antibacterial property, even against *S. aureus* (Sotelo-Boyás et al. 2017).

Encapsulation using cyclodextrin has been employed for different components of cumin oil. Essential oil can be simply blended with cyclodextrin in water (Anaya-Castro et al. 2017; Kfoury et al. 2015) and further dried using advanced drying methods (freeze drying, which generally follows freezing and lyophilization of the essential oil and β-cyclodextrin complex (Mangolim et al. 2014; Rakmai et al. 2017; Tao et al. 2014)) for maximum retention of the essential oils; however, other complex and advanced methods of sample preparation and production have also been used. Figure 9.5 shows the process of spray drying encapsulation. In 2003, Manolikar and Sawant introduced the "kneading" method for encapsulation of essential oil with cyclodextrins by mixing using

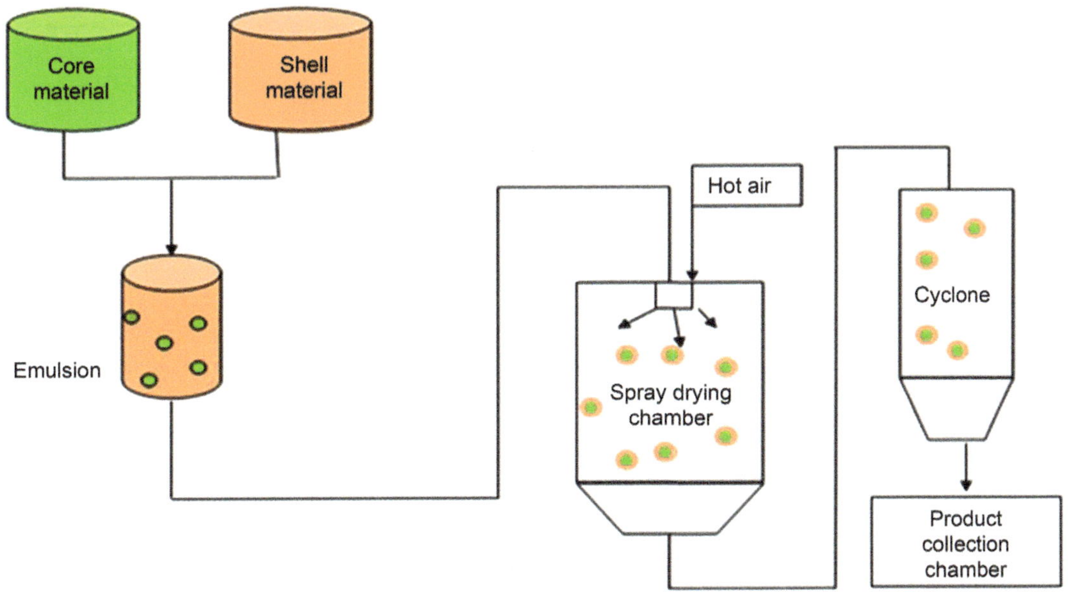

Figure 9.5 Spray-drying encapsulation process.

a mortar and kneading along with addition of small quantities of ethanol to obtain a paste. Then the paste was dried using a vacuum dryer at –20°C to produce a fine powder (Manolikar and Sawant 2003). By applying the same method, Santos and his team demonstrated that the kneading complexes of β-cyclodextrin and carvacrol (a component of cumin extract) resulted in significant antimicrobial property against *Salmonella typhimurium* and *E. coli* with a great entrapment efficiency. The complex had good storage stability in the presence of regular light (Santos et al. 2015; Tao et al. 2014). In a different study, Köse et al. (2021) demonstrated "simultaneous isolation and selective encapsulation of volatile compounds", including cuminaldehyde from essential oils through elctrospraying of β-cyclodextrin, which can be further exploited for the application in the case of other significant compounds.

Sachharomeyces cerevisiae base microcarriers have also been recently employed for the encapsulation of black cumin oil (Karaman, 2020), where the stability of black cumin oil was improved as a significant decrease in thymoquinone concentration was observed at 65°C. It was observed that the seed oil quality was preserved in the plasmolysed yeast encapsules (Karaman, 2020). Nanophytosome has also been proposed as a promising carrier for improving the beneficial properties of cumin essential oils (Mohammadi et al. 2021), which can be further explored for different applications.

9.6.2 Characterization of Bioactive Compounds

Characterization of cumin essential oils is usually done according to the specific guidelines and laws imposed by the exporting and importing countries. The criteria of characterization are explained by International Organization for Standardization (ISO) or French AFNOR (French Association of Normalization). Therefore, the physical and organoleptic properties such as colour, odour, solubility, refraction, and flash point, apart from the chemical properties like acid and ester indices, are administered (Boughendjioua and Djeddi 2018). According to Boughendjioua (2019), physicochemical properties of the essential oil have a standard refractive index (20°C) 1.48, density (20°C) 0.91, alcohol solubility (80% v/v) 1.1, aldehyde percentage (on the basis of cuminaldehyde) 50%, acidity (on the basis of cuminic acid) 1, alcohol percentage (on the basis of cuminol) 3.50%, steric index 19.24, and carbonyl index 9.32.

9.7 OTHER APPLICATIONS OF CUMIN

Apart from the culinary uses, cumin also has many applications in different sectors including the packaging sector and most importantly the pharmaceutical industry.

9.7.1 Applications of Essential Oil in Food Packaging

Essential oil extracted from cumin comes under permitted additives and has efficient preservation properties attributed to the presence of bioactive compounds leading to antioxidant and antimicrobial properties. Antimicrobial properties can delay the food spoilage by hindering the microbial activity by foodborne pathogens and their toxins. Generally chemical food additives are used for controlling the microbial growth and the corresponding effects, but replacing those chemically synthesized additives with natural additives is always a better choice. Essential oils are considered to be alternatives of those chemically synthesized additives (Carocho et al. 2014; Ribeiro-Santos et al. 2017). Their application in active packaging generally shows a better result, attributed to the release of biological components in packaged food leading to increased shelf-life (Ribeiro-Santos et al. 2017). Apart from that, the use of essential oils in active packaging can improve the water vapour barrier properties of the packaging film along within corporation of antioxidant activity.

9.7.1.1 Antimicrobial Activity in Packaging

Several research studies have been carried out for improving the safety and quality of food products through the application of natural additives instead of chemical additives to maintain the organoleptic and nutritional quality and hindering the growth of foodborne pathogens during storage; however, proper packaging plays an important role in the maintenance of the food quality. It is observed that the use of essential oils in biodegradable or edible packaging minimizes the use of chemical additives (Jafarzadeh et al. 2020; Rehman et al. 2020; Ribeiro-Santos et al. 2017). According to a research carried out by Shetty et al. (1994), the yeasts (*Saccharomyces* and *Candida* spp.) and fungi (*Aspergillus* and *Penicillium* spp.) were highly affected by essential oils of cumin and cuminaldehyde as compared to the bacteria. Fungi had microbial inhibitory concentration values 10–20 times lower than bacteria. Among the different microorganisms investigated, *E. coli* were the most sensitive and *Pseudomonas aeruginosa* was the most resistant among all the gram-negative bacteria. Furthermore, among all the gram-positive bacteria, *S. aureus* had microbial inhibitory concentration almost double than the others tested.

9.7.1.2 Antioxidant Activity in Packaging

Food spoilage can also occur due to the presence of oxygen attributed to the microbial activity (Gómez-Estaca et al. 2014). Oxidative deterioration can happen due to the interaction of reactive oxygen species (ROS) with the components, which are highly vulnerable to be oxidization leading to the structural changes. Hydroxyl radicals (OH^-), superoxide radicals (O_2^{2-}), and hydrogen peroxide (H_2O_2) are categorized as ROS (Shankar and Mehendale 2014). ROS can be responsible for various undesirable changes in foods, which include minimization of the shelf-life and corresponding effects on the nutritional value, colour, and flavour of the food along with rancidity (Shahidi and Ambigaipalan 2015). The oxidation and peroxidation of fatty acids and other fats are mainly responsible for the rancidity of foods. In packaged foods, this process is generally considered to be initiated by change in metals' oxidation state already present in food and/or exposure to light containing ultraviolet (UV) (such as sunlight) can cause the changes in fat globules (Lee 2014). Toxic aldehydes and ketones may be produced due to this process (Gómez-Estaca et al. 2014). Other factors such as moisture content and other molecules such as sugars and short chain fatty acids have also been found to be responsible for deteriorative reaction, because they are easily accessed by the microorganisms. In a similar process, foods get subjected to an enzymatic reaction and ROS liberation. That's why the microbial action and oxidative reaction are the most important reasons for the spoilage of foods (Ribeiro-Santos et al. 2017). To avoid the interaction between ROS and foods, oxygen can be achieved by thermoplastic films (Atarés and Chiralt 2016). Two major problems include the degradability of thermoplastic films and challenges replace the synthetic antioxidant. In packaging films fortified with essential oils, both of these problems can be overcome. Thus, incorporation of essential oils into the packaging films could be a great decision.

Cumin essential oils have been incorporated in active packages developed for meat products including chicken breast meat (Takma and Korel 2019) and lamb loins (Hasani-Javanmardi et al. 2021). Active polyethylene terephthalate (PET) films were prepared using chitosan and alginate coatings incorporating black cumin oil, which were fabricated using a layer-by-layer technique by Takma and Korel (2019). Chicken meat samples stored in these packages demonstrated lower total aerobic mesophilic and psychrotrophic bacteria count along with significant antimicrobial activity against *S. aureus* and *E. coli*. In a different study, safflower oil nano-emulsion and oxygen absorber were combined with essential oil for developing packaging materials (Hasani-Javanmardi et al. 2021) and evaluating the corresponding effects on the storage quality of lamb loins during the 20 days period under chilled conditions. The treatment delayed the growth of some microorganisms (mesophilic and psychrotrophic bacteria, *Enterobacteriaceae*, and lactic acid bacteria). Several other positive physiochemical aspects were also observed, including decrease in lipid and protein oxidation and colour and sensory deterioration in different treatment groups (Hasani-Javanmardi et al. 2021). Hence, overall, it can be deduced that the application of essential oils has enhanced the quality of packaging films along with making the process more sustainable and devoid of synthetic chemicals.

9.7.2 Pharmacological Uses of Cumin

Cumin and its active constituents have been used as an antibacterial, antifungal, anti-inflammatory, antioxidant, astringent, and blood thinner. The spice and its different varieties and forms have been found to be effective for general health maintenance and as a general stimulant. These have also been active against atherosclerosis (hardening of the arteries), bone loss, cancer, cardiovascular disease, carpal tunnel syndrome, cataracts (eye disease), cavities, dental plaque, diabetes, digestion, diuretic (improves urine flow), ear infections, gastrointestinal disorders, high cholesterol, immune modulation (affects the immune system), and low blood sugar. The extracts of cumin have been also used as a menstrual flow stimulant, for promoting flow of breast milk, relaxation, and against seizures/epilepsy, ulcers, and weight loss. The pharmacological activities or corresponding mechanism of uses of cumin have been mentioned in detail in Section 9.4.

9.8 CONCLUSION

Cumin has been extensively utilized as an ingredient to enhance the flavour of foods and beverages, where it also provides a unique pungent taste. However, attributed to its beneficial properties, it has been widely exploited through different extraction and encapsulation techniques to preserve and efficiently utilize its health beneficial effects. However, with advanced technologies, further improvement in the activities of corresponding extracts and formulations can be achieved. Also, there is a huge scope of exploitation of cumin as a source of healthy dietary fibres and proteins, which should be further explored.

REFERENCES

Agrawal, S., and Gour, H. (1992). Anti-fungal substance in the essential oil of cumin. Paper presented at the *proceedings of national seminar on exploration of indigenous raw materials for essential oil industry*.

Ali, A., Wu, H., Ponnampalam, E. N., Cottrell, J. J., Dunshea, F. R., and Suleria, H. A. 2021. Comprehensive profiling of most widely used spices for their phenolic compounds through LC-ESI-QToF-MS2 and their antioxidant potential. *Antioxidants*, 10(5): 721.

Ali, B., and Blunden, G. 2003. Pharmacological and toxicological properties of *Nigella sativa*. *Phytotherapy Research*, 17(4): 299–305.

Amiri, A., Mousakhani-Ganjeh, A., Amiri, Z., Guo, Y.-G., Singh, A. P., and Kenari, R. E. 2020. Fabrication of cumin loaded-chitosan particles: Characterized by molecular, morphological, thermal, antioxidant and anticancer properties as well as its utilization in food system. *Food Chemistry*, 310: 125821.

Anandharamakrishnan, C. (2014). *Techniques for nanoencapsulation of food ingredients* (Vol. 8): Springer.

Anaya-Castro, M. A., Ayala-Zavala, J. F., Muñoz-Castellanos, L., Hernández-Ochoa, L., Peydecastaing, J., and Durrieu, V. 2017. β-Cyclodextrin inclusion complexes containing clove (*Eugenia caryophyllata*) and Mexican oregano (*Lippia berlandieri*) essential oils: Preparation, physicochemical and antimicrobial characterization. *Food Packaging and Shelf Life*, 14: 96–101.

Andallu, B., and Ramya, V. 2007. Antihyperglycemic, cholesterol-lowering and HDL-raising effects of cumin (*Cuminum cyminum*) seeds in type-2 diabetes. *Journal of Natural Remedies*, 7(1): 142–149.

Atarés, L., and Chiralt, A. 2016. Essential oils as additives in biodegradable films and coatings for active food packaging. *Trends in Food Science & Technology*, 48: 51–62.

Balassa, L. L., Fanger, G. O., and Wurzburg, O. B. 1971. Microencapsulation in the food industry. *Critical Reviews in Food Science & Nutrition*, 2(2): 245–265.

Banerjee, S., Chattopadhyay, P., Ghosh, A., Goyary, D., Karmakar, S., and Veer, V. 2013. Influence of process variables on essential oil microcapsule properties by carbohydrate polymer–protein blends. *Carbohydrate Polymers*, 93(2): 691–697.

Behbahani, B. A., Noshad, M., and Falah, F. 2019. Cumin essential oil: Phytochemical analysis, antimicrobial activity and investigation of its mechanism of action through scanning electron microscopy. *Microbial Pathogenesis*, 136: 103716.

Benmoussa, H., Elfalleh, W., He, S., Romdhane, M., Benhamou, A., and Chawech, R. 2018. Microwave hydrodiffusion and gravity for rapid extraction of essential oil from Tunisian cumin (*Cuminum cyminum* L.) seeds: Optimization by response surface methodology. *Industrial Crops and Products*, 124: 633–642.

Bhandari, B. R., Patel, K. C., and Chen, X. D. 2008. Spray drying of food materials – Process and product characteristics. *Drying Technologies in Food Processing*, 4: 113–157.

Boughendjioua, H. 2019. Characterization of aroma active compounds of cumin (*Cuminum cyminum*, L.) seed essential oil. *Modified Applied Nioequivalence Bioavailability*, 4: 1–5.

Boughendjioua, H., and Djeddi, S. 2018. Study of the organoleptic and physicochemical properties of cinnamon essential oil (*Cinnamomum zeylanicum*). *American Journal of Life Science Researches*, 6(3): 123–130.

Boyraz, N., and Özcan, M. 2005. Antifungal effect of some spice hydrosols. *Fitoterapia*, 76(7–8): 661–665.

Carocho, M., Barreiro, M. F., Morales, P., and Ferreira, I. C. 2014. Adding molecules to food, pros and cons: A review on synthetic and natural food additives. *Comprehensive Reviews in Food Science and Food Safety*, 13(4): 377–399.

Chauhan, P. S., Satti, N. K., Suri, K. A., Amina, M., and Bani, S. 2010. Stimulatory effects of *Cuminum cyminum* and flavonoid glycoside on cyclosporine-A and restraint stress induced immune-suppression in Swiss albino mice. *Chemico-biological Interactions*, 185(1): 66–72.

Cheikh-Rouhou, S., Besbes, S., Lognay, G., Blecker, C., Deroanne, C., and Attia, H. 2008. Sterol composition of black cumin (*Nigella sativa* L.) and Aleppo pine (*Pinus halepensis* Mill.) seed oils. *Journal of Food Composition and Analysis*, 21(2): 162–168.

Chen, X.-G., Lee, C. M., and Park, H.-J. 2003. O/W emulsification for the self-aggregation and nanoparticle formation of linoleic acid modified chitosan in the aqueous system. *Journal of Agricultural and Food Chemistry*, 51(10): 3135–3139.

De Martino, L., De Feo, V., Fratianni, F., and Nazzaro, F. 2009. Chemistry, antioxidant, antibacterial and antifungal activities of volatile oils and their components. *Natural Product Communications*, 4(12): 1934578X0900401226.

de Oliveira, E. F., Paula, H. C., and de Paula, R. C. 2014. Alginate/cashew gum nanoparticles for essential oil encapsulation. *Colloids and Surfaces B: Biointerfaces*, 113: 146–151.

Deng, C., N. Yao, B. Wang, and X. Zhang. 2006. Development of microwave-assisted extraction followed by headspace single-drop microextraction for fast determination of paeonol in traditional Chinese medicines. *Journal of Chromatography A* 1103: 15–21. doi: 10.1016/j.chroma.2005.11.023

Derakhshan, S., Sattari, M., and Bigdeli, M. 2007. P2081 evaluation of antibacterial activity and biofilm formation in *Klebsiella pneumoniae* in contact with essential oil and alcoholic extract of cumin seed (*Cuminum cyminum*). *International Journal of Antimicrobial Agents*, 29: S601.

Derakhshan, S., Sattari, M., and Bigdeli, M. 2010. Effect of cumin (*Cuminum cyminum*) seed essential oil on biofilm formation and plasmid Integrity of *Klebsiella pneumoniae*. *Pharmacognosy Magazine*, 6(21): 57.

Desai, K. G. H., and Jin Park, H. 2005. Recent developments in microencapsulation of food ingredients. *Drying Technology*, 23(7): 1361–1394.

Dua, A., Gupta, S., Mittal, A., and Mahajan, R. 2012. A study of antioxidant properties and antioxidant compounds of cumin (*Cuminum cyminum*). *International Journal of Pharmaceutical and Biological Science Archive*, 3(5): 1110–1116.

Eikani, M. H., Golmohammad, F., Mirza, M., and Rowshanzamir, S. 2007. Extraction of volatile oil from cumin (*Cuminum cyminum* L.) with superheated water. *Journal of Food Process Engineering*, 30(2): 255–266.

Eikani, M. H., Goodarznia, I., and Mirza, M. 1999. Supercritical carbon dioxide extraction of cumin seeds (*Cuminum cyminum* L.). *Flavour and Fragrance Journal*, 14(1): 29–31.

Esmaeili, A., and Asgari, A. 2015. In vitro release and biological activities of *Carum copticum* essential oil (CEO) loaded chitosan nanoparticles. *International Journal of Biological Macromolecules*, 81: 283–290.

Farag, R., Daw, Z., and Abo-Raya, S. 1989a. Influence of some spice essential oils on *Aspergillus parasiticus* growth and production of aflatoxins in a synthetic medium. *Journal of Food Science*, 54(1): 74–76.

Farag, R., Daw, Z., Hewedi, F., and El-Baroty, G. 1989b. Antimicrobial activity of some Egyptian spice essential oils. *Journal of Food Protection*, 52(9): 665–667.

Fioramonti, S. A., Martinez, M. J., Pilosof, A. M., Rubiolo, A. C., and Santiago, L. G. 2015. Multilayer emulsions as a strategy for linseed oil microencapsulation: Effect of pH and alginate concentration. *Food Hydrocolloids*, 43: 8–17.

Freiberger, E. B., Kaufmann, K. C., Bona, E., de Araújo, P. H. H., Sayer, C., Leimann, F. V., and Goncalves, O. H. 2015. Encapsulation of roasted coffee oil in biocompatible nanoparticles. *LWT – Food Science and Technology*, 64(1): 381–389.

Gagandeep, Dhanalakshmi, S., Mendiz, E., Rao, A. R., and Kale, R. K. 2003. Chemopreventive effects of *Cuminum cyminum* in chemically induced forestomach and uterine cervix tumors in murine model systems. *Nutrition and Cancer*, 47(2): 171–180.

Getachew, A. T., and B. S. Chun. 2016. Influence of hydrothermal process on bioactive compounds extraction from green coffee bean. *Innovative Food Science & Emerging Technologies* 18(A): 24–31. doi: 10.1016/j.ifset.2016.09.006

Ghaderi-Ghahfarokhi, M., Barzegar, M., Sahari, M., Gavlighi, H. A., and Gardini, F. 2017. Chitosan-cinnamon essential oil nano-formulation: Application as a novel additive for controlled release and shelf life extension of beef patties. *International Journal of Biological Macromolecules*, 102: 19–28.

Gholamian, S., Nourani, M., and Bakhshi, N. 2021. Formation and characterization of calcium alginate hydrogel beads filled with cumin seeds essential oil. *Food Chemistry*, 338: 128143.

Gironi, F., and Maschietti, M. 2008. Continuous countercurrent deterpenation of lemon essential oil by means of supercritical carbon dioxide: Experimental data and process modelling. *Chemical Engineering Science*, 63(3): 651–661.

Glišić, S. B., Mišić, D. R., Stamenić, M. D., Zizovic, I. T., Ašanin, R. M., and Skala, D. U. 2007. Supercritical carbon dioxide extraction of carrot fruit essential oil: Chemical composition and antimicrobial activity. *Food Chemistry*, 105(1): 346–352.

Gómez-Estaca, J., López-de-Dicastillo, C., Hernández-Muñoz, P., Catalá, R., and Gavara, R. 2014. Advances in antioxidant active food packaging. *Trends in Food Science & Technology*, 35(1): 42–51.

Hajlaoui, H., Mighri, H., Noumi, E., Snoussi, M., Trabelsi, N., Ksouri, R., and Bakhrouf, A. 2010. Chemical composition and biological activities of Tunisian *Cuminum cyminum* L. essential oil: A high effectiveness against *Vibrio* spp. strains. *Food and Chemical Toxicology*, 48(8–9): 2186–2192.

Hammad, A. A. I., and Youssef, B. M. 1995. Growth inhibition of some pharmacological molds by garlic and cumin essential oils. *Egyptian Journal of Microbiology*, 29: 115–120.

Hasani-Javanmardi, M., Fallah, A. A., and Abbasvali, M. 2021. Effect of safflower oil nanoemulsion and cumin essential oil combined with oxygen absorber packaging on the quality and shelf-life of refrigerated lamb loins. *LWT – Food Science and Technology*, 147: 111557.

Hawthorne, S. B., Rickkola, M.-L., Screnius, K., Holm, Y., Hiltunen, R., and Hartonen, K. 1993. Comparison of hydrodistillation and supercritical fluid extraction for the determination of essential oils in aromatic plants. *Journal of Chromatography A*, 634(2): 297–308.

Hosseini, S. F., Zandi, M., Rezaei, M., and Farahmandghavi, F. 2013a. Two-step method for encapsulation of oregano essential oil in chitosan nanoparticles: Preparation, characterization and in vitro release study. *Carbohydrate Polymers*, 95(1): 50–56.

Hosseini, S. M., Hosseini, H., Mohammadifar, M. A., Mortazavian, A. M., Mohammadi, A., Khosravi-Darani, K., Shojaee-Aliabadi, S., Dehghan, S., and Khaksar, R. 2013b. Incorporation of essential oil in alginate microparticles by multiple emulsion/ionic gelation process. *International Journal of Biological Macromolecules*, 62: 582–588.

Jafarzadeh, S., Jafari, S. M., Salehabadi, A., Nafchi, A. M., Kumar, U. S. U., and Khalil, H. A. 2020. Biodegradable green packaging with antimicrobial functions based on the bioactive compounds from tropical plants and their by-products. *Trends in Food Science & Technology*, 100: 262–277.

Jimenez-Carmona, M., Ubera, J., and De Castro, M. L. 1999. Comparison of continuous subcritical water extraction and hydrodistillation of marjoram essential oil. *Journal of Chromatography A*, 855(2): 625–632.

Kalaivani, P., Saranya, R. B., Ramakrishnan, G., Ranju, V., Sathiya, S., Gayathri, V., Thiyagarajan, L. K., Venkhatesh, J. R., Babu, C. S., and Thanikachalam, S. 2013. *Cuminum cyminum*, a dietary spice, attenuates hypertension via endothelial nitric oxide synthase and NO pathway in renovascular hypertensive rats. *Clinical and Experimental Hypertension*, 35(7): 534–542.

Kanakdande, D., Bhosale, R., and Singhal, R. S. 2007. Stability of cumin oleoresin microencapsulated in different combination of gum arabic, maltodextrin and modified starch. *Carbohydrate Polymers*, 67(4): 536–541.

Kanter, M. 2009. Effects of *Nigella sativa* seed extract on ameliorating lung tissue damage in rats after experimental pulmonary aspirations. *Acta Histochemica*, 111(5): 393–403.

Karaman, K. 2020. Characterization of *Saccharomyces cerevisiae* based microcarriers for encapsulation of black cumin seed oil: Stability of thymoquinone and bioactive properties. *Food Chemistry*, 313: 126129.

Karnic, C. 1991. A clinical trial of a composite herbal drug in treatment of diabetic mellitus (madhumeha). *Aryavaidyan*, 5: 36–46.

Kaur, G., Upadhyay, N., Tharappel, L. J. P., and Invally, M. 2019. Pharmacodynamic interaction of cumin seeds (*Cuminum cyminum* L.) with glyburide in diabetes. *Journal of Complementary and Integrative Medicine*, 16(4). doi:10.1515/jcim-2018-0080.

Keerthiga, N., Anitha, R., Rajeshkumar, S., and Lakshmi, T. 2019. Antioxidant activity of cumin oil mediated silver nanoparticles. *Pharmacognosy Journal*, 11(4): 787–789.

Kfoury, M., Auezova, L., Greige-Gerges, H., and Fourmentin, S. 2015. Promising applications of cyclodextrins in food: Improvement of essential oils retention, controlled release and antiradical activity. *Carbohydrate Polymers*, 131: 264–272.

Khalil, N., Ashour, M., Fikry, S., Singab, A. N., and Salama, O. 2018. Chemical composition and antimicrobial activity of the essential oils of selected Apiaceous fruits. *Future Journal of Pharmaceutical Sciences*, 4(1): 88–92.

Kiralan, M., Özkan, G., Bayrak, A., and Ramadan, M. F. 2014. Physicochemical properties and stability of black cumin (*Nigella sativa*) seed oil as affected by different extraction methods. *Industrial Crops and Products*, 57: 52–58.

Köse, M. D., Tekin, B. N., and Bayraktar, O. 2021. Simultaneous isolation and selective encapsulation of volatile compounds from essential oil during electrospraying of β-cyclodextrin. *Carbohydrate Polymers*, 258: 117673.

Krishnakantha, T., and Lokesh, B. R. 1993. Scavenging of superoxide anions by spice principles. *Indian Journal of Biochemistry & Biophysics*, 30(2): 133–134.

Kubo, I., and Kinst-Hori, I. 1998. Tyrosinase inhibitors from cumin. *Journal of Agricultural and Food Chemistry*, 46(12): 5338–5341.

Kuhn, K. R., de Silva, F. G. D., Netto, F. M., and da Cunha, R. L. 2014. Assessing the potential of flaxseed protein as an emulsifier combined with whey protein isolate. *Food Research International*, 58: 89–97.

Kunzemann, J., and Herrmann, K. 1977. Isolation and identification of flavon(ol)-*O*-glycosides in caraway (*Carum carvi* L.), fennel (*Foeniculum vulgare* Mill.), anise (*Pimpinella anisum* L.), and coriander (*Coriandrum sativum* L.), and of flavon-*C*-glycosides in anise. I. Phenolics of spices (author's transl.). *Zeitschrift fur Lebensmittel-untersuchung und-Forschung*, 164(3): 194–200.

Lal, G., Saran, P. L., Gangadevi, D., and Rishi, R. (2014). Production technology of cumin (*Cuminum cyminum* L.). In *Advances in vegetable agronomy* (pp. 223–231): Post Graduate School Indian Agricultural Research Institute New Delhi, India and Indian Council of Agricultural Research Department of Agricultural Research and Education Ministry of Agriculture, Government of India New Delhi, India.

Lee, D. S. (2014). Antioxidative packaging system. In *Innovations in food packaging* (pp. 111–131): Elsevier.

Lee, H.-S. 2005. Cuminaldehyde: aldose reductase and α-glucosidase inhibitor derived from *Cuminum cyminum* L. seeds. *Journal of Agricultural and Food Chemistry*, 53(7): 2446–2450.

Leung, A. Y. (1980). *Encyclopedia of common natural ingredients used in food, drugs, and cosmetics.*

Li, X. H., Yun, J., Xing, Y. G., Xiao, Y., and Tang, Y. (2011). Complexation of cinnamon essential oil by β-cyclodextrin and its release characteristics at high temperature. Paper presented at the *advanced materials research.*

Majdalawieh, A. F., and Fayyad, M. W. 2015. Immunomodulatory and anti-inflammatory action of *Nigella sativa* and thymoquinone: A comprehensive review. *International Immunopharmacology*, 28(1): 295–304.

Malini, T., and Vanithakumari, G. 1987. Estrogenic activity of *Cuminum cyminum* in rats. *Indian Journal of Experimental Biology*, 25(7): 442–444.

Mangolim, C. S., Moriwaki, C., Nogueira, A. C., Sato, F., Baesso, M. L., Neto, A. M., and Matioli, G. 2014. Curcumin–β-cyclodextrin inclusion complex: Stability, solubility, characterisation by FT-IR, FT-Raman, X-ray diffraction and photoacoustic spectroscopy, and food application. *Food Chemistry*, 153: 361–370.

Manolikar, M., and Sawant, M. 2003. Study of solubility of isoproturon by its complexation with β-cyclodextrin. *Chemosphere*, 51(8): 811–816.

Mehdi, R., Masoomch, S., Mohammad, B., Kamkar, J., Soheil, A., and Reza, S. 2009. Chemical composition and antiaflatoxicogenic activity of *Carum carvi* L., Thymus vulgaris, and *Citrus aurantifolia* essential oils. *Food Control*, 20(11): 1018–1024.

Mehta, B., Verma, M., and Gupta, M. 2008. Novel lipid constituents identified in seeds of *Nigella sativa* (Linn). *Journal of the Brazilian Chemical Society*, 19(3): 458–462.

Mékaoui, R., Benkaci-Ali, F., Scholl, G., and Eppe, G. 2016. Effect of the extraction technique, heating time and cryogenic grinding (N_2 at –196°C) on the composition of cumin seeds volatile oil. *Journal of Essential Oil Bearing Plants*, 19(8): 1903–1919.

Mohammadi, A., Hashemi, M., and Hosseini, S. M. 2015. Nanoencapsulation of *Zataria multiflora* essential oil preparation and characterization with enhanced antifungal activity for controlling *Botrytis cinerea*, the causal agent of gray mould disease. *Innovative Food Science & Emerging Technologies*, 28: 73–80.

Mohammadi, M., Hamishehkar, H., and Piruzifard, M. K. 2021. Nanophytosome as a promising carrier for improving cumin essential oil properties. *Food Bioscience*, 42: 101079.

Najda, A., Dyduch, J., and Brzozowski, N. 2008. Flavonoid content and antioxidant activity of caraway roots (Carum carvi L.). *Vegetable Crops Research Bulletin*, 68: 127.

Nalini, N., Manju, V., and Menon, V. 2006. Effect of spices on lipid metabolism in 1,2-dimethylhydrazine-induced rat colon carcinogenesis. *Journal of Medicinal Food*, 9(2): 237–245.

Nalini, N., Sabitha, K., Viswanathan, P., and Menon, V. 1998. Influence of spices on the bacterial (enzyme) activity in experimental colon cancer. *Journal of Ethnopharmacology*, 62(1): 15–24.

Nickavar, B., Mojab, F., Javidnia, K., and Amoli, M. A. R. 2003. Chemical composition of the fixed and volatile oils of *Nigella sativa* L. from Iran. *Zeitschrift für Naturforschung C*, 58(9–10): 629–631.

Noppakundilograt, S., Piboon, P., Graisuwan, W., Nuisin, R., and Kiatkamjornwong, S. 2015. Encapsulated eucalyptus oil in ionically cross-linked alginate microcapsules and its controlled release. *Carbohydrate Polymers*, 131: 23–33.

Oetjen, G.-W., and Haseley, P. (2004). *Freeze-drying*: John Wiley & Sons.

Pande, C., and Goswami, L. 2000. Composition of essential oil from seeds of *Cuminum cyminum* L. *Indian Perfumer*, 44(4): 265–266.

Petrović, G. M., Stojanović, G. S., and Radulović, N. S. 2013. Encapsulation of cinnamon oil in b-cyclodextrin. *Journal of Medicinal Plants Research*, 4(14): 1382–1390.

Piornos, J. A., Burgos-Díaz, C., Morales, E., Rubilar, M., and Acevedo, F. 2017. Highly efficient encapsulation of linseed oil into alginate/lupin protein beads: Optimization of the emulsion formulation. *Food Hydrocolloids*, 63: 139–148.

Platel, K., and Srinivasan, K. 2000a. Influence of dietary spices and their active principles on pancreatic digestive enzymes in albino rats. *Food/Nahrung*, 44(1): 42–46.

Platel, K., and Srinivasan, K. 2000b. Stimulatory influence of select spices on bile secretion in rats. *Nutrition Research*, 20(10): 1493–1503.

Platel, K., and Srinivasan, K. 2001. Studies on the influence of dietary spices on food transit time in experimental rats. *Nutrition Research*, 21(9): 1309–1314.

Pourmortazavi, S. M., Ghadiri, M., and Hajimirsadeghi, S. S. 2005. Supercritical fluid extraction of volatile components from *Bunium persicum* Boiss. (black cumin) and *Mespilus germanica* L. (medlar) seeds. *Journal of Food Composition and Analysis*, 18(5): 439–446.

Rakmai, J., Cheirsilp, B., Mejuto, J. C., Torrado-Agrasar, A., and Simal-Gándara, J. 2017. Physico-chemical characterization and evaluation of bio-efficacies of black pepper essential oil encapsulated in hydroxy-propyl-beta-cyclodextrin. *Food Hydrocolloids*, 65: 157–164.

Razzaghi-Abyaneh, M., Shams-Ghahfarokhi, M., Rezaee, M.-B., Jaimand, K., Alinezhad, S., Saberi, R., and Yoshinari, T. 2009. Chemical composition and antiaflatoxigenic activity of *Carum carvi* L., *Thymus vulgaris* and *Citrus aurantifolia* essential oils. *Food Control*, 20(11): 1018–1024.

Rebey, I. B., Bourgou, S., Debez, I. B. S., Karoui, I. J., Sellami, I. H., Msaada, K., Limam, F., and Marzouk, B. 2012. Effects of extraction solvents and provenances on phenolic contents and antioxidant activities of cumin (*Cuminum cyminum* L.) seeds. *Food and Bioprocess Technology*, 5(7): 2827–2836.

Rehman, A., Jafari, S. M., Aadil, R. M., Assadpour, E., Randhawa, M. A., and Mahmood, S. 2020. Development of active food packaging via incorporation of biopolymeric nanocarriers containing essential oils. *Trends in Food Science & Technology*, 101: 106–121.

Ribeiro-Santos, R., Andrade, M., de Melo, N. R., and Sanches-Silva, A. 2017. Use of essential oils in active food packaging: Recent advances and future trends. *Trends in Food Science & Technology*, 61: 132–140.

Rodov, V., Vinokur, Y., Gogia, N., and Chkhikvishvili, I. 2010. Hydrophilic and lipophilic antioxidant capacities of Georgian spices for meat and their possible health implications. *Georgian Medical News*, 179: 61–66.

Romagnoli, C., Andreotti, E., Maietti, S., Mahendra, R., and Mares, D. 2010. Antifungal activity of essential oil from fruits of Indian *Cuminum cyminum*. *Pharmaceutical Biology*, 48(7): 834–838.

Roman-Ramos, R., Flores-Saenz, J., and Alarcon-Aguilar, F. 1995. Anti-hyperglycemic effect of some edible plants. *Journal of Ethnopharmacology*, 48(1): 25–32.

Ruberto, G., and Baratta, M. T. 2000. Antioxidant activity of selected essential oil components in two lipid model systems. *Food Chemistry*, 69(2): 167–174.

Sachin, B., Monica, P., Sharma, S., Satti, N., Tikoo, M., Tikoo, A., Suri, K., Gupta, B., and Johri, R. 2009. Pharmacokinetic interaction of some antitubercular drugs with caraway: Implications in the enhancement of drug bioavailability. *Human & Experimental Toxicology*, 28(4): 175–184.

Sachin, B., Sharma, S., Sethi, S., Tasduq, S., Tikoo, M., Tikoo, A., Satti, N., Gupta, B., Suri, K., and Johri, R. 2007. Herbal modulation of drug bioavailability: Enhancement of rifampicin levels in plasma by herbal products and a flavonoid glycoside derived from *Cuminum cyminum*. *Phytotherapy Research*, 21(2): 157–163.

Samojlik, I., Lakic, N., Mimica-Dukic, N., Đaković-Švajcer, K., and Bozin, B. 2010. Antioxidant and hepatoprotective potential of essential oils of coriander (*Coriandrum sativum* L.) and caraway (*Carum carvi* L.) (Apiaceae). *Journal of Agricultural and Food Chemistry*, 58(15): 8848–8853.

Santos, E. H., Kamimura, J. A., Hill, L. E., and Gomes, C. L. 2015. Characterization of carvacrol beta-cyclodextrin inclusion complexes as delivery systems for antibacterial and antioxidant applications. *LWT – Food Science and Technology*, 60(1): 583–592.

Sarkar, C., Jamaddar, S., Islam, T., Mondal, M., Islam, M. T., and Mubarak, M. S. 2021. Therapeutic perspectives of the black cumin component thymoquinone: A review. *Food & Function*, 12: 6167–6213.

Sayyah, M., Mahboubi, A., and Kamalinejad, M. 2002. Anticonvulsant effect of the fruit essential oil of *Cuminum cyminum* in mice. *Pharmaceutical Biology*, 40(6): 478–480.

Senorans, F., Ibanez, E., Cavero, S., Tabera, J., and Reglero, G. 2000. Liquid chromatographic–mass spectrometric analysis of supercritical-fluid extracts of rosemary plants. *Journal of Chromatography A*, 870(1–2): 491–499.

Shahidi, F., and Ambigaipalan, P. 2015. Phenolics and polyphenolics in foods, beverages and spices: Antioxidant activity and health effects – A review. *Journal of Functional Foods*, 18: 820–897.

Shankar, K., and Mehendale, H. M. (2014). Oxidative stress. In *Encyclopedia of toxicology*, 3rd ed. (pp. 735–737).

Shankaracharya, N. B. (1974). The spice in India. *Symposium on spice industry in India* (pp. 24–26).

Shayegh, S., Rasooli, I., Taghizadeh, M., and Alipoor Astaneh, S. D. 2008. Phytotherapeutic inhibition of supragingival dental plaque. *Natural Product Research*, 22(5): 428–439.

Shetty, R., Singhal, R., and Kulkarni, P. 1994. Antimicrobial properties of cumin. *World Journal of Microbiology and Biotechnology*, 10(2): 232–233.

Shrestha, M., Ho, T. M., and Bhandari, B. R. 2017. Encapsulation of tea tree oil by amorphous beta-cyclodextrin powder. *Food Chemistry*, 221: 1474–1483.

Singh, R. P., Gangadharappa, H., and Mruthunjaya, K. 2017. *Cuminum cyminum* – A popular spice: An updated review. *Pharmacognosy Journal*, 9(3): 292–301.

Skrinjar, M., Mandic, A., Misan, A., Sakac, M., Saric, L. C., and Zec, M. 2009. Effect of Mint (*Mentha piperita* L.) and Caraway (*Carum carvi* L.) on the growth of some toxigenic *Aspergillus* species and aflatoxin B1 production. *Matica Srpska proceedings for natural sciences*.

Sotelo-Boyás, M., Correa-Pacheco, Z., Bautista-Baños, S., and y Gómez, Y. G. 2017. Release study and inhibitory activity of thyme essential oil-loaded chitosan nanoparticles and nanocapsules against foodborne bacteria. *International Journal of Biological Macromolecules*, 103: 409–414.

Sowbhagya, H., Srinivas, P., Purnima, K. T., and Krishnamurthy, N. 2011. Enzyme-assisted extraction of volatiles from cumin (*Cuminum cyminum* L.) seeds. *Food Chemistry*, 127(4): 1856–1861.

Srinivasan, K. 2018. Cumin (*Cuminum cyminum*) and black cumin (*Nigella sativa*) seeds: Traditional uses, chemical constituents, and nutraceutical effects. *Food Quality and Safety*, 2(1): 1–16.

Srivastava, K. 1989. Extracts from two frequently consumed spices – cumin (*Cuminum cyminum*) and turmeric (*Curcuma longa*) – inhibit platelet aggregation and alter eicosanoid biosynthesis in human blood platelets. *Prostaglandins, Leukotrienes and Essential Fatty Acids*, 37(1): 57–64.

Surya, D., Vijayakumar, R. S., and Nalini, N. 2005. Oxidative stress and the role of cumin (*Cuminum cyminum* Linn.) in alloxan-induced diabetic rats. *Journal of Herbs, Spices & Medicinal Plants*, 11(3): 127–139.

Takma, D. K., and Korel, F. 2019. Active packaging films as a carrier of black cumin essential oil: Development and effect on quality and shelf-life of chicken breast meat. *Food Packaging and Shelf Life*, 19: 210–217.

Tamnak, S., Mirhosseini, H., Tan, C. P., Amid, B. T., Kazemi, M., and Hedayatnia, S. 2016. Encapsulation properties, release behavior and physicochemical characteristics of water-in-oil-in-water (W/O/W) emulsion stabilized with pectin–pea protein isolate conjugate and Tween 80. *Food Hydrocolloids*, 61: 599–608.

Tao, F., Hill, L. E., Peng, Y., and Gomes, C. L. 2014. Synthesis and characterization of β-cyclodextrin inclusion complexes of thymol and thyme oil for antimicrobial delivery applications. *LWT – Food Science and Technology*, 59(1): 247–255.

Tayman, C., Cekmez, F., Kafa, I. M., Canpolat, F. E., Cetinkaya, M., Uysal, S., Tunc, T., and Sarıcı, S. U. 2012. Beneficial effects of *Nigella sativa* oil on intestinal damage in necrotizing enterocolitis. *Journal of Investigative Surgery*, 25(5): 286–294.

Vignesh, S., Anitha, R., Kumar, S. R., and Lakshmi, T. 2019. Evaluation of the antimicrobial activity of cumin oil mediated silver nanoparticles on oral microbes. *Research Journal of Pharmacy and Technology*, 12(8): 3709–3712.

Wanner, J., Bail, S., Jirovetz, L., Buchbauer, G., Schmidt, E., Gochev, V., Girova, T., Atanasova, T., and Stoyanova, A. 2010. Chemical composition and antimicrobial activity of cumin oil (*Cuminum cyminum*, Apiaceae). *Natural Product Communications*, 5(9): 1934578X1000500904.

Willatgamuwa, S., Platel, K., Saraswathi, G., and Srinivasan, K. 1998. Antidiabetic influence of dietary cumin seeds (*Cuminum cyminum*) in streptozotocin induced diabetic rats. *Nutrition Research*, 18(1): 131–142.

Xiao, J., Ke, Z. P., Shi, Y., Zeng, Q., and Cao, Z. 2018. The cardioprotective effect of thymoquinone on ischemia-reperfusion injury in isolated rat heart via regulation of apoptosis and autophagy. *Journal of Cellular Biochemistry*, 119(9): 7212–7217.

Zhao, Y., Wang, P., Zheng, W., Yu, G., Li, Z., She, Y., and Lee, M. 2019. Three-stage microwave extraction of cumin (*Cuminum cyminum* L.) seed essential oil with natural deep eutectic solvents. *Industrial Crops and Products*, 140: 111660.

CHAPTER 10

Bioactive Compounds in Nigella Seeds

Kanchan Suri and Shelja Sareen

CONTENTS

DOI: 10.1201/9781003215387-10

ABBREVIATIONS

ABTS	2,2′-Azino-bis-(3-ethylbenzothiazoline-6-sulfonic acid) diammonium salt
ASE	Accelerated solvent extraction
BMI	Body mass index
BW	Body weight
CRPSerum	C-reactive protein
DPPH	1,1-Diphenyl-2-picrylhydrazyl
DTQ	Dithymoquinone
FRAP	Ferric-reducing antioxidant power
FTIR	Fourier transform infrared spectroscopy
GC-FID	Gas chromatography coupled with flame ionization detector
GC-MS	Gas chromatography with mass spectroscopy
GC-O	Gas chromatography olfactory
HD	Hydro-distillation
HDL	High-density lipoprotein
HPLC	High performance liquid chromatography
HS-SPME	Headspace-solid-phase micro-extraction
LC-ESI-QTOF/MS	Liquid chromatography coupled with electrospray ionization-quadrupole time-of-flight and mass spectrometry
LDL	Low-density lipoprotein
MAE	Microwave-assisted extraction
MDA	Malondialdehyde
NMR	Nuclear magnetic resonance
Ns-PoC	Nigella seed post-conditioning
PCOS	Polycystic ovarian syndrome
PEF	Pulsed electric field

SD	Steam distillation
SE	Solvent extraction
SFE	Supercritical fluid extraction
SPME	Solid-phase micro-extraction
T2D	Type-2 diabetes
TFC	Total flavonoid content
THQ	Thymohydroquinone
TLC	Thin-layer chromatography
TPC	Total phenolic content
TQ	Thymoquinone
UAE	Ultrasound-assisted extraction
UHPLC	MS-ultra-high-performance liquid chromatography tandem mass spectrometry
WC	Waist circumference

10.1 INTRODUCTION

The plant kingdom has myriad of medicinal plants, spices, and phytoconstituents, which have shown promising therapeutic potential since traditional times (Manoharan et al. 2021). The therapeutic potential about their application and utilization govern through their success rate as evidenced through satisfactory clinical efficacy and also through least toxicity impact (Ijaz et al. 2017). Nigella seeds, a common spice obtained from annual herb *Nigella sativa*, which belong to *Ranunculaceae* family, have extensive medicinal uses (Dubey et al. 2016; Suri et al. 2019). Its seeds are small in size and are commonly known by variety of names, that is, black caraway, black seed, black cumin seed, kalonji, Iranian black cumin, habbatulbarakah, seed of blessing, Al Habbah Al Sawda, and qazheshuniz (Tembhurne et al. 2014). Nigella word is derived from the *Latin* word "niger" meaning black, suggesting the colour of its seed. The bitterness of the seed is due to the presence of nigellin, a seed protein (Kalidasu et al. 2017). Its seeds are among the highly valued medicinal seeds in history with reference in Bible too. Islamic prophet Mohammad quoted that it "can heal every disease except death" (Naz 2011). Its seeds are native to Mediterranean countries like Southwest Asia, Southern Europe, and Northern Africa. Nigella seeds are also grown in Asian countries like India, Italy, Indonesia, Afghanistan, and Pakistan. In India, it is particularly grown in the northern and eastern parts spanning states of Himachal Pradesh, Madhya Pradesh, Punjab, Gangetic plains of Assam, Bengal, and Maharashtra (Botnick et al. 2012; Dubey et al. 2016).

Since ancient times nigella seeds have been used as herbal medicine for wide range of diseases in Prophetic, Arabic, Unani, and Indian medicines (Ahmad et al. 2013; Koshak et al. 2017a). They are used for various diseases including asthma, back pain, fever, bronchitis, cough, chest congestion, dizziness, paralysis, chronic headache, inflammation, infertility, and other gastrointestinal (GIT) disorders such as flatulence, dysentery, diarrhoea, and dyspepsia (Gholamnezhad et al. 2016; Nasir et al. 2014). Seeds have potential to be an anodyne, liver tonic, purgative, and appetite stimulant (Tavakkoli et al. 2017). Moreover, the seeds are also employed to prepare high-cost nutritive oil which is further utilized as a remedy against many diseases like swollen joints, cysts, rhinal ulcer, testicular inflammation and dermatitis. Further, Nigella oil along with honey is panacea for bronchospasms, asthma, and chest congestion (Gholamnezhad et al. 2016; Nasir et al. 2014). Besides their use for remedial purposes, these seeds are also considered important among the non-conventional sources of edible oils and also in use as condiments and spices in baked products including cakes, breads, pastries, curries, and pickles. Seeds also act as an ingredient in various seasonings due to their characteristic aroma and bitter peppery taste (Tariq 2008).

10.2 CHARACTERISTICS AND BIOACTIVE COMPOUNDS

10.2.1 Characteristics

Nigella is a bushy, self-branching plant containing myriad coloured flowers such as pink, white, yellow, or pale blue (Darakhshan et al. 2015). It is cultivated as a winter crop with flowering and fruiting from January to April (Dubey et al. 2016). Its plant forms fruit capsule with 3–7 fused follicles by self-reproduction, each of which contains many trigonal seeds (Darakhshan et al. 2015). Its seeds are dicotyledonous, small sized (1–5 mm), and are dark greyish or blackish in colour from outside and white from inside. Its seeds also have slightly aromatic odour and peppery bitter taste with a crunchy texture (Al-Gaby 1998).

10.2.2 Bioactive Compounds

Chemically nigella seeds are highly nutrient rich and till now 100 bioactive constituents have been extracted from seeds, of which some of the components have not been structurally identified and pharmacologically tested (Al-Ghamdi 2001). Nigella seeds contain various components including oil (fixed and volatile/essential oil), vitamins, minerals, amino acids, and various bioactive compounds. Plants produce various types of bioactive constituents, generally known as phytochemical components. These include carbohydrates, phenolic compounds, flavonoids, alkaloids, tannins, terpenoids, and steroids, which are responsible for numerous pharmacological uses of the plants and their derived products (Yessuf 2015). These compounds can be further divided into various categories according to their biochemical pathways and chemical classes such as glycosides, flavonoids, tannins, terpenoids, lignans, and alkaloids (Bernhoft 2010). Some secondary metabolites like alkaloids (10.11 mg/100 g), flavonoids (3.78 mg/100 g), saponins (7.58 mg/100 g), and tannins (21 mg/100 g) are also present in nigella seeds which contribute to the majority of seed phytochemicals (Agradi et al. 2001; Mamun and Absar 2018). Chemical structures of few common bioactive compounds are given in Figure 10.1.

10.2.2.1 Oil (Fixed and Volatile)

Nigella seeds contain oil as a major component and the yield of the same varies from 30% to 40%, under stress conditions such as salt stress, alkalinity stress, and low temperature; the yield declines to 13–23% (Gharby et al. 2015). Its seeds contain two types of oils, that is, fixed oil and volatile oil, whose content varied from 32% to 40% and 0.4% to 0.5%, respectively (Tembhurne et al. 2014). The oil yield and composition of seeds also vary with the change in geographical conditions, storage conditions, and the method used for oil extraction (Darakhshan et al. 2015). Fixed oil constitutes myriad of fatty acids such as linoleic acid (58.38%), oleic acid (23.39%), palmitic acid (11.24%), stearic acid (2.87%), behenic acid (2.25%), linolenic acid (0.32%), arachidic acid (0.19%), and myristic acid (0.17%) (Al Juhaimi et al. 2018). Its essential or volatile oil is regarded as an important bioactive ingredient of nigella seeds (Liu et al. 2011).

10.2.2.2 Vitamins and Minerals

Nigella seeds from different geographical locations have reported the minerals such as potassium (675–1498 mg/100 g), phosphorus (378–543 mg/100 g), calcium (188–570 mg/100 g), sodium (17.6–412.50 mg/100 g), magnesium (138.17–355.2 mg/100 g), manganese (3.1–265 mg/100 g), and iron (9.70–57.5 mg/100 g), wherein potassium and phosphorus were present in the highest amount. Its seeds also contain zinc (4.43–6.7 mg/100 g) and copper (1.5–3.17 mg/100 g) as trace elements (Ashraf et al. 2006; Nergiz and Otleş 1993; Sultan et al. 2009). These seeds are considered to be

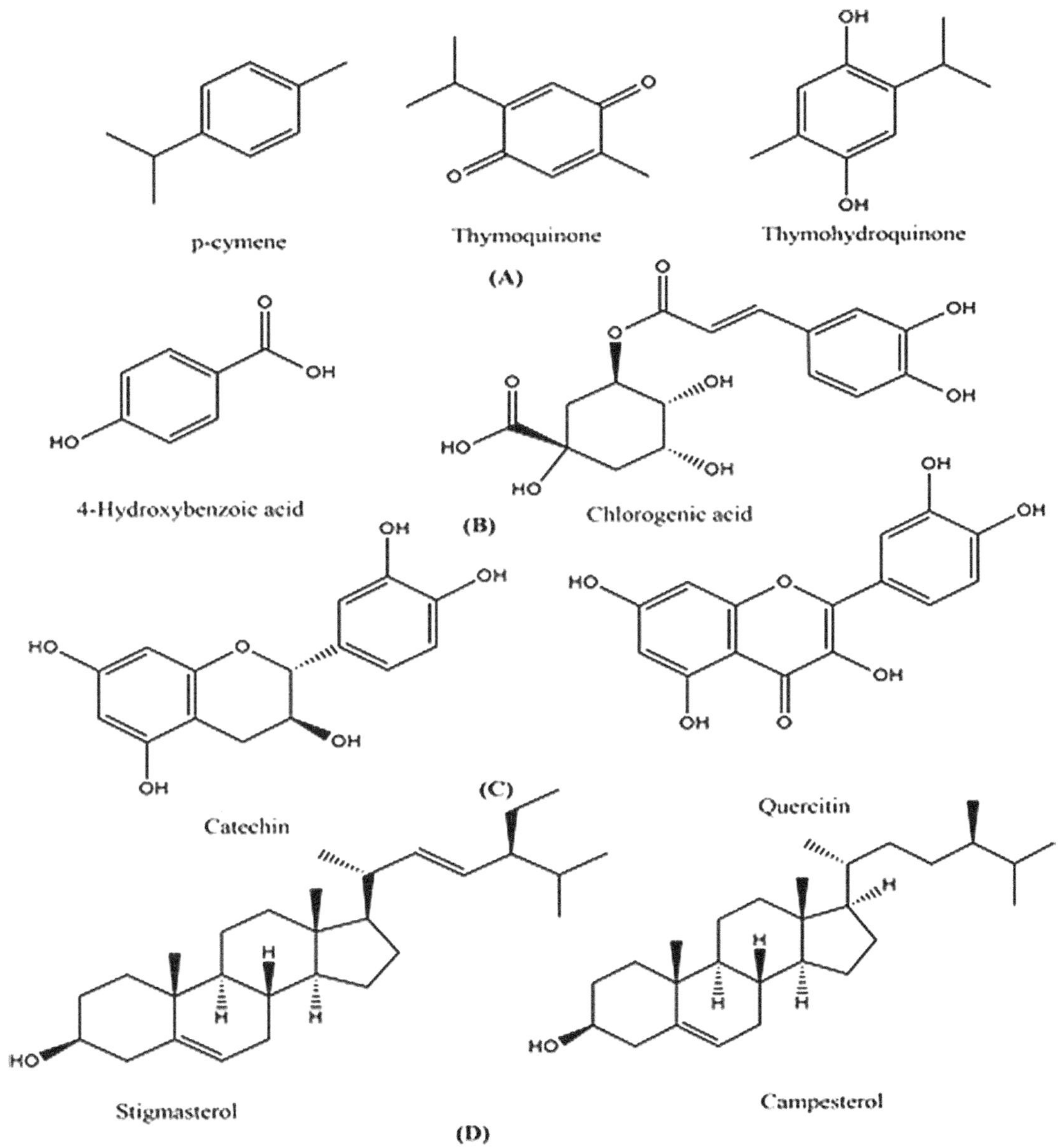

Figure 10.1 Bioactive compounds: chemical structures (A), terpenes/terpenoids (B), phenolic acids (C), flavonoids (D), phytosterols.

an abundant source of both water-soluble vitamins such as thiamine (831 μg/100 g), riboflavin (63 μg/100 g), pyridoxine (789 μg/100 g), niacin (6311 μg/100 g), folic acid (42 μg/100 g), and fat-soluble vitamins like retinol (0.15–0.42 μg/g), vitamin D2 (1.38 μg/g), μg/g), vitamin K1 (1.85 μg/g), vitamin K2 (2.15 μg/g), and tocopherols (451 mg/kg) (Nergiz and Otleş 1993; Vatansev et al. 2013).

10.2.2.3 Amino Acids

Nigella seed proteins also contain decent amount of essential amino acids such as leucine (6.00%), valine (4.77%), lysine (3.86%), threonine (3.86%), phenylalanine (3.70%), isoleucine (3.69%), histidine

(2.79%), methionine (1.88%), and tryptophan (1.50%) and non-essential amino acids like glutamic acid (23.95%), aspartic acid (9.32%), arginine (8.90%), glycine (6.11%), proline (4.88%), serine (4.50%), alanine (4.39%), tyrosine (3.43%), and cysteine (2.47%) (Kabir et al. 2019).

10.2.2.4 Terpenes/Terpenoids

Nigella seed contains number of terpenes including p-cymene, thymoquinone (TQ), dithymoquinone (DTQ), thymohydoquinone (THQ), α-pinene, γ-terpinene, α-thujene, α-terpinene, limonene, β-pinene, carvacrol, longifolene, thymol, sabinene bornyl acetate, β-elemene, trans-anethole, and paeoniflorinand β-myrcene (Benkaci-Ali et al. 2007; Feng et al. 2020; Salehi et al. 2021). Two compounds, including 1,8-cineole and p-cymen-8-ol, were reported by Bourgou et al. (2010). Kazemi (2014) also reported the presence of α-phellandrene, terpinolene, camphor, carvone, longicyclene, and borneol. Further, Ghosheh et al. (1999) and Naz (2011) also reported the presence of 4-terpineol and citronellol. Further, two another terpenoids named cycloart-3-one-7,22-diene-24-ol and cycloart-23-methyl-7,20,22-triene-3b, 30-diol were identified by Mehta et al. (2008).

10.2.2.5 Phenolic Acids

It includes hydro-cinnamic acids and hydroxybenzoic acids as two groups. Phenolic acids such as caffeic acid, vanillic acid, gallic acid, sinapic acid, syringic acid, chlorogenic acid, p-coumaric acid, ferulic acid, protocatechuic acid, 4-hydroxybenzoic acid, trans-cinnamic acid, 5-hydroxyferulic acid, 3-hydroxybenzoic acid, rosmarinic acid, and salicylic acid were reported in earlier studies (Bourgou et al. 2008; Topcagic et al. 2017). Further, Feng et al. (2020) reported hippuric acid, caffeic acid 3-*o*-glucuronide, p-coumaroyl tartaric acid, caffeic acid-3-sulphate, ferulic acid 4-*O*-glucoside, 4-*o*-methylgallic acid, and vanillic acid 4-sulphate as phenolic acids derived from nigella seeds.

10.2.2.6 Flavonoids

Flavonoids are secondary plant metabolites, having a polyphenolic structure and imparts antioxidant, anticarcinogenic, antimutagenic and anti-inflammatory properties to the plants. Nigella seeds contain various flavonoids such as quercetin, quercetin-3-galactosyl glucoside, trigillin quercetin-3-glucoside, kaempferol-3-glucosyl galactosyl glucoside, and (+)-catechin, myricetin, epicathechin, rutin, quercitrin dihydro, naringine, apigenin, naringenin, pinocembrin, galangin, and quercimeritrin(Hameed et al. 2019; Merfort et al. 1997; Topcagic et al. 2017). Feng et al. (2020) mentioned different flavonoids which include flavonols, flavones, and isoflavonoids. In flavonols, compounds such as myricetin 3-*o*-rhamnoside, kaempferol 3,7,4′-*o*-triglucoside, kaempferol 3,7-*o*-diglucoside, and patuletin 3-*O*-glucosyl-(1->6)-[apiosyl (1->2)]-glucoside) were dominant. In flavones, apigenin 6, diosmin, and 8-di-C-glucoside were present, while in isoflavonoids compounds such as kaempferol, quercetin, and 3′,4′,7-trihydroxyisoflavanone were present.

10.2.2.7 Alkaloids

The alkaloids are heterocyclic compounds containing nitrogen in their structure and they are bitter in taste (Bernhoft 2010). Nigella seeds comprise alkaloids such as nigellicine, nigellimine-nigellidine, and nigellimine-*N*-oxide which were reported by Malik et al. (1985, 1995), Malik and Zaman (1992), and Liu et al. (2011). Further, eight dolabellane-type diterpene alkaloids including nigellamine A_1, nigellamine A_2, nigellamine B_1, nigellamine B_2 and nigellamines A_3, A_4, A_5, and C were isolated from methanolic seed extracts (Morikawa et al. 2004a, 2004b). Ali et al. (2008) reported presence of sulphated indazole-type alkaloid named nigellidine-4-*O*-sulphite from seeds of nigella. Besides these, five indazole-containing alkaloids such as nigelanoid, nigeglanine, 17-*O*-(β-D-glucopyranosyl)-4-*O*-methylnigellidine, 4-*O*-methylnigellidine, and 4-*O*-methylnigeglanine were also reported by Yuan et al. (2014).

10.2.2.8 Saponins and Coumarins

Saponins are a heterogeneous group of glycosides, which includes a steroidal or triterpenoid aglycone and one or more saccharide chains in their structure (Güçlü-Üstündağ and Mazza 2007). Kumara and Huat (2001) isolated a triterpene saponin named α-hederin, from nigella seeds. Another two saponins extracted from nigella seeds include terylglucosides and acetyl-steryl-glucoside (Sudhir et al. 2016). Another saponin identified as 3-*O*-[β-D-xylopyranosyl(1→3)-α-L-rhamnopyranosyl(1→4)-β-D-glucopyranosyl]-11-methoxy-16-hydroxy-17-acetoxy hederagenin was reported by Mehta et al. (2008). Taskin et al. (2005) isolated three triterpene saponins from nigella seeds which were 3-*O*-[β-D-xylopyranosyl-(1→3)-α-L-rhamnopyranosyl-(1→2)-α-L-arabinopyranosyl]-28-*O*-[α-L-rhamnopyranosyl-(1→4)-β-D-glucopyranosyl-(1→6)-β-D-glucopyranosyl]-hederagenin, 3-*O*-[α-L-rhamnopyranosyl-(1→2)-α-L-arabinopyranosyl]- 28-*O*-[α-L-rhamnopyranosyl-(1→4)-β-D-glucopyranosyl-(1→6)-β-D-glucopyranosyl]-hederagenin, and 3-*O*-[β-D-xylopyranosyl-(1→3)-α-L-rhamnopyranosyl-(1→2)-α-L-arabinopyranosyl]-hederagenin. Elbandy et al. (2009) isolated two saponins from nigella seeds named sativosides A and B. Its seed also contain coumarins such as 6-methoxy-coumarin, 7-hydroxy-coumarin, and 7-oxy-coumarin (Tembhurne et al. 2014).

10.2.2.9 Phytosterols

Phytosterols are a significant class of compounds and are gaining more attention these days due to nutraceutical properties. They are known to reduce LDL (low-density lipoprotein) and thus lowering blood cholesterol (Yimer et al. 2019). Nigella seeds contain eight phytosterols such as β-sitosterol, stigmasterol, campesterol, Δ^5-avenasterol, cholesterol, Δ^7-cholesterol, γ-sitosterol, and desmosterol (Benkaci-Ali et al. 2012). Total tocopherols in seed oil ranged between 9.15 and 27.92 mg/100 g wherein α- and γ-tocopherol along with β-tocotrienol were dominant (Matthaus and Özcan 2011).

10.2.2.10 Minor Compounds

Feng et al. (2020) reported the presence of minor compounds in nigella seeds including anthocyanins, lignans, tyrosols, and stilbenes. Among them anthocyanins include compounds such as peonidin 3-*o*-rutinoside, delphinidin 3-*o*-glucosyl-glucoside, and peonidin 3-*o*-sambubioside-5-*o*-glucoside. On the other hand, lignans include compounds such as cyclolariciresinol and schisandrin while tyrosols include 3,4-DHPEA-AC and hydroxytyrosol 4-*o*-glucoside. Stilbenes include 3′-hydroxy-3,4,5,4′-tetramethoxystilbene and resveratrol. Nigella seeds also contain some other compounds including glycerolipids such as monoacylglycerols, diacylglycerols, and triacylglycerols along with phospholipids such as phosphatidylcholine, phosphatidylglycerol, and phosphatidylinositol. Further, carbohydrates such as, xylose, rhamnose, and arabinose along with some alkane hydrocarbons including n-nonane, 8-heptadecene, 2-undecanone, and n-octyl isobutyrateare also present in nigella seeds (Hannan et al. 2021).

10.3 EXTRACTION METHODS OF BIOACTIVE COMPOUNDS

Extracting bioactive components is the preliminary step in an experimental study which influences both qualitative and quantitative aspects. Thus, choice of extraction method becomes crucial to accomplish adequate separation, identification, and characterization of bioactive constituents

(Azmir et al. 2013). Extraction involves basic steps such as grinding of seeds in order to reduce the particle size of material followed by homogenization. The homogenized material is then subjected to filtration using Whatman filter paper or centrifugation. Following centrifugation, the filtrate is dried using rotary vacuum evaporator in order to obtain crude extract containing bioactive compounds. Generally, both traditional and non-conventional methods are utilized for the extraction of bioactive compounds. Traditional extraction methods include solid liquid extraction, that is, Soxhlet method (Khattak and Simpson 2008; Kiralan et al. 2014; Solati et al. 2014), maceration (Yessuf 2015), and hydro-distillation (HD) or steam distillation (SD) (Benkaci-Ali et al. 2007; Hadi et al. 2016) while non-conventional methods include solid-phase micro-extraction (SPME) (Farag et al. 2017; Kiralan et al. 2014), supercritical fluid extraction (SFE) (Ghahramanloo et al. 2017; Machmudah et al. 2005), microwave-assisted extraction (MAE) (Altemimi et al. 2017; Liu et al. 2013), ultrasonic-assisted extraction (UAE) (Alam et al. 2021; Moghimi et al. 2018), pulsed electric-field-assisted extraction (Bakhshabadi et al. (2017), and accelerated solvent extraction (ASE) (Liu et al. 2012).

10.3.1 Traditional Extraction Methods

10.3.1.1 Solvent Extraction Method

In this method, powdered sample for extraction purpose is placed in thimble and is mixed with different solvents under mild heat application followed by drying using rotary vacuum evaporator to obtain solvent free extract. Different solvents such as water, hexane, ethanol, petroleum ether, methanol, and acetone are utilized for extraction of bioactive compounds. Khattak and Simpson (2008) utilized Soxhlet method for preparation of water, hexane, methanol, and acetone extracts of nigella seeds followed by their estimation for total phenolic content (TPC) and free radical scavenging assay using 1,1-diphenyl-2-picrylhydrazyl (DPPH). The results were promising for methanolic extracts which showed highest TPC and DPPH radical scavenging capacity. Solati et al. (2014) utilized petroleum ether in Soxhlet extractor for extraction of bioactive constituents from nigella seeds. Kiralan et al. (2014) employed the same method to extract oil from nigella seeds.

10.3.1.2 Maceration

It is quite simple and inexpensive method to extract bioactive compounds from seed. Its basic steps include grinding of seeds followed by mixing with solvent with occasional shaking of the material, after which filtration is done to separate solid residue from liquid (Azmir et al. 2013). Yessuf (2015) performed the maceration for 24 h to obtain petroleum ether extracts. Further, Hadi et al. (2016) identified 28 bioactive compounds from methanolic extracts using maceration.

10.3.1.3 Hydro-Distillation or Steam Distillation

Another traditional method utilized for extraction of volatile bioactive compounds includes HD or SD using a Clevenger-type or Likens-Nickerson apparatus (Liu et al. 2011). In this process, initially seeds are filled in a still compartment and water is added in adequate quantity, followed by boiling them together, and in SD method, steam is directly infused into the seed material. Further, vapour mixture containing water and oil is condensed by indirect cooling of water. Then resultant mixture flows to separator, where oil containing bioactive constituents gets separated from water (Silva et al. 2005).

Stoyanova et al. (2003) extracted essential oil from nigella seeds using HD, SD, and hexane extracts and reported that main components from HD extracted oil were α-thujene (17.3%) and p-cymene (56.4%) while the oil obtained by SD and the hexane extract were found to be rich in TQ

(48.2 and 42.8%) and p-cymene (30.1% and 33.9%), respectively. Benkaci-Ali et al. (2007) utilized HD method for the extraction of essential volatile oil from nigella seeds and reported that resultant oil contains 76.3% monoterpene hydrocarbons, 8.2% oxygenated monoterpenes, and 5.69% of sesquiterpenes hydrocarbons.

10.3.2 Non-Conventional Extraction Methods

Traditional extraction methods were clouded by several limitations such as larger volumes of pure solvent requirement, longer extraction time, removal of solvents, lower extraction selectivity, and degradation of heat labile compounds. Hence, in order to overcome these limitations, new extraction methods needed to be introduced (Iqbal et al. 2018). For this reason, various novel extraction techniques such as SFE, MAE, ultrasound, and enzyme-assisted extraction either alone or in combination have presented noteworthy benefits in comparison to conventional methods. Moreover, the yield was also reported to be increased with minimum effect on the bioactive compounds (Wen et al. 2020).

10.3.2.1 Solid-Phase Micro-Extraction

It is a modern extraction technique developed in early 1990s and practised for extraction of volatile compounds due to its simplicity, rapidity, efficiency, elimination of solvents cost, less toxicity, and no heat application (Abo-Atya et al. 2021). Additionally, SPME enables the enrichment of volatile compounds from liquid or gaseous samples, over a fused-silica fibre followed by their subsequent desorption, leading to identification of even those volatiles which are present in negligible amounts (Handa 2008). Kiralan et al. (2014) utilized headspace (HS)-SPME for determination of changes in nigella seed volatiles during accelerated oxidation conditions (60 and 100°C). Farag et al. (2017) practised HS-SPME to extract volatile compounds from 12 nigella accessions and reported 34 volatile components with TQ being the highest in concentration.

10.3.2.2 Supercritical Fluid Extraction

It is an efficient method to extract bioactive compounds from plant-based seeds in comparison to traditional methods due to several advantages like higher selectivity, shorter extraction time, and use of non-flammable, safe, environment-friendly solvents such as CO_2, propane, and compressed liquefied petroleum gas (Wrona et al. 2017). Several studies have done in order to examine the effect of SFE on the yield, antioxidant activity, and volatile compounds of essential oil. Machmudah et al. (2007) extracted nigella seed oil using supercritical CO_2 at low and high separation pressures of 20–50 MPa at 313 K temperature using 2 mL/min CO_2 flow rate for 170 min extraction time. Analysis of extracts was done using GC-MS and antioxidant activity was measured. Ghahramanloo et al. (2017) done comparative analysis of nigella seed oil obtained by SFE and solvent extraction (SE) and reported that oil extracted with SFE gives higher yield and more linoleic acid and TQ in comparison to oil extracted by SE.

10.3.2.3 Microwave-Assisted Extraction

This method is employed due to reduced extraction time and lesser solvent volume required. Basically, in this extraction, microwaves help to heat the least proportion of solvent and able to quench out the bioactive compounds of many types from them (Altemimi et al. 2017). In a study carried out by Liu et al. (2013), MAE was used to extract bioactive compounds from nigella seeds at 700-W power with 2450 microwave frequency for 10 min extraction time. Further authors identified 32 volatile compounds using GC-FID and GC-MS from the extract and reported TQ, p-cymene,

4-Isopropyl-6-methoxy-1-methyl-1-cyclohexene and longifolene, α-thujene, carvacrol, limonene and β-pinene, α-longipinene, and β-cyclocitral as major volatile compounds.

10.3.2.4 Ultrasound-Assisted Extraction

It is another green extraction technology used in recent years for the extraction of bioactive constituents from seeds due to benefits like higher yield, short processing time, and low maintenance cost (Wen et al. 2018). One more key point about UAE is its mechanical impact, which disrupts the cell walls due to acoustical cavitation, hence improved the efficiency of process by increasing the penetration of solvent into the seed materials (Toma et al. 2001). Moghimi et al. (2018) extracted oil from nigella seeds using various ultrasound power and irradiation time. Further, they evaluated oils for various physicochemical parameters and total phenolic content and reported that ultrasound treatment significantly increased the TPC of extracted seed oils. Alam et al. (2021) also carried out UAE of essential oil from nigella seeds and reported increase in yield with reduced extraction time.

10.3.2.5 Pulsed Electric Field Extraction

This is a non-thermal promising way for the extraction of high valued seed oils. In this plant, seeds are placed between two electrodes at ambient temperature or slightly higher temperature which induces critical electrical potential across the cell membrane and cause structural damage to the cell membrane via electroporation (Kaseke et al. 2021). Bakhshabadi et al. (2018) conducted PEF treatment of nigella seeds and compared PEF treatment with MAE. They analysed the obtained oils for yield, chemical composition, and oxidative stability and reported that PEF significantly increased the oil extraction yield and oxidative stability of extracts. Further, they stated that PEF treatment extracted greater amount of TQ and cymene in comparison to MAE.

10.3.2.6 Accelerated Solvent Extraction

ASE is also known as pressurized liquid extraction (PLE), pressurized solvent extraction, or subcritical solvent extraction. When pure water is utilized as a solvent, PLE is commonly known as superheated water extraction, pressurized low polarity water extraction or subcritical water extraction, or pressurized hot water extraction (Pronyk and Mazza 2009). Liu et al. (2012) characterized secondary volatile compounds from nigella seeds using ASE and reported 25 compounds from the same. Ahmad et al. (2020) employed ASE for the TQ extraction from nigella seeds and compared it with traditionally operated extraction tactics. The authors reported higher extraction yield of TQ with ASE in competition with other primitive ones.

10.4 IDENTIFICATION AND CHARACTERIZATION OF BIOACTIVE COMPOUNDS

Following extraction, the next step is to identify and characterize bioactive constituents. Identification of compounds is carried out by comparing retention times and spectral characteristics of their peaks with those of standards. In some cases, spiking the sample with standards is also performed to identify the desired compound and its confirmation (Benkaci-Ali et al. 2012). Further, compounds can be characterized structurally by precise analytical techniques such as high-performance liquid chromatography (HPLC), ultra-high performance liquid chromatography tandem mass spectrometry (UPLC-MS), gas chromatography (GC), GC-MS, and LC-Q-TOF/MS and other spectral techniques like Fourier transform infrared spectroscopy (FTIR) and nuclear magnetic resonance (NMR) while biochemical characterization can be done by using clinical studies, *in-vivo* evaluation, and toxicity assays (Sasidharan et al. 2011). In recent years, various researchers have carried out numerous studies and a range of compounds have been identified and characterized from nigella seeds using various tools and techniques.

10.4.1 Column and Thin Layer Chromatography

Column chromatography and TLC can be used for the isolation and purification of the bioactive compounds due to their economy, convenience, and availability in wide range of stationary phases (Altemimi et al. 2017). Yessuf (2015) utilized TLC and column chromatography for screening of alkaloids, phenols, flavonoids, and steroids from methanolic extract of nigella seeds. Few studies reported use of TLC assay and UV spectroscopy for the quantification of TQ from nigella seeds (Habib and Choudhary 2021; Mohammed et al. 2019). Tiji et al. (2021) performed the qualitative screening of nigella seed fractions using different solvents for phytochemicals, namely, alkaloids, steroids/terpenes, flavonoids, polyphenols, saponins, and coumarins using TLC and UV lamp.

10.4.2 Liquid Chromatography

HPLC is extensively used for the isolation and separation of constituents due to its versatile nature, robustness, and reliable results. It uses ultraviolet-visible (UV-vis) diode array detector and in some cases mass spectroscopy to provide information about structural elucidation of the compounds (Sridharan et al. 2011). Hameed et al. (2019) performed characterization of nigella seed extract for various bioactive compounds using HPLC and reported caffeic acid, kaempferol, and chlorogenic acid as major phenolic acids. In addition to this, isolation and identification of eight phenolic compounds from nigella seeds using reverse phase HPLC with diode array (UV-vis) detector was also carried out. The identified polyphenol compounds were flavonoids including (+)-catechin, epicathechin, rutin, quercitrin dihydro, naringine, quercimeritrin, quercetin, and callistephin, respectively. Topcagic et al. (2017) identified 23 phenolic compounds using UPLC-MS/MS and reported sinapic acid as major phenolic acid while kaempferol as major flavonoid from nigella seed extracts. Few studies reported the use of RP-HPLC for the estimation of TQ content from nigella seeds (Gad and El-Ahmady 2018; Habib and Choudhry 2021).

These days, liquid chromatography coupled with electrospray ionization-quadrupole time-of-flight and mass spectrometry (LC-ESI-QTOF/MS) is another advanced analytical technique, used for identification and characterization of diverse range of polyphenols (Spínola et al. 2015). Feng et al. (2020) did identification and characterization of nigella seed phenolics using LC-ESI-QTOF/MS and categorized them as hydro-cinnamic acids, hydroxybenzoic acids, anthocyanins, flavonols, flavones, isoflavonoids, lignans, tyrosols, and stilbenes. Yuan et al. (2014) also carried out the characterization of chemical constituents from nigella seeds using UPLC-Q-TOF/MS followed by HPLC fingerprint analysis.

10.4.3 Gas Chromatography

GC with flame ionization detector (GC-FID) or tandem mass spectroscopy (GC-MS) is utilized for the separation and characterization of compounds of volatile nature (Dinagaran et al. 2016). Burits and Bucar (2000) analysed essential oil from nigella seeds using GC-MS and reported TQ, t-anethole, r-cymene, carvacrol, longifoline, and 4-terpineol as major compounds. Hadi et al. (2016) also evaluated bioactive compounds of nigella seeds using GC-MS and identified 28 bioactive compounds. Liu et al. (2012) carried out identification and quantification of 25 bioactive compounds using GC-FID/GC-MS. Various researchers also employed GC-FID for the identification of fatty acids and sterol from nigella seeds and extracted oils (Atta 2003; Saxena et al. 2017; Suri et al 2019). Recent studies used GC-MS and (gas chromatography-olfactory) GC-O for comparing volatile compounds from Bangladesh and Indian nigella seeds (Kabir et al. 2020).

10.4.4 Fourier Transform Infrared Spectroscopy and Nuclear Magnetic Resonance

FTIR is important technique for the identification and characterization of functional groups (chemical bonds) or chemical constituents of unknown mixture of seed extracts (Sasidharan et al. 2011). NMR spectroscopy is one of the major techniques generally used for elucidation of 3D structure of biological compounds (Hameed et al. 2017). Venkatachallam et al. (2010) characterized nigella seed extracts obtained by SFE for quinones and phenolics. The analysis of phenolic compounds was performed by GC and GC-MS and their confirmation using ^{1}H and ^{13}C NMR spectroscopy. They reported 47 volatile compounds with TQ, DTQ, and THQ and thymol as major phenolic constituents.

Merfort et al. (1997) isolated and identified three flavonoid glycosides such as kaempferol 3-glucosyl (1-2) galactosyl (1-2) glucoside, quercetin, and quercetin 3-(6-feruloylglucosyl) (1-2) galactosyl (1-2) glucoside from nigella seeds. The structures have been elucidated by using spectral techniques such as UV-vis and mass spectroscopy with ^{1}H and ^{13}C NMR which were further elucidated by 2D NMR. Mohammed et al. (2019) have done structural characterization of bioactive components from nigella seed oil using FTIR and reported presence of TQ, DTQ, thymol, and THQ. Maulidiani et al. (2015) utilized ^{1}H NMR spectroscopy for the identification of phenolics, flavonoids, carbohydrates, terpenoids, and fatty acid components from nigella seed extracts. Further, they carried out confirmation of some compounds including thymol, TQ, fatty acids, p-cymene, thymol-β-glucopyranoside, and sucrose by using 2D NMR. Mehta et al. (2008) isolated and characterized various compounds from nigella seeds using ^{1}H and ^{13}C NMR spectroscopy. FTIR characterization of nigella seeds indicated the presence of carbonyl C=O, C=C group, and C–H stretching while ^{1}H and ^{13}C NMR spectroscopy revealed the presence of chemical shifts corresponding to different functional groups (Krishnaveni and Saranya 2016).

10.4.5 Combinational Approach

Butt et al. (2019) carried out isolation of TQ from nigella seeds using HPLC followed by identification using FTIR spectroscopy. Farag et al. (2014) did annotation of 52 compounds including 18 fatty acids, 10 flavonoids, 10 alkaloids, 6 phenolics, and 8 saponins from nigella seed extracts using UPLC-Q-TOF-MS and GC-MS analysis. Krishnaveni and Saranya (2016) carried out identification of 19 compounds from nigella seeds using GC-MS and structural characterization was performed using FTIR, ^{1}H, and ^{13}C NMR spectroscopy to attain an insight about functional groups and chemical shifts, respectively. Further, Trigui et al. (2018) isolated water-soluble polysaccharides from nigella seeds and done their characterization by using FTIR, NMR spectroscopy, and X-ray diffraction. Ahmad et al. (2020) identified nigella seed bioactives using UHPLC, FTIR, and NMR spectroscopy and reported the presence of TQ and dihydroxy TQ as quinones and quercetin and kaempferol as flavonoids along with coumarins and alkaloids.

10.5 PROPERTIES OF NIGELLA SEEDS

10.5.1 Physicochemical Properties

Nigella seeds contain (4.6–7.0%) moisture, (30–40%) oil, (13.5–22.80%) protein, (3.7–6%) ash, (5.5–7.9%) crude fibre, and (5.5–7.9%) total carbohydrate content according to previously reported literature (Ahmad et al. 2013; Ashraf et al. 2006; Dubey et al. 2016; Nergiz and Otleş 1993; Sultan et al. 2009; Tembhurne et al. 2014). The physical properties are of considerable importance in order to design proper processing equipment, maintenance of adequate storage, and processing conditions (Singh et al. 2015). Some of the physical properties of seeds are given in Table 10.1.

Table 10.1 Various Physical Properties of Nigella Seeds

Physical Characteristics	Reported Values	References
Length (mm)	2.63	Singh et al. (2015)
Width (mm)	1.42	Singh et al. (2015)
Thickness (mm)	0.99	Singh et al. (2015)
Sphericity	0.59	Singh et al. (2015)
Porosity (g/100 g)	50.37	Singh et al. (2015)
Surface area (mm²)	1.09	Singh et al. (2015)
Seed volume (cm³)	2.76	Atta (2003)
Thousand seed weight	2.21–2.39	Atta (2003); Singh et al. (2015)
True density (kg/m³)	1113.43	Singh et al. (2015)
Bulk density (kg/m³)	552.20	Singh et al. (2015)
Angle of repose	24.4	Singh et al. (2015)

Alrashidi et al. (2020) investigated physicochemical properties of nigella seeds extracted by different extraction methods. They reported oil yield of seeds varied from 21.16% to 40.16% while other physicochemical properties of nigella seed oil like free fatty acid content ranged from 7.84% to 19.28% (as % oleic acid)and peroxide value varied from 8.14 to 9.24 meq O_2/kg. Saponification value and iodine value for these oils varied from 188.27–211.75 and 87–121 g I_2/100 g oil, respectively. Refractive index and viscosity of the oils were 1.454–1.476 and 50–76 mPa s, respectively. Further, Kiralan et al. (2014) reported chlorophylls content varied from 0.30 to 0.81 mg/kg and carotenoids content from 0.18 to 0.40 m/kg in nigella seed oils extracted with different extraction procedures.

10.5.2 Functional Properties

Functional properties of seed/flour help us to evaluate how new proteins, fat, fibre, and carbohydrates may behave in certain systems and is there impact of environmental variables on their determination (Kumar et al. 2017). Jan et al. (2019) determined various functional properties of nigella seed flour. The water absorption and oil absorption capacity were 22.40% and 12.85%, respectively, while foaming capacity was 3.20% and foaming stability was reported to be 0.52%. The emulsion capacity and emulsion stability were 4.3% and 2.26%, respectively.

10.5.3 Antioxidant Properties

Numerous studies have testified the antioxidant properties of nigella seeds and its oil (Burits and Bucar 2000; Meziti et al. 2012; Solati et al. 2014). The active compound TQ exhibits strong antioxidant property and thus acts as strong scavenger of superoxide, singlet molecular oxygen, and hydroxyl radicals (Alenzi et al. 2013). Besides TQ, presence of various polyphenolic compounds also contributes to the antioxidant property of nigella seeds (Feng et al. 2020). Antioxidant capacity of nigella seed or oils was estimated with different *in-vitro* methods including FRAP (ferric reducing antioxidant power), ABTS (2,2′-Azino-bis (3-ethylbenzothiazoline-6-sulfonic acid) diammonium salt), DPPH radical scavenging assay, deoxyribose degradation, and non-enzymatic lipid peroxidation to consider site-specific reactions (Thilakarathne et al. 2018; Iqbal et al. 2019). Burits and Bucar (2000) while studying found that four components of nigella seed essential oil including TQ, carvacrol, 4-terpineol, and anethole exhibited strong antioxidant activity. Thilakarathne et al. (2018) reported that whole nigella seeds have highest antioxidant activity followed by seed coat and cotyledon as confirmed through various *in vitro* assays like DPPH, ABTS, and TPC. Ahmed et al. (2018) mentioned in earlier studies that nigella seed fraction extracted with 40% ethanol showed the highest TPC of 138.85 mg pyrocatechol equivalent/g extract while the highest total flavonoid content (TFC) of 140.11 mg rutin equivalent/g extract was observed with 20% ethanol extracts.

10.5.4 Pharmacological Properties

In recent years, numerous bioactive constituents including α-hederin, carvacrol, TQ, DTQ, THQ, thymol, nigellidine, nigellicine, nigellimine-N-oxide, and nigellone have been isolated from nigella seeds and its oil which were further tested on human and animal models for both *in vivo* and *in vitro* studies. These studies have revealed that nigella seeds and its active constituents have innumerable pharmacological potential of human concern, thus presented in Table 10.2.

10.5.5 Toxicological Properties

Many studies in the past have estimated the toxicological effects of nigella seeds and TQ isolated from it. In earlier literature, Badary et al. (1998) assayed the acute and oral toxicity of TQ in mice model and observed the LD_{50} value of TQ to be 2.4 g/kg while supplementation with high doses (2–3 g/kg) was found to cause hypoactivity and respiratory issues. After 24 h of TQ administration, a decline in glutathione level was observed in heart, liver, and kidney along with

Table 10.2 Pharmacological Properties of Nigella Seeds

Pharmacological Properties	Active Components	References
Anti-cancer	TQ and nigella seed extracts	Almatroudi et al. (2020); Khan et al. (2011); Shoieb et al. (2003)
Anti-bacterial	TQ, nigella seed extract, and nigella seed oil extracts	Al Attas et al. (2016); Aljabre et al. (2015); Ashraf et al. (2018); Topcagic et al. (2017)
Anti-fungal	TQ, thymol, and ether extracts,	Aljabre et al. (2005); Moghim et al. (2015); Taha et al. (2010)
Anti-diabetic	Nigella seeds and TQ	Abdelmeguid et al. (2010); Bamosa et al. (2010)
Anti-inflammatory	Nigella oil and TQ	Nieto et al. (2000)
Analgesic	TQ and nigella seed extracts	Alemi et al. (2013); Majdalawieh and Fayyad (2015)
Immunomodulatory	Nigella oil and TQ	Chakrabarty et al. (2003); Majdalawieh and Fayyad (2015)
Gastro-protective	Nigella oil and TQ	El-Dakhakhny et al. (2000); Kanter et al. (2006); Salem et al. (2010)
Nephroprotective	TQ	Awad et al. (2011); Benhelima et al. (2016); Canayakin et al. (2016); Darakhshan et al. (2015)
Anti-histaminic	Nigellone and nigella seeds	Chakravarty (1993); Boskabady et al. (2004)
Hepatoprotective	TQ	Kapoor (2009); Saheb et al. (2016); Shreelakshmidevi et al. (2019)
Pulmonary protective	Nigellone and TQ	Hossein et al. (2008); Wienkötter et al. (2008)
Neuroprotective	TQ and nigella seeds	Sayeed et al. (2013)
Antioxitocic	Nigella seed oil and nigella hexane extracts	Mahmoud and Shaheen (1996)
Testicular protective	TQ	Gokçe et al. (2011)
Anti-fertility	Nigella seed oil	Mahdavi et al. (2015); Mohamed et al. (2009)
Anti-viral	Nigella oil and nigella seed extracts	Barakat et al. (2013); Forouzanfar et al. (2014); Khan (1999)
Anti-atherogenic	Nigella oil and TQ	Ali and Blunden (2003); Asgary et al. (2013)
Anti-hypertensive	Nigella oil	Fallah Huseini et al. (2013); Tasar et al. (2012)
Anthelminitic	Nigella seeds and nigella oil	Akhtar and Riffat (1991); Mohamed et al. (2005)

significant reduction in weight of these organs; however, heightened enzymatic activity of lactate dehydrogenase, alanine amino transferase, and creatine phosphokinase was observed. Meanwhile, sub-chronic study revealed that inclusion of TQ in drinking water for 90 days at 0.01–0.03% concentration in mice caused reduction in fasting sugar with no signs of toxicity (Badary et al. 1998). Mashayekhi-Sardoo et al. (2020) reviewed *in-vivo* toxicological profile of TQ and concluded that the degree and pattern of TQ toxicity were influenced by factors such as route of administration where LD_{50} of TQ administrated orally was significantly lower in comparison to intra-peritoneal mode of administration. For example, the LD_{50} values of TQ after intra-peritoneal administration were 10–15 times more, and for oral gavages, it was 100–150 times more as compared to its effective doses for antioxidant, anti-cancer, and anti-inflammatory activities. Another influential factor is animal model chosen for the study. The LD_{50} in rat model was 57.5 mg/kg when administered intra-peritoneally and 794.3 mg/kg when administered orally. On the other hand, in mice model LD_{50} of TQ was found to be 104.7 mg/kg and 870.9 mg/kg, respectively, for intra-peritoneal and oral administration (Al-Ali et al. 2008). Besides these two factors, TQ toxicity also depends on dosage size as reported by Harzallah et al. (2012) that TQ at 40 mg/kg is safer than dosage of 80 mg/kg as higher dosage causes significant DNA damage and chromosomal aberrations in the liver and kidney while no such anomalies were observed in mice fed at 40 mg/kg dosage. Bamosa et al. (2002) also analysed effects of TQ administration on blood lipid parameters (HDL, LDL, cholesterol, and triglycerides) in albino rats for two dosage sizes, namely, high dose (8 mg/kg/day) and low dose (6 mg/kg/day for 2 weeks). They concluded that low dosage size didn't produce any signs of toxicity in rats. On the contrary, most of the rats fed at high dose were dead at the end of the first week of TQ administration and the once who survived after first week showed signs of peritonitis.

Abukhader (2012) assessed the effect of gender on TQ-based acute toxicity in rats at 0, 20, 30, and 40 mg/kg for intra-peritoneal administration and 0, 200, 300, and 500 mg/kg concentrations for oral administration. The results indicated that for the intra-peritoneal administration, the maximum dose tolerance was different for both males (22.5 mg/kg) and females (15 mg/kg), whereas for oral ingestion it was observed same for both male and female rats (250 mg/kg). The study also reported that signs of toxicity varied with the TQ administration mode. It was observed that rats with intra-peritoneal injection showed acute pancreatitis while rats with oral administration of TQ ingestion showed transient toxicity signs.

10.6 HEALTH BENEFITS OF NIGELLA SEEDS IN VARIOUS DISEASES

Nearby 80% of the world's population are reliant on ancient medicines in one or other way (Ali and Blunden 2003). Since ancient times, plants are the richest sources of effective and safe medicines and have been successfully exploited for their therapeutic uses against various diseases.

10.6.1 Bacterial, Viral, and Fungal Infections

According to previously published data, nigella seeds exhibit antimicrobial activities and provides protection against certain bacterial, viral, and fungal pathogens. Nigella seed oil was found to be efficient against Gram (–) bacteria, while diethyl-ether extracts of nigella seeds were effective against both Gram (+) and Gram (–) bacteria. It indicated synergistic effect with certain antibiotics like gentamycin and streptomycin while it showed additive effect with other antibiotics like ampicillin, erythromycin, tobramycin, spectinomycin, chloramphenicol, lincomycin, doxycycline, co-trimoxazole, and nalidixic acid (Aljabre et al. 2015). Several authors have reported the positive effects of nigella seeds, its oil, and extracts in providing protection against certain bacterial infections (Al Attas et al. 2016; Ashraf et al. 2018; Topcagic et al. 2017). The efficiency of nigella seeds and its derived components are also reported to provide protection against certain viral infections

(Barakat et al. 2013; Forouzanfar et al. 2014; Khan 1999). The results from several studies suggested that anti-viral activity of nigella seeds could be due to increase in natural killer cell activity and ratio of T_H cell to T_S cell. Further, the number and function of M-phi and CD4-positive T cells also increased along with more production of IFN-γ with nigella seed oil. Several authors had reported the anti-fungal activity of nigella seeds (Aljabre et al. 2005; Moghim et al. 2015). Islam and colleagues (2019) reviewed that nigella seeds were beneficial against *Madurella mycetomatis* and *Candida albicans* while TQ derived from its seed was effective against other pathogens such as *Fusarium solani*, *Aspergillus niger*, and *Scopulariopsis brevicaulis*, *Microsporum* spp., *Epidermophyton* spp., and *Trichophyton* spp. In another study by Nadaf et al. (2015), nigella seed oil was found to be effective at concentration of 10–200 μg/mL against *Saccharomyces cerevisiae and Candida utilis*. Various nigella seed components including TQ, thymol, and THQ at 1 mg/mL concentration also exhibited strong anti-fungal activity against some moulds, yeasts, and dermatophytes (Taha et al. 2010).

10.6.2 Cancer

Cancer is an ailment symbolized by uncontrolled cell growth, generally caused due to somatic alterations in genes or carried genetically which ultimately leads to the growth of tumour in an individual (Shipitsin and Polyak 2008). Nigella seed is known for its anti-cancerous activity according to earlier reported data which states that TQ found in nigella seeds possess potent bioactivities such as antioxidant (Badary et al. 2003), antiproliferative (El-Mahdy et al. 2005) chemo-preventive (Khan and Sultana 2005), anti-mutagenic and anti-carcinogenic (Bourgou et al. 2008), and anti-neoplastic and pro-apoptotic (Gali-Muhtasib et al. 2004). In recent study, silver-coated nanoparticles of nigella seed extracts exhibited cytotoxicity effect in HCC712 cancer cell lines and they can be utilized in anti-cancer therapy (Almatroudi et al. 2020). Khan et al. (2011) reviewed the potential of nigella seeds and its derived components including TQ, seed extracts, and fixed and volatile oil against various type of cancers such as leukaemia, breast, colon, liver, pancreatic, renal, prostate, cervical, lung, and skin cancer. The study also elaborated on the possible mechanism of TQ action by which it controls the cancer. First, TQ induces the apoptotic cell death via upregulating the expression of apoptotic genes (Bcl.) and downregulating the expression of ant-apoptotic genes. Second, TQ suppresses Akt activation via dephosphorylation and thus haults cancer cell survival. In addition to this, TQ also deactivates the NF-κB pathway, and thus exercise control over expression of oncogenes. Afterwards, TQ increases the antioxidant enzymatic activity and thus confer protection against cancer. Besides these, TQ also provides protection to normal cells against ionizing radiation damage along with imparting protection to CYP450 enzymes from environmental damage (Khan et al. 2011).

10.6.3 Respiratory Disorders

Asthma is a respiratory disorder affecting more than 0.3 billion population across the globe. Nigella seeds are used in different forms such as oil, herbal tea, capsules, or boiled extracts in order to treat acute and chronic asthma. Furthermore, several studies carried out on both animals and humans proved that nigella seeds significantly reduced the histamine release and inflammation of airways along with improved bronchodilatory responses (Koshak et al. 2017a, 2017b; Majeed et al. 2021; Salem et al. 2017). Nigella seed oil was also found helpful in the treatment of asthma and bronchitis due to the presence of nigellone compound which exhibits anti-asthmatic activity and causes mast cells degranulation resulting in release of mediators such as histamine. Previous studies also reported enhancement in peak expiratory flow rate and reduction in pulmonary index with the consumption of nigella seed oil in asthma patients (Mazaheri et al. 2019). Koshak and Koshak (2020) reviewed some *in-silico* studies carried out recently on severe acute respiratory syndrome coronavirus-2 (SARS-CoV-2). They reported that several nigella seed compound, including TQ, nigelledine, THQ, hederagenin, and α-hederin, has high-to-moderate affinity with proteins and enzymes of SARS-CoV-2 and hence inhibits the replication of virus and its adherence to host cell receptor. Mahboubi (2018) studied the therapeutical potential of nigella seed oil in controlling the sinusitis and mentioned that nigella seed oil imparts various properties including analgesic, antioxidant, anti-inflammatory, immune-modulator, anti-microbial, and anti-histaminic. Due to these, nigella seeds inhibit the inflammation of respiratory airways and sinuses along with reduction in microbial infections, thereby helps in reducing the sinusitis symptoms.

10.6.4 Neurodegenerative Disorders

Nigella seed and its derived products such as oil, TQ, and extracts exhibits several neuroprotective functions and prevent from certain neurodegenerative disorders like Schizophrenia, epilepsy, Alzheimer's and Parkinson's disease along with depression, stress, and anxiety (Hannan et al. 2021). Yimer et al. (2019) reviewed the effects of nigella seeds and TQ in the treatment of epilepsy, Alzheimer's, and Parkinson's disease. They reported that in case of Alzheimer's disease, TQ reduces thiobarbituric acid reactive substances and 5-LOX levels, improves activity of enzymes (superoxide dismutase, glutathione peroxidase), prevents neurons declining and depolarization of mitochondrial membrane, causes disaggregation of beta amyloid peptides, and slows down the degeneration of cognitive thinking and formation of reactive oxygen species. In case of Parkinson's disease, TQ at dose of 5 and 10 mg/kg significantly reduces the malondialdehyde content and prevents the loss of dopaminergic neurons in *substantia nigra* due to its antioxidant and anti-inflammatory activity. TQ when given in duration of 4 weeks caused significant reduction infrequency of seizure and epilepsy. In another study carried out on stressed mice, TQ at concentration of 20 mg/kg showed the therapeutic effects in decreasing anxiety with the reduction in plasma nitrite and rise in GABA content of brain (Gilhotra and Dhingra 2011). Asrar et al. (2020) reported in their study that nigella seed oil showed significant positive results in reducing anxiety, improving memory, and increasing prefrontal and hippocampal noradrenalin levels in rats fed with ethanol, thereby reducing the ill effects of ethanol consumption. In clinical trials, consumption of nigella seed capsules (500 mg) for 4 weeks significantly decreased the depression and ameliorates the memory positivity, reduced anxiety and stabilized the disturbed mood (Sayeed et al. 2014), and improved the neuroprotective function (Sayeed et al. 2013).

10.6.5 Skin Disorders

Nigella seed, oil, extracts, and its derived components were found to be effective against skin diseases such as psoriasis, vitiligo, and acne vulgaris. For example, Dwarampudi et al. (2012) determined the effect of ethanolic seed extracts against psoriasis, both *in-vivo* and *in-vitro* mode. For *in-vivo* studies, researchers used mouse tail model, and for *in-vitro* studies, they carried out SRB assay employing HaCaT cell lines. They concluded from their undertaken researches that nigella seed extracts proved 65.01% activity in the mouse tail model for psoriasis and increases the relative epidermal thickness. Recently, a clinical study was carried out with 60 patients to determine the effect of TQ-based hydrogel for the treatment of acne vulgaris. After 2 months of application, patients reported reduced number of papules, pustules, and comedones without any side effects. Moreover, acne disability index was also decreased to 63.49% indicating that nigella seed derived TQ-based hydrogel can effectively minimize the symptoms of *acne vulgaris* (Soleymani et al. 2020). Nawarathne et al. (2019) also reported from their studies that application of three topical gels formulated with ethyl acetate extract of nigella seeds significantly inhibited the growth of acne causing micro-organisms (*Staphylococcus aureus* and *Propionibacterium acnes*) in patients when treated for 30 days. A randomized, double blind clinical study was carried out with 52 patients to compare the effectiveness of nigella seed oil and fish oil in the treatment of vitiligo. After 6 months, authors reported that upon topical application of oils for twice a day, nigella seed oil-treated patients showed more improvement in skin lesions when compared to fish oil-treated patients (Ghorbanibirgani et al. 2014). Moreover, in recent study carried out by Sarac et al. (2019), topical application of nigella seed oil showed significant repigmentation of vitiligo-affected areas, especially face, hands, and genital region of patients. Wound healing ability of nigella seeds and oil was also explored by various researchers (Aljabre et al. 2015; Javadi et al. 2018).

10.6.6 Arthritis and Bone Regeneration

Nigella seed oil and TQ plays useful role in the treatment of rheumatoid arthritis and inflammatory diseases by inhibiting the eicosanoids such as leukotriene B4 and thromboxane B2 (Mazaheri et al. 2019). In another study, nigella seed extract was supplemented in collagen-induced arthritis mice and significant reduction in the neutrophil infiltration and inflammation score was observed (Saidin et al. 2020). Further, nigella seeds and its derived products also exhibited significant contribution in bone healing as reported by Ezirganli et al. (2016). The authors observed that systemic application of nigella seed extract significantly increased the osteoblast cells and hence promotes bone healing. Arslan et al. (2017) assessed the systemic effect of TQ on bone healing in a rat model and reported that TQ administration accelerated the new bone formation and could be helpful for bone healing. Abd Elrahman et al. (2019) also reported that more active bone formation, thick trabeculae with highly vascular bone marrow and bone remodelling along with increased number of osteocytes in rabbit models upon application of nigella seed extracts.

10.6.7 Liver and Kidney Diseases

Nigella seeds and its constituents impart protection to liver by numerous ways such as by increasing the level of antioxidant enzymes, total thiol, and glutathione, inhibiting the oxidative stress and lipid peroxidation, reducing fat accumulation, by preventing the inflammation and other histopathological changes in the liver (Hannan et al. 2021). In another study carried out by Saheb et al. (2016), treatment with nigella seed powder significantly exhibited the hepatoprotective effect in streptozotocin induced diabetic rats. Nigella seeds also showed significant hepatoprotective effect against lead acetate-induced oxidative damage in rabbits (Ali et al. 2017). Shreelakshmidevi et al. (2019) discussed beneficial effect of nigella seed extracts against isoniazid and rifampicin-induced hepatotoxicity. Canayakin et al. (2016) also evaluated the effect of ethanolic nigella seed extracts in paracetamol-induced nephrotoxicity in rats and reported that administration of nigella seed extracts intensely decreased the serum creatinine and urea levels. This decrease might be due to the presence of compounds such as TQ, kaempferol, and quercetin which delivers their anti-inflammatory and antioxidant activities, thereby protects the rats from paracetamol induced acute nephrotoxicity. Further, Ansari et al. (2017) stated protective role of nigella seeds in diabetic nephropathy and reported decrease in serum creatinine, blood glucose, and blood urea while improvement in glomerular filtration rate, haemoglobin, and total urinary volume. Hannan et al. (2021) reviewed the lung protective effects of nigella seeds and its derived products, that is, TQ, oil, and seed extracts.

10.6.8 Cardiovascular Disorders and Hypolipidemic Effect

Bashir et al. (2014) evaluated the effect of ethanolic nigella seed extracts on reducing the level of serum C-reactive protein (CRP) and reported that nigella seeds possess ability to significantly decrease the CRP content. Ghoreyshi et al. (2020) studied the effect of nigella seed post-conditioning (Ns-PoC) on oxidative stress induced injury during ischemic-reperfusion in rat heart and reported that Ns-PoC significantly improved the cardiac functions as indicated by enhancement in parameters such as rate pressure product, left ventricular developed pressure, and thiol groups along with improvement in the enzyme activities (superoxide anion dismutase, and catalase). In addition to this, Ns-PoC also declined the levels of MDA and 4-hydroxynonenal. Bader (2015) also reported the protective anti-ischemic effects of nigella seeds against different kinds of vascular ischemia. Further, Shoaei-Hagh et al. (2021) carried out a randomized double-blind clinical trial to analyse the effect of nigella seed oil in reducing the cardiovascular risks and reported that significant reduction in LDL, total cholesterol, diastolic blood pressure, MDA, and fasting blood sugar. Moreover, a significant increase in HDL and glutathione reductase level was also observed. Similarly, Razmpoosh et al. (2021) carried out a double-blind clinical trial on obese and overweight women to evaluate the effect of nigella seed supplementation on reducing the cardiovascular risk factors. From the study, they reported that nigella seeds significantly enhanced the HDL level and reduced the level of LDL and serum glutamic-oxaloacetic transaminase. Atherogenic index and systolic blood pressure were also decreased. Daryabeygi-Khotbehsara and colleagues (2017) in their meta-analysis review described the role of nigella seeds oil and powder supplementation on serum lipids. Reduction in triglycerides level, total cholesterol, and LDL level along with improvement in HDL cholesterol levels was observed.

10.6.9 Obesity and Hypertension

Nigella seed and its derived products such as TQ, seed extracts, and oil reduce obesity and hypertension. For example, Namazi et al. (2018) reviewed the earlier published data in context of determining the effect of nigella seed and its derived products supplementation on anthropometric indices, that is, body mass index (BMI), waist circumference (WC), and body weight (BW) in adult humans, and reported that nigella seed supplementation significantly decreased the BW, BMI, and WC in human studies with no side effects. Mahdavi et al. (2015) also studied the influence of nigella seed oil supplementation with a low-calorie diet for 8 weeks and reported significant reduction in risk of cardiometabolic factors in obese women. Anti-hypertensive activity of nigella seed oil is due to certain constituents, that is, TQ, polyphenols, and flavonoids (Mazaheri et al. 2019). The antioxidant activity of TQ is due to its free radical scavenging activity which helps in the treatment of hypertension. Some studies also depicted positive effect of nigella seed and its derived products including oil, TQ, and seed extracts on reducing the elevated level of blood pressure by reducing oxidative stress (Shoaei-Hagh et al. 2021).

10.6.10 Diabetes

Daryabeygi-Khotbehsara and colleagues (2017) reviewed the effectiveness of nigella seeds and their derived products, that is, oil, TQ, and seed extracts against the type-2 diabetes (T2D) and stated that nigella seeds improve fasting blood sugar, glycosylated haemoglobin along with improvement in total cholesterol, and LDL in T2D patients. Pari and Sankaranarayanan (2009) carried out a study to estimate the anti-hyperglycemic effect of TQ on the activities of enzymes (glucose 6-phosphate dehydrogenase and hexokinase) involved in carbohydrate metabolism in case of streptozotocin-nicotinamide-induced diabetic rats. Authors reported from their study that oral administration of TQ (80 mg/kg BW for 45 days) efficiently control T2D and also restores the enzyme activity close to their normal range thus increasing the glucose uptake by tissues. Further, Hammad-Shafiq et al. (2014) proposed the mechanism of anti-diabetic effect of nigella seeds and stated that TQ or nigella seeds decreased the oxidative stress on liver cells followed by improvement in enzyme activities and reduction in lipid peroxidation which enhances the beta cell activity causing it to produce more insulin, thereby reducing the incidence of type-1 diabetes.

10.6.11 Reproductive Health

Mahdavi et al. (2015) reviewed the effect of nigella seeds on male fertility and stated that nigella seeds significantly influenced the sperm parameters including leydig cells, semen, sexual hormones, and overall health of male reproductive organs. They also correlated TQ and unsaturated fatty acids in improving the male sterility due to antioxidant activity of these two compounds. Nigella seeds also increases milk production in lactating women as stated by Ali and Blunden (2003). Nigella seed reportedly reduced menopausal symptoms in perimenopausal women (Latiff et al. 2014). Recent studies reportedly found that oral administration of nigella seed oil for 45 days enhanced the semen parameters, testicular spermatogenesis, and seminal vesicle development in rats (Mosbah et al. 2018). Naeimi et al. (2020) carried out double-blinded clinical study on 84 patients and administered them 500 mg nigella seed oil capsule for 16 weeks. From the study, they reported that nigella seed oil significantly improved the polycystic ovarian syndrome (PCOS) and menstrual irregularities in women.

10.7 CONCLUSION

Existing research supports nigella seed benefits in practically every physiological system. In this current chapter, an attempt is made to get an insight about the bioactive compounds present in seeds along with their extraction, isolation, and characterization methods. This chapter has been made comprehensive by highlighting various assays to testify antioxidant properties, *in-vitro* and *in-vivo* studies to explore pharmacological properties and toxicological effects of the TQ and seeds itself in the animal models. The review according to best of our knowledge is first of its kind to provide in depth insight on both traditional as well as non-conventional extraction methods. Although plenty of work has been done to explore the extraction methods but the biggest knowledge gap here is to look for the best extraction method for a specific bioactive compound. Another road block is to identify and characterize the compounds as many of the compounds are still needed to be identified and characterized. Regardless of the innumerable benefits nigella seeds have to offer, as evident from published literature, this chapter is focussing upon exploring the presence of various bioactive compounds apart from TQ, such as phenols, alkaloids, flavonoids, phytosterols, saponins, and coumarins, which lately has received much attention from scientific community owing to their various health benefits. Most of the clinical trials are done for TQ, but in near future, more trials are needed for the compounds other than TQ to establish the clinical evidences. Further to assess toxicological properties, limited numbers of studies are done on animal model which are of short-term duration. Moreover, many studies exist to present vast and global pharmacological characterization of nigella seeds, but only a few non-clinical safety investigations have been reported so far. Undoubtedly nigella seeds have many health benefits as obvious from the published literature, the need of the time is to explore the pharmacology of its constituents, standardize the contents of dosage forms and establish pharmacokinetics to build a safety profile.

CONFLICT OF INTEREST

Authors declared no conflict of interest.

REFERENCES

Abd Elrahman, S.M., Younes, S.A. and Kawana, K.Y., 2019. Evaluation of *Nigella sativa* on socket healing in rabbits. *Alexandria Dental Journal* 44:60–64.

Abdelmeguid, N.E., Fakhoury, R., Kamal, S.M. and Al Wafai, R.J., 2010. Effects of *Nigella sativa* and thymoquinone on biochemical and subcellular changes in pancreatic β-cells of streptozotocin-induced diabetic rats. *Journal of Diabetes* 2:256–266.

Abo-Atya, D.M., El-Mallah, M.F., El-Seedi, H.R. and Farag, M.A., 2021. Novel prospective of *N. sativa* essential oil analysis, culinary and medicinal uses. *Black cumin (Nigella sativa) seeds: Chemistry, technology, functionality, and applications*, 97–129. Springer, Cham.

Abukhader, M.M., 2012. The effect of route of administration in thymoquinone toxicity in male and female rats. *Indian Journal of Pharmaceutical Sciences* 74:195.

Agradi, E., Fico, G., Cillo, F., Francisci, C. and Tomè, F., 2001. Estrogenic activity of phenolic compounds from *Nigella damascena* evaluated using a recombinant yeast screen. *Planta Medica* 67:553–555.

Ahmad, A., Husain, A., Mujeeb, M., et al. 2013. A review on therapeutic potential of *Nigella sativa*: A miracle herb. *Asian Pacific Journal of Tropical Biomedicine* 3:337–352.

Ahmed, A.F., He, N., Xia, Z.Y. and Kang, W.Y., 2018. Total phenolic and flavonoid content and antioxidant properties of *Nigella sativa* L. seeds. *Current Topics in Nutraceutical Research* 16:147–153.

Ahmad, R., Ahmad, N., Mohd, A.M.I.R., et al. 2020. Variation in *Nigella sativa* quality and its standardization via instrumental analysis: A study based on geographical origin. *Notulae Botanicae Horti Agrobotanici Cluj-Napoca* 48:1141–1154.

Akhtar, M.S. and Riffat, S., 1991. Field trial of *Saussurea lappa* roots against nematodes and *Nigella sativa* seeds against cestodes in children. *Journal of the Pakistan Medical Association* 41:185–187.

Al Juhaimi, F., Uslu, N. and Özcan, M.M., 2018. The effect of preultrasonic process on oil content and fatty acid composition of hazelnut, peanut and black cumin seeds. *Journal of Food Processing and Preservation* 42:e13335.

Al-Ali, A., Alkhawajah, A.A., Randhawa, M.A. and Shaikh, N.A., 2008. Oral and intraperitoneal LD50 of thymoquinone, an active principle of *Nigella sativa*, in mice and rats. *Journal of Ayub Medical College Abbottabad* 20:25–27.

Alam, I., Shahi, N.C., Lohani, U.C., Kumar, A. and Prakash, O., 2021. Ultrasound assisted extraction of oil from black cumin (*Nigella sativa* L.). *International Journal of Chemical Sciences* 9:87–91.

Al Attas, S.A., Fat'heya, M.Z. and Turkistany, S.A., 2016. *Nigella sativa* and its active constituent thymoquinone in oral health. *Saudi Medical Journal* 37:235.

Alemi, M., Sabouni, F., Sanjarian, F., Haghbeen, K. and Ansari, S., 2013. Anti-inflammatory effect of seeds and callus of *Nigella sativa* L. extracts on mix glial cells with regard to their thymoquinone content. *AAPS Pharmscitech* 14:160–167.

Alenzi, F.Q., Altamimi, M.A., Kujan, O., et al. 2013. Antioxidant properties of *Nigella sativa*. *Journal of Molecular and Genetic Medicine* 7:1–5.

Al-Gaby, A.M.A., 1998. Amino acid composition and biological effects of supplementing broad bean and corn proteins with *Nigella sativa* (black cumin) cake protein. *Food/Nahrung* 42:290–294.

Al-Ghamdi, M.S., 2001. The anti-inflammatory, analgesic and antipyretic activity of *Nigella sativa*. *Journal of Ethnopharmacology* 76:45–48.

Ali, B.H. and Blunden, G., 2003. Pharmacological and toxicological properties of *Nigella sativa*. *Phytotherapy Research* 17:299–305.

Ali, Z., Ferreira, D., Carvalho, P., Avery, M.A. and Khan, I.A., 2008. Nigellidine-4-*O*-sulfite, the first sulfated indazole-type alkaloid from the seeds of *Nigella sativa*. *Journal of Natural Products* 71:1111–1112.

Ali, H., Korshom, M.A., Mandour, A.A., El-Bessoumy, A.A. and El-Sayed, Y.S., 2017. Hepatoprotective efficacy of *Nigella sativa* seeds dietary supplementation against lead acetate-induced oxidative damage in rabbit–purification and characterization of glutathione peroxidase. *Biomedicine and Pharmacotherapy* 89:711–718.

Aljabre, S.H., Alakloby, O.M. and Randhawa, M.A., 2015. Dermatological effects of *Nigella sativa*. *Journal of Dermatology and Dermatologic Surgery* 19:92–98.

Aljabre, S.H.M., Randhawa, M.A., Akhtar, N. et al. 2005. Antidermatophyte activity of ether extract of *Nigella sativa* and its active principle, thymoquinone. *Journal of Ethnopharmacology* 101:116–119.

Al-Jassir, M.S., 1992. Chemical composition and microflora of black cumin (*Nigella sativa* L.) seeds growing in Saudi Arabia. *Food Chemistry* 45:239–242.

Almatroudi, A., Khadri, H., Azam, M., et al. 2020. Antibacterial, antibiofilm and anticancer activity of biologically synthesized silver nanoparticles using seed extract of *Nigella sativa*. *Processes* 8:388.

Alrashidi, M., Derawi, D., Salimon, J. and Yusoff, M.F., 2020. An investigation of physicochemical properties of *Nigella sativa* L. seed oil from Al-Qassim by different extraction methods. *Journal of King Saud University-Science* 32:3337–3342.

Altemimi, A., Lakhssassi, N., Baharlouei, A., Watson, D.G. and Lightfoot, D.A., 2017. Phytochemicals: Extraction, isolation, and identification of bioactive compounds from plant extracts. *Plants* 6:42.

Ansari, Z.M., Nasiruddin, M., Khan, R.A. and Haque, S.F., 2017. Protective role of *Nigella sativa* in diabetic nephropathy: A randomized clinical trial. *Saudi Journal of Kidney Diseases and Transplantation* 28:9.

Arslan, A.H., Tomruk, C.Ö., Meydanlı, E.G., et al. 2017. Histopathological evaluation of the effect of systemic thymoquinone administration on healing of bone defects in rat tibia. *Biotechnology and Biotechnological Equipment* 31:175–181.

Asgary, S., Ghannadi, A., Dashti, G., et al. 2013. *Nigella sativa* L. improves lipid profile and prevents atherosclerosis: Evidence from an experimental study on hypercholesterolemic rabbits. *Journal of Functional Foods* 5:228–234.

Ashraf, M., Ali, Q. and Iqbal, Z., 2006. Effect of nitrogen application rate on the content and composition of oil, essential oil and minerals in black cumin (*Nigella sativa* L.) seeds. *Journal of the Science of Food and Agriculture* 86:871–876.

Ashraf, S., Anjum, A.A., Ahmad, A., et al. 2018. In vitro activity of *Nigella sativa* against antibiotic resistant *Salmonella enterica*. *Environmental Toxicology and Pharmacology* 58:54–58.

Asrar, B., Khan, I., Homberg, J.R. and Haleem, D.J., 2020. *Nigella sativa* oil ameliorates chronic ethanol induced anxiety and impaired spatial memory by modulating noradrenaline levels. *Pakistan Veterinary Journal* 40:350–354.

Atta, M.B., 2003. Some characteristics of nigella (*Nigella sativa* L.) seed cultivated in Egypt and its lipid profile. *Food Chemistry* 83:63–68.

Awad, A.S., Kamel, R. and Sherief, M.A.E., 2011. Effect of thymoquinone on hepatorenal dysfunction and alteration of CYP3A1 and spermidine/spermine N-1-acetyl-transferase gene expression induced by renal ischaemia–reperfusion in rats. *Journal of Pharmacy and Pharmacology* 63:1037–1042.

Azmir, J., Zaidul, I.S.M., Rahman, M.M., et al. 2013. Techniques for extraction of bioactive compounds from plant materials: A review. *Journal of Food Engineering* 117:426–436.

Badary, O.A., Taha, R.A., Gamal El-Din, A.M., and Abdel-Wahab, M.H. 2003. Thymoquinone is a potent superoxide anion scavenger. *Drug and Chemical Toxicology*, 26:87–98.

Badary, O.A., Al-Shabanah, O.A., Nagi, M.N., Al-Bekairi, A.M. and Elmazar, M., 1998. Acute and subchronic toxicity of thymoquinone in mice. *Drug Development Research* 44:56–61.

Bader, A.A., 2015. Anti-ischemic Properties of *Nigella sativa* against cardiac and non-cardiovascular ischemia. *International Journal of Pharmacology and Toxicology* 5:53–61.

Bakhshabadi, H., Mirzaei, H., Ghodsvali, A., et al. 2017. Effects of pulsed electric field and microwave pretreatments on some the black cumin seeds oil characterises. *Innovative Food Technologies* 4:21–29.

Bakhshabadi, H., Mirzaei, H., Ghodsvali, A., et al. 2018. The influence of pulsed electric fields and microwave pretreatments on some selected physicochemical properties of oil extracted from black cumin seed. *Food Science and Nutrition* 6:111–118.

Bamosa, A., Ali, B.A. and Al-Hawsawi, Z.A., 2002. The effect of thymoquinone on blood lipids in rats. *Indian Journal of Physiology and Pharmacology* 46:195–201.

Bamosa, A.O., Kaatabi, H., Lebdaa, F.M., Elq, A.M. and Al-Sultanb, A., 2010. Effect of *Nigella sativa* seeds on the glycemic control of patients with type 2 diabetes mellitus. *Indian Journal of Physiology and Pharmacology* 54:344–54.

Barakat, E.M.F., El Wakeel, L.M. and Hagag, R.S., 2013. Effects of *Nigella sativa* on outcome of hepatitis C in Egypt. *World Journal of Gastroenterology* 19:2529.

Bashir, M.U., Qureshi, H.J. and Mumtaz, U., 2014. Effect of *Nigella sativa* seeds extract on serum C-reactive protein in albino rats. *Annals of King Edward Medical University* 20:35–35.

Benhelima, A., Kaid-Omar, Z., Hemida, H., Benmahdi, T. and Addou, A., 2016. Nephroprotective and diuretic effect of *Nigella sativa* L. seeds oil on Lithiasic Wistar rats. *African Journal of Traditional, Complementary and Alternative Medicines* 13:204–214.

Benkaci-Ali, F., Baaliouamer, A., Wathelet, J.P. and Marlier, M., 2012. Chemical composition and physicochemical characteristics of fixed oils from Algerian *Nigella sativa* seeds. *Chemistry of Natural Compounds* 47:925–931.

Benkaci-Ali, F., Baaliouamer, A., Meklati, B.Y. and Chemat, F. 2007. Chemical composition of seed essential oils from Algerian Nigella sativa extracted by microwave and hydrodistillation. *Flavour and Fragrance Journal* 22:148–153.

Bernhoft, A., 2010. A brief review on bioactive compounds in plants. *Bioactive Compounds in Plants-Benefits and Risks for Man and Animals* 50:11–17.

Boskabady, M.H., Shirmohammadi, B., Jandaghi, P. and Kiani, S., 2004. Possible mechanism(s) for relaxant effect of aqueous and macerated extracts from *Nigella sativa* on tracheal chains of guinea pig. *BMC Pharmacology* 4:1–6.

Botnick, I., Xue, W., Bar, E., et al. 2012. Distribution of primary and specialized metabolites in *Nigella sativa* seeds, a spice with vast traditional and historical uses. *Molecules* 17:10159–10177.

Bourgou, S., Ksouri, R., Bellila, A., et al. 2008. Phenolic composition and biological activities of Tunisian *Nigella sativa* L. shoots and roots. *Comptes Rendus Biologies* 331:48–55.

Bourgou, S., Pichette, A., Marzouk, B. and Legault, J., 2010. Bioactivities of black cumin essential oil and its main terpenes from Tunisia. *South African Journal of Botany* 76:210–216.

Burits, M. and Bucar, F., 2000. Antioxidant activity of *Nigella sativa* essential oil. *Phytotherapy Research* 14:323–328.

Butt, A.S., Nisar, N., Mughal, T.A., Ghani, N., and Altaf, I. 2019. Anti-oxidative and anti-proliferative activities of extracted phytochemical compound thymoquinone. *Journal of Pakistan Medical Association* 6:1479–1485.

Canayakin, D., Bayir, Y., Kilic Baygutalp, N., et al. 2016. Paracetamol-induced nephrotoxicity and oxidative stress in rats: The protective role of *Nigella sativa*. *Pharmaceutical Biology* 54:2082–2091.

Chakrabarty, A., Emerson, M.R. and LeVine, S.M., 2003. Hemeoxygenase-1 in SJL mice with experimental allergic encephalomyelitis. *Multiple Sclerosis Journal* 9:372–381.

Chakravarty, N., 1993. Inhibition of histamine release from mast cells by nigellone. *Annals of Allergy* 70:237–242.

Darakhshan, S., Tahvilian, R., Colagar, A.H. and Babolsar, I., 2015. *Nigella sativa*: A plant with multiple therapeutic implications. *International Journal of Pharmacognosy* 2:190–14.

Daryabeygi-Khotbehsara, R., Golzarand, M., Ghaffari, M.P. and Djafarian, K., 2017. *Nigella sativa* improves glucose homeostasis and serum lipids in type 2 diabetes: A systematic review and meta-analysis. *Complementary Therapies in Medicine* 35:6–13.

Dinagaran, S., Sridhar, S. and Eganathan, P., 2016. Chemical composition and antioxidant activities of black seed oil (*Nigella sativa* L.). *International Journal of Pharmaceutical Sciences and Research* 7:4473.

Dubey, P.N., Singh, B., Mishra, B.K., Kant, K. and Solanki, R.K., 2016. Nigella (*Nigella sativa*): A high value seed spice with immense medicinal potential. *Indian Journal of Agricultural Sciences* 86:967–979.

Dwarampudi, L.P., Palaniswamy, D. and Muruganantham Nithyanantham, P.S., 2012. Antipsoriatic activity and cytotoxicity of ethanolic extract of *Nigella sativa* seeds. *Pharmacognosy Magazine* 8:268.

Elbandy, M., Kang, O.H., Kwon, D.Y. and Rho, J.R., 2009. Two new antiinflammatory triterpene saponins from the Egyptian medicinal food black cumin (Seeds of *Nigella sativa*). *Bulletin of the Korean Chemical Society* 30:1811–1816.

El-Dakhakhny, M., Barakat, M., Abd El-Halim, M. and Aly, S.M., 2000. Effects of *Nigella sativa* oil on gastric secretion and ethanol induced ulcer in rats. *Journal of Ethnopharmacology* 72:299–304.

El-Mahdy, M.A., Zhu, Q., Wang, Q.E., Wani, G. and Wani, A.A., 2005. Thymoquinone induces apoptosis through activation of caspase-8 and mitochondrial events in p53-null myeloblastic leukemia HL-60 cells. *International Journal of Cancer* 117:409–417.

Ezirganli, S., Kazancioglu, H.O., Ozdemir, H., Inan, D.S. and Tek, M., 2016. The effects of *Nigella sativa* seed extract on bone healing in an experimental model. *Journal of Craniofacial Surgery* 27:1905–1909.

Fallah Huseini, H., Amini, M., Mohtashami, R., Ghamarchehre, M.E., et al. 2013. Blood pressure lowering effect of *Nigella sativa* L. seed oil in healthy volunteers: A randomized, double-blind, placebo-controlled clinical trial. *Phytotherapy Research* 27:1849–1853.

Farag, M.A., El-Kersh, D.M., Rasheed, D.M. and Heiss, A.G., 2017. Volatiles distribution in *Nigella* species (black cumin seeds) and in response to roasting as analyzed via solid-phase microextraction (SPME) coupled to chemometrics. *Industrial Crops and Products* 108:564–571.

Farag, M.A., Gad, H.A., Heiss, A.G. and Wessjohann, L.A., 2014. Metabolomics driven analysis of six *Nigella* species seeds via UPLC-qTOF-MS and GC–MS coupled to chemometrics. *Food Chemistry* 151:333–342.

Feng, Y., Dunshea, F.R. and Suleria, H.A., 2020. LC–ESI–QTOF/MS characterization of bioactive compounds from black spices and their potential antioxidant activities. *Journal of Food Science and Technology* 57:4671–4687.

Forouzanfar, F., Bazzaz, B.S.F. and Hosseinzadeh, H., 2014. Black cumin (*Nigella sativa*) and its constituent (thymoquinone): A review on antimicrobial effects. *Iranian Journal of Basic Medical Sciences* 17:929.

Gad, H.A. and El-Ahmady, S.H., 2018. Prediction of thymoquinone content in black seed oil using multivariate analysis: An efficient model for its quality assessment. *Industrial Crops and Products* 124:626–632.

Gali-Muhtasib, H., Diab-Assaf, M., Boltze, C., et al. 2004. Thymoquinone extracted from black seed triggers apoptotic cell death in human colorectal cancer cells via a p53-dependent mechanism. *International Journal of Oncology* 25:857–866.

Ghahramanloo, K.H., Kamalidehghan, B., Javar, H.A. et al., 2017. Comparative analysis of essential oil composition of Iranian and Indian *Nigella sativa* L. extracted using supercritical fluid extraction and solvent extraction. *Drug Design, Development and Therapy* 11:2221.

Gharby, S., Harhar, H., Guillaume, D., et al., 2015. Chemical investigation of Nigella sativa L. seed oil produced in Morocco. *Journal of the Saudi Society of Agricultural Sciences* 14:172–177.

Gholamnezhad, Z., Havakhah, S. and Boskabady, M.H., 2016. Preclinical and clinical effects of *Nigella sativa* and its constituent, thymoquinone: A review. *Journal of Ethnopharmacology* 190:372–386.

Ghorbanibirgani, A., Khalili, A. and Rokhafrooz, D., 2014. Comparing *Nigella sativa* oil and fish oil in treatment of vitiligo. *Iranian Red Crescent Medical Journal* 16:e4515.

Ghoreyshi, M., Mahmoudabady, M., Bafadam, S. and Niazmand, S., 2020. The protective effects of pharmacologic postconditioning of hydroalcoholic extract of *Nigella sativa* on functional activities and oxidative stress injury during ischemia–reperfusion in isolated rat heart. *Cardiovascular Toxicology* 20:130–138.

Ghosheh, O.A., Houdi, A.A. and Crooks, P.A., 1999. High performance liquid chromatographic analysis of the pharmacologically active quinones and related compounds in the oil of the black seed (*Nigella sativa* L.). *Journal of Pharmaceutical and Biomedical Analysis* 19:757–762.

Gilhotra, N. and Dhingra, D., 2011. Thymoquinone produced antianxiety-like effects in mice through modulation of GABA and NO levels. *Pharmacological Reports* 63:660–669.

Gokçe, A., Oktar, S., Koc, A. and Yonden, Z., 2011. Protective effects of thymoquinone against methotrexate-induced testicular injury. *Human and Experimental Toxicology* 30:897–903.

Güçlü-Üstündağ, Ö. and Mazza, G., 2007. Saponins: Properties, applications and processing. *Critical Reviews in Food Science and Nutrition* 47:231–258.

Habib, N., and Choudhry, S., 2021. HPLC quantification of thymoquinone extracted from Nigella sativa L. (Ranunculaceae) seeds and antibacterial activity of its extracts against Bacillus species. *Evidence-Based Complementary and Alternative Medicin* 2021:6645680.

Hadi, M.Y., Mohammed, G.J. and Hameed, I.H., 2016. Analysis of bioactive chemical compounds of *Nigella sativa* using gas chromatography-mass spectrometry. *Journal of Pharmacognosy and Phytotherapy* 8:8–24.

Hameed, I.H., Al-Rubaye, A.F. and Kadhim, M.J., 2017. Uses of nuclear magnetic resonance spectroscopy technique in pharmaceutical analysis: A review. *International Journal of Current Pharmaceutical Review and Research* 8:79–84.

Hameed, S., Imran, A., Nisa, M.U., et al. 2019. Characterization of extracted phenolics from black cumin (*Nigella sativa* Linn.), coriander seed (*Coriandrum sativum* L.), and fenugreek seed (*Trigonella foenum-graecum*). *International Journal of Food Properties* 22:714–726.

Hammad-Shafiq, A.A., Masud, T. and Kaleem, M., 2014. Cardio-protective and anti-cancer therapeutic potential of *Nigella sativa. Iranian Journal of Basic Medical Sciences* 17:967.

Handa, S.S., 2008. An overview of extraction techniques for medicinal and aromatic plants. *Extraction Technologies for Medicinal and Aromatic Plants* 1:21–40.

Hannan, M., Rahman, M., Sohag, A.A.M., et al. 2021. Black cumin (*Nigella sativa* L.): A comprehensive review on phytochemistry, health benefits, molecular pharmacology, and safety. *Nutrients* 13:1784.

Harzallah, H.J., Grayaa, R., Kharoubi, W., et al. 2012. Thymoquinone, the *Nigella sativa* bioactive compound, prevents circulatory oxidative stress caused by 1,2-dimethylhydrazine in erythrocyte during colon postinitiation carcinogenesis. *Oxidative Medicine and Cellular Longevity* 2012:854065.

Hossein, B.M., Nasim, V. and Sediqa, A., 2008. The protective effect of *Nigella sativa* on lung injury of sulfur mustard–exposed guinea pigs. *Experimental Lung Research* 34:183–194.

Ijaz, H., Tulain, U.R., Qureshi, J., et al. 2017. *Nigella sativa* (prophetic medicine): A review. *Pakistan Journal of Pharmaceutical Sciences* 30:229–234.

Iqbal, M.J., Butt, M.S. and Suleria, H.A.R., 2018. Black cumin: A review of phytochemistry, antioxidant potential, extraction techniques, and therapeutic perspectives. *Plant-and Marine-Based Phytochemicals for Human Health*, 12:151–188.

Iqbal, M.J., Butt, M.S., Sohail, M. and Suleria, H.A., 2019. The antioxidant potential of black cumin (*Nigella sativa* L.) extracts through different extraction methods. *Current Bioactive Compounds* 15:623–630.

Islam, M.T., Khan, M.R. and Mishra, S.K., 2019. An updated literature-based review: Phytochemistry, pharmacology and therapeutic promises of *Nigella sativa* L. *Oriental Pharmacy and Experimental Medicine* 19:115–129.

Jan, K., Ahmad, M., Rehman, S., Gani, A. and Khaqan, K., 2019. Effect of roasting on physicochemical and antioxidant properties of kalonji (*Nigella sativa*) seed flour. *Journal of Food Measurement and Characterization* 13:1364–1372.

Javadi, S.M.R., Hashemi, M., Mohammadi, Y., et al. 2018. Synergistic effect of honey and *Nigella sativa* on wound healing in rats. *Acta Cirurgica Brasileira* 33:518–523.

Kabir, Y., Akasaka-Hashimoto, Y., Kubota, K. and Komai, M., 2020. Volatile compounds of black cumin (*Nigella sativa* L.) seeds cultivated in Bangladesh and India. *Heliyon* 6:e05343.

Kabir, Y., Shirakawa, H. and Komai, M., 2019. Nutritional composition of the indigenous cultivar of black cumin seeds from Bangladesh. *Progress in Nutrition* 21:428–434.

Kalidasu, G., Reddy, G.S., Kumari, S.S., Kumari, A.L. and Sivasankar, A., 2017. Secondary volatiles and metabolites from *Nigella sativa* L. seed. *Indian Journal of Natural Products and Resources* 8:151–158.

Kanter, M., Coskun, O. and Uysal, H., 2006. The antioxidative and antihistaminic effect of *Nigella sativa* and its major constituent, thymoquinone on ethanol-induced gastric mucosal damage. *Archives of Toxicology* 80:217–224.

Kapoor, S., 2009. Emerging clinical and therapeutic applications of *Nigella sativa* in gastroenterology. *World Journal of Gastroenterology* 15:2170.

Kaseke, T., Opara, U.L. and Fawole, O.A., 2021. Novel seeds pretreatment techniques: Effect on oil quality and antioxidant properties: A review. *Journal of Food Science and Technology* 58:4451–4464.

Kazemi, M., 2014. Phytochemical composition, antioxidant, anti-inflammatory and antimicrobial activity of *Nigella sativa* L. essential oil. *Journal of Essential Oil Bearing Plants* 17:1002–1011.

Keshri, G., Singh, M.M., Lakshmi, V. and Kamboj, V.P., 1995. Post-coital contraceptive efficacy of the seeds of *Nigella sativa* in rats. *Indian Journal of Physiology and Pharmacology* 39:59–59.

Khan, M.A., 1999. Chemical composition and medicinal properties of *Nigella sativa* Linn. *Inflammopharmacology* 7:15–35.

Khan, N. and Sultana, S., 2005. Inhibition of two stage renal carcinogenesis, oxidative damage and hyperproliferative response by *Nigella sativa*. *European Journal of Cancer Prevention* 14:159–168.

Khan, A., Chen, H.C., Tania, M. and Zhang, D.Z., 2011. Anticancer activities of *Nigella sativa* (black cumin). *African Journal of Traditional, Complementary and Alternative Medicines* 8:226.

Khattak, K.F. and Simpson, T.J., 2008. Effect of gamma irradiation on the extraction yield, total phenolic content and free radical-scavenging activity of *Nigella sativa* seed. *Food Chemistry* 110:967–972.

Kiralan, M., Özkan, G., Bayrak, A. and Ramadan, M.F., 2014. Physicochemical properties and stability of black cumin (*Nigella sativa*) seed oil as affected by different extraction methods. *Industrial Crops and Products* 57:52–58.

Koshak, A.E. and Koshak, E.A., 2020. *Nigella sativa* L. as a potential phytotherapy for COVID-19: A mini-review of in-silico studies. *Current Therapeutic Research* 93:100602.

Koshak, A., Koshak, E. and Heinrich, M., 2017a. Medicinal benefits of *Nigella sativa* in bronchial asthma: A literature review. *Saudi Pharmaceutical Journal* 25:1130–1136.

Koshak, A., Wei, L., Koshak, E., et al. 2017b. *Nigella sativa* supplementation improves asthma control and biomarkers: A randomized, double-blind, placebo-controlled trial. *Phytotherapy Research* 31:403–409.

Krishnaveni, M. and Saranya, S., 2016. Phytoconstituent analysis of *Nigella sativa* seeds using analytical techniques. *Bulletin of Environment, Pharmacology and Life Sciences* 5:25–38.

Kumar, K., Chandra, S., Samsher, N.C., Singh, J. and Kumar, M., 2017. Functional properties of food commodities (wheat, kidney bean, cowpea, turnip, cauliflower) flours. *International Journal of Chemical Studies* 5:1199–1202.

Kumara, S.S.M. and Huat, B.T.K., 2001. Extraction, isolation and characterisation of antitumor principle, α-hederin, from the seeds of *Nigella sativa*. *Planta Medica* 67:29–32.

Latiff, L.A., Parhizkar, S., Dollah, M.A. and Hassan, S.T.S., 2014. Alternative supplement for enhancement of reproductive health and metabolic profile among perimenopausal women: A novel role of *Nigella sativa*. *Iranian Journal of Basic Medical Sciences* 17:980.

Liu, X., M Abd El-Aty, A. and Shim, J.H., 2011. Various extraction and analytical techniques for isolation and identification of secondary metabolites from *Nigella sativa* seeds. *Mini Reviews in Medicinal Chemistry* 11:947–955.

Liu, X., Abd El-Aty, A.M., Cho, S.K., et al. 2012. Characterization of secondary volatile profiles in *Nigella sativa* seeds from two different origins using accelerated solvent extraction and gas chromatography-mass spectrometry. *Biomedical Chromatography* 26:1157–1162.

Liu, X., Park, J.H., Abd El-Aty, A.M., et al. 2013. Isolation of volatiles from *Nigella sativa* seeds using microwave-assisted extraction: Effect of whole extracts on canine and murine CYP1A. *Biomedical Chromatography* 27:938–945.

Machmudah, S., Shiramizu, Y., Goto, M., Sasaki, M. and Hirose, T., 2005. Extraction of *Nigella sativa* L. using supercritical CO_2: A study of antioxidant activity of the extract. *Separation Science and Technology* 40:1267–1275.

Machmudah, S., Kawahito, Y., Sasaki, M., and Goto, M., 2007. Supercritical CO_2 extraction of rosehip seed oil: Fatty acids composition and process optimization. *The Journal of Supercritical Fluids* 41:421–428.

Mahboubi, M., 2018. Natural therapeutic approach of *Nigella sativa* (black seed) fixed oil in management of Sinusitis. *Integrative Medicine Research* 7:27–32.

Mahdavi, R., Heshmati, J. and Namazi, N., 2015. Effects of black seeds (*Nigella sativa*) on male infertility: A systematic review. *Journal of Herbal Medicine* 5:133–139.

Majdalawieh, A.F. and Fayyad, M.W., 2015. Immunomodulatory and anti-inflammatory action of *Nigella sativa* and thymoquinone: A comprehensive review. *International Immunopharmacology* 28:295–304.

Majeed, A., Muhammad, Z., Ahmad, H., et al. 2021. *Nigella sativa* L.: Uses in traditional and contemporary medicines – An overview. *Acta Ecologica Sinica* 41:253–258.

Malik, S. and Zaman, K., 1992. Nigellimine: A new isoquinoline alkaloid from the seeds of *Nigella sativa*. *Journal of Natural Products* 55:676–678.

Malik, S., Cun-Heng, H. and Clardy, J., 1985. Isolation and structure determination of nigellicine, a novel alkaloid from the seeds of *Nigella sativa*. *Tetrahedron Letters* 26:2759–2762.

Malik, S., Hasan, S.S., Choudhary, M.I., Ni, C.Z. and Clardy, J., 1995. Nigellidine – A new indazole alkaloid from the seeds of *Nigella sativa*. *Tetrahedron Letters* 36:1993–1996.

Mamun, M.A. and Absar, N., 2018. Major nutritional compositions of black cumin seeds – cultivated in Bangladesh and the physicochemical characteristics of its oil. *International Food Research Journal* 25:2634–2639.

Manoharan, N., Jayamurali, D., Parasuraman, R. and Govindarajulu, S.N., 2021. Phytochemical composition, therapeutical and pharmacological potential of *Nigella sativa*: A review. *Traditional Medicine Research* 6:32.

Mashayekhi-Sardoo, H., Rezaee, R. and Karimi, G., 2020. An overview of in vivo toxicological profile of thymoquinone. *Toxin Reviews* 39:115–122.

Matthaus, B. and Özcan, M.M., 2011. Fatty acids, tocopherol, and sterol contents of some *Nigella* species seed oil. *Czech Journal of Food Sciences* 29:145–150.

Maulidiani, M., Sheikh, B.Y., Mediani, A., et al. 2015. Differentiation of *Nigella sativa* seeds from four different origins and their bioactivity correlations based on NMR-metabolomics approach. *Phytochemistry Letters* 13:308–318.

Mazaheri, Y., Torbati, M., Azadmard-Damirchi, S. and Savage, G.P., 2019. A comprehensive review of the physicochemical, quality and nutritional properties of *Nigella sativa* oil. *Food Reviews International* 35:342–362.

Mehta, B.K., Sharma, U., Agrawal, S., et al. 2008. Isolation and characterization of new compounds from seeds of *Nigella sativa*. *Medicinal Chemistry Research* 17:462–473.

Merfort, I., Wray, V., Barakat, H.H., et al. 1997. Flavonol triglycosides from seeds of *Nigella sativa*. *Phytochemistry* 46:359–363.

Meziti, A., Meziti, H., Boudiaf, K., Mustapha, B. and Bouriche, H., 2012. Polyphenolic profile and antioxidant activities of *Nigella sativa* seed extracts in vitro and in vivo. *International Journal of Biotechnology and Bioengineering* 6:109–117.

Moghim, H., Taghipoor, S., Shahinfard, N., Kheiri, S. and Panahi, R., 2015. Antifungal effects of *Zataria multiflora* and *Nigella sativa* extracts against *Candida albicans*. *Journal of Herbmed Pharmacology* 4:138–141.

Moghimi, M., Farzaneh, V. and Bakhshabadi, H., 2018. The effect of ultrasound pretreatment on some selected physicochemical properties of black cumin (*Nigella sativa*). *Nutrire* 43:1–8.

Mohamed, A.M., Metwally, N.M. and Mahmoud, S.S., 2005. *Sativa* seeds against *Schistosoma mansoni* different stages. *Memórias do Instituto Oswaldo Cruz* 100:205–211.

Mohamed, A., Waris, H.M., Ramadan, H., Quereshi, M. and Kalra, J., 2009. Amelioration of chronic relapsing experimental autoimmune encephalomyelitis (CR-EAE) using thymoquinone-biomed 2009. *Biomedical Sciences Instrumentation* 45:274–279.

Mohammed, S.J., Amin, H.H.H., Aziz, et al., 2019. Structural characterization, antimicrobial activity, and in vitro cytotoxicity effect of black seed oil. *Evidence-based Complementary and Alternative Medicine* 2019:6515671–6515671.

Morikawa, T., Xu, F., Ninomiya, K., Matsuda, H. and Yoshikawa, M., 2004a. Nigellamines A3, A4, A5, and C, new dolabellane-type diterpene alkaloids, with lipid metabolism-promoting activities from the Egyptian medicinal food black cumin. *Chemical and Pharmaceutical Bulletin* 52:494–497.

Morikawa, T., Xu, F., Kashima, Y., et al., 2004b. Novel dolabellane-type diterpene alkaloids with lipid metabolism promoting activities from the seeds of *Nigella sativa*. *Organic Letters* 6:869–872.

Mosbah, R., Djerrou, Z. and Mantovani, A., 2018. Protective effect of *Nigella sativa* oil against acetamiprid induced reproductive toxicity in male rats. *Drug and Chemical Toxicology* 41:206–212.

Nadaf, N.H., Gawade, S.S., Muniv, A.S., et al., 2015. Exploring anti-yeast activity of *Nigella sativa* seed extracts. *Industrial Crops and Products* 77:624–630.

Naeimi, S.A., Hajimehdipoor, H. and Saber, S., 2020. Comparing the effect of *Nigella sativa* oil soft gel and placebo on oligomenorrhea, amenorrhea and laboratory characteristics in patients with polycystic ovarian syndrome, a randomized clinical trial. *Research Journal of Pharmacognosy* 7:49–59.

Namazi, N., Larijani, B., Ayati, M.H. and Abdollahi, M., 2018. The effects of *Nigella sativa* L. on obesity: A systematic review and meta-analysis. *Journal of Ethnopharmacology* 219:173–181.

Nasir, A., Siddiqui, M.Y. and Mohsin, M., 2014. Therapeutic uses of Shoneez (*Nigella sativa* Linn.) mentioned in Unani system of medicine – A review. *International Journal of Pharmacy and Phytopharmacology Research* 4:47–49.

Nawarathne, N.W., Wijesekera, K., Wijayaratne, W.M.D.G.B. and Napagoda, M., 2019. Development of novel topical cosmeceutical formulations from *Nigella sativa* L. with antimicrobial activity against acne-causing microorganisms. *The Scientific World Journal* 2019:5985207.

Naz, H., 2011. *Nigella sativa*: The miraculous herb. *Pakistan Journal of Biochemistry and Molecular Biology* 44:44–48.

Nergiz, C. and Otleş, S., 1993. Chemical composition of *Nigella sativa* L. seeds. *Food Chemistry* 48:259–261.

Nieto, N., Torres, M.I., Fernandez, M.I., et al. 2000. Experimental ulcerative colitis impairs antioxidant defense system in rat intestine. *Digestive Diseases and Sciences* 45:1820–1827.

Panche, A. N., Diwan, A. D., and Chandra, S. R., 2016. Flavonoids: An overview. *Journal of Nutritional Science*-5:1–15.

Pari, L. and Sankaranarayanan, C., 2009. Beneficial effects of thymoquinone on hepatic key enzymes in streptozotocin–nicotinamide induced diabetic rats. *Life Sciences* 85:830–834.

Pronyk, C. and Mazza, G., 2009. Design and scale-up of pressurized fluid extractors for food and bioproducts. *Journal of Food Engineering* 95:215–226.

Ramadan, M.F., 2007. Nutritional value, functional properties and nutraceutical applications of black cumin (*Nigella sativa* L.): An overview. *International Journal of Food Science and Technology* 42:1208–1218.

Razmpoosh, E., Safi, S., Nadjarzadeh, A., Fallahzadeh, H., Abdollahi, N., Mazaheri, M., Nazari, M. and Salehi-Abargouei, A., 2021. The effect of *Nigella sativa* supplementation on cardiovascular risk factors in obese and overweight women: A crossover, double-blind, placebo-controlled randomized clinical trial. *European Journal of Nutrition* 60:1863–1874.

Saheb, S.H., Desai, S.D., Das, K.K. and Haseena, S., 2016. Hepatoprotective effect of *Nigella sativa* seed in sterptozotocine induced diabetic albino rats: Histological observations. *International Journal of Anatomy and Research* 4:2459–2463.

Saidin, K.S., Jais, M.R., Ismail, E.N. and Ishak, R., 2020. The effect of *Nigella sativa* and *Eucheuma cottonii* in collagen-induced arthritis mice. *Research Journal of Pharmacy and Technology* 13:1319–1323.

Salehi, B., Quispe, C., Imran, M., et al. 2021. Nigella plants – Traditional uses, bioactive phytoconstituents, preclinical and clinical studies. *Frontiers in Pharmacology* 12:417.

Salem, E.M., Yar, T., Bamosa, A.O., et al. 2010. Comparative study of *Nigella sativa* and triple therapy in eradication of *Helicobacter pylori* in patients with non-ulcer dyspepsia. *Saudi Journal of Gastroenterology* 16:207.

Salem, A.M., Bamosa, A.O., Qutub, H.O., et al. 2017. Effect of *Nigella sativa* supplementation on lung function and inflammatory mediators in partly controlled asthma: A randomized controlled trial. *Annals of Saudi medicine* 37:64–71.

Sarac, G., Kapicioglu, Y., Sener, S., et al. 2019. Effectiveness of topical *Nigella sativa* for vitiligo treatment. *Dermatologic Therapy* 32:e12949.

Sasidharan, S., Chen, Y., Saravanan, D., Sundram, K.M. and Latha, L.Y., 2011. Extraction, isolation and characterization of bioactive compounds from plants' extracts. *African Journal of Traditional, Complementary and Alternative Medicines* 8:1–10.

Saxena, S.N., Rathore, S.S., Diwakar, Y., et al. 2017. Genetic diversity in fatty acid composition and antioxidant capacity of *Nigella sativa* L. genotypes. *LWT* 78:198–207.

Sayeed, M.S.B., Asaduzzaman, M., Morshed, H., Hossain, M.M., Kadir, M.F. and Rahman, M.R., 2013. The effect of *Nigella sativa* Linn. seed on memory, attention and cognition in healthy human volunteers. *Journal of Ethnopharmacology* 148:780–786.

Sayeed, M.S.B., Shams, T., Hossain, S.F., et al. 2014. *Nigella sativa* L. seeds modulate mood, anxiety and cognition in healthy adolescent males. *Journal of Ethnopharmacology* 152:156–162.

Shipitsin, M. and Polyak, K., 2008. The cancer stem cell hypothesis: In search of definitions, markers, and relevance. *Laboratory Investigation* 88:459–463.

Shoaei-Hagh, P., Kamelan Kafi, F., Najafi, S., et al. 2021. A randomized, double-blind, placebo-controlled, clinical trial to evaluate the benefits of *Nigella sativa* seeds oil in reducing cardiovascular risks in hypertensive patients. *Phytotherapy Research* 35:4388–4400.

Shoieb, A.M., Elgayyar, M., Dudrick, P.S., Bell, J.L. and Tithof, P.K., 2003. In vitro inhibition of growth and induction of apoptosis in cancer cell lines by thymoquinone. *International Journal of Oncology* 22:107–113.

Shreelakshmidevi, S., Jaikumar, S. and Pammy, S., 2019. Hepatoprotective effect of *Nigella sativa* seed extract against isoniazid and rifampacin induced hepatotoxicity. *Indian Journal of Public Health Research and Development* 10:344–349.

Silva, L.V., Nelson, D.L., Drummond, M.F.B., Dufossé, L. and Glória, M.B.A., 2005. Comparison of hydrodistillation methods for the deodorization of turmeric. *Food Research International* 38:1087–1096.

Singh, R.K., Vishwakarma, R.K., Vishal, M.K., Singh, S.K. and Saharan, V.K., 2015. Moisture dependent physical properties of nigella seeds. *African Journal of Agricultural Research* 10:58–66.

Solati, Z., Baharin, B.S. and Bagheri, H., 2014. Antioxidant property, thymoquinone content and chemical characteristics of different extracts from *Nigella sativa* L. seeds. *Journal of the American Oil Chemists' Society* 91:295–300.

Soleymani, S., Zargaran, A., Farzaei, M.H., et al. 2020. The effect of a hydrogel made by *Nigella sativa* L. on acne vulgaris: A randomized double-blind clinical trial. *Phytotherapy Research* 34:3052–3062.

Spínola, V., Pinto, J. and Castilho, P.C., 2015. Identification and quantification of phenolic compounds of selected fruits from Madeira Island by HPLC-DAD–ESI-MSn and screening for their antioxidant activity. *Food Chemistry* 173:14–30.

Sridharan, V., Suryavanshi, P.A. and Menendez, J.C., 2011. Advances in the chemistry of tetrahydroquinolines. *Chemical Reviews* 111:7157–7259.

Stoyanova, A., Georgiev, E., Wajs, A. and Kalemba, D., 2003. A comparative investigation on the composition of the volatiles from seeds of *Nigella sativa* L. from Bulgaria. *Journal of Essential Oil Bearing Plants* 6:207–209.

Sudhir, S.P., Deshmukh, V.O. and Verma, H.N., 2016. *Nigella sativa* seed, a novel beauty care ingredient: A review. *International Journal of Pharmaceutical Sciences and Research* 7:3185.

Sultan, M.T., Butt, M.S., Anjum, F.M., et al. 2009. Nutritional profile of indigenous cultivar of black cumin seeds and antioxidant potential of its fixed and essential oil. *Pakistan Journal of Botany* 41:1321–1330.

Suri, K., Singh, B., Kaur, A., Yadav, M.P. and Singh, N., 2019. Impact of infrared and dry air roasting on the oxidative stability, fatty acid composition, Maillard reaction products and other chemical properties of black cumin (*Nigella sativa* L.) seed oil. *Food Chemistry* 295:537–547.

Taha, M., Azeiz, A.Z. and Saudi, W., 2010. Antifungal effect of thymol, thymoquinone and thymohydroquinone against yeasts, dermatophytes and non-dermatophyte molds isolated from skin and nails fungal infections. *Egyptian Journal of Biochemistry and Molecular Biology* 28:109–126.

Tariq, M., 2008. *Nigella sativa* seeds: Folklore treatment in modern day medicine. *Saudi Journal of Gastroenterology* 14:105.

Tasar, N., Sehirli, O., Yigner, O., Leymanoglu, S. and Yegen, B., 2012. Protective effects of *Nigella sativa* against hypertension-induced oxidative stress and cardiovascular dysfunction in rats. *Marmara Pharmaceutical Journal* 16:141–149.

Taşkin, M.K., Çalişkan, Ö.A., Anil, H., et al. 2005. Triterpene saponins from *Nigella sativa* L. *Turkish Journal of Chemistry* 29:561–569.

Tavakkoli, A., Mahdian, V., Razavi, B.M. and Hosseinzadeh, H., 2017. Review on clinical trials of black seed (*Nigella sativa*) and its active constituent, thymoquinone. *Journal of Pharmacopuncture* 20:179.

Tembhurne, S.V., Feroz, S., More, B.H. and Sakarkar, D.M., 2014. A review on therapeutic potential of *Nigella sativa* (kalonji) seeds. *Journal of Medicinal Plants Research* 8:167–177.

Thilakarathne, R.C.N., Madushanka, G.D.M.P. and Navaratne, S.B., 2018. Phytochemical analysis of Indian and Ethiopian black cumin seeds (*Nigella sativa*). *Agricultural Research and Technology* 17:556011.

Tiji, S., Benayad, O., Berrabah, M., El Mounsi, I. and Mimouni, M., 2021. Phytochemical profile and antioxidant activity of *Nigella sativa* L. growing in Morocco. *The Scientific World Journal* 2021:6623609.

Toma, M., Vinatoru, M., Paniwnyk, L. and Mason, T.J., 2001. Investigation of the effects of ultrasound on vegetal tissues during solvent extraction. *Ultrasonics Sonochemistry* 8:137–142.

Topcagic, A., Zeljkovic, S.C., Karalija, E., Galijasevic, S. and Sofic, E., 2017. Evaluation of phenolic profile, enzyme inhibitory and antimicrobial activities of *Nigella sativa* L. seed extracts. *Bosnian Journal of Basic Medical Sciences* 17:286.

Trigui, I., Yaich, H., Sila, A., et al. 2018. Physicochemical properties of water-soluble polysaccharides from black cumin seeds. *International Journal of Biological Macromolecules* 117:937–946.

Vatansev, H., Ciftci, H., Ozkaya, A., et al. 2013. Chemical composition of *Nigella sativa* L. seeds used as a medical aromatic plant from East Anatolia Region, Turkey. *Asian Journal of Chemistry* 25:5490–5492.

Venkatachallam, S.K.T., Pattekhan, H., Divakar, S. and Kadimi, U.S., 2010. Chemical composition of *Nigella sativa* L. seed extracts obtained by supercritical carbon dioxide. *Journal of Food Science and Technology* 47:598–605.

Wen, C., Zhang, J., Zhang, H., et al. 2018. Advances in ultrasound assisted extraction of bioactive compounds from cash crops – A review. *Ultrasonics Sonochemistry* 48:538–549.

Wen, L., Zhang, Z., Sun, D.W., Sivagnanam, S.P. and Tiwari, B.K., 2020. Combination of emerging technologies for the extraction of bioactive compounds. *Critical Reviews in Food Science and Nutrition* 60:1826–1841.

Wienkötter, N., Höpner, D., Schütte, U., et al. 2008. The effect of nigellone and thymoquinone on inhibiting trachea contraction and mucociliary clearance. *Planta Medica* 74:105–108.

Wrona, O., Rafińska, K., Możeński, C. and Buszewski, B., 2017. Supercritical fluid extraction of bioactive compounds from plant materials. *Journal of AOAC International* 100:1624–1635.

Yessuf, A.M., 2015. Phytochemical extraction and screening of bioactive compounds from black cumin (*Nigella sativa*) seeds extract. *American Journal of Life Sciences* 3:358–364.

Yimer, E.M., Tuem, K.B., Karim, A., Ur-Rehman, N. and Anwar, F., 2019. *Nigella sativa* L. (black cumin): A promising natural remedy for wide range of illnesses. *Evidence-Based Complementary and Alternative Medicine: Ecam* 2019:1528635–1528635.

Yuan, T., Nahar, P., Sharma, M., et al. 2014. Indazole-type alkaloids from *Nigella sativa* seeds exhibit antihyperglycemic effects via AMPK activation in vitro. *Journal of Natural Products* 77:2316–2320.

CHAPTER 11

Bioactive Components in Saffron

Vandana Sablania, Suradeep Basak, and Vidhu Yadav

CONTENTS

DOI: 10.1201/9781003215387-11

11.1 INTRODUCTION

Crocus sativus L. (Saffron) is a perennial plant, which belongs to Iridaceae family. Saffron is one of the most important and highly priced spices which is also known by various names like "Plant gold" and "Red gold". *C. sativus* L. is a stemless shrub with 3–5 cm globule-shaped corm under the ground. During growth period, each and every corm produces 6–9 long, narrow grass-like leaves along with purple-coloured tubular flower which consists of 4–6 tepals emerging out of the flower. Central part of the flower is known as pistil which contains ovary and leads to the growth of stalk-like structure known as style. As the style grows, it splits yellow-orange-red-colour-lobed stigma of length around 2.5–3 cm. Drying of style portion along with stigma leads to the production of pure saffron (Sampathu et al. 1984). Saffron tepal is rich in protein, fibre, fats, and minerals like potassium, calcium, and phosphorus which is necessary for the growth of animals. Thus, it may be used as an animal fodder (Fahim et al. 2012). Tepals of *C. sativus* L. are supplements of biologically active compounds such as quercetin, kaempferol, rutin, lutein, hesperidin, flavonoids, tannins, and anthocyanins (Srivastava et al. 2010).

It is being suggested to pluck the flowers before sunrise, when they are in budding approach and the grown tepals protect stigma from physical and environmental damage to maintain the quality of saffron. Saffron is being cultivated in different parts of the world such as Spain, India, Greece, Turkey, Iran, and China (Ordoubi et al. 2017). Iran contributes about 90% of the total worldwide saffron production, India contributes 5% production, and remaining 5% contribution is specified by some Mediterranean countries such as Spain, Morocco, Italy, and Turkey (Fernandez et al. 2004). The soil used for cultivation of saffron should have neutral pH with loamy texture, free from stones, and must be having high water retention capacity. Possibly, dry and moderate humid conditions with rainfall 30–40 cm have been advisable for ideal growth of flowers (Sampathu et al. 1984). Saffron is a triploid geophyte due to which it is unable to produce seed. It was observed that 70% of its ovule contains polygamous sac-like structure which corresponds to low pollen viability and reduces the chances of germination or fertilization due to meiotic abnormalities. Therefore, saffron undergoes for self-pollination to increase the rate of pollen viability (Gresta et al. 2008).

The characteristics of saffron influenced by various pre-harvesting factors like cultivation time, rate of irrigation, temperature, type of soil, fertilizers, and post-harvesting factors such as duration of harvesting, storage conditions, and drying conditions (Negbi 2003). As per ISO-3632-2-2003 (2003), saffron is also known as "the stigmas of *Crocus sativus* L. dried, dark red in colour and trumpet-shaped, serrated, or indented at the distal end. The length is between 20 mm and

40 mm. The stigmas may be isolated or joined in pairs or threes at the end of the portion of the style, which is white/yellow in colour". There are different varieties of saffron available globally depending on their grading methods. In India, Mongra and Lachcha are two varieties of saffron produced by the cultivators. Mongra variety is of red colour and contains mainly broken stigma, whereas Lachcha variety contains few yellow colour stigma along with stigma and pieces of tepals (Sampathu et al. 1984).

Saffron is composed of carbohydrate in the form of pectin, dextrins, pentosans, gums, sugar (63%), protein (12%), crude fibre (5%), minerals (5%), and fat (5%) (Shafiee et al. 2017). It has been also added into various cereal products (pasta), desserts (chocolate soup, yogurt, and jam), and beverages (vermouth, herbal tea) to increase consumer's acceptability (Armellini et al. 2018; Almodovar et al. 2020).

Saffron is commonly used as seasoning, colouring, and fragrant agent in different commodities due to its robust taste, odour, and red colour (Zeka et al. 2015). Even saffron extract can be infused into various cosmetics like cream, lotion, shampoo, face wash, and body wash to protect them from sun exposure; rather it reduces the rate of ageing process due to the presence of bioactive compounds in saffron (Hosseinzadeh and Nassiri-Asl 2010). It has also been used for the development of traditional medicines for curing cramps, depression, liver disease, pain, and tumours. Similarly, there are some pharmacological applications of saffron due to which it can be used for the treatment of several metabolic disorders such as cardiovascular diseases, hypertension, dyslipidaemia, diabetes, and stoutness (Shafiee et al. 2017). Due to the potential functions of saffron, the market demand has been increased over time. Hence, this chapter deals with the extraction of various bioactive components and their biological, chemical, and pharmacological properties with different methods of extraction for further utilization in various food products.

11.2 BIOACTIVE COMPONENTS IN *C. SATIVUS*

Various researchers have reported numerous activities of *C. sativus* L. which is due to the presence of bioactive components in its stigma portion (Figure 11.1). These components are picrocrocin, safranal, crocetin, and crocin which are attributed to several pharmacological, therapeutic, and antioxidant properties and culinary importance (Zeka et al. 2015; Baba et al. 2015). Carotenoids, monoterpenoids, flavonoid, and anthocyanins are the most important components of stigma, accountable for specific colour of the saffron. They are composed of both fat-soluble carotenoids (lycopene, α-carotene and β-carotene, zeaxanthin) and water-soluble carotenoids (apocarotenoid, crocetin) with "cis" and "trans" configuration of carotenoids (Gresta et al. 2008). Moreover, different hues (yellow, orange, red) of stigma correspond to the presence of hydrophilic carotenoids which is also known as crocin, and its presence is abundant in saffron plant. Crocin is mono or di-glycosyl esters of dicarboxylic acid known as crocetin which is derived from zeaxanthin with molecular weight 976.96 g/mol. The change in sugar moieties of crocin at R1 and R2 position leads to the formation of different metabolites like digentiobioside crocetin (crocin 1), gentioglucoside crocetin (crocin 2), gentiobioside crocetin (crocin 3), glucoside crocetin (crocin 4), and diglucoside crocetin (crocin 5) which makes them highly soluble in water and increases the antioxidant activity; rather it also plays a significant role in plant reproduction (Serrano-Díaz et al. 2011; Samarghandian and Borji 2014). The melting point of crocin is 186°C and crocin-1 may contribute >10% mass of dry saffron which can be observed at 440 nm (Singla and Bhat 2011). According to the International Union of Pure and Applied Chemistry (IUPAC), crocetin is named as 8,8′-diapo 8,8′-carotenoic acid having molecular weight of 328.4 g/mol. The melting point of crocetin is 285°C (Samarghandian and Borji 2014). Shahi et al. (2016) reviewed from various articles that 94% of total crocetin present in the form of glycosides and remaining 6% as free crocetin. Zeaxanthin acts as a precursor which undergoes oxidative degradation and produces various metabolites of crocetin, for example, α-crocetin,

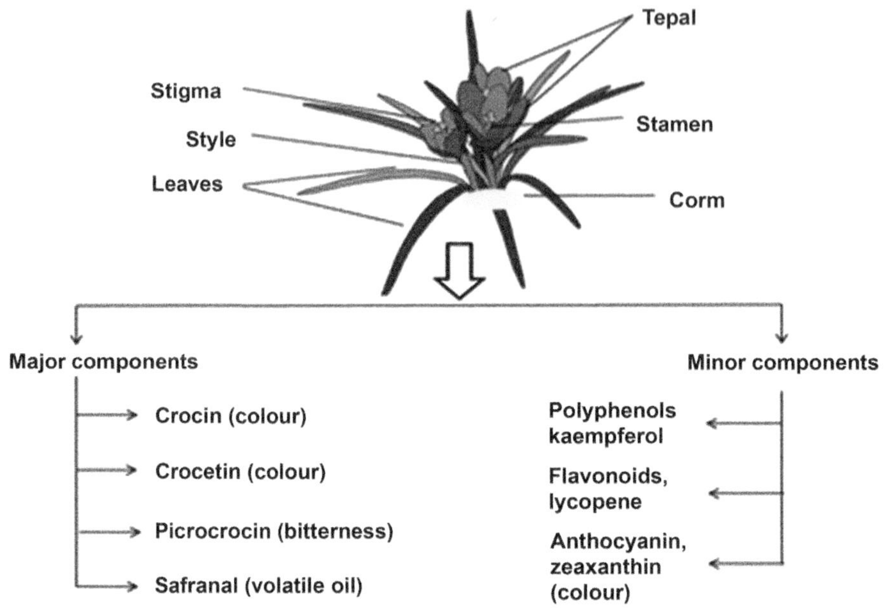

Figure 11.1 Schematic diagram and bioactive components in *C. sativus* L.

β-crocetin, and γ-crocetin which possess many pharmaceutical applications. The chemical structure of various bioactive components is detailed in Table 11.1.

Picrocrocin corresponds to the bitter taste of saffron and found to be second most abundant component (1–13% of saffron's dry weight) which is being produced as a metabolite during crocin production pathway with molecular weight of 330.37 g/mol. Picrocrocin is a monoterpene glycoside and its IUPAC name is 4-(b-D-glucopyranosyloxy)-2,6,6-trimethyl-1 cyclohexene-1-carboxaldehyde. The absorbance of picrocrocin can be observed at 254 nm (Shahi et al. 2016). In picrocrocin pathway, loss of one glucose unit transforms picrocrocin into oxysafranal, which consequently acts as a precursor of safranal production. Safranal is a monoterpene aldehyde and aglycon of picrocrocin with molecular weight of 150.21 g/mol. Safranal is one of the most important volatile components in saffron with IUPAC name 2,6,6-trimethyl-1,3 cyclohexadiene-1-carboxaldehyde. Seventy per cent of the total volatile fraction is contributed by safranal and its different derivatives (Esfanjani et al. 2015). During the dehydration of stigma, safranal is being produced from β-glucosidase of picrocrocin due to the activation of enzymes at higher temperature. The maximum absorption of safranal can be obtained at 330 nm (Shahi et al. 2016). Similarly, several metabolites in the form of polyphenolic compounds have been studied by researchers in which major components are beta-carotene, mangicrocin, alkaloids, terpenoids, flavonoids, anthocyanin, and saponins (Samarghandian and Borji 2014). Gresta et al. (2008) have also reviewed that polyphenolic compounds such as terpenoids, kaempferol, alkaloids, and saponins are considered as a precursor of volatile compounds which probably react through picrocrocin to impart unpleasant taste in saffron.

11.3 CHEMISTRY OF SAFFRON

Saffron (*C. sativus* L.) is an integral part of herbal medicine as it possesses wide range of biological activities, such as antioxidant, antimicrobial, anti-depressant, anti-carcinogenic, anti-asthma, anti-obesity, anti-diabetic, anti-apoptotic, anti-Alzheimer, and aphrodisiac to name a few (Kakouri

Table 11.1 Classification of Bioactive Components of *Crocus sativus* L.

Components and Their Derivatives	Melting Point (°C)	Structure
α-Crocin	176–182	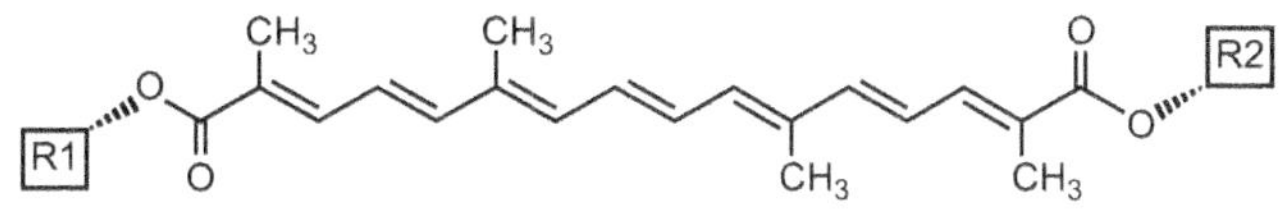
Crocetin	275–285	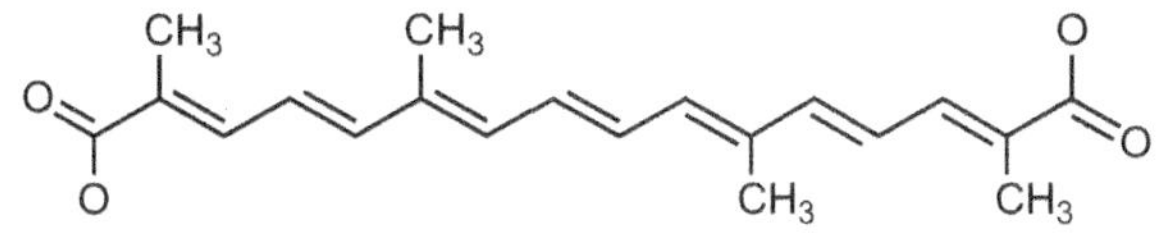
4-Hydroxyl-2,6,6 trimethyl-1-tetrahydrobenzaladehyde	<150	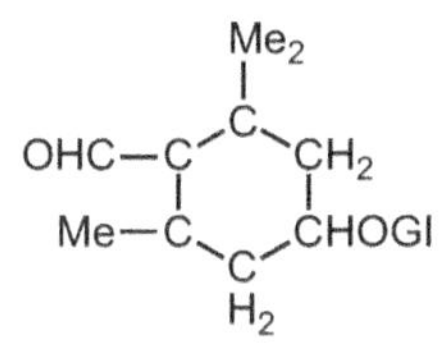
β-D-Gentibiose	180–195	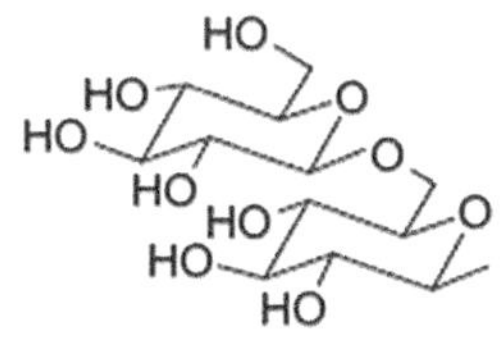
Picrocrocin	156	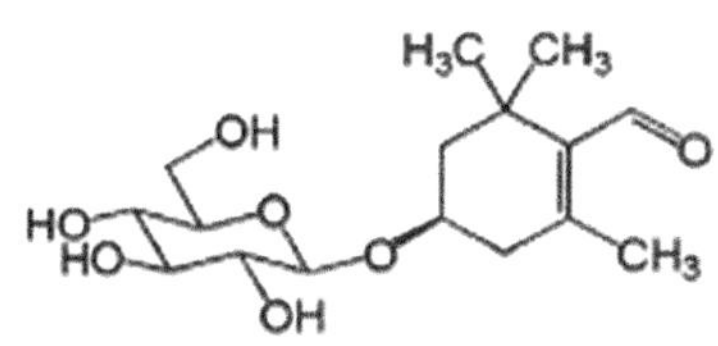
Isophorone	–6 to –8	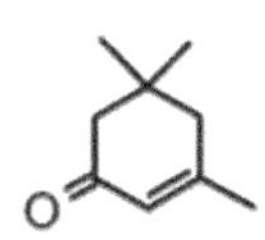
Saffranal	106–110	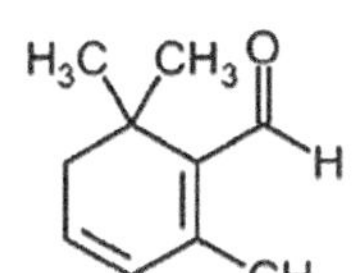
Zeaxanthin	210–220	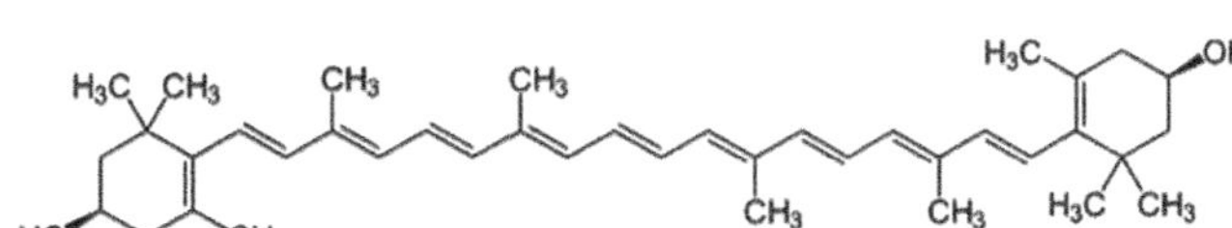

Source: Sampathu et al. (1984); Gresta et al. (2008).

et al. 2017; Pandey et al. 2020). With the emergence of green consumerism, the use of saffron at levels of 1–260 ppm as food-colouring agent has picked up in bakery and confectionary and beverage sector of food industries (Selim et al. 2000). The dried elongated stigma and yellowish style of *C. sativus* flower encompass the commercial saffron, and all studies on saffron are cited to the stigma. Based on the present scientific literature, it can be adjudged that there are numerous volatile, non-volatile, and aromatic chemical compounds present in saffron. The proximate composition (% w/w) of saffron suggested 63% of total carbohydrates that includes starch, 20% reducing sugars (viz., glucose, fructose, gentiobiose, xylose, and ramnose), 6–7% of pentosans, 9–10% of gums, dextrin, and pectin; 5% of crude fibre; 12–13% of protein, 5–8% of fat, 5–7% of minerals (mostly calcium, sodium, zinc, phosphorous, potassium, and manganese in trace amount, and 10–12% of moisture, and vitamin particularly riboflavin and thiamine (Sampathu et al. 1984; Rios et al. 1996; Melnyk et al. 2010)). The secondary metabolites of saffron include flavanoids, terpenoids, terpene alcohols, and their esters apocarotenoids (Pandey et al. 2020). Based on their structure or precursor, these volatile and aromatic compounds are divided into two groups, namely, C_{13}-norisoprenoids and C_9 and C_{10} group (Shahi et al. 2016). Oxidation of carotenoids results in release of apocarotenoids that operate as signalling molecules which eventually ends up being part of several pathways responsible for biological activity of the spice. Chemical analysis of extract of dry stigma has revealed apocarotenoids, namely, crocin (mono-glycosyl or di-glycosyl polyene ester responsible for colour), safranal (product of zeaxanthin degradation and responsible for aroma), picrocrocin (monoterperne glycoside precursor of safranal), and crocetin (produces upon hydrolysis of crocin) as the backbone of sensory and biological properties of saffron (Sampathu et al. 1984; Shahi et al. 2016).

The apocarotenoids and volatile oils of saffron, which is mainly composed of cineole, pinene, phenethenol, greaniol, limonene, borneol, etc., are responsible for colour, flavour, and aroma, whereas the non-volatile portion includes compounds like zeaxanthin, β- and γ-crocetin, lycopene, tetrahydrolycopene, phytoene, phytofluene and several α- and β-carotenes, vitamins particularly riboflavin and thiamine, etc., which add up to its biological activities (Samarghandian and Borji 2014; Shahi et al. 2016; Pandey et al. 2020). The concentration of phenolic compounds and carotenoids in saffron changes under the influence of varying environmental conditions, such as drought, altitude, temperature, and UV radiation (Zarinkamar et al. 2011).

11.3.1 Crocin

Crocin is glycoside carotenoid which is responsible for the distinct bright yellow-red colour of the spice (Christodoulou et al. 2015). More precisely, it is monoglycosyl or diglycosyl esters of polyene dicarboxylic acid crocetin with elementary composition and molecular weight of $C_{44}H_{64}O_{24}$ and 976.96 g/mol, respectively, and about 6–16% of total dry matter of saffron is composed of crocin (Tarantilis et al. 1998; Samarghandian and Borji 2014). Sugar moieties, such as gentiobiose, glucose, triglucose, or neapolitanose, present in crocin are found in cis- or trans-configuration (Sanchez et al. 2009). There are different types of crocins, namely, α-crocin or crocin-I or digentiobioside crocetin or crocetin digentibiosyl ester, crocin-II or tricrocin or gentioglucoside crocetin or crocetin digentibiosyl glucosyl ester, crocin-III or gentiobioside crocetin or crocetin monogentibiosyl ester, crocin-IV or glucoside crocetin, and crocin-V or diglucoside crocetin, and they are all stable at ambient temperature (Christodoulou et al. 2015; Pandey et al. 2020). Among these, crocin-I constitutes 10% of total saffron dry weight with blends of golden-yellow-orange colour (Shahi et al. 2016). Unlike other carotenoids, crocin is water soluble; hence, it is hugely popular colouring agent in food and traditional medicine (Basker and Negbi 1983; Lage and Cantrell 2009).

11.3.2 Crocetin

Crocetin is lipophilic, low molecular weight (328.4 g/mol) natural carotenoid having elementary composition of $C_{20}H_{24}O_4$ with seven unsaturation, four methyl groups, and carboxylic acid group at each end of the molecule (Christodoulou et al. 2015). Esterification of crocetin forms crocins; therefore, it forms core of crocin and also responsible for the distinct colour and accounts 14% of dry matter of the spice (Samarghandian and Borji 2014). Crocetin is basically aglycon of crocin, and in saffron, it is derived from oxidative degradation of zeaxanthin precursor (Shahi et al. 2016). It has a melting point of 285°C and upon crystallization it forms red needles, whereas in solution its colour remains yellow (Samarghandian and Borji 2014). It is sparingly soluble in basic aqueous solution (20 μM at pH 8) but readily gets dissolved with organic bases, namely, pyridine (Christodoulou et al. 2015). Crocetin has wide biological activity that includes antioxidant, anti-cancer, anti-depressant, anti-apoptotic, cardiovascular improver among a few (Wang et al. 2014).

11.3.3 Picrocrocin

Picrocrocin is colourless, water-soluble monoterpene glycoside that constitute 1–13% of dry matter of the spice and also responsible for the bitter taste (Alonso et al. 2000). It has an elementary formula of $C_{16}H_{26}O_7$ and molecular weight of 330.37 g/mol, and precursor of aromatic compounds of saffron (Lage and Cantrell 2009; Samarghandian and Borji 2014). Upon action by β-glucosidase, picrocrocin breaks down into a molecule of glucose and an aglycone (4-hydroxy-β-cyclocitral), which on dehydration yields dehydro-β-cyclocitral or safranal (Christodoulou et al. 2015).

11.3.4 Safranal

Safranal is the major component that constitutes up to 70% of volatile fraction of the essential oil of saffron and is responsible for the distinct aroma (Rodel and Petrzika 1991; Lage and Cantrell 2009). It is monoterpene aldehyde with chemical formula of $C_{10}H_{14}O$ and molecular weight of 150.21 g/mol (Christodoulou et al. 2015). Safranal is not present in the fresh stigma and is formed from picrocrocin only during postharvest processing (enzymatic or acid/alkali treatment) of the spice (Carmona et al. 2007).

11.4 EXTRACTION METHOD USED FOR EXTRACTION OF BIOACTIVE COMPONENTS OF SAFFRON

Various techniques have been developed owing to the demand to extract biological compounds having maximum quality and extraction efficacy. Extraction of bioactive components from various plants can be done by traditional as well as conventional methods. Traditional methods include maceration, hydro-distillation, and Soxhlet extraction methods whereas novel techniques include sonication, microwave-assisted extraction (MAE), supercritical fluid extraction (SCFE), high-pressure extraction, high-performance liquid chromatography (HPLC), ultrasound-assisted extraction (UAE), and ultrafiltration (UF) (Azwanida 2015). Extraction is defined as the separation of biologically active components, metabolites, and residues from the plants by using definite amount of solvents depending on their polarity. Extraction basically involves the separation of required components from the substrate which further depends on several factors such extraction time, temperature, nature of solvent, concentration of solvent, and particle size of sample (Handa et al. 2008). So far, a large number of solvents have been utilized to release bioactive glycosides of saffron (Kyriakoudi et al. 2012). In contrast to the traditional methods, novel techniques are usually preferred to control elution of unnecessary components, to reduce the extraction period and energy, and loss of solvents. Extraction done by using different traditional and novel extraction methods has been summarized in Table 11.2.

11.4.1 Conventional Extraction Techniques

11.4.1.1 Soxhlet Extraction

Mostly it is used for the extraction of fat, bioactive compounds, and essential oils. Saffron is kept in porous cellulose thimble and then set the Soxhlet apparatus with optimum amount of solvent (ethanol, methanol, diethyl ether, chloroform, etc.). Let heat the sample by maintaining optimum

Table 11.2 Enlisting Various Extraction Methods for the Extraction of Bioactive Components from *Crocus sativus* L.

Extraction Method	Extraction Parameters	Extracted Component	References
Maceration	Sample: 0.05 g Ethanol: 33.33% Extraction time: 2 h Extraction temperature: 85°C	Best suited for the extraction of picrocrocin and safranal	Sarfarazi et al. (2015)
	Sample: 1 g Ethanol: 33.33% Extraction temperature: 85°C Extraction time: 2 h	Anthocyanin, hydroxyl trimethyl cyclohexene Carboxaldehyde	Khazaei et al. (2016)
	Solvent: hexane and methanol Drying temperature including oven-drying and freeze-drying at 87°C for 20 min and 43°C for 70 min, respectively	All bioactive components High drying temperature resulted in greater retention of crocin as compared to the intermediate temperature (46–58°C)	Gregory et al. (2005)
Soxhlet apparatus	Sample: 5 g Methanol: 100 mL Extraction time: 6 h at room temperature	Maximum extraction yield (51.25%). Higher amount of anthocyanins (0.67 mg/g) in tuber followed by flavones (2.52 mg/g) in flower	Senizza et al. (2020)
	Sample: 10 g Methanol: 150 mL Extraction time: 8 h at room temperature	Recovery of safranal (12.24 mg/g) and crocin (2.04 mg/g)	Goleroudbary and Ghoreishi (2016)
	Sample: 20 g Methanol: 100 mL Extraction time: 5–6 h at room temperature	All bioactive components of saffron	Tarantilis et al. (1994)
Hydro-distillation	Sample: 5 g 50 mL of deionised water, 5 mL of diethyl ether, Methanol: 100 mL Extraction time: 2 h Condensation at −10°C by using a solution of water and glycol	Extraction of essential oil of saffron	Kanakis et al. (2004)
Microwave-assisted extraction	Sample: 1 g 50:50 ethanol:water Extraction time: 18 min Power: 200 W	Strong antioxidant activity in terms of DPPH, ABTS, and FRAP due to the presence of polyphenols	Gallo et al. (2010)
	Sample: 1 g Power: 800 W Temperature: 66°C Solvent: water:ethanol (41:59) Time: 0.5, 1, 2, 3, 4, 5 min	Total polyphenols and anthocyanin	Da Porto and Natolino (2018)
	Sample: 1 g 50:50 ethanol:water Extraction time: 18 min Power: 1000 W Frequency: 2.45 GHz Temperature: 95.91°C Time: 30 min Ethanol: 59.59%	Extraction of picrocrocin, crocin, and safranal	Sarfarazi et al. (2019)
Ultrasound-assisted extraction	Sample: 3 g 30 mL of 50:50 Ethanol: water solvent Sonication time: 30 min Frequency: 33 kHz	Extraction of polyphenols	Gallo et al. (2010)

(Continued)

Table 11.2 Enlisting Various Extraction Methods for the Extraction of Bioactive Components from *Crocus sativus* L. (*Continued*)

Extraction Method	Extraction Parameters	Extracted Component	References
	Sample: 4 g Temperature: 25°C Frequency: 35 kHz 40 mL of diethyl ether Sonication time: 30 min	Extraction of safranal and other major compounds like 4-hydroxy-2,6,6-trimethyl-1-cyclohexen-1-carboxaldehyde and 3,5,5-trimethyl-2-cyclohexene-1-one	Maggi et al. (2011)
	Sample: 2 g of tepal Temperature: 25°C Frequency: 35 kHz Petroleum ether: 75 mL	Bioactives compounds from tepal like aglycon flavonoids, and determination antioxidant activity	Kakouri et al. (2017)
Supercritical fluid extraction	Sample: 0.5 g Temperature: 80°C CO_2 pump rate: 3 mL/min Entrainer pump rate (methanol/water): 0.2 mL/min Pressure: 30 MPa/40 MPa	Crocin, picrocrocin, and α-crocin (30 MPa) Safranal, and 4-hydroxy-2,6,6-trimethyl-1-cyclohexen-1-carboxaldehyde (40 MPa) HTCC, picrocrocin, safranal (methanol) α-Crocin and deglycosylatedcrocin (water)	Nerome et al. (2016)
	Sample: 1 g Temperature: 62°C Plunger pump rate: 100 mL/min Solvent: 30 mL of 5% ethanol Pressure: 164 MPa Time: 45 min	Phenolic and flavonoid content	Ahmadian-Kouchaksaraie and Niazmand (2017)
	Sample: 1.5 g Temperature: 92°C/44°C Flow rate: 1 cm^3 Solvent: 50% methanol Pressure: 21.3 MPa/19.3 MPa Time: 122 min/110 min	Safranal recovery (91.76%) Crocin recovery (32.67%)	Goleroudbary and Ghoreishi (2016)
Ohmic-assisted extraction	Sample: 5 g Time: 45 min NaCl for conducting electricity: 0.3 g Voltage: 225 V Distilled water: 100 mL	Total phenols, flavonoids, and antioxidant activity	Gahruie et al. (2020)
High hydrostatic pressure extraction	Sample: 1 g Temperature: 50°C Solvent: 100 mL water Pressure: 580 MPa Time: 5 min	Increase in extraction quality of crocin (52–63%), picrocrocin (54–85 %), safranal (55–62%)	Shinwari and Rao (2018)
Enzyme-assisted extraction	Pectinase enzyme: 7 U/g dry matter β-Glucosidase: 25 U/g dry matter Time: 1 h Temperature: 50°C pH: 4.8 Solvent: ethanol	Anthocyanins	Lotfi et al. (2015)
Ultrafiltration	Sample: 0.5 g of extracted sample from 20 mL of cyclohexane Solvent: 1 L of water Centrifugation: 7000 rpm, 20 min Filter membrane pore size: 0.8, 0.45, and 0.2 μm	Crocetin ester, picrocrocin, and kaempferol glycoside	Sanchez et al. (2008)

(Continued)

Table 11.2 Enlisting Various Extraction Methods for the Extraction of Bioactive Components from *Crocus sativus* L. (*Continued*)

Extraction Method	Extraction Parameters	Extracted Component	References
Solid-phase extraction	Sample: 5 g Solvent: 30 mL of cyclohexane Centrifugation: 4000 rpm, 10 min Elution: 90 mL of 10% acetonitrile/ water Time: 4 h	Picrocrocin	Sanchez et al. (2009)
Solid-phase extraction	Sample: 5 g Solvent: 30 mL of cyclohexane Centrifugation: 4000 rpm, 10 min Elution: 90 mL of 10% acetonitrile/ water Time: 4 h	Picrocrocin	Sanchez et al. (2009)
Emulsion liquid-membrane extraction	Surfactant: 2.5% *n*-decane Agitating speed: 300 rpm Internal aqueous phase to membrane ratio: 0.8 Emulsion to external aqueous emulsion ratio: 0.3	Safranal, picrocrocin, and crocin	Mokhtari and Pourabdollah (2013)
Pulsed electric field extraction	Sample: 2 mg/mL of water Temperature: room temperature Whatman filter paper: 0.7 µm Frequency: 0.1–10 Hz Pulse width: 10–100 µs Voltage: 5 kV Electric current: 33 A	Crocin, safranal, and picrocrocin	Pourzaki et al. (2013)
	Sample: 0.2 g water: 150 mL Temperature: 25°C Whatman filter paper: 0.7 µm Electric field: 2 kV/cm Voltage: 8 kV Electric current: 23 A	Polyphenolic compounds, crocin, flavonoids, and antioxidant activity	Neri et al. (2021)

temperature for 5–6 h on the basis of boiling temperature of the solvents. This apparatus works by siphon process which is responsible for the release of components into the solvent involving the principle of "like dissolves like" and this can be separated by the heating in hot air oven or by using rotary evaporator (Azwanida 2015). Senizza et al. (2020) conducted an experiment by extracting various phytochemical components from different parts of the *Colchicum triphyllum* plant (Meadow saffron) like flower, leaves, and tubers by taking 100 mL of methanol. The reported results have shown significant difference in anthocyanins content in which tuber showed higher value as compared to the flower of meadow saffron. Aqueous extract of saffron was obtained from 15 g of powdered petal, mixed with 400 mL of distilled water, and run in Soxhlet extractor for 18 h. The prepared saffron extract was concentrated to 100 mL by using rotatory evaporator under low pressure and then filtered through a 0.2-mm filter paper. The semi-dried extract was utilized to determine the presence of metabolites which causes lung cancer (Samarghandian et al. 2013).

11.4.1.2 Maceration

It is frequently used method for the extraction of different components from plants. According to Azwanida (2015), maceration involved the soaking of plant sample and would be kept at room temperature for a period of at least 2–3 days with intermittent agitation. Soaking of plant sample for certain time period leads to the breakdown of plant cell wall to increase the infusion of components into the solvent. Sometimes, heat is also allowed to pass through the system to facilitate the extraction process depending on the nature of solvent added during soaking. Sometimes, wide mixture of solvents including water, organic solvents, and other solvents has been involved for the extraction of saffron bioactive compounds. Sarfarazi et al. (2015) have performed the statistical analysis by response surface methodology to optimize extraction parameters for the separation of bioactive components from saffron. The optimization parameters involved ethanol concentration (0–100%), extraction time (2–7 h), and temperature (5–85°C). He reported that increasing temperature for a reduced time period resulted in higher picrocrocin content sequentially while decreased during higher extraction time. The relationship between ethanol concentration and temperature showed an interactive effect on picrocrocin extraction. It was found that the amount of carotenoids from saffron extracted by chloroform was higher than acetone due to its high solubility in it (Castillo et al. 2005).

11.4.1.3 Hydro-Distillation/Steam-Distillation

Hydro- or steam-distillation technique can be used for the extraction of bioactives and essential oils from various plants. It may allocate with water, steam, or the mixture of both along with some solvents like diethyl ether. For extraction, it is important to soak the stigma in water and then heating should be done by selecting any mode of heat treatment aforementioned. Steam and hot water are considered as an essential factor for release of bioactive compounds from plant tissues. The extraction should be deliberately done at high extraction temperature to remove the unnecessary compounds in bioactive contents. This method has some of the disadvantages like difficult to recover heat-sensitive bioactive compounds, long time process, loss of polar components, etc. (Azmir et al. 2013). To overcome such losses, new and updated version of hydro-distillation technique has been introduced by the name of turbo-distillation to improve the efficiency of the process by re-using the water source during condensation stage. Turbo-distillation is one of the important methods to increase the withdrawal of biological active complexes (El Asbahani et al. 2015). It was found that Kanakis et al. (2004) have extracted safranal and 4-hydroxy-2,6,6-trimethyl-1-cyclohexene-1-carboxaldehyde (HTCC) from saffron by using hydro-distillation extraction method. The amount of safranal was reported in the range of 288.1–687.9 mg/100 g of saffron.

11.4.2 Novel Extraction Techniques

11.4.2.1 Microwave-Assisted Extraction

In this extraction method, utilization of solvents is assisted by microwave heating which results in improved extraction efficiency in reduced extraction time. The electromagnetic waves are diffused at a frequency ranged between 300 MHz and 300 GHz, whereas in food processing, food industry, pharmaceuticals, and microwaves are used in the range of 0.915–2.45 GHz. In this method, the rate of heat and mass transfer occurs in the same direction (Mason et al. 2011). The samples are taken into optimum concentration and allowed to heat up to its boiling temperature. The diffusion of solvents in plant cells softens the tissues and releases the dissolved components into the solvent. Muzaffar et al. (2015) have used MAE for the elution of bioactive components from saffron and effective release of crocin, safranal picrocrocin, and anthocyanin was noted. Microwave assisted extraction of *C. sativus* showed antioxidant activity of 73.94 mmol of Fe^{2+}/100 g of sample due to the higher affinity of crocetin in water which easily adsorb microwave energy during extraction (Gallo et al. 2010). Sarfarazi et al. (2020) have accomplished response surface methodology to improve different variables such as extraction time (10–30 min), extraction temperature (45–125°C), and concentration of ethanol (0–100%) for the extraction of bioactive components from saffron.

The obtained responses have shown significant difference in terms of influence of temperature and solvent concentration.

11.4.2.2 Ultrasound-Assisted Extraction/Sonication-Assisted Extraction

This method involves the use of sound wave in the range of 20 kHz–100 MHz. The ultrasound waves are responsible for promoting the surface interaction between the solvent and sample, thereby facilitating the release of components. This method is quite simple and cost-effective to produce components in both small and large-scale production area with low consumption of solvent in less time period (Azwanida 2015). Ultrasound waves have been established to release entangled organic and inorganic metabolites from plants by intensifying the mass transfer and accelerate the interaction of solvent to the target compounds (Garavand et al. 2019). Numerous components of saffron like polyphenols (kaempferol, quercetin, and myricetin), crocetin, picrocrocin, and safranal have been examined by UAE due to no severe effect of heat production with short extraction time period (Garavand et al. 2019).

11.4.2.3 Supercritical Fluid Extraction

It is an extraction method that utilizes carbon dioxide as supercritical fluid which exists in both liquid and gaseous form at its critical point. The density of carbon dioxide, low viscosity, high extraction efficiency, and high diffusivity, makes it suitable for the extraction of low polar molecules, especially safranal (Azwanida 2015). Though the supercritical fluid extraction (SCFE) has low permeability towards the polar analytes, therefore the addition of aqueous ethanol and methanol was actively used for the separation of those polar elements like crocin and crocetin from saffron. The solubility efficiency and water absorption capacity are the important criteria which need to consider while using aqueous solvents in SCFE method (Goleroudbary and Ghoreishi 2016). During the extraction of components from saffron by SCFE method, Nerome et al. (2016) reported that increasing temperature and pressure up to 80°C and 30 MPa, respectively, resulted in increased extract concentration that might be due to improved polarity of carbon dioxide which further enhances the solubility of extracted analytes such as deglycosylated crocin, HTCC, and safranal. Box-Behnken design was successfully implied to optimize extraction time, temperature, and pressure for the extraction of bioactive components of saffron which illustrates enhanced antioxidant activity and flavanoid content at optimized condition (Ahmadian-Kouchaksaraie and Niazmand 2017).

11.4.2.4 Ohmic-Assisted Extraction

Ohmic-assisted extraction (OAE) method is one of the important methods used for the extraction of phytochemicals, bioactive components, and essential oils by modifying various extraction factors (Gahruie et al. 2020). In this method, an alternating current is allowed to pass through the sample which leads to generation of electric resistance into the material, thus improves the extraction efficiency with qualitative components from the sample in shorter period and less consumption of energy. It has been quoted as "green technology" which utilizes less amount of energy and can also work by using renewable source of energy by producing electricity with reduced production of carbon dioxide (Khajehei et al. 2017). Gahruie et al. (2020) have optimized the extraction parameters such as time (15–45 min) and voltage (75–225 V) for the abstraction of phytochemicals from *C. sativus* (Table 11.2) by using OAE method. The obtained extract resulted in inhibition concentration (IC_{50}), phenolic content, and flavonoid content of 178 μg/mL, 928 mg/100 g, and 148 μg/mL, respectively. Therefore, it was concluded that OAE method showed high bioactive content with maximum antimicrobial and lipase enzyme inhibiting activity as compared to UAE and MAE method.

11.4.2.5 Enzyme-Assisted Extraction

Enzyme-assisted extraction (EAE) technique is becoming popular method for the extraction of phytochemicals. This method involves the use of enzymes which act as catalyst to promote the abstraction of components and synthesis of complex polyphenols based on their inherent ability. Enzymes are known for their ability to degrade the cell wall and cell membrane of various components for their sufficient release (Puri et al. 2012). EAE methods have been used for the extraction of polysaccharides, protein, oils, flavours, and pigments with reduced energy consumption and processing time. A large number of enzymes such as cellulase, hemicellulase, pectinase, pectinesterase, cellobiase, protease, and α-amylase have been employed for rapid extraction of polyphenolic compounds (Sowbhagya and Chitra 2010). Enzymes with better hydrolytic properties can only be used to degrade the esterified sides of various components. Pectinase is one of the most important exploited enzymes due to its numerous properties in food industry such as extract yield, digestibility, filterability, and reducing viscosity (Danalache et al. 2018). Lotfi et al. (2015) have performed EAE method for the separation of anthocyanin from saffron. Parameters used for the extraction of pigment have been mentioned in Table 11.2.

11.4.2.6 High Hydrostatic Pressure Extraction

High hydrostatic pressure extraction (HHPE) is considered as one of the non-thermal and alternative methods for the extraction of bioactive components from numerous plants. The pressure can be ranged from 100 to 1000 MPa for the effective release of components from plants (Prasad et al. 2009. High hydrostatic pressure instrument controls the pressure and temperature by using sensor valve. Ethylene glycol is usually used to enhance the pressure of fluid due to its higher viscosity than water. Hence, it remains in liquid form at sub-zero (0 to –10°C) temperature for efficient working of instrument at lower temperature (Buzrul et al. 2008). It was described that high-pressure inclusion for the extraction process reduces processing time, improves the mass transfer rate, diffusion rate of components in solvent, and also increases the extraction rate (Prasad et al. 2009). The pressure difference between the inner and outer membrane of the commodity significantly increases the permeation rate of components. Shinwari and Rao (2018) have extracted picrocrocin, crocin, and safranal from saffron by means of HHPE method with a pressure and temperature range of 100–600 MPa and 30–70°C, respectively. However, positive pressure exhibited increased crocin content. Increasing temperature resulted in reduced crocin content and increased content of safranal up to 50°C. Moreover, the extract also showed anti-cancer activity against the cells of colon cancer with 50% inhibition concentration of 0.96 mg/mL. Therefore, HHPE method is advised for the extraction of biologically active components from *C. sativus* for its utilization in various nutraceutical products.

11.4.2.7 Ultrafiltration

UF is a membrane separation process which is highly selective and accountable for the segregation of compounds depending on their molecular weight, colloidal particles, and other polymeric components. These polymeric membranes are made up of cellulose acetate, polypropylene, etc. In this method, solids particles are separated from the liquid and further can be concentrated by reverse osmosis or de-acidification treatment (Hernandez et al. 1995). Since a long time, UF is being used for the separation of biologically active components such polyphenols, flavonoids, carotenoids, polysaccharides, etc. It depends on multiple factors like particles size of components, thickness, and pore size of filtering aid, cross-sectional area of membrane, and flow rate of solutes. The separation of phytochemicals from *C. sativus* was done by using 1 g of saffron powder in 20 mL of cyclohexane and incubated for 5 h at room temperature. Furthermore, extract was centrifuged at 7000 rpm for 20 min and the obtained extract was additionally mixed with water for extracting bioactive components by passing it through different size membranes. Therefore, the combination of centrifugation along with UF has shown a significant release of biological components by selecting appropriate design of membrane (Sanchez et al. 2008). Not much studies have been done for the extraction of saffron's bioactive components by using UF method but it can be used in future for the extraction of numerous components.

11.4.2.8 Solid-Phase Extraction

Solid-phase extraction (SPE) is commonly used less expansive extraction and purification technique. This method involves separation of constituents depending on their solid and liquid phase. It assists the constituents to get adsorbed over the surface of solid phase (fibre coating) and later on it will be fractionated depending on their particle size and concentration by using appropriate amount of solvents (Sanchez et al. 2009). SPE depends on the interaction of both sorbent and analytes, hydrophobic group, ionic, and non-ionic interactions. The components of interest are mostly polar in nature which gets absorbed on the sorbent surface due to their strong carbon hydrogen interface, whereas non-polar components create hindrance between the absorbed component and sorbent. Furthermore, it can be used for the elution of components from the sorbent by reversed phase SPE (Zwir-Ferenec and Biziuk 2006). Sanchez et al. (2009) worked on extraction of picrocrocin from saffron by SPE by optimizing different parameters which is mentioned in Table 11.2.

11.4.2.9 Solid-Liquid Extraction

Solid-liquid extraction (SLE) is relatively a new technique for the extraction of polyphenolic compounds under the influence of repeated pressure. In this method, the sample is initially immersed into the solvent and thus pressure is applied for a certain period of time at room temperature. The Naviglio extractor used for this purpose consists of steel piston and porous septa which allows the permeation of solvent across it. The extraction chambers are linked with pressure valve which need to be closed during extraction and should be maintained throughout the extraction. During this phase, equilibrium sustained between the solid (sample) and liquid (solvent) which causes the separation of bioactive components from the sample and solvent will get release as soon as the pressure valve opens resulting in creation of negative pressure gradient between inside and outside of the sample matrix. Therefore, the remained residues over the solid matrix are physically removed and mixed with solvent for further release of bioactive components (Ferrara et al. 2014). Ferrara et al. (2014) have compared the ultrasound and SLE method for the separation of biological components from saffron and it was reported that SLE method was best suited for the extraction these metabolites.

11.4.2.10 Emulsion Liquid-Membrane Extraction

Emulsion liquid-membrane extraction (ELME) is the method in which the extraction occurs at one side and exposed at the other side, whereas in conventional methods, extraction and stripping are done within the same apparatus. ELME is known for its double emulsion phase in which one aqueous phase contains stripping agents and other aqueous phase contains eluted compounds like polyphenols, metals, organic compounds, etc. These two emulsion phases are separated by organic membrane which embraces diluents, surfactants, and carriers (Garavand et al. 2019). The extraction of bioactive compounds from saffron was done by ELME method and reported 90% abstraction of safranal, picrocrocin, and crocin (Mokhtari and Pourabdollah 2013).

11.4.2.11 Subcritical Water Extraction

Subcritical water extraction (SCWE) works under the influence of high pressure and high temperature to keep it in liquid phase for accomplishing quick extraction. Water is known as distinctive solvent due to its large number of hydrogen bonds. The critical temperature and pressure of water is 374°C and 22 MPa, respectively, at 218 atm whereas at subcritical point the temperature varies from 100 to 374°C under high pressure to sustain water in its liquid form. Elevation in temperature alters the dielectric constant which thus helps in maintaining the polarity of liquid for conducting selective extraction of poly-phenolic components (Sarfarazi et al. 2019; Rodriguez-Meizoso et al. 2010). SCWE method was selected for the extraction of safranal, crocin, and crocetin from saffron. The parameters like extraction time (5–15 min) and temperature (105–125°C) were optimized for the same by using response surface methodology (Sarfarazi et al. 2019). The obtained results showed moderate time and low temperature extraction has significant impact over the extraction efficiency of safranal and crocin which means they are thermally more stable as compared to picrocrocin.

11.4.2.12 Pulsed Electric Field Extraction

Pulsed electric field is a non-thermal extraction method whose working efficiency depends on electric pulses generated during release of electric power. An electric power of 20–80 kV/cm can be used for the generation of pulsed electric field. Pulsed electric field extraction (PEFE) method works at low electric power, which generates electric field around the cell membrane or cell wall of the plant and causes the partial or complete loss of membrane to increase its permeability for the liberation of bioactive components. PEFE is responsible for the stimulation of thermo-labile components and also avoids their losses during extraction. PEFE method found to have improved mass transfer, better extraction efficiency, and energy saving and takes less time for the extraction (Barba et al. 2015). Neri et al. (2021) have performed an experiment to determine the effect of pulsed electric field as a pretreatment to evaluate the functional and quality attributes of *C. sativus* which showed relatively improved properties in respect to antioxidant activity and phenolic content. It was also observed that PEFE-treated saffron enhanced the leaching of safranal and crocin into the treatment media.

11.4.2.13 Stir-Bar Sorptive Extraction

It is one of the most efficient microextraction methods which requires less time for extraction with less solvent. In this method, the desired sample is taken into immiscible solvent and a polymer-coated bar is added into it under the controlled conditions such as time, temperature, and solvent concentration. After extraction, the stir bar is removed and rinsed with distilled water to remove sugar, salts, and protein metabolites. Rinsing does not cause the loss of polyphenolic components which is being absorbed in it. Commonly used stir bar is a non-polar polydimethylsiloxane which act as sorptive phase for the extracting residues due to its thermal stability and high diffusion ability (Prieto et al. 2010), whereas it is considered as one of the disadvantages due to non-efficient adsorption of polar components over polydimethylsiloxane stir bar. The obtained analytes from stir bar are analysed by gas chromatography-mass spectra or mass spectra.

11.5 IDENTIFICATION AND QUANTIFICATION OF SAFFRON BIOACTIVE COMPONENTS BY VARIOUS METHODS

11.5.1 High-Performance Liquid Chromatography

HPLC analysis is commonly employed method for the quantification of non-volatile bioactive compounds from different extract of plants, fruits, and vegetables. Almodovar et al. (2020) have performed an experiment to determine the bioaccessibility of saffron extract in which the identification and quantification were done by HPLC-PAD/MS having mass range 50–1500 amu, drying temperature, and drying flow rate (340°C, 10 L/min), along with capillary tension of 2000 V. Only crocetin was detected in concentrated amount as compared to other high molecular weight carotenoids. Sarfarazi et al. (2020) have performed HPLC analysis for the microwave-assisted extract of saffron by using C18 column. A solvent concentration of methanol:water in the ratio of 20:80 to 80:20 was taken at a constant flow rate of 1 mL/min with an elution time of 60 min at 25°C. Twenty microlitres of saffron extract was injected and the obtained peaks were reported at 440 nm along with reference compound (crocin). The identified areas under the curve were analysed and compared. Subsequently, the obtained area was reported similar to the area of reference compounds. Further, gas chromatography-mass spectrometry (GC-MS) analysis was also done which quantified safranal compound with a molecular weight of 150 mol and retention time of 6.719 min. More specifically, Kyriakoudi et al. (2012) have performed reverse-phase HPLC-DAD analysis for the saffron extract obtained by UAE method for the determination of crocin and picrocrocin. A UV6000 LP DAD was used and separation was carried out by C18 column (128 mm × 4 mm) with a gradient mixture of (A) water:acetic acid (1%) and (B) acetonitrile (20–100%) at a flow rate of 0.5 mL/min. The sample was injected at the rate of 20 μL and monitored at the wavelength of 200–500 nm, whereas the identification of crocin and picrocrocin was reported at 440 nm and 250 nm, respectively. The quantification of these two components was determined by constructing a calibration curve of crocin and crocetin standards by using aqueous methanolic solution. The reverse-phase HPLC-DAD analysis was employed to obtain maximum recovery of glycosides from saffron extract. Solvent concentration in the ratio of 1:1 methanol and water is highly recommended for the extraction of these components. Even the purity of picrocrocin was analysed by Sanchez et al. (2009) with column chromatography through C18 adsorbent at 250 nm under the obtained peak area and was found to be 96%. Goleroudbary and Ghoreishi (2016) have quantified the content of crocin and safranal through HPLC by using C18 column (250 mm × 4.6 mm × 10 μm) with an aliquot sample of 20 μL with a mobile phase (methanol) at a flow rate of 0.7 mL/min and was detected in UV-vis at 310 and 440 nm for safranal and crocin, respectively. Lotfi et al. (2015) injected 50 μL of saffron extract into C18 column (12.5 mm × 5 mm × 5 μm). The mobile phase included two different solvents in which A corresponds to 2.5% acetic acid in water and B corresponds to 2.5% of acetic acid in methanol. The solvent gradients were taken at different concentration such as 100%, 90%, and 50% for a time period of 5, 15, and 45 min, respectively, and was analysed for the determination of anthocyanins from saffron tepals which showed a better results in terms of stability, decomposition process, and colour values as compared to the one which was extracted with methanol. Hence, these compounds can be used as better source of antioxidant activity.

11.5.2 Gas Chromatography-Mass Spectrometry

GC-MS is mostly used for the quantification of volatile compounds. Due to high heat treatment, there are chances for loss of heat-sensitive components such as safranal which is volatile in nature; therefore, it is preferably quantified by GC-MS. The author has analysed the aromatic extract obtained by different techniques such as SPE, liquid–liquid extraction, solvent-assisted flavour evaporation, and simultaneous distillation extraction. Around 28 aromatic compounds were reported in saffron extract obtained from solvent-assisted flavour evaporation method in which predominantly volatiles were ketones followed by aldehydes and acids. Gas chromatography was coupled with flame ionization diode and mass spectrometry for the quantification of numerous volatile components of saffron. In this method, 3 μL of saffron extract was injected to separate aromatic compounds on DB-Wax (30 m × 0.25 mm × 0.5 μm thickness) and set at a temperature of 280°C. The flow rate of helium gas was maintained at 1.5 mL/min. The mass spectra were operated with electronic ionization energy of 70 eV. GC-MS and ionization energy source were maintained at 250°C and 180°C, respectively, and identification was done in a mass range of *m/z* 30–300 at 2.0 scan per minute whereas quantification was performed by comparing the peak of pure compounds and resulted in Amanpour et al. (2015). The coupling of GC with MS has been modified in various ways regarding the characteristics of column, capillary chromatography, temperature of oven, and sample preparation and quantification is typically done by referring the peaks of standard of high purity. Liu et al. (2018) have performed GC-MS analysis of saffron extract obtained by ultrasonication extraction method. GC/MS-QP equipped with quadrupole mass spectrometer with optima 5 capillary column (30 m × 0.32 mm × 0.25 μm thickness) and the flow rate of helium (carrier gas) was 1.50 mL/min. The temperature was maintained from 80 to 150°C at the rate of 5°C/min and then 250°C at 25°C/min with electronic ionization energy of 70 eV. Identification and quantification were performed with the full scan in mass range of *m/z* 20–600. This combination resulted in higher content of picrocrocin and safranal but lower content of kaempferol-3-*O*-sophoroside and isophorone.

11.5.3 Liquid Chromatography-Mass Spectroscopy

Liquid chromatography-mass spectroscopy (LC-MS) analysis is used for the determination of bioactive components of various plants. LC-MS and GC-MS are almost similar analysis methods; only difference is between their mobile phases. In case of LC-MS solvents and in GC-MS helium gas were used as a mobile phase for the identification and quantification of polyphenolic compounds. Kakouri et al. (2017) have focussed on the quantitative analysis of bioactive compounds of *C. sativus* extract by LC-DAD/MS by injecting 10 μL extract of tepals on C18 column at 35°C. Solvent A (99.99% acetic acid) and solvent B (99.99% acetonitrile) were operated at various combinations of solvents and temperature gradient started with 5–30% of solvent B for 0–74 min up to 90% B to a time period up to 120 min. The flow rate was maintained at 0.4 mL/min and voltage was set at 1.85 with scan range 100–900 *m/z* having scan speed of 4000 amu/s. The obtained results showed derivatives of kaempherol and glycosides such as myricetin and quercetin. Moreover, Zalacain et al. (2005) have also quantified the amount of crocetin and picrocrocin in saffron extract by using LC-DAD-MS.

11.5.4 UV-vis Spectrometry

Various bioactive components extracted from saffron were analysed by ISO 3632 trade standard (ISO/TS3632, 2003). The obtained extract of *C. sativus* was measured for its pigment (crocin), aroma (safranal), and bitter taste (picrocrocin) at a wavelength of 440, 330, and 257 nm, respectively. The observed absorbance values were used for the determination of crocin, picrocrocin, and safranal by using the following formula (Lage and Cantrell 2009):

$$E_{1\,cm}^{1\%} = \frac{D \times 10,000}{\{m \times (100 - H)\}}$$

where D is the absorbance at different wavelength (440, 330, and 257 nm), m is the initial weight, and H is the mass fraction. Instead of using a specific wavelength for zero order derivatives, it is also suggested to explore second-order derivatives for the quantification of various components to avoid the overlapping of peaks and to obtain appropriate resolution of those peaks (Rajabi et al. 2015). Orfanou and Tsimidou (1996) have analysed the colour strength of saffron by using UV-vis spectra having wavelength range of 200–600 nm using quartz cuvette. The experimental conditions such as scaling factor and wavelength difference used for the first and second-order derivatives were ×10 and ×400, and 5 nm and 20 nm, respectively. The obtained results for first-order derivative showed improved image of overlapped peaks around the wavelength of 400 nm, whereas the inverse of second derivative spectrum was reported as more suitable for the quantification of colouring compounds from saffron at wavelength range of 400–500 nm. UV-vis diode array detector is mostly used for the quantification of non-volatile compounds. Masoum et al. (2015) have explored the second-order derivative of UV-vis spectra to detect the presence of tartrazine and sunset yellow in adulterated sample of saffron. The absorbance spectrum was generated at 5 nm intervals ranged between 250 and 600 nm for each extract of saffron. Saffron extract was prepared by numerous researchers by using different extraction techniques, in which Maggi et al. (2011) have prepared saffron extract through UAE method by using diethyl ether or chloroform as a solvent. The absorbance of the extract was determined at 330 nm via UV-vis spectrometry and the resultant peaks of safranal were further analysed by GC-MS for better results.

11.5.5 Near-Infrared Spectrometry

Near-infrared spectrometry (NIR) absorption specifically contains large number of bands having wave number ranged between 12,820 and 3959 cm^{-1}. They are related to large number of major vibrations of –OH, –CH, –SH, and –NH groups. It comprises a light source (tungsten lamp), monochromater, sample holder, and detector (lead sulphide, silicon, indium gallium arsenide) for the transmittance of light. NIR analysis has been used for the identification of various bioactive components from extracts which could be subsequently used in pharmaceutical industry (Reich 2005). Hence, Zalacain et al. (2005) took the extract of saffron and placed it into the quartz plates having 3 cm diameter. NIR was coupled with NIR reflectance adjunct at wavelength range between 4000 and 10,000 cm^{-1} and the resolution was maintained at 16 cm^{-1} for the detection of bioactive components in it. The identified peaks were compared with pure crocetin and confirm the presence of cis-crocetin glycosides at wavenumber of 6500 cm^{-1}. Therefore, NIR can also be used for the determination of biological compounds without any further hesitation.

11.5.6 Fourier Transform Infrared Spectroscopy/Raman Spectroscopy Analysis

Fourier transform infrared spectroscopy (FTIR) analysis depends on interaction between electromagnetic radiations and samples. FTIR has been used in combination with various techniques for the identification and quantification of bioactive compounds. The extract of *Curcuma longa* was analysed by using FTIR in association with principal component regression analysis for the determination of curcuminoid in it (Rohman et al. 2015). FTIR and Raman spectra of saffron extract were analysed between 200 and 2000 cm^{-1} which showed C=C and C–C stretching vibrations at 1540 and 1167 cm^{-1}, respectively, which corresponded to crocetin (major pigment present in saffron). In FTIR, the bands were reported at 3291 cm^{-1} which is allocated to stretching vibrations of –OH group, bands at 2923 and 2854 cm^{-1} corresponded to the presence of symmetric and asymmetric stretching vibrations, whereas the band reported at 1732 cm^{-1} corresponded to carbonyl group which was related to crocin structure in saffron (Er et al. 2017). Hence, these two methods are rapid techniques which can be used for measuring colour pigments in various samples in lesser time period without compromising their chemical structure.

11.5.7 Electronic Nose Method

Sometimes, conventional mode of analysis is not suitable for quality control of various components; therefore, "green methods" such as E-nose have been brought into work place for their better utilization. E-nose is a rapid and fast-operating technique used for identification of volatile components. It consists of multiple sensors which are able to impersonate the numerous odours by means of an array of gas sensors. These sensors are treated with various aromatic compounds to produce peculiar odour which is known as smell print. The patterns created by sensors after treating with different chemicals could be used as database to identify the unknown samples (Peris and Escuder-Gilabert 2016). Various researchers have already used E-nose for different products such as milk, olive oil, honey, and tea (Jafari et al. 2020). Carmona et al. (2005) have established a relation between gas sensors and principal component analysis to identify the odour of saffron collected from various geographical areas. Moreover, the similar combination was used by Heidarbeigi et al. (2015) to determine adulteration in collected saffron samples. Kiani et al. (2017) have also quantified aromatic compounds from 33 types of saffron collected from various regions. Multilayer perception and radial basis function neural model were applied for the prediction of E-nose results and further correlated with principal component analysis for the identification of aromatic compounds. Therefore, E-nose could be used for determination of sensitive components without causing any damage to them.

11.6 BIOLOGICAL AND PHARMACEUTICAL PROPERTIES OF SAFFRON

Saffron is one of the most important food additives, colorant, and flavouring compounds used in many food industries, though the stigma of saffron is commercially being used in pharmaceutical industry due to the presence of numerous bioactive compounds which has been already discussed in previous sections. The most stated bioactive components in saffron are crocin, crocetin, and apocarotenoids which effectively fight against diseases like coronary artery diseases, neurodegenerative disorders, asthma, bronchitis, immune-modulatory ailments, colds, fever, anti-depressant, anti-tumour, and anti-carcinogenic activities (Shahi et al. 2016). Many researchers have done in-vivo and in-vitro studies of saffron to determine the rate of these activities for their pharmacological applications.

11.6.1 Anti-Tumour and Anti-Cancer Activities

Cancer has become one of the common diseases worldwide and the cases are constantly rising since a decade. Bioactive components of saffron have been reported as potential ingredient which can fight against cancerous cells. The extract of saffron acts against cellular DNA, RNA synthesis, and free radicals by reducing the growth of infected cells (Bhandari 2015). In this regard, various scientists have performed experiments to utilize saffron extract for the treatment of different cancers such as colorectal cancer (Haraldsdottir et al. 2016), hepatic cancer (Lim and Torresi 2014), lung cancer (Samarghandian et al. 2013), prostate cancer (Festuccia et al. 2014), and pancreatic cancer (Bakshi et al. 2010). Zhao et al. (2008) suggested that crocin acts as a major component to control the anti-proliferation activity of colorectal cancer cells. Moreover, others researchers have also mentioned that saffron can play role in gene expression and apoptosis in cancer cells by reducing the growth of bladder cancer TCCB T24 cell line. Crocetin has also showed an effect on cancerous cells by inhibiting their growth. This effect could be observed due to reduced synthesis of DNA, RNA, and protein in neoplastic cells, by inhibiting the activity of RNA polymerase II, and their interaction with histone H1 and H1-DNA structures. Safranal could also be used to reduce the growth of tumour cells with low toxicity (Milajerdi et al. 2016). Therefore, the rate of anti-cancer activity of saffron extract can also be studied on other cancers as well for identification of its constituent's effect.

11.6.2 Antioxidant Activity

Saffron has been found to have beneficial health effect in terms of its antioxidant properties. The activity is inherited due to the presence of α-tocopherol, water-soluble carotenoid, crocin, saffranal, and crocin in saffron (Chen et al. 2008). These components are able to quench the produced free radicals and thus are used for the treatment of diseases such as cardiovascular and neurological disorders. Researchers have suggested that crocetin is an important constituent which deals with free radicals to scavenge them and reduces the lipid peroxidation activity. Ochiai et al. (2007) have reported increased peroxidation rate in semi-permeable membrane due to lack of aldohexose activity which further reduces the enzyme activity. Therefore, co-treatment of cells with crocin reduces the oxidative rate significantly as compared to α-tocopherol present in semi-permeable membrane. Likewise, aqueous extract of saffron was infused in vascular endothelial cells of human being under stress condition. The reported results revealed that saffron is not only showing antioxidant effect but also acts as a modulator in micro-RNAs under stress conditions and delivers a therapeutic importance in the treatment of oxidative stress-mediated diseases. Therefore, it is suggested to consume saffron extract as an adjunct in foods products and also in cosmetics and drugs (Rodriguez-Ruiz et al. 2016).

11.6.3 Anti-Diabetic Activity

The incidence of diabetes mellitus is increasing globally. The overall frequency to this metabolic disorder will increase 33% more as per the current scenario. Uncontrolled diabetes can lead to improper working efficiency of tissues and organs by promoting the activity of apoptosis, protein kinas, oxidative stress, and inflammation (Boyle et al. 2010). Various synthetic anti-diabetic agents such as insulin and dipeptidyl peptidase inhibitors have been used to deal with it but they have a number of side effects. Thus, it becomes essential to use natural sources to fight against diabetes mellitus disorder. Saffron is found to be effective against insulin sensitizing, hypoglycemic effects, and by improving insulin-signalling pathway to control the formation of β cells. Components like crocin, safranal, β-carotenes, and crocetin possess hypoglycemic effects in insulin-dependent tissues via stimulation of insulin sensitivity (Yaribeygi et al. 2019). Some of the studies reported that saffron extracts could inhibit the pathophysiological mechanism responsible for the destruction of β cells (Xi et al. 2005). Subsequently, hyperglycaemia has toxic effect on β cells that may be reduced by using saffron extract which prevents β-cell toxicity by inducing insulin sensitivity and reduction in blood glucose. Crocetin is an important component for improving β-cell function and helps in the prevention of dexamethasone-dependent insulin-resistance activity and hyperglycaemia (Xi et al. 2005). This suggests that crocetin could be used for the treatment of insulin-resistance disorder.

11.6.4 Anti-Depressant and Anxiety

Saffron plays a crucial role to treat depression and acts as a stress reliever, due to the presence of its volatile compound which further corresponds to neuron modulatory effect and antioxidant activity. Administration of saffron extract is associated with neurological cells which reduce the intensity of anxiety-induced during post-trauma situation (Hashtjini et al. 2018). Anti-depressant property of saffron was found to be more effective than placebo for the treatment of depression by optimizing its dose concentration and time period of treatment, whereas the combination of curcumin and saffron was also effective against the depression and anxiolytic disorder (Lopresti and Drummond 2017). Amin et al. (2015) have evaluated the anti-depressant effects of crocin and crocetin by infusing it into the mice in two concentrations of acute and sub-acute administration. The obtained results showed crocetin was more effective than crocin in treatment of depression as higher doses of crocin have shown comparatively reduced anti-depressant effects. Asrari et al. (2018) performed an experiment on rats by administering saffron extract (40 and 80 mg/kg/day) and imipramine (10 mg/kg/day) in their intra-peritoneal cavity for 21 days. Prolonged administration resulted in anti-depressant activity which was associated with enhanced phosphorylation of cAMP response which was comparable with synthetic drug (imipramine).

11.6.5 Cardiovascular Activities

Bioactive components in saffron like crocin, crocetin, and safranal possess cardio-protective characteristics which can reduce the malonaldehyde content and lipid peroxidation in heart tissues. It was also reported that these components reduce the histopathological changes in heart tissue by decreasing CK-MB and LDH activities in serum of Wistar rats (Mehdizadeh et al. 2013). Crocetin from saffron extract helps in repairing the cardial damage. Crocetin was injected at the rate of 50 mg/kg/day in rats and was observed for their protein and gene expression which indicated that crocetin can fight against myocardial ischemia-reperfusion injury (MIRI) in rats by inhibiting reactive oxygen species ROS) production and myocardium apoptosis (Shahi et al. 2016). Administration of crocetin was reported to increase the mitochondrial potential in cardiac myocyte treated with noradrenaline and reduce the activity of lactate dehydrogenase. Isoproterenol prompted to control cardiotoxicity contains reduced systolic, diastolic, and average arterial blood pressures, additionally to improve the level of oxidative enzymes (Goyal et al. 2010). In quails, the rate of atherosclerosis decreased after injecting with crocin due to uptake of oxidized low density lipoprotein (LDL) (He et al. 2005). Moreover, crocin (100 mg/kg/day) also reduces the total triglycerides, LDL, and very low density lipoprotein (VLDL) cholesterol in rat's serum by maximum release of fat in their faecal waste (Shen and Qian 2006).

11.6.6 Anti-Alzheimer's Activity

Alzheimer's disease is a cerebrovascular impairment with degenerated mental disabilities and loss of memory periodically. To overcome with this disability, extract of saffron has been induced into different animals to analyse their efficacy. Crocin as a potent structure used to treat animal from this impairment (Asadi et al. 2015). Naghizadeh et al. (2013) also reported enhanced cerebral performance in rats which were treated with streptozotocin followed by crocin administration and hence suggested valuable effects of crocin in the treatment of neurodegenerative disorders. It has also been observed that saffron employs synergistic effects with other nutraceutical supplements on cognitive disabilities. Similarly, administration of apocarotenoid-enriched saffron extract has resulted in tightness of blood–brain cells which enhances the transport of amlyloid β cells in Alzheimer's disease. Consumption of saffron 30 mg/day significantly reduces help in treatment of Alzheimer's disease (Batarseh et al. 2017).

11.6.7 Anti-Inflammatory Activity

A large number of studies have been done in support of role of inflammation and immune dysfunction like depression and anxiety disorders. The extract of *C. sativus* was taken for in-vitro and in-vivo study. For in-vitro human blood sample, whereas in in-vivo study rats were taken and extract was infused at the rate of 400 mg/mL and they were compared with samples treated with diclofenac (10 mg/mL). Not much significant results were obtained for these two different anti-inflammatory agents (Kumar et al. 2012). Various studies were taken into consideration to determine the efficacy of saffron extract over inflammatory activity such as arthritis (Ding et al. 2013), asthma (Mahmoudabady et al. 2013), and colitis (Kawabata et al. 2012). Crocin was effectively used to neutralize the level of inflammation due to its antioxidant nature. Anti-inflammatory test was performed in mice to protect the ear oedema which was tentatively due to the presence of flavonoids, tannins, anthocyanins, alkaloids, and saponin content in the saffron flower (Hemshekhar et al. 2012).

11.6.8 Neuroprotective Activity

Our brain contains a large number of neuroprotective cells which is essential for the growth of cells, their differentiation, and ultimately decline in their growth period. These cells play a crucial role to protect a living organism from anxiety and depression. Shafiee et al. (2018) have included a study who reported a significant low amount of these neuroprotective cells in person who were suffering from severe depression as compared to healthy individual. Thus, these neuroprotective cells act as an anti-depressant agent. There are some studies which say saffron can be used as one of the neuroprotective agents due to its crocin content that increases the binding efficiency of cyclic monophosphate with protein and neuroprotective cells present in hippocampus portion of brain (Ghasemi et al. 2015). Crocin can also reduce some chemical and psychological effects in neurons produced by malathion as neuroprotective effect on brain derived neurotrophic factor (Dorri et al. 2015). Dementia is one of the neurological disorders which is linked with oxidative stress disorder which releases number of reactive oxygen species. These reactive substances regulate the formation of nitric oxide and malondialdehyde leads to reduction in antioxidant activity. Thus, it produces a large number of oxidant inside the brain cells and subsequently leads to necrosis. To control such activities, many studies were done by utilizing synthetic drug which has its own harmful effects; therefore, it was tried to reduce these activities by the influence of natural sources. Safrnal, one of the important constituents in saffron, showed protective effects against the oxidative damage in rats. Moreover, crocin, crocetin, and picrocrocin have also shown beneficial effect in numerous pathological studies (Hosseinzadeh and Sadeghnia 2005).

11.6.9 Anti-Gastric Activity

Saffron has been reported as potential anti-gastric agent in mice. Saffron is commonly used for the ailment of ulcer problem since ancient time. Safranal and crocin exhibit antioxidant properties which reduce the formation of indomethacin—gastric complex in intestinal mucosa layer by increasing glutathione level, thus prevents lipid oxidation (Kianbakht and Mozafari 2009). Xu et al. (2009) also found that crocin can help in the prevention of mucosal layer depletion in rats which were treated with indomethacin. In another study, mice were treated with 100 mg/kg of saffron which resulted in significant high percentage of gastric acid in their excreta. This could be due to the activation of nitric oxide synthetase which promote the release of histamine from the cells and further increase the release of gastric in excretions (Nabavizadeh et al. 2009). The efficacy of saffron as an anti-gastric agent was comparable with the synthetic drug omeprazole which also possesses antioxidant property (Kianbakht and Mozafari 2009).

11.7 CONCLUSION

This chapter demonstrated the work of several scientists in updating and modernizing means of extraction, analysis, and identification of bioactive compounds of saffron. The consumption of saffron correlates with a lower risk of many types of cancers and other diseases that are attributed to a large number of phytochemicals present in saffron. Among these phytochemicals, crocin, crocetin, and safranal were considered the most bioactive ingredients in a variety of studies. Various mechanisms have been identified to describe protective or therapeutic effects of saffron. However, there is a need to investigate a clear mechanism of action of saffron and its individual components, to check toxicity effects more extensively.

REFERENCES

Ahmadian-Kouchaksaraie, Z., and Niazmand, R. 2017. Supercritical carbon dioxide extraction of antioxidants from *Crocus sativus* petals of saffron industry residues: Optimization using response surface methodology. *The Journal of Supercritical Fluids* 121: 19–31.

Almodovar, P., Briskey, D., Rao, A., Prodanov, M., and Inarejos-García, A. M. 2020. Bioaccessibility and pharmacokinetics of a commercial saffron (*Crocus sativus* L.) extract. *Evidence-Based Complementary and Alternative*. https://doi.org/10.1155/2020/1575730

Alonso, G. L., Salinas, R. M., Garijo, J., and Sánchez-fernández, M. A. 2000. Composition of crocins and picrocrocin from Spanish saffron (*Crocus satnus* L.). *Journal of Food Quality* 24: 219–233.

Amanpour, A., Sonmezdag, A. S., Kelebek, H., and Selli, S. 2015. GC-MS-olfactometric characterization of the most aroma-active components in a representative aromatic extract from Iranian saffron (*Crocus sativus* L.). *Food Chemistry* 182: 251–256.

Amin, B., Nakhsaz, A., and Hosseinzadeh, H. 2015. Evaluation of the antidepressant-like effects of acute and sub-acute administration of crocin and crocetin in mice. *Avicenna Journal of Phytomedicine* 5(5): 458.

Armellini, R., Peinado, I., Pittia, P. et al. 2018. Effect of saffron (*Crocus sativus* L.) enrichment on antioxidant and sensorial properties of wheat flour pasta. *Food Chemistry* 254: 55–63.

Asadi, F., Jamshidi, A. H., Khodagholi, F., Yans, A., Azimi, L., Faizi, M., and Sharifzadeh, M. 2015. Reversal effects of crocin on amyloid β-induced memory deficit: Modification of autophagy or apoptosis markers. *Pharmacology Biochemistry and Behavior* 139: 47–58.

Asrari, N., Yazdian-Robati, R., Abnous, K., Razavi, B. M., Rashednia, M., Hasani, F. V., and Hosseinzadeh, H. 2018. Antidepressant effects of aqueous extract of saffron and its effects on CREB, P-CREB, BDNF, and VGF proteins in rat cerebellum. *Journal of Pharmacopuncture* 21(1): 35.

Azmir, J., Zaidul, I. S. M., Rahman, M. M. et al. 2013. Techniques for extraction of bioactive compounds from plant materials: A review. *Journal of Food Engineering* 117(4): 426–436.

Azwanida, N. N. 2015. A review on the extraction methods use in medicinal plants, principle, strength and limitation. *Journal of Medical Aromatic Plants* 4(196): 2167–0412.

Baba, S. A., Malik, A. H., Wani, Z. A., Mohiuddin, T., Shah, Z., Abbas, N., Ashraf, N. 2015. Phytochemical analysis and antioxidant activity of different tissue types of *Crocus sativus* and oxidative stress alleviating potential of saffron extract in plants, bacteria, and yeast. *South African Journal of Botany* 99: 80–87.

Bakshi, H., Sam, S., Rozati, R., et al. 2010. DNA fragmentation and cell cycle arrest: A hallmark of apoptosis induced by crocin from Kashmiri saffron in a human pancreatic cancer cell line. *Asian Pacific Journal of Cancer Prevention* 11(3): 675–679.

Barba, F. J., Parniakov, O., Pereira, S. A. et al. 2015. Current applications and new opportunities for the use of pulsed electric fields in food science and industry. *Food Research International* 77: 773–798.

Basker, D., and Negbi, M. 1983. Uses of Saffron. *Economic Botany* 37(2): 228–236.

Batarseh, Y. S., Bharate, S. S., Kumar, V., Kumar, A., Vishwakarma, R. A., Bharate, S. B., and Kaddoumi, A. 2017. *Crocus sativus* extract tightens the blood-brain barrier, reduces amyloid β load and related toxicity in 5XFAD mice. *ACS Chemical Neuroscience* 8(8): 1756–1766.

Bhandari, P. R. 2015. *Crocus sativus* L. (saffron) for cancer chemoprevention: A mini review. *Journal of Traditional and Complementary Medicine* 5(2): 81–87.

Boyle, J. P., Thompson, T. J., Gregg, E. W., Barker, L. E., and Williamson, D. F. 2010. Projection of the year 2050 burden of diabetes in the US adult population: Dynamic modeling of incidence, mortality, and prediabetes prevalence. *Population Health Metrics* 8(1): 29.

Buzrul, S., Alpas, H., Largeteau, A., and Demazeau, G. 2008. Inactivation of *Escherichia coli* and *Listeria innocua* in kiwifruit and pineapple juices by high hydrostatic pressure. *International Journal of Food Microbiology* 124(3): 275–278.

Carmona, M., Zalacain, A., Pardo, J. E., López, E., Alvarruiz, A., and Alonso, G. L. 2005. Influence of different drying and aging conditions on saffron constituents. *Journal of Agricultural and Food Chemistry* 53(10): 3974–3979.

Carmona, M, Zalacain, A., Salinas, M. R., and Alonso, G. L. 2007. A new approach to saffron aroma. *Critical Reviews in Food Science and Nutrition* 47: 145–159.

Castillo, R., Fernández, J. A., and Gómez-Gómez, L. 2005. Implications of carotenoid biosynthetic genes in apocarotenoid formation during the stigma development of *Crocus sativus* and its closer relatives. *Plant Physiology* 139(2): 674–689.

Chen, Y., Zhang, H., Tian, X., Zhao, C., Cai, L., Liu, Y., and Chen, C. 2008. Antioxidant potential of crocins and ethanol extracts of *Gardenia jasminoides* ELLIS and *Crocus sativus* L.: A relationship investigation between antioxidant activity and crocin contents. *Food Chemistry* 109(3): 484–492.

Christodoulou, E., Kadoglou, N. P., Kostomitsopoulos, N., and Valsami, G. 2015. Saffron: A natural product with potential pharmaceutical applications. *Journal of Pharmacy and Pharmacology* 67(12): 1634–1649.

Da Porto, C., and Natolino, A. 2018. Extraction kinetic modelling of total polyphenols and total anthocyanins from saffron floral bio-residues: Comparison of extraction methods. *Food Chemistry* 258: 137–143.

Danalache, F., Mata, P., Alves, V. D., and Moldão-Martins, M. 2018. Enzyme-assisted extraction of fruit juices. In *Fruit Juices*. Academic Press, ISBN- 978-0-12-802230-6: 183–200.

Ding, Q., Zhong, H., Qi, Y., Cheng, Y., Li, W., Yan, S., and Wang, X. 2013. Anti-arthritic effects of crocin in interleukin-1β-treated articular chondrocytes and cartilage in a rabbit osteoarthritic model. *Inflammation Research* 62(1): 17–25.

Dorri, S. A., Hosseinzadeh, H., Abnous, K., et al. 2015. Involvement of brain-derived neurotrophic factor (BDNF) on malathion induced depressive-like behavior in subacute exposure and protective effects of crocin. *Iranian Journal of Basic Medical Sciences* 18(10): 958.

El Asbahani, A., Miladi, K., Badri, W., Sala, M., Addi, E. A., Casabianca, H., Elaissari, A. 2015. Essential oils: From extraction to encapsulation. *International Journal of Pharmaceutics* 483(1–2), 220–243.

Er, S. V., Eksi-Kocak, H., Yetim, H., and Boyaci, I. H. 2017. Novel spectroscopic method for determination and quantification of saffron adulteration. *Food Analytical Methods* 10(5): 1547–1555.

Esfanjani, A. F., Jafari, S. M., Assadpoor, E., and Mohammadi, A. 2015. Nano-encapsulation of saffron extract through double-layered multiple emulsions of pectin and whey protein concentrate. *Journal of Food Engineering* 165: 149–155.

Fahim, N. K., Janati, S. F., and Feizy, J. 2012. Chemical composition of agriproduct saffron (*Crocus sativus* L.) petals and its considerations as animal feed. *GIAD Journal of Food* 37: 197–201.

Ferrara, L., Naviglio, D., and Gallo, M. 2014. Extraction of bioactive compounds of saffron (*Crocus sativus* L.) by ultrasound assisted extraction (UAE) and by rapid solid-liquid dynamic extraction (RSLDE). *European Scientific Journal* 10(3): ISSN-1857-7431.

Fernandez, J. A. 2004. Biology, biotechnology and biomedicine of saffron. *Recent Research Developments in Plant Science* 127(2): 159.

Festuccia, C., Mancini, A., Gravina, G. L., et al. 2014. Antitumor effects of saffron-derived carotenoids in prostate cancer cell models. *BioMedical Research International* 135048. doi:10.1155/2014/135048.

Gahruie, H. H., Parastouei, K., Mokhtarian, M., et al. 2020. Application of innovative processing methods for the extraction of bioactive compounds from saffron (*Crocus sativus*) petals. *Journal of Applied Research on Medicinal and Aromatic Plants* 19: 100264.

Gallo, M., Ferracane, R., Graziani, G., Ritieni, A., and Fogliano, V. 2010. Microwave assisted extraction of phenolic compounds from four different spices. *Molecules* 15(9): 6365–6374.

Garavand, F., Rahaee, S., Vahedikia, N., and Jafari, S. M. 2019. Different techniques for extraction and micro/nanoencapsulation of saffron bioactive ingredients. *Trends in Food Science and Technology* 89: 26–44.

Ghasemi, T., Abnous, K., Vahdati, F., Mehri, S., Razavi, B. M., and Hosseinzadeh, H. 2015. Antidepressant effect of *Crocus sativus* aqueous extract and its effect on CREB, BDNF, and VGF transcript and protein levels in rat hippocampus. *Drug Research* 65(07): 337–343.

Goleroudbary, M. G., and Ghoreishi, S. M. 2016. Response surface optimization of safranal and crocin extraction from *Crocus sativus* L. via supercritical fluid technology. *The Journal of Supercritical Fluids* 108: 136–144.

Goyal, S. N., Arora, S., Sharma, A. K., Joshi, S., Ray, R., Bhatia, J., and Arya, D. S. 2010. Preventive effect of crocin of *Crocus sativus* on hemodynamic, biochemical, histopathological and ultrastuctural alterations in isoproterenol-induced cardiotoxicity in rats. *Phytomedicine* 17(3–4): 227–232.

Gregory, M. J., Menary, R. C., and Davies, N. W. 2005. Effect of drying temperature and air flow on the production and retention of secondary metabolites in saffron. *Journal of Agricultural and Food Chemistry* 53(15): 5969–5975.

Gresta, F., Lombardo, G. M., Siracusa, L., and Ruberto, G. 2008. Saffron, an alternative crop for sustainable agricultural systems: A review. *Agronomy for Sustainable Development* 28(1): 95–112.

Handa, S. S., Khanuja, S. P. S., Longo, G., and Rakesh, D. D. 2008. *Extraction Technologies for Medicinal and Aromatic Plants. Earth, Environmental and Marine Sciences, and Technologies*, United Nations Industrial Development Organization.

Haraldsdottir, S., Roth, R., Pearlman, R., et al. 2016. Mismatch repair deficiency concordance between primary colorectal cancer and corresponding metastasis. *Familial Cancer* 15(2): 253–260.

Hashtjini, M. M., Jahromi, G. P., and Meftahi, G. H. 2018. Aqueous extract of saffron administration along with amygdala deep brain stimulation promoted alleviation of symptoms in post-traumatic stress disorder (PTSD) in rats. *Avicenna Journal of Phytomedicine* 8(4): 358.

He, S. Y., Qian, Z. Y., Tang, F. T., Wen, N., Xu, G. L., and Sheng, L. 2005. Effect of crocin on experimental atherosclerosis in quails and its mechanisms. *Life Sciences* 77(8): 907–921.

Heidarbeigi, K., Mohtasebi, S. S., Foroughirad, A., et al. 2015. Detection of adulteration in saffron samples using electronic nose. *International Journal of Food Properties* 18(7): 1391–1401.

Hemshekhar, M., Santhosh, M. S., Sunitha, K., Thushara, R. M., Kemparaju, K., Rangappa, K. S., and Girish, K. S. 2012. A dietary colorant crocin mitigates arthritis and associated secondary complications by modulating cartilage deteriorating enzymes, inflammatory mediators and antioxidant status. *Biochimie* 94(12): 2723–2733.

Hernandez, E., Chen, C. S., Johnson, J., and Carter, R. D. 1995. Viscosity changes in orange juice after ultrafiltration and evaporation. *Journal of Food Engineering* 25(3): 387–396.

Hosseinzadeh, H., and Sadeghnia, H. R. 2005. Safranal, a constituent of *Crocus sativus* (saffron), attenuated cerebral ischemia induced oxidative damage in rat hippocampus. *Journal of Pharmacy and Pharmceutical Sciences* 8(3): 394–399.

Hosseinzadeh, H., and Nassiri-Asl, M. 2013. Avicenna's (IbnSina) the canon of medicine and saffron (*Crocus sativus*): A review. *Phytotherapy Research* 27(4): 475–483.

ISO-3632-2-2003. 2003. *Part 1: Specification, Part 2: Test Methods*. International Organisation for Standardization, Geneva.

Jafari, S. M., Tsimidou, M. Z., Rajabi, H., and Kyriakoudi, A. 2020. Bioactive ingredients of saffron: Extraction, analysis, applications. *Saffron Science Technology and Health*, Woodhead Publishing Series in Food Science, Technology and Nutrition, pp. 261–290.

Kakouri, E., Daferera, D., Paramithiotis, S., et al. 2017. *Crocus sativus* L. tepals: The natural source of antioxidant and antimicrobial factors. *Journal of Applied Research on Medicinal and Aromatic Plants* 4: 66–74.

Kanakis, C. D., Daferera, D. J., Tarantilis, P. A., and Polissiou, M. G. 2004. Qualitative determination of volatile compounds and quantitative evaluation of safranal and 4-hydroxy-2, 6, 6-trimethyl-1-cyclohexene-1-carboxaldehyde (HTCC) in Greek saffron. *Journal of Agricultural and Food Chemistry* 52(14): 4515–4521.

Kawabata, K., Tung, N. H., Shoyama, Y., Sugie, S., Mori, T., and Tanaka, T. 2012. Dietary crocin inhibits colitis and colitis-associated colorectal carcinogenesis in male ICR mice. *Evidence-Based Complementary and Alternative Medicine*. 820415. doi:10.1155/2012/820415.

Khajehei, F., Niakousari, M., SeidiDamyeh, M., et al. 2017. Impact of ohmic-assisted decoction on bioactive components extracted from yacon (*Smallanthus sonchifolius* Poepp.) leaves: Comparison with conventional decoction. *Molecules* 22(12): 2043.

Khazaei, K. M., Jafari, S. M., Ghorbani, M., Kakhki, A. H., and Sarfarazi, M. 2016. Optimization of anthocyanin extraction from saffron petals with response surface methodology. *Food Analytical Methods* 9(7): 1993–2001.

Kianbakht, S., and Mozafari, K. 2009. Effects of saffron and its active constituents, crocin and safranal, on prevention of indomethacin induced gastric ulcers in diabetic and nondiabetic rats. *Journal of Medical Plants* 8(5): 30–38.

Kiani, S., Minaei, S., Ghasemi-Varnamkhasti, M., and Ayyari, M. 2017. An original approach for the quantitative characterization of saffron aroma strength using electronic nose. *International Journal of Food Properties* 20: S673–S683.

Kumar, V., Bhat, Z. A., Kumar, D., et al. 2012. Evaluation of anti-inflammatory potential of petal extracts of *Crocus sativus* "cashmerianus". *International Journal of Phytopharmacy* 3: 27–31.

Kyriakoudi, A., Chrysanthou, A., Mantzouridou, F., and Tsimidou, M. Z. 2012. Revisiting extraction of bioactive apocarotenoids from *Crocus sativus* L. dry stigmas (saffron). *Analytica Chimica Acta* 755: 77–85.

Lage, M., and Cantrell, C. L. 2009. Quantification of saffron (*Crocus sativus* L.) metabolites crocins, picrocrocin and safranal for quality determination of the spice grown under different environmental Moroccan conditions. *Scientia Horticulturae* 121(3): 366–373.

Lim, E. J., and Torresi, J. 2014. Prevention of hepatitis C virus infection and liver cancer. *Viruses and Human Cancer* 193: 113–133.

Liu, J., Chen, N., Yang, J., Yang, B., Ouyang, Z., Wu, C., and Chen, M. 2018. An integrated approach combining HPLC, GC/MS, NIRS, and chemometrics for the geographical discrimination and commercial categorization of saffron. *Food Chemistry* 253: 284–292.

Lopresti, A. L., and Drummond, P. D. 2017. Efficacy of curcumin, and a saffron/curcumin combination for the treatment of major depression: A randomised, double-blind, placebo-controlled study. *Journal of Affective Disorders* 207: 188–196.

Lotfi, L., Kalbasi-Ashtari, A., Hamedi, M., and Ghorbani, F. 2015. Effects of enzymatic extraction on anthocyanins yield of saffron tepals (*Crocos sativus*) along with its color properties and structural stability. *Journal of Food and Drug Analysis* 23(2): 210–218.
Mahmoudabady, M., Neamati, A., Vosooghi, S., and Aghababa, H. 2013. Hydroalcoholic extract of *Crocus sativus* effects on bronchial inflammatory cells in ovalbumin sensitized rats. *Avicenna Journal of Phytomedicine* 3(4): 356.
Maggi, L., Sánchez, A. M., Carmona, M., et al. 2011. Rapid determination of safranal in the quality control of saffron spice (*Crocus sativus* L.). *Food Chemistry* 127(1): 369–373.
Masoum, S., Gholami, A., Hemmesi, M., and Abbasi, S. 2015. Quality assessment of the saffron samples using second-order spectrophotometric data assisted by three-way chemometric methods via quantitative analysis of synthetic colorants in adulterated saffron. *Spectrochimica Acta Part A: Molecular and Biomolecular Spectroscopy* 148: 389–395.
Mason, J. T., Chemat, F., and Vinatoru, M. 2011. The extraction of natural products using ultrasound or microwaves. *Current Organic Chemistry* 15(2): 237–247.
Mehdizadeh, R., Parizadeh, M. R., Khooei, A. R., Mehri, S., and Hosseinzadeh, H. 2013. Cardioprotective effect of saffron extract and safranal in isoproterenol-induced myocardial infarction in Wistar rats. *Iranian Journal of Basic Medical Sciences* 16(1): 56.
Milajerdi, A., Djafarian, K., and Hosseini, B. 2016. The toxicity of saffron (*Crocus sativus* L.) and its constituents against normal and cancer cells. *Journal of Nutrition and Intermediary Metabolism* 3: 23–32.
Mokhtari, B., and Pourabdollah, K. 2013. Extraction of saffron ingredients and its fingerprinting by nanoemulsion membranes. *International Journal of Chemical Technology* 20: 222–228.
Muzaffar, S., Khan, K. Z., Riaz, M., Mir, J. A., and Ahmed, A. 2015. High performance liquid chromatography for validation method of oxolamine citrate from bulk drug and pharmaceutical formulation. *Journal of Chemical and Pharmaceutical Research* 7(1): 111–115.
Nabavizadeh, F., Salimi, E., Sadroleslami, Z., Karimian, S. M., and Vahedian, J. 2009. Saffron (*Crocus sativus*) increases gastric acid and pepsin secretions in rats: Role of nitric oxide (NO). *African Journal of Pharmacy and Pharmacology* 3(5): 181–184.
Naghizadeh, B., Mansouri, M. T., Ghorbanzadeh, B., Farbood, Y., and Sarkaki, A. 2013. Protective effects of oral crocin against intracerebroventricular streptozotocin-induced spatial memory deficit and oxidative stress in rats. *Phytomedicine* 20(6): 537–542.
Neri, L., Giancaterino, M., Rocchi, R., et al. 2021. Pulsed electric fields (PEF) as hot air drying pre-treatment: Effect on quality and functional properties of saffron (*Crocus sativus* L.). *Innovative Food Science and Emerging Technologies* 67: 102592.
Negbi, M. (2003). Saffron cultivation: Past, present and future prospects. In: Negbi, M. (Ed.), *Saffron: Crocus sativus L.* Harwood Academic Publishers, Australia, pp. 14–28.
Nerome, H., Ito, M., Machmudah, S., Kand a, H., and Goto, M. 2016. Extraction of phytochemicals from saffron by supercritical carbon dioxide with water and methanol as entrainer. *The Journal of Supercritical Fluids* 107: 377–383.
Ochiai, T., Shimeno, H., Mishima, K. I. et al. 2007. Protective effects of carotenoids from saffron on neuronal injury in vitro and in vivo. *Biochimica et Biophysica Acta (BBA) – General Subjects* 1770(4): 578–584.
Ordoubi, S. A., Staikidou, C., Kyriakoudi, A. 2017. A stepwise approach for the detection of carminic acid in saffron with regard to religious food certification. *Food Chemistry* 267: 410–419.
Orfanou, O., and Tsimidou, M. 1996. Evaluation of the colouring strength of saffron spice by UV-vis spectrometry. *Food Chemistry* 57(3): 463–469.
Pandey, D. K., Nandy, S., Mukherjee, A., and Dey, A. 2020. Advances in bioactive compounds from *Crocus sativus* (saffron): Structure, bioactivity and biotechnology. *Studies in Natural Products Chemistry* 66: 273–304.
Peris, M., and Escuder-Gilabert, L. 2016. Electronic noses and tongues to assess food authenticity and adulteration. *Trends in Food Science and Technology* 58: 40–54.
Puri, M., Sharma, D., and Barrow, C. J. 2012. Enzyme-assisted extraction of bioactives from plants. *Trends in Biotechnology* 30(1): 37–44.
Prasad, K. N., Yang, E., Yi, C., Zhao, M., and Jiang, Y. 2009. Effects of high pressure extraction on the extraction yield, total phenolic content and antioxidant activity of Longan fruit pericarp. *Innovative Food Science and Emerging Technologies* 10(2): 155–159.

Pourzaki, A., Mirzaee, H., and HemmatiKakhi, A. B. B. A. S. 2013. Using pulsed electric field for improvement of components extraction of saffron (*Crocus sativus*) stigma and its pomace. *Journal of Food Processing and Preservation* 37(5): 1008–1013.

Prieto, A., Basauri, O., Rodil, R., et al. 2010. Stir-bar sorptive extraction: A view on method optimisation, novel applications, limitations and potential solutions. *Journal of Chromatography A* 1217(16): 2642–2666.

Rajabi, H., Ghorbani, M., Jafari, S. M., Mahoonak, A. S., and Rajabzadeh, G. 2015. Retention of saffron bioactive components by spray drying encapsulation using maltodextrin, gum Arabic and gelatin as wall materials. *Food Hydrocolloids* 51: 327–337.

Reich, G. 2005. Near-infrared spectroscopy and imaging: Basic principles and pharmaceutical applications. *Advanced Drug Delivery Reviews* 57(8): 1109–1143.

Rios, J. L., Recio, M. C., Giner, R. M., and Meñez, S. 1996. An update review of saffron and its active constituents. *Phytotherapy Research* 10(3): 189–193.

Rodel, W., and Petrzika, M. 1991. Analysis of the volatile components of saffron. *Journal of High Resolution Chromatography* 14: 771–774.

Rodriguez-Meizoso, I., Jaime, L., Santoyo, S., et al. 2010. Subcritical water extraction and characterization of bioactive compounds from *Haematococcus pluvialis* microalga. *Journal of Pharmaceutical and Biomedical Analysis* 51(2): 456–463.

Rodriguez-Ruiz, V., Barzegari, A., Zuluaga, M., et al. 2016. Potential of aqueous extract of saffron (*Crocus sativus* L.) in blocking the oxidative stress by modulation of signal transduction in human vascular endothelial cells. *Journal of Functional Foods* 26: 123–134.

Rohman, A., Ramadhani, D., and Nugroho, A. 2015. Analysis of curcumin in *Curcuma longa* and *Curcuma xanthorriza* using FTIR spectroscopy and chemometrics. *Research Journal of Medicinal Plant* 9(4): 179–186.

Sanchez, A. M., Carmona, M., Prodanov, M., and Alonso, G. L. 2008. Effect of centrifugal ultrafiltration on the composition of aqueous extracts of saffron spice (*Crocus sativus* L.). *Journal of Agricultural and Food Chemistry* 56(16): 7293–7301.

Sanchez, A. M., Carmona, M., del Campo, C. P., and Alonso, G. L. 2009. Solid-phase extraction for picrocrocin determination in the quality control of saffron spice (*Crocus sativus* L.). *Food Chemistry* 116(3): 792–798.

Sarfarazi, M., Jafari, S. M., and Rajabzadeh, G. 2015. Extraction optimization of saffron nutraceuticals through response surface methodology. *Food Analytical Methods* 8(9): 2273–2285.

Sarfarazi, M., Jafari, S. M., Rajabzadeh, G., and Feizi, J. 2019. Development of an environmentally-friendly solvent-free extraction of saffron bioactives using subcritical water. *LWT* 114: 108428.

Samarghandian, S., and Borji, A. 2014. Anticarcinogenic effect of saffron (*Crocus sativus* L.) and its ingredients. *Pharmacognosy Research* 6(2): 99.

Samarghandian, S., Borji, A., Farahmand, S. K., Afshari, R., and Davoodi, S. 2013. *Crocus sativus* L. (saffron) stigma aqueous extract induces apoptosis in alveolar human lung cancer cells through caspase-dependent pathways activation. *BioMed Research International*. doi: 10.1155/2013/417928.

Sampathu, S. R., Shivashankar, S., Lewis, Y. S., and Wood, A. B. 1984. Saffron (*Crocus sativus* Linn.) – Cultivation, processing, chemistry and standardization. *Critical Reviews in Food Science and Nutrition* 20(2): 123–157.

Sarfarazi, M., Jafari, S. M., Rajabzadeh, G., and Galanakis, C. M. 2020. Evaluation of microwave-assisted extraction technology for separation of bioactive components of saffron (*Crocus sativus* L.). *Industrial Crops and Products* 145: 111978.

Selim, K., Tsimidou, M., and Biliaderis, C. G. 2000. Kinetic studies of degradation of saffron carotenoids encapsulated in amorphous polymer matrices. *Food Chemistry* 71: 199–206.

Senizza, B., Rocchetti, G., Okur, M. A., et al. 2020. Phytochemical profile and biological properties of *Colchicum triphyllum* (meadow saffron). *Foods* 9(4): 457.

Serrano-Díaz, J., Sánchez, A. M., Maggi, L., Carmona, M., and Alonso, G. L. 2011. Synergic effect of water-soluble components on the coloring strength of saffron spice. *Journal of Food Composition and Analysis* 24(6): 873–879.

Shafiee, M., AghiliMoghaddam, N. S., Nosrati, M., et al. 2017. Saffron against components of metabolic syndrome: Current status and prospective. *Journal of Agricultural and Food Chemistry* 65(50): 10837–10843.

Shafiee, M., Arekhi, S., Omranzadeh, A., and Sahebkar, A. 2018. Saffron in the treatment of depression, anxiety and other mental disorders: Current evidence and potential mechanisms of action. *Journal of Affective Disorders* 227: 330–337.

Shahi, T., Assadpour, E., and Jafari, S. M. 2016. Main chemical compounds and pharmacological activities of stigmas and tepals of 'red gold'; saffron. *Trends in Food Science and Technology* 58: 69–78.

Shen, X. C., and Qian, Z. Y. 2006. Effects of crocetin on antioxidant enzymatic activities in cardiac hypertrophy induced by norepinephrine in rats. *Die Pharmazie – An International Journal of Pharmaceutical Sciences* 61(4): 348–352.

Shinwari, K. J., and Rao, P. S. 2018. Stability of bioactive compounds in fruit jam and jelly during processing and storage: A review. *Trends in Food Science and Technology* 75: 181–193.

Singla, R. K., and Bhat, G. V. 2011. Crocin: An overview. *Indo Global Journal of Pharmaceutical Sciences* 1(4): 281–286.

Sowbhagya, H. B., and Chitra, V. N. 2010. Enzyme-assisted extraction of flavorings and colorants from plant materials. *Critical Reviews in Food Science and Nutrition* 50(2): 146–161.

Srivastava, R., Ahmed, H., and Saraf, S. A. 2010. *Crocus sativus* L.: A comprehensive review. *Pharmacognosy Reviews* 4(8): 200–208.

Tarantilis, P. A., Polissiou, M., and Manfait, M. 1994. Separation of picrocrocin, cis-trans-crocins and safranal of saffron using high-performance liquid chromatography with photodiode-array detection. *Journal of Chromatography A* 664(1): 55–61.

Tarantilis, P. A., Beljebbar, A., Manfait, M., and Polissiou, M. 1998. FT-IR, FT-Raman spectroscopic study of carotenoids from saffron (*Crocus sativus* L.) and some derivatives. *Spectrochimica Acta – Part A: Molecular and Biomolecular Spectroscopy* 54: 651–657.

Wang, Y., Sun, J., Liu, C., and Fang, C. 2014. Protective effects of crocetin pretreatment on myocardial injury in an ischemia/reperfusion rat model. *European Journal of Pharmacology* 741: 290–296.

Xi, L., Qian, Z., Shen, X., Wen, N., and Zhang, Y. 2005. Crocetin prevents dexamethasone-induced insulin resistance in rats. *Planta Medica* 71(10): 917–922.

Xu, G. L., Li, G., Ma, H. P., Zhong, H., Liu, F., and Ao, G. Z. 2009. Preventive effect of crocin in inflamed animals and in LPS-challenged RAW 264.7 cells. *Journal of Agricultural and Food Chemistry* 57(18): 8325–8330.

Yaribeygi, H., Zare, V., Butler, A. E., Barreto, G. E., and Sahebkar, A. 2019. Antidiabetic potential of saffron and its active constituents. *Journal of Cellular Physiology* 234(6): 8610–8617.

Zalacain, A., Ordoudi, S. A., Díaz-Plaza, E. M., et al. 2005. Near-infrared spectroscopy in saffron quality control: Determination of chemical composition and geographical origin. *Journal of Agricultural and Food Chemistry* 53(24): 9337–9341.

Zarinkamar, F., Tajik, S., and Soleimanpour, S. 2011. Effects of altitude on anatomy and concentration of crocin, picrocrocin and safranal in *Crocus sativus* L. *Australian Journal of Crop Science* 5(7): 831–838.

Zeka, K., Ruparelia, K. C., Continenza, M. A. et al. 2015. Petals of *Crocus sativus* L. as a potential source of the antioxidants crocin and kaempferol. *Fitoterapia* 107: 128–134.

Zhao, P., Luo, C. L., Wu, X. H., Hu, H. B., Lv, C. F., and Ji, H. Y. 2008. Proliferation apoptotic influence of crocin on human bladder cancer T24 cell line. *Zhongguo Zhongyaozazhi Zhongguozhongyaozazhi. China Journal of Chinese Materiamedica* 33(15): 1869–1873.

Zwir-Ferenc, A., and Biziuk, M. 2006. Solid Phase Extraction Technique – Trends, Opportunities and Applications. *Polish Journal of Environmental Studies* 15(5).

CHAPTER 12

Bioactive Compounds in Mint

Shilpi Ahluwalia and Navneet Kaur

CONTENTS

DOI: 10.1201/9781003215387-12

12.1 INTRODUCTION

Plants are also referred to as "*Biosynthetic Laboratories*" or "*Chemical Goldmines*" as they are source of various bioactive secondary metabolites with immense therapeutic potential (Ahad et al. 2021; Kaur and Ahmed 2021). Among plants, herbs and spices belong to the category of medicinal plants which are quite popular and are used from antiquity in food to impart flavour and enhance organoleptic characteristics (Juana et al. 2012). Additionally, being rich in volatile compounds and antioxidants, they are widely recognized in ancient medicine for exerting beneficial effect on human health. They are also used as preservatives being anti-microbial in nature (Yashin et al. 2017). Herbs and spices can be utilized in any form either as whole plant or plant part such as fresh or dried seeds (cumin, fennel, cardamom, etc.), bulbs (onion, garlic, turmeric, etc.), roots and rhizomes (ginger, etc.), stems (celery, balm, basil, etc.), leaves (mint, dill, coriander, etc.), buds (clove, etc.), fruits (black pepper, star anise, etc.), bark (cinnamon, etc.), and other plant parts (Asfaw 2019).

Mint (Figure 12.1) is an aromatic herb rich in essential oils (EOs), belonging to genus *Mentha* and family Lamiaceae. It commonly grows in temperate and tropical or subtropical regions of world. The name mint has been supposed to be derived from Greek legend of the nymph Minthe, which was turned into a herb by Hades in order to save her from his wife who wanted to kill her. As per recent taxonomical classification, genus *Mentha* consists of nearly 61 species with about 100 varieties and cultivars. Out of these, more than 10 species are found in the Indian subcontinent making it the richest source of *Mentha* species. All the species contain various EOs and most of them are cultivated as industrial crops (Taneja and Chandra 2012). This chapter provides an insight into

Figure 12.1 Mint leaves.

comprehensive description of genus *Mentha*, that is, its taxonomical classification, various commercially important species, bioactive compounds in mint along with their extraction and identification methods, therapeutic properties exhibited by those compounds, their pharmacological significance, and lastly the toxicological and safety studies.

12.2 TAXONOMIC CLASSIFICATION AND IMPORTANT SPECIES

Genus *Mentha* includes perennials with leafy runners, stolons, or underground rootstocks. It comprises more than 60 species. However, the taxonomy of genus *Mentha* is quite confusing as more than 3000 names have been published since 1753. Most of these names are either synonyms of each other or invalid and illegitimate or belong to intraspecific taxa. The major reason for this ambiguity is frequent hyberdization taking place between wild and cultivated species. Other factors responsible for this confusion are genodiocy favoured outcrossing, concominant polyploidy, and stable novel forms produced by vegetative propagation. Genus *Mentha* is generally classified into five sections, namely, *Audibertia*, *Eriodontes*, *Mentha*, *Preslia*, and *Pulegium*. Within section *Mentha*, five basic species have been proposed which include *Mentha arvensis* L., *Mentha aquatica* L., *Mentha spicata* L., *Mentha longifolia* (L.) Huds, and *Mentha suaveolens* Ehrh. It is suggested that these five species have been responsible for the origin of 11 natural and named hybrids (Brahmi et al. 2018). However, according to present literature reports, *Mentha* section is divided into three basic lines based on the inflorescence characters while another classification has been made on the basis of major monoterpene present in EOs given in Table 12.1 (Saric-Kundalic et al. 2009). According to dominant monoterpene, three main pathways have been suggested, menthol pathway (menthol, menthone, and menthofuran), carvon pathway (carvon, dihydrocarvone, and caveol), and linalool pathway (linalool and linalylacetate). The species are placed under these pathways on the basis of major monoterpene they are producing.

In the plant list database, 1086 species of *Mentha* L. with scientific names have been recorded. Out of which only 38 are accepted:

(http://www.theplantlist.org/browse/A/Lamiaceae/Mentha/).

The genus *Mentha* L. is extensively present in almost all continents of world and grows best in damp soil and humid environment. Mints are generally 10–150 cm in height, fast growing, and very invasive. Though they prefer cool and moist shady areas to grow, they can tolerate even full sun conditions. Different species produce various coloured flowers from white, pink to lilac. Among various mint species and hybrids, the commonly available and commercially useful varieties are as follows.

12.2.1 *Mentha arvensis* L.

M. arvensis L. is also known as Japanese mint or corn mint. It is cultivated in Japan since antiquity but also grown in Brazil, China, and India these days. It is an erect plant, generally branched and up to 60 cm tall. The plant has mauve-coloured flowers. It is generally used as food seasoning or in salads and as household remedy for the treatment of indigestion, upset stomach, diarrhoea, and dysentery. The plant has also been employed in folk medicine for relieving symptoms of various diseases like asthma, jaundice, liver and spleen disease, rheumatic pains, arthritis, hypertension, inflamed joints, and ischemia. The cornmint EO being rich source of menthol is utilized in industries, namely, pharmaceutical, food, and cosmetic. Menthol has several therapeutic properties also as it acts as antiseptic, carminative, and diuretic agent. The EO obtained from cornmint is used as a remedy to treat several skin infections like scabies and ringworm infections and also relieves itching, rashes, burns, and even pain (Kokkini et al. 2003; Thawkar et al. 2018).

Table 12.1 Classification of Section *Mentha* on the Basis of Inflorescence Characters and the Dominant Monoterpene Present in Essential Oils

Based on Inflorescence Characters

Line	Characters	Type Species	Other Species
Capitatae	Compact, head like inflorescence	*M. aquatica* L.	*Mentha dahurica* Fisch. ex. Benth *Mentha japonica* (Miq.) Makino
Spicatae	Spikes	*M. spicata* L.	*M. longifolia* (L.) L. *M. suaveolens* Ehrh. *Mentha* × *piperita* L. Huds. *Mentha* × *maximilianea* F. W. Schultz *Mentha* × *maximilianea* *Mentha* × *villosa* Huds
Verticillatae	Inflorescence vertically partitioned into whorls	*M. arvensis* L.	*Mentha canadensis* L. *Mentha* × *verticillata* L. *Mentha* × *gracilis* Sole *Mentha* × *dalmatica* Tausch *Mentha* × *carinthiaca* Host *Mentha* × *smithiana* R. Graham

Based on Dominant Monoterpene Compounds in Essential Oil

Pathway	Dominant Monoterpene Compound	Species
Menthol pathway	Menthol, menthone, and menthofuran	*M. aquatica* L. *M. arvensis* L. *Mentha cervina* L. *M. canadensis* L. *Mentha dahurica* Fisch. ex. Benth. *Mentha diemenica Spreng.* *Mentha* × *dumetorum Schultes* *Mentha gattefossei Maire* *M.* × *gracilis* Sole (syn. *M.* × *gentilis* auct non. L.) *Mentha grandiflora* Benth *Mentha japonica* (Miq.) Makino *M. longifolia* (L.) *Mentha longifolia* subsp. *typhoides* (Briq.) Harley *M.* × *maximilianea* F. W. Schultz *M.* × *piperita* L. *Mentha satureoides* R. Br. *Mentha suaveolens* Ehrh. (syn. *M. rotundifolia* L.) *M.* × *verticillata* L. *Mentha* × *villosa* Huds. *Mentha repens* (Hook. f.) Briq. *Mentha requienii* Benth. *Mentha pulegium* L.
Carvon pathway	Carvon, dihydrocarvone, and carveol	*Mentha* × *dalmatica Tausch* *M. longifolia* L. *Mentha* × *smithiana R. Graham* *M. suaveolens* Ehrh. (syn. *Mentha* × *rotundifolia* L.) *M. spicata* L. *Mentha* × *villosa Huds.*
Linalool pathway	Linalool and linalylacetate	*M. aquatica* var. *citrata* (Ehrh.) Frensen. *M.* × *piperita* var. *citrata* (Ehrh.) *M.* × *piperita* (L.) Huds. var. *lavanduliodora* ined. *M. longifolia* L. *var. lavanduliodora* ined.

Source: Adapted from Saric-Kundalic et al. (2009).

12.2.2 *Mentha piperita* L.

M. piperita L. is commonly called peppermint, black mint, candy mint, or Mitcham in various languages. It is a natural and sterile hybrid cultivated by crossing between *M. aquatica* L. or water mint and *M. spicata* L. or spear mint. Botanist John Ray in 1696 first recognized peppermint as distinct species while *London Pharmacopoeia* formally accepted it as species in 1721. Peppermint is native to Europe and some varieties are also indigenous to North America and Australia. It consists of three varieties, namely, *vulgaris* sole (Mitcham mint), *sylvestris* sole (Hungarian mint), and *officinalis* sole (Black mint). Mitcham mint is extensively found throughout the world while the substantially cultivated variety is black mint as it yields more volatile oil in comparison to other two varieties. Due to wide array of applications in food, perfumery, therapeutic, and medicinal sectors, peppermint is widely cultivated throughout the world but the major cultivators are Bulgaria, Greece, Spain, and the United States (Pushpangadan and Tewari 2006; Masomeh et al. 2017).

The plants of peppermint are 30–90 cm in height and have lilac flowers. The leaves are dark green leaves with reddish veins and possess pungent and strong aroma. Peppermint leaves characteristic cooling sensation in mouth as after taste due to the presence of menthol and menthone. Due to this property, peppermint oil is used in chewing gums, toothpastes, and mouthwashes as flavouring agent. Peppermint leaves being stimulant and carminative in action have found significant place in folk medicine and consumed in the form of mint tea or water decoction to treat sore throat, headache, common cold, fever, insomnia, and morning sickness. They are also used as decongestant and possess antiseptic properties. Their use has also been widely recognized in relieving symptoms of gastric disorders such as gas, colic, and irritable bowel syndrome (Pushpangadan and Tewari 2006).

12.2.3 *Mentha spicata* L.

M. spicata L. is also known by other vernacular names like spearmint, garden mint, lady's mint, brown mint, sage of Bethlehem, etc. The plant is 50–56 cm tall, perennial, and mainly propagate from underground stolons. The flowers are pink to white in colour. The leaves exhibit distinctive aroma and pungent taste but lack after cooling sensation unlike peppermint and Japanese mint. It is extensively grown all over the world due to its carvone and limonene rich EO. The major producers of spearmint are Australia, China, Germany, Russia, and the United States (Patra and Kumar 2006).

Spearmint is added to various food items as it imparts characteristic flavour and aroma, and due to its anti-microbial action, it also has role in food preservation. It is widely employed in cosmetics and in oral preparations. In traditional herbal therapies, it has been used to combat gastric problems, cold and flu, fever, etc. It also exhibits antioxidant, carminative, and anti-spasmodic activities (AyKee et al. 2017).

12.2.4 *Mentha longifolia* L.

M. longifolia is known by different names like wild mint and horse mint. It is found in abundance in Mediterranean regions, Australia, Europe, and North America. The plant is 40–120 cm tall, perennial, and possesses peppermint-like aroma. Leaves are generally green-greyish in colour, being green from above while white from below. The flower colour varies from lilac, purple to white (Mikaili et al. 2013). The EO mainly consists of limonene and carvone as major constituents. The medicinal and therapeutic properties of wild mint are due to various phytochemicals and EO components. Different parts of mint, namely, leave, flowers, and stem, have been employed extensively in conventional pharmacopeia for the treatment of numerous infectious diseases (Farzaei et al. 2017).

12.2.5 *Mentha pulegium*

M. pulegium is popularly known as pennyroyal, pennyrile, sqawmint, mosquito plant, and pudding grass. The plants are extensively grown in north and central Europe and Mediterranean regions. The EO obtained from pennyroyal is rich source of pulegone, which is a highly toxic organic compound having deleterious effects on liver and uterus. Though being toxic even at smaller concentrations, its use as culinary herb is still prevalent. It was used in traditional medicine as abortifacient because of its ability to stimulate menstrual flow (Yadi et al. 2017).

12.3 APPLICATIONS OF MINT IN FOOD AND COSMETICS

12.3.1 Uses in Food

Mint is widely used in food from pre-historic times majorly for flavour and aroma purposes such as in soups, herbal teas, lemonades, cheese, paneer, breads, and salads. Herbal teas incorporated with mint are widely consumed all over the world because of their exclusive taste and therapeutic effects like colic relief, sore throat, flatulence relief, etc. It is used as spice and condiment in various food delicacies. In India, mint chutney is consumed along with food to aid digestion. It is added as condiment to vegetables and salads in China. It is also used to flavour wines and sauces in Roman and Egyptian cultures. Commercially it is exploited by food industry in confectionary industry as ingredient in flavouring agent in candies and chewing gums, in alcoholic beverages, and as preservative in various foods and different types of edible packages (Mamadalieva et al. 2020).

12.3.2 Uses in Cosmetics

The EO obtained from mint is extensively employed in personal care products due to its strong and pleasant odour along with various therapeutic effects. They are widely added to creams, gels, ointments, body talc's, and perfumes due to the cooling and soothing effect of menthol present in mint (Sarkic and Stappen 2017).

The EOs are also used in aromatherapy as it is believed that the aroma compounds present in EOs have curative effect on body, mind, and soul. They are administered in body via inhalation using candles or by massaging them on the surface of skin. Due to numerous biological activities exhibited by mint EO such as analgesic, anti-inflammatory, antiseptic, anti-fungal, anti-viral, anti-bacterial, carminative, diuretic, spasmolytic, and stimulatory effects, they are used in aromatherapy. Mint EO is used to relieve pain, arthritis, headaches, and in treatment of skin diseases, irritable bowel syndrome, sinus, and lung congestion (Ali et al. 2015).

12.4 EXTRACTION AND IDENTIFICATION METHODS FOR BIOACTIVE COMPOUNDS IN MINT

The basic strategy for isolation, extraction, and identification of bioactive constituents from any natural source involves three steps (Banu and Catherine 2015; Chavez-Gonzalez 2020a):

a. Pre-treatment or preparation of sample: In this stage, samples are generally dried, centrifuged, or filtered and prepared for extraction.
b. Extraction and isolation of bioactive compounds: Different processes are used for extraction and isolation of bioactive constituents which involves conventional methods like maceration, Soxhlet extraction, decoction, serial exhaustive extraction, water infusion, steam, or hydro-distillation and advanced methods like microwave-assisted extraction (MAE), ultrasound-assisted extraction

(UAE), enzyme-assisted extraction (EAE), supercritical fluid extraction (SFE), and solid microphase extraction.

c. Identification and quantification of bioactive compounds: This step involves various chromatographic and mass spectrophotometric techniques such as gas chromatography and mass spectroscopy (GC-MS), high-performance liquid chromatography (HPLC), high-performance thin-layer chromatography, and electrospray ionization mass spectroscopy (ESI-MS).

The two major groups of bioactive compounds present in various mint species are essential oils (EOs) and phenolic compounds (Mamadalieva et al. 2020). Various extraction methods have been used to derive high concentration of these fractions so far. Research communities are continuously working on both conventional and advanced methods to improve extraction yield and concentration of bioactive compounds in mint extracts. Some of the widely used approaches for extraction of bioactive components from mint are discussed below.

12.4.1 Extraction Methods

12.4.1.1 Steam Distillation or Hydro-Distillation

The process of steam distillation (SD) involves use of heat either from steam or water to rupture the oil glands present in plants. Then, the oil is vaporized, condensed, and lastly separated from water. The method has been traditionally employed to separate EO from peppermint. Among various techniques, it is most widely accepted for extraction of EOs at industrial scale. Though water distillation is more common than SD due to economics involved in the process, the latter method gives more yield along with high-quality EO. The method also prevents direct contact between water and plant material (Radwan et al. 2020). In spite of all these merits, the process is not free of certain demerits such as it is both time consuming and requires more fuel. Moreover, it also results in flavour modification and hence overall quality deterioration of EO (Dai et al. 2010).

A comparative study to check the efficiency and qualitative properties of peppermint oil obtained by hydro-distillation (HD) and SD was carried out by Gavahian and co-workers (2015). They concluded in their work that EO obtained by both the methods was similar in properties and components as analysed using GC-MS. Hence, both methods can be utilized as qualitative reference methods for extracting EO from peppermint.

12.4.1.2 Soxhlet and Solid-Liquid Extraction

Soxhlet and solid-liquid extractions are old-style methods which use organic solvents for extraction of polyphenolic compounds and EOs (Mussatto 2015). The method is commonly employed for the extraction of mint oil (Al-Marzougi et al. 2013). Among various solvents, hexane is widely used for attaining considerable yield of oil from mint. The process of Soxhlet extraction operates under milder process conditions and high temperature is generally avoided. However, the process has some limitations, being time consuming, requires greater quantities of solvent, residues of solvent can contaminate extracted product leading to low quality, and it is not suitable for thermolabile fractions (Dai et al. 2010; Almeida et al. 2012). Despite these limitations, the process has still been used by researchers worldwide and recently by Almeida and company (2012) and Hossain and co-workers (2014) for extraction of secondary metabolites from *M. piperita* L. and *M. spicata* L. due to simplicity of operation and good extraction results.

12.4.1.3 Microwave-Assisted Extraction

MAE comes under advanced methods for the isolation of biologically active phytochemicals. The process is accomplished under controlled conditions of temperature and pressure while keeping solvent and sample mixed in a close vessel. Microwaves are used to heat the solvent. The technique offers various advantages over conventional methods of extraction. Some of the merits include decreased time of extraction, greater yields along with low energy, and solvent consumption (Mussatto 2015).

Dai and co-workers (2009) investigated the effect of different extraction techniques like room temperature extraction, reflux temperature extraction, UAE, and MAE. The effect of techniques was investigated on extraction yield and extracted compounds (menthol, menthone, and methofuran) from peppermint oil. They favoured the use of MAS as the process offered greater yield in shorter process time and by utilizing low-solvent quantity.

Recently, similar study has been carried out by Radivojac et al. (2021) in which the researchers compared HD and microwave-assisted hydro-distillation (MAHD) for extraction of peppermint EO while comparison between Soxhlet extraction, MAE, and UAE was made for extraction of lipids from peppermint. They concluded that out of HD and MAHD, MAHD is better technique for extraction of EO as it improved the yield significantly with low energy consumption. On the other hand for extraction of lipids, UAE and MAE were found better than conventional Soxhlet extraction in terms of solvent utilization and time consumption but the extraction yield was found better with Soxhlet extraction.

12.4.1.4 Ultrasound-Assisted Extraction

UAE requires the use of ultrasonic shock waves which travel through the solvent to rupture the cellular membrane of solid plant tissue and extract desired bioactive components. The method is quite appealing as it is cheap and requires few types of equipment in the form of ultrasonic probe system and ultrasonic bath (Mussatto 2015).

A comparison was made between classical hot water extraction method and UAE for isolation of secondary metabolites from peppermint. Out of the two methods, UAE was found much more effective as it extracted compounds in short span of time and even elevated their bioavailability and bioactivity (Dogan and Tornuk 2019).

12.4.1.5 Supercritical Fluid Extraction

Supercritical fluid may be defined as the fluid which is compelled at temperature and pressure above its thermodynamic critical point. This changes the properties of the supercritical fluid which acts as solvent leading to high diffusivity and low viscosity, thereby facilitating the extraction of desired bioactive components. The process generally utilizes carbon dioxide as solvent as it is cheap, easily available, non-inflammable, and free of toxicity. Additionally, separation of CO_2 from mixture is quite easy and free of any chemical residues (Chavez-Gonzalez et al. 2020b).

The method has gained immense popularity recently because of several advantages offered in contrast to other methods of extraction which includes low temperature of operation which maintains integrity of thermolabile fractions, low waste generation due to high volatility of solvent used, and easy separation of both solvent and extracted compounds. The only limitation associated with the technique is high initial investment and operating cost (Mussatto 2015).

Almeida and company (2012) extracted mint EO using different extraction techniques, namely, HD, Soxhlet extraction and SFE with and without modifiers using different process conditions. Among all the methods, studied SFE was regarded as best method to obtain phytochemicals and antioxidants from mint EO as the method resulted in highest quality oil with bioactive compounds of proven therapeutic potential. The use of ethanol as co-solvent in SFE increased the extraction yield.

Similar study was carried out in which comparison was made between SD and SFE for studying the extraction yield and compositional quality of *M. spicata* L. EO. The researchers found SFE as a superior method than SD (Al-Marzougi et al. 2013).

12.4.1.6 Enzyme-Assisted Extraction

EAE is also called micro-organism-assisted extraction. The method comes under clean and green technology and gives similar results as MAE, UAE, SFE, etc. for extraction of polyphenolic compounds. It involves culturing micro-organisms, mainly fungal species, vegetable extracts, or animal organs or direct addition of commercially available enzymes. In case of micro-organisms, during growth various enzymes such as pectinases, hemicellulases, cellulases, and glucanases are produced depending upon type of micro-organism involved. These enzymes disintegrate the cell wall of plant material and hence assist in release of bioactive compounds. The method being environment friendly has many merits like specificity due to enzyme action, mild operating conditions, no solvent requirement, reduced extraction time, increased extraction efficiency, and better compositional quality of metabolites (Chavez-Gonzalez et al. 2020b).

In a recent study by Shimotori and co-workers (2020), different polysaccharide degrading enzymes, namely, cellulase A, cellulase T, and hemicellulase, were used either alone or in combination for the extraction of phytochemicals from Japanese mint (*M. arvensis* L.). Maximum extraction of phytochemicals was obtained using combination of cellulase and hemicellulase. The EAE also resulted in improved yield and extraction efficiency of Japanese mint EO.

12.4.2 Identification Methods

After efficient extraction of bioactive, another major step is their identification and quantification using right analytical technique. Recently several approaches including chromatographic (column chromatography, thin-layer chromatography, etc.) and spectroscopic methods (UV-visible, infrared (IR), nuclear magnetic resonance (NMR), and mass spectroscopy) and hyphenated techniques including combination of both chromatographic and spectrophotometric methods such as high-performance liquid chromatography-diode array detection (HPLC-DAD) and high-performance liquid chromatography-mass spectrometry (HPLC-MS) have been used by researchers all over the world for detection of biomolecules. The goal behind identification of these bioactive compounds is to screen the metabolites responsible for bioactivity possessed by the natural plant material such as antioxidant, anti-microbial, or anti-tumour activity along with ease, specificity, and in less time (Donno et al. 2020). Among various methods employed for quantification and identification of bioactive molecules, the most widely used techniques for detection of mint phytochemicals include GC-MS, Fourier transform infrared (FTIR), and LC-MS which are described below in detail.

12.4.2.1 Gas Chromatography and Mass Spectroscopy

GC-MS involves vaporization of metabolites to be analysed. These analytes move along with carrier gas through the silica column. The method requires derivatization of volatile chemicals before analysis. Electron ionization (EI) technique is mainly employed in GC-MS as it produces intense metabolite fragmentation. Various mass spectral libraries are available which can be used to detect the peaks generated by GC-MS. Though it can be used for the detection of sugars, organic acids, amines, amino acids, and fatty acids, it is highly suitable for identification of terpenoids and EOs (Panda et al. 2018). Consequently, it has been employed by various researchers for the detection of mint EO constituents and bioactive molecules such as Hossain et al. (2014), Tambe and Tumane (2017), and Shimotori and co-workers (2020).

12.4.2.2 Fourier Transform Infrared Spectroscopy

FTIR is a high-resolution analytical method used for the detection of chemical components along with their structure elucidation. The technique offers several advantages such as rapid sample preparation and ease of use. It is a non-destructive method employed for fingerprinting herbal extracts and powders. It involves the use of infra-red radiations which are passed through the sample. Some of the frequencies are absorbed by the sample while some are transmitted without any absorption. Thus, FTIR is based on vibrational changes that occur inside the molecules of analyte after its exposure to infra-red light. Hence, it is also called vibrational spectroscopy (Singh et al. 2011). Pramila and company (2012) assessed the IR spectrum of peppermint leaf extract to determine the existence of functional groups and detected tannins and flavonoids in the extract.

12.4.2.3 Liquid Chromatography-Mass Spectroscopy

LC-MS is extensively employed for the detection of secondary metabolites mainly alkaloids, saponins, and phenylpropanoids along with lipids. The coupling of HPLC to MS enhances both its speed and accuracy. It is mainly used for the detection of low-, moderate-, and high-molecular-weight secondary metabolites and for compounds sensitive to heat as derivitization is not required in LC-MS as in GC-MS. However, the major drawback involves limited mass spectral libraries for identification of biomolecules which makes it less efficient (Panda et al. 2018). Several workers like Brown et al. (2019) used HPLC/ESI-MS/MS for detection of phenolic compounds from mint. Similarly, Alexa et al. (2018) employed the technique for the detection of hydroxycinnamic acids in *M. piperita* L.

12.5 BIOACTIVE COMPOUNDS FROM MINT AND THEIR THERAPEUTIC PROPERTIES

Bioactive compounds or bioactive secondary metabolites are valuable phytochemicals produced by plants which have multiple applications in various industries such as food, cosmetics, perfumery, agriculture, and pharmaceutical. They are also popularly termed as "*Chemical Goldmines*". These secondary metabolites are synthesized as by-products or intermediates of primary biosynthetic pathway (Kaur and Ahmed 2021). Species of genus *Mentha* or commonly known as mint have been widely used throughout the history for culinary and medicinal purposes. The medicinal properties of mint may be attributed to two major groups of secondary metabolites, that is, volatile compounds which include EOs and non-volatile compounds which involve phenolic compounds (Mamadalieva et al. 2020).

12.5.1 Volatile Compounds or Essential Oils

The volatile components of plants having strong odour or essence are called EOs. They usually evaporate at room temperature when comes in contact with air and are volatile under steam. They are complex liquids containing low-molecular-weight compounds mainly synthesized by plants as secondary metabolites. The extraction of these compounds is mainly done by SD, HD, or Soxhlet extraction. Almost all plant parts such as leaves, flowers, stems, fruits, roots, seeds, buds, and bark produce EOs and store them in glands or trichomes, oil ducts, or resin ducts. The EOs are mainly used for flavouring purposes in food items, in perfume and cosmetic industry, and in pharmaceutical sector (Chavez-Gonzalez et al. 2020a).

Mentha L. or mint comes second after *Citrus* L. in producing EO worldwide (Mucciarelli et al. 2001). The EO is secreted by grandular trichomes present in both vegetative and reproductive organs of mint plant. Even the reason behind worldwide cultivation of different species of mint is their EO. Among most widely cultivated mint species are Japanese mint (*M. cardensis* L.), peppermint (*M. piperita* L.), and spearmint (*M. spicata* L.), out of these corn mint is mainly cultivated for its

EO. Peppermint oil is however extensively used in pharmaceutical and oral preparations. The major producers of mint oil in world are North America for peppermint and spearmint oil while India and China are leading producers of corn mint oil. The amount of EO produced by different mint species is variable in composition due to different chemotypes present, environmental and growth conditions, time of collection, plant part used for extraction of oil, and method of isolation and analysis followed for EO (Singh and Pandey 2018).

The EO mainly consists of terpenes, that is, monoterpenes, diterpenes, and sesquiterpenes along with alkaloid and phenolic substances. Other compounds present in minor quantity include phenylpropanoids, polyacetylenes, and some fatty acids. The major monoterpenes produced by different mint species are C3-oxygenated p-menthane-type like menthone, menthol, and pulegone or C4-oxygentaed p-menthane such as carvone. The other major compounds present in mint oil are alcohols like elemol, geraniol, linalool, menthol, neomenthol, 3-octanol, cis/trans-sabinene hydrate, *a*-terpineol, terpinen-4-ol, viridoflorol; esters like decyl acetate, dihydrocarvyl acetate, 1,2-epoxyneomenthyl acetate, menthyl acetate, neoisomenthyl acetate, neomenthyl acetate, 3-octyl acetate, *a*-terpinyl acetate; hydrocarbons, namely, *b*-caryophyllene, germacrene-D, limonene; ketones, namely, carvone, cis-/trans-dihydrocarvone, isomenthone, menthone, 3-octanone, pulegone, piperitenone, piperitone; and ethers like 1,8-cineole, menthofuran, caryophyllene oxide, piperitenone oxide, and piperitone oxide. The structures of principal compounds produced in various mint species are given in Figure 12.2 while the major compounds along with the therapeutic properties exhibited by the particular EO are given in Table 12.2.

The EO from *M. arvensis* L. has been found to exhibit least number of chemical constituents than any other species of genus *Mentha*. In various studies performed on EO composition of *M. arvensis* L., the most abundant compounds extracted were menthol and isomenthone. Similarly the major constituent of *M. canadensis* L. EO was menthol. Studies have been made in which salt stress has been related to decreased menthol content in *M. canadensis* L. while this led to increase in menthone and pulegone content (Salehi et al. 2018). The structures of major compounds found in important species of genus *Mentha* are given in Figure 12.2.

Regarding compositional studies performed on *M. spicata* L. EO, carvone and limonene were observed as abundant compounds. However, pulegone, piperitenone oxide, and 1,8-cineol (Figure 12.2) represented major EO components in other important studies. In research studies performed on composition analysis of EO of *M. pulegium* L., pulegone was reported as major compound. Additionally, other compounds present in comparatively high quantity were menthone, limonene, isomenthone, piperitone, and piperitenone (Figure 12.2) (Salehi et al. 2018).

M. longifolia L. (wild mint) EO has also been studied extensively due to the presence of high morphological variability within the species. However, the dominant compounds extracted were oxygenated monoterpenes, that is, pulegone, piperitenone oxide, and 1,8-cineol. Besides these compounds, other compounds found in noteworthy quantities were α-pinene, β-pinene, β-caryphyllene, borneol, carvone, dihydrocarvone, eucalyptol, isomenthone, isopulegone, menthol, menthone, menthofuran, and piperitenone (Figure 12.2) (Salehi et al. 2018).

In comparison to all the species, *M. piperita* L. EO due to presence of numerous chemotypes represented great variability in EO composition. In a study performed on chocolate mint variety of *M. piperita*, methofuran, menthone-d-neoisomenthol and pulegone (Figure 12.2) were found in highest quantity, while in grapefruit mint variety, linalyl acetate along with linalool (Figure 12.2) existed in major amount. In similar studies, variation in chemical composition of EO from *M. piperita* L. grown in different countries has also been observed. In Indian variety of peppermint, menthol, isomenthone, and limonene (Figure 12.2) prevailed in abundance while in Serbia and Morocco, menthol, menthyl acetate, and menthone (Figure 12.2) were found as major constituents. In Brazil, China, Iran, Saudi Arabia, and Taiwan, menthol dominated the composition of EO as major constituent (Salehi et al. 2018).

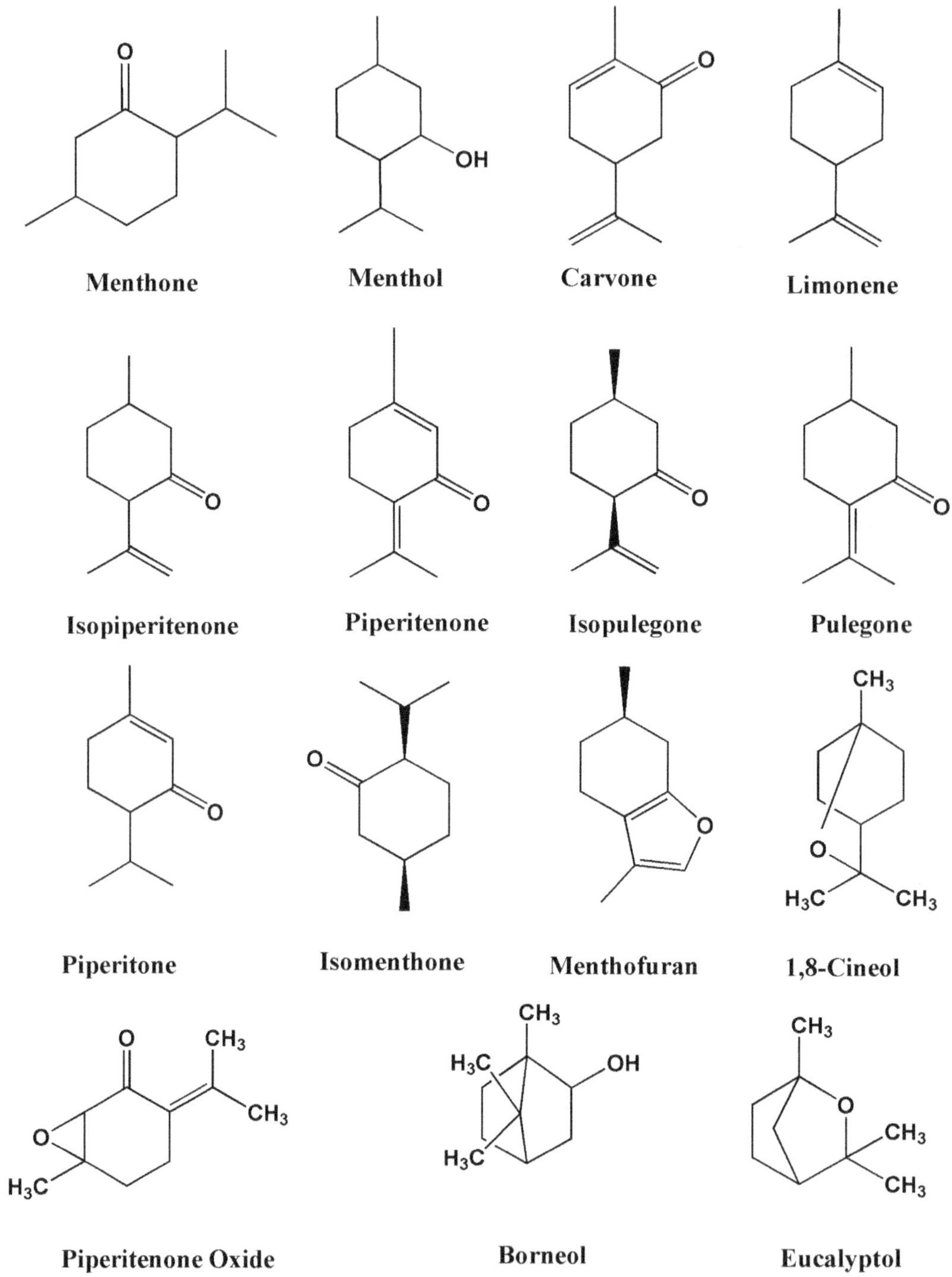

Figure 12.2 Principal compounds present in mint EO. (*Continued*)

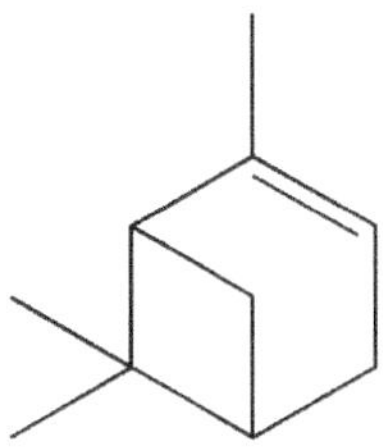

Alpha-pinene

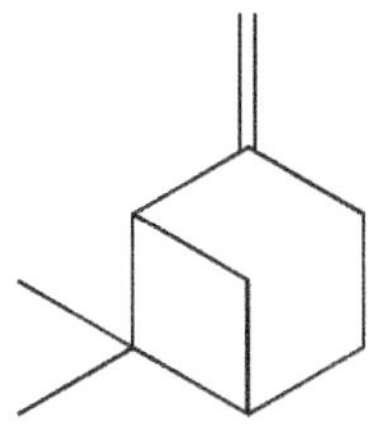

Beta-pinene

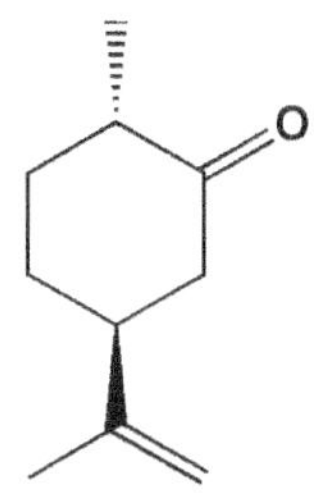

Dihydrocarvone

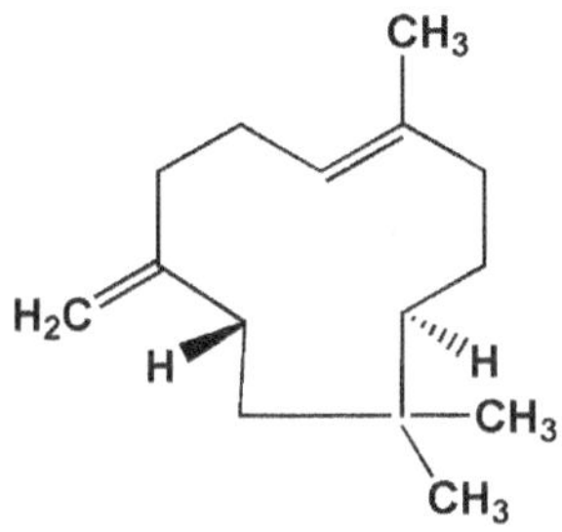

Beta-caryophyllene

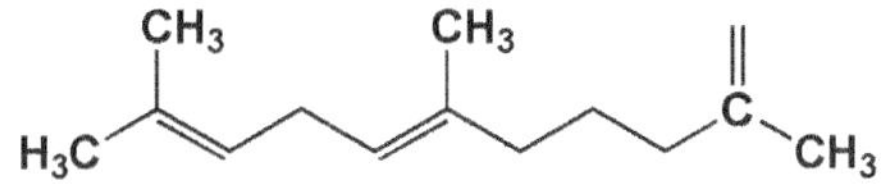

Linanyl acetate

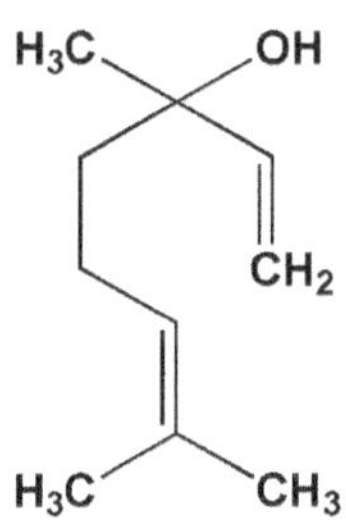

Linalool

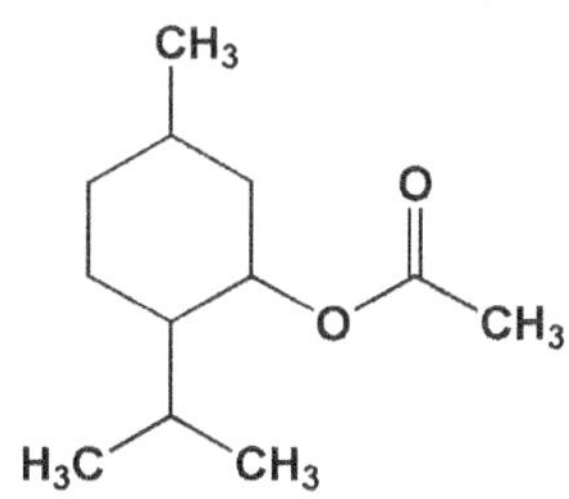

Menthyl acetate

Figure 12.2 *(Continued)*

12.5.2 Non-Volatile Compounds or Phenolic Compounds

Different species belonging to genus *Mentha* have been reported to contain different classes of phenolic or non-volatile compounds. The major classes along with various compounds are (Salehi et al. 2018):

a. *Anthocyanidins* such as cyanidin, delphinidin, luteolinidin, pelargonidin, and petunidin.
b. *Coumarins* including esculetin and scopoletin.
c. *Flavanols*, that is, catechin and epicatechin.
d. *Flavanones* like eriocitrin, eriodictyol, hesperidin, naringenin, and narirutin.

e. *Flavones* such as apigenin, diosmetin, diosmin, luteolin, luteolin-*O*-glucuronide, gardenin B, luteolin-*O*-glucoside, pebrellin, salvigenin, thymusin, and thymonin.
f. *Flavonols* namely quercetin, kaempferol, and rutin.
g. *Phenolic acids* which include cinnamic acid, its analogs such as hydroxybenzoic acid, p-coumaric acid, ferulic acid, caffeic acid, sinapic acid, rosmarinic acid, salvianolic acid, isosalvianolic acid, didehydrosalvianolic acid and lithospermic acids, nepetoidin A/B and glycosides of phenolic acids like caffeic acid glucuronide, chlorogenic, caftaric, gallic, syringic, and vanillic acids.
h. Other compounds such as reveretrol, tyrosol, and oleanolic acid.

The therapeutic activity exhibited by various classes of compounds along with basic structure is given in Table 12.3.

Table 12.2 Major Compounds Present in EO of Commercially Important Mint Species and Therapeutic Properties of Oil

S. No.	Mint Species	Major Bioactive Compounds Present in EO	Therapeutic Properties of EO	References
1.	*M. arvensis* L. (Cornmint)	Menthol, pulegone, p-cymene, isomenthone, menthyl acetate, limonene, isopulegol, linalyl acetate, linalool, 1,8-cineol	Anti-spasmodic, antioxidant, antiseptics, analgesic, anti-microbial, and carminative effect	Makkar et al. (2018); Salehi et al. (2018)
2.	*M. candensis* L. (Japanese cornmint)	Menthol, menthone, isomenthone, isopulegol, pulegone, α-pinene, α-terpineol, limonene	Antioxidant, anti-microbial, anti-nociceptive, anti-hyper-analgesic, cooling sore throats, anti-inflammatory, anti-diabetic, anti-viral, anti-cancer and gastroprotective activity	Vieira et al. (2018); Salehi et al. (2018); Mamadielieva et al. (2020)
3.	*M. longifolia* L. (wild mint)	Menthone, pulegone, piperitone, dihydrocarvone, 1,8-cineol, 3-terpinolenone, limonene, germacrene D, α-pinene, limonene, β-pinene, α-pinene, caryophyllene, piperitone oxide, sabinene, borneol, myrcene, thymol, carvone, eucalyptol, caryophyllene oxide, trans-caryophyllene, cis-caryophyllene	Anti-spasmodic, antioxidant, anti-bacterial, anti-headache, sedative, aphrodisiac, blood purifier	Mikaili et al. (2013); Salehi et al. (2018)
4.	*M. piperita* L. (peppermint)	Menthol, menthofuran, pulegone, menthone, 1,8-cineol, menthyl acetate, limonene, isomenthol, α-pinene, limonene, β-pinene, β-cineol, linalool, linalyl acetate, D-carvone	Antioxidant, anti-microbial, anti-spasmodic, ulcer healing, anti-obesity, anti-cancer, immunomodulatory, gastroprotective, anti-headache, anti-bloating effect	Masomeh et al. (2017); Salehi et al. (2018)
5.	*M. pulegium* L. (pennyroyal)	Pulegone, piperitenone, methone, limonene, eucalyptol, humulene, α-pinene, isomenthone, piperitone	Abortifacient, acaricidal, anti-bacterial, anti-myometrium effect, antioxidant, anti-spasmodic effect	Hadi et al. (2017); Salehi et al. (2018)
6.	*M. spicata* L. (spearmint)	Carvone, cis-carveol, limonene, piperitone, piperitenone, 1,8-cineol, piperitone oxide, β-pinene, cis-dihydrocarvone, dihydrocarveol, camphor, germacrene, α-terpeniol, menthone	Treat smooth muscle spasms, cramps, morning sickness, increase blood circulation, relieve pain, antiseptic, nausea, menstrual cramps, flatulence, flu, and vomiting	Alu'datt et al. (2018); Salehi et al. (2018)

Table 12.3 Therapeutic Activities of Various Classes of Secondary Metabolites Present in Genus *Mentha*

S. No.	Class of Phenolics	Basic Structure	Therapeutic Properties
1.	Anthocyanidins	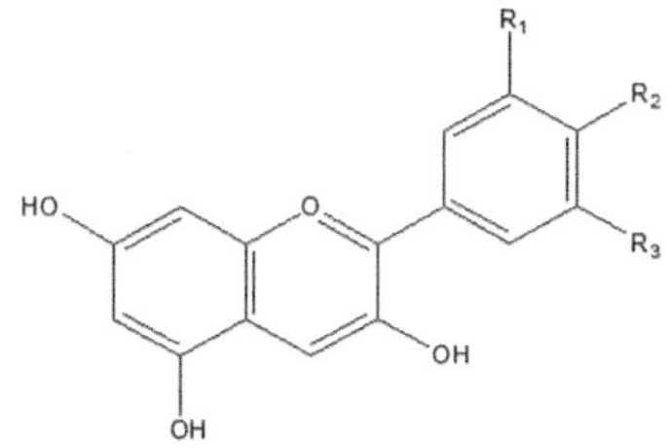	Anti-diabetic, anti-microbial, anti-atherosclerosis, anti-cancer, anti-inflammatory, and anti-obesity
2.	Coumarins	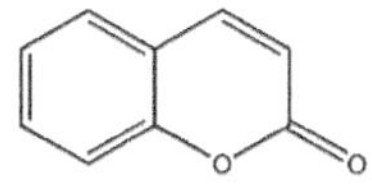	Antioxidant, analgesic, anti-microbial, anti-inflammatory, anti-cancerous, anti-coagulant, anti-diabetic, and anti-neurodegenerative agent
3.	Flavanols	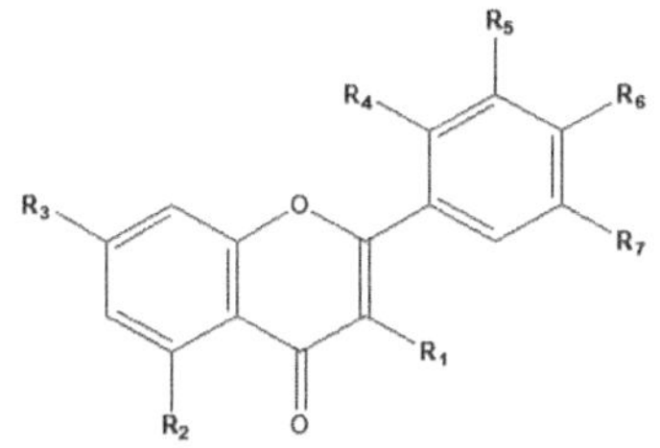	Anti-cancer, antioxidant, anti-atherosclerosis, and neuroprotective activities
4.	Flavanones	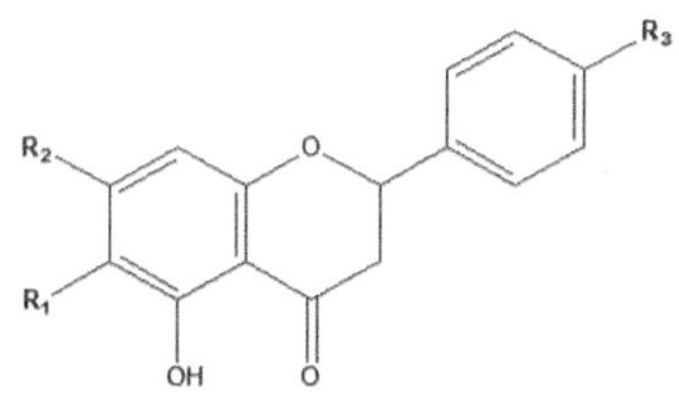	Antioxidant, anti-cancer, anti-inflammatory, anti-hypertensive, anti-cholesteromic, anti-diabetic, and anti-lipidemic
5.	Flavones	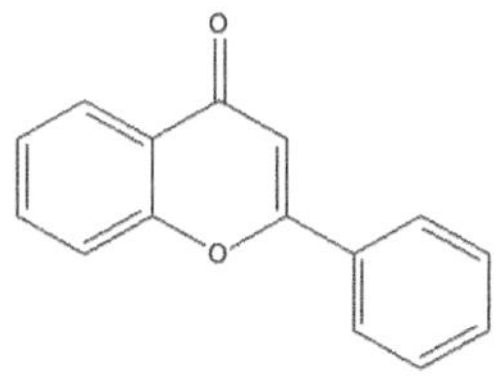	Antioxidant, anti-tumour, anti-microbial, and anti-inflammatory activity
6.	Flavonols	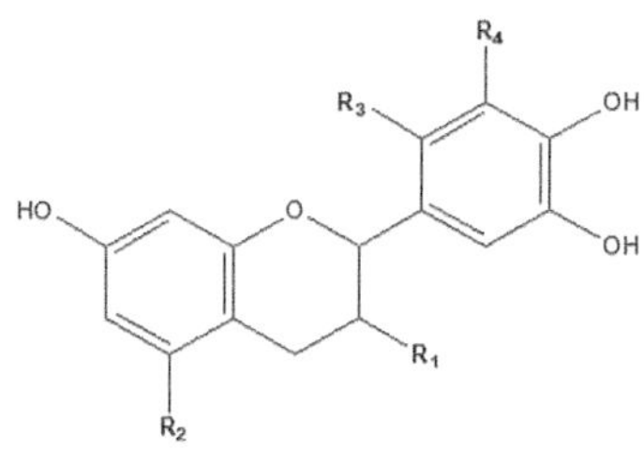	Antioxidant, anti-tumour, vasodilating, and cardio-protective activity, anti-thrombotic and anti-inflammatory
7.	Phenolic acids	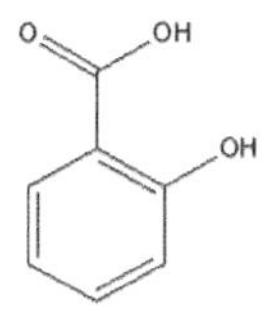	Antioxidant, anti-allergy, anti-inflammatory, anti-tumour, anti-diabetic, anti-microbial, and anti-viral activity

Source: Kaur and Ahmed (2021).

12.6 PHARMACOLOGICAL STUDIES IN MINT

Due to the presence of wide array of bioactive compounds in the form of EOs and phenolic compounds, the genus *Mentha* L. has been widely studied and compilation of extensive data on pharmacological potential is quite arduous. Its bioactive constituents possess various therapeutic properties like analgesic, antioxidant, anti-inflammatory, anti-allergic, anti-microbial, anti-cancer, anti-viral, anti-diabetic, anti-thrombotic, anti-atheroseclerosis, cardio-protective, anti-neurodegenerative, etc. which have been clearly discussed in the previous section. Thus, the authors have made an attempt to compile information on some of the major pharmacological studies carried out on mint in recent years.

12.6.1 Anti-Microbial Activity

The inefficacy of antibiotics as drugs against infection-causing micro-organisms has led to emergence of antibiotic resistant strains worldwide. This has provoked the researchers to recognize novel anti-microbial agents with potent activity and minor side effects (Kaur and Ahmed 2021). In this context, various species of genus *Mentha* have been studied against pathogenic and disease-causing micro-organisms. The EOs from mint have shown to possess anti-bacterial, anti-fungal, and anti-viral activity. Moreover, the constituents of EOs are safer, economical, and exhibit lesser side effects in comparison to synthetic drugs (Zaidi and Dahiya 2015).

Recently, a study was performed to investigate the efficacy of EOs and water extracts of various *Mentha* species, namely, *M. piperita*, *M. spicata*, *M.* × *villosa*, *M.* × *gracilis*, *M. aquatica*, *M. arvensis*, *M. longifolia*, *M. suaveolens*, and their different chemotypes against pathogenic *Escherichia coli* and *Staphylococcus aureus*. The researchers concluded that the EO exhibited anti-bacterial activity against both *E. coli* and *S. aureus* while water extracts were only potent against *S. aureus*. However, some of the samples were found to be bacteriocidal also (Kapp et al. 2020).

In another study implemented by Mejdoub and co-workers (2019), anti-fungal activity of EO from *M. spicata*, *M. pulegium*, *M. piperita*, and *M. rotundifulias* was studied against *Aspergillus niger*, *Aspergillus flavus*, *Alternaria* spp., and *Penicillium* spp. The researchers concluded that all the tested species exhibited substantial anti-fungal activity while the EO from *M. pulegium* was found to be most effective by showing complete mortality at 15 μL/mL air dose. The other major studies carried out on different mint species are given in Table 12.4.

12.6.2 Anti-Cancer Activity

Cancer is mainly caused by abnormal growth of any kind of cells in the body. It is distinguished into two types, benign and malignant tumours. Benign tumours generally do not spread to the surrounding cells or tissues and stays at the original site. On the other hand, malignant tumour is quite dangerous as it invades the surrounding normal cells and tissues. The metastasis caused by the malignant cancer cells spreads throughout the body and finally results in death if not taken care at proper time. The medicines used for the treatment of cancer cells are both costly and unsafe. Additionally, they are associated with undesirable side effects. This all has led the scientific community to find alternative therapy in the form of herbal drugs to treat cancer in an effective and safe manner (Kaur and Ahmed 2021).

The role of *Mentha* species in controlling and treating different types of cancers is widely studied by researchers due to the cytotoxic and anti-proliferative activities exhibited by phytochemicals present in EO and crude extracts of *Mentha*. The anti-cancer activity of various mint species such as *M. piperita*, *M. spicata*, *M. longifolia*, *M. aquatica*, *M. pulegium*, *M. arvensis*, and *M. crispa* has been studied on different cancer cell lines, namely, HeLa (cervical cancer), HEP-2 (human larynx epithelioma cancer), human monocytic U937 (human monocytic leukemic) cells, MCF-7 (Michigan

Table 12.4 In-Vitro Studies Carried on Various Mint Species against Human Pathogenic Micro-Organisms in Last Five Years (2015–2020)

S. No.	Mint Species Used	Extract/EO Used	Tested Strain	Reference
1.	*M. spicata* and *M. piperita*	EO	*Escherichia coli*, *Salmonella typhi*, *Salmonella paratyphi*, *Staphylococcus aureus*, *Klebsiella pneumoniae*, *Pseudomonas aeruginosa*, and *Acinetobacter* spp.	Zaidi and Dahiya (2015)
2.	*M. piperita*	Methanolic extract	*Bacillus subtilis* and *Staphylococcus aureus Escherichia coli* and *Pseudomonas aeruginosa*, *Trichophyton mentagrophytes*, *Trichophyton verrucosum*, and *Candida albicans*	Farzaei et al. (2015)
3.	*M. piperita*	EO	*Staphylococus aureus*, *Escherichia coli*, *Pseudomonas aeruginosa*, and *Bacillus cereus*	Grulova et al. (2016)
4.	*M. piperita* (var. chocolate mint, peppermint and citrate), *M. spicata* (var. apple mint and pineapple mint), *M. pulegium*, *M. longifolia*, *M. aquatic*, and *M. arvensis*	Ethanol, methanol, hexane, ethyl acetate, diethyl ether	*E. coli*, *Staphylococccus haemolyticus*, *Aeromona shydrophila*, *Coronobacter sakazakii*, *Aeromonas salmonicida*	Park et al. (2016)
5.	*M. piperita*	Leave EO and Stem EO	*E. coli*, *Pseudomonas aeroginosa*, *Micrococcus luteus*, *S. aureus*, *B. subtilis*, *Salmonella typhimurium*, and *B. cereus*	Marva et al. (2017)
6.	*M. spicata*	Leaf EO	*E. coli*, *Kluyvera* spp., *Kleibsiela* spp., *Salmonella* spp., and *Serratia* spp.	Selles et al. (2018)
7.	*M. spicata*, *M. pulegium*, *M. piperita*, and *M. rotundifulias*	EO	*Aspergillus niger*, *Aspergillus flavus*, *Alternaria* spp., and *Penicillium* spp.	Mejdoub et al. (2019)
8.	*M. piperita*, *M. spicata*, and *M. × villosa*, *M. × gracilis*, *M. aquatica*, *M. arvensis*, *M. longifolia*, *M. suaveolens*	EO and water extracts	*E. coli* and *S. aureus*	Kapp et al. (2020)

Cancer Foundation-7), Jurkat (immortalized human T-lymphocyte cells), T24 (human urinary bladder carcinoma), HT-29 (human colorectal adenocarcinoma cell line), and MIA PaCa-2 (human pancreatic cancer cell line) (Sharma et al. 2014).

In a study taken out by Yang and company (2017), cytotoxicity of platinum-based nano-particles (PtNPs) entrapping polyphenols from *M. piperita* aqueous leaf extract was tested against human colon cancer cell line (HCT 116). The researchers deduced that the biosynthesized PtNPs significantly reduced the survival of cancer cells at a minor concentration of IC50 20 μg/mL.

In a latest study, crude methanol extract of *M. longifolia* was tested against adrenocortical tumour cell models, namely, H295R and SW13. The study suggested the role of methanolic extract of *M. longifolia* in reducing cell viability and vitality of both the cell lines particularly in SW13 cancer cell line (Patti et al. 2020).

12.6.3 Anti-Diabetic Activity

Diabetes mellitus is a common metabolic disorder which is increasing at an alarming rate with 25% of total world population affected from this disorder. It is basically characterized by elevated levels of glucose in the blood along with altered metabolism of proteins and lipids. It occurs in two major types, namely, type-I diabetes mellitus or insulin-dependent diabetes resulting due to poor production of insulin by the pancreas and another is type-II diabetes mellitus or insulin-independent diabetes, as the name suggests, in this type though the body produce insulin but somehow it becomes ineffective in utilizing the produced insulin and hence results in enhanced blood glucose levels after immediate meals. The latter type, that is, type-II diabetes mellitus, occurs more frequently in developed and developing countries than type-I disorders and it commonly affects population with age above 30 years. Hence, it is also called adult-onset diabetes (Reddy and Manoharbabu 2019; Kaur and Ahmed 2021).

Several mint species such as *M. arvensis*, *M. aquatica*, *M. pulegium*, *M. piperita*, and *M. spicata* have been studied for their anti-diabetic potential and it is one of the recognized traditional herbs known for its wide medicinal properties (Bayania et al. 2017; Farid et al. 2018).

The anti-diabetic efficacy of methanolic extract of *M. arvensis* has been studied in both in-vivo and in-vitro conditions using key enzymes α-amylase and α-glucosidase linked to type-II diabetes mellitus in in-vitro model while starch-induced diabetic Wistar rats for in-vivo model. The results presented by study were astonishing as the extract showed exceptional results by inhibiting both key enzymes. It also effectively repressed postprandial rise in blood glucose level in Wistar rats. Thus, the researchers supported the use of *M. arvensis* in regulation of type-II diabetes mellitus (Agawane et al. 2019).

Similarly, the effect of *M. pulegium* aqueous extract from aerial parts was tested in vivo for anti-diabetic activity in streptozotocin-induced diabetic and normal rats. The study concluded that the extract significantly reduced blood glucose levels at a dose of 20 mg/kg. The histopathology of liver and pancreas also supported the use of extract with beneficial effects seen on cellular population. Lastly, oral glucose tolerance test conducted on normal rats also showed improved glucose tolerance (Farid et al. 2019).

12.6.4 Analgesic/Anti-Nociceptive Activity

As per the International Organization of Pain, it may be defined as, "Unpleasant sensory and emotional experience associated with actual and potential tissue damage". Various synthetic drugs are available in market to relieve and calm pain such as opiod analgesics and non-steroidal anti-inflammatory drugs. Though they are quite effective as pain killers, they are associated with number of side effects such as gastrointestinal disturbance in the form of vomiting and constipation, other adverse reactions like itching and nausea, and increased risk of kidney problems (Nandhini et al. 2018). Thus, the researchers are taking interest in herbal plants with numerous effective bioactive components which offer great alternative to synthetic pain relievers. *Mentha* species is well known in traditional medicine for its analgesic and anti-nociceptive properties.

Biswas and company (2014) studied the analgesic potential of *M. arvensis* in acetic acid-induced writhing test on Swiss albino mice at doses 250 and 500 mg/kg body weight and the extract demonstrated potent activity against managing pain sensation in mice as number of writhes decreased considerably with increased dose of extract.

Analgesic potential of *M. piperita* has also been widely studied among all *Mentha* species. The latest study has been reported using aqueous extract of *M. piperita* at different doses, that is, 200, 400, and 600 mg/kg bw. in Swiss albino mice by Eddy's hot plate method and tail clip method. The extract exhibited analgesic activity at all three doses, but at dose of 600 mg/kg, it was found to be most effective with analgesic activity similar to that of reference drug aspirin used in the study (Prasanna et al. 2018).

12.6.5 Anti-Inflammatory Activity

Inflammation is an immune response initiated by body when it encounters any physical injury or pathogens such as germs, bacteria, fungi, and viruses. Though it is a protective mechanism, it results in hotness, redness, and swelling at the site of infection due to release of several inflammatory mediators. Inflammation could be acute or chronic depending upon the condition. In chronic inflammatory responses such as Rheumatoid arthritis and Crohn's disease, the severity of inflammation is much more (IQWig 2006). *Mentha* extracts and EOs possess various phytochemicals having anti-inflammatory response and several studies have been carried out in past to check the efficacy of mint as powerful anti-inflammatory agent.

In an in-vivo study, the EO from three *Mentha* species, namely, *M. piperita* L., *M. spicata* L., and *M. suaveolens* Ehrh. was investigated for anti-inflammatory activity using rat-paw oedema test induced by γ-carrageenan. All the studied species exhibited anti-inflammatory activity but significant results were observed in rat groups treated with *M. spicata* EO as it gave superior results in comparison to recommended drug diclofenac. The EO of *M. spicata* also caused notable decrease in prostaglandin concentration at the site of infected area (Mogosan et al. 2017).

Similar results were obtained by Kehili et al. (2020) also who recently performed in-vivo study to check potential of *M. spicata* EO as anti-inflammatory agent in Swiss NMRI male mice, carrageenan-induced paw oedema, and xyline-induced ear oedema models. Histological examination was also performed on inflamed and non-inflamed tissues. The EO markedly inhibited paw volume by 77.24%. In addition, topical administration of EO also notably decreased xylene-induced ear oedema similar to control ketoprofen. Histological examination revealed role of EO in reduced cellular infiltration and subcutaneous oedema.

12.7 TOXICOLOGICAL AND SAFETY STUDIES

Herbal products are extensively used for the treatment of numerous diseases, but the laws regarding their toxicity and safety issues are still in their infancy. Many studies are present in literature which impart confirmation for the existence of bioactives that are boon to human health. However, it should be an important concern to ascertain the dangerous effects of some of the phytochemicals present in the plants (Jothy et al. 2011).

Toxicity may be defined as declaration of being poisonous or lethal, representing the circumstances of detrimental changes caused by the interaction between poisonous substance and cells. This interaction generally depends on different factors such as the chemical nature of the poisonous material and the cell membrane. The results of toxicity can occur before the attachment of the venomous material to critical organs such as liver and kidneys. Thus, assessment of lethal nature of a substance is pivotal when scrutinizing its use in human health as introduction to chemicals can be dangerous and may lead to unfavourable effects on health and well-being. In general, toxicity is estimated by observing acute, sub-chronic, chronic, carcinogenic, and reproductive effects (Jothy et al. 2011).

Mentha species is generally considered safe because of its numerous culinary uses and hence its chronic toxicity studies involving human subjects are quite limited. Few clinical studies have been reported on safety of *M. spicata* oil consumption and it has been regarded as safe with no side effects at 0.24 mL daily dose for continuously for about 3 weeks. However, the leaf EO of *M. spicata* used in toothpaste flavouring has been associated with contact cheilitis, a form of contact allergy. Other side effects linked to spearmint EO include flushing, headaches, heartburn, nausea, and vomiting. Neuronal toxicity has also been displayed by *M. spicata* extracts at high doses (Tafrihi et al. 2021).

Similar adverse effects have been noticed in case of *M. piperita* as excessive intake of peppermint tea has been linked to low iron absorption by the body leading to anaemia. Other side effects include abdominal pain, bradycardia, cheilitis, dermatitis, mouth ulcers, nausea, tremors, and vomiting. At high doses, peppermint oil can cause transient shock, hematuria along with other complications such as kidney and lung failure (Tafrihi et al. 2021).

Certain bioactive compounds and their metabolites present in different species of *Mentha* have also been reported to possess adverse effects when consumed or inhaled such as carvone, limonene, menthol, pulegone, methofuran, and menthone. Menthol is dominant compound present in mint EO. Though it is widely used in oral preparations, candies, etc. and regarded as safe with no adverse toxicity reports present in literature, consumption of pure menthol can be fatal. Even overexposure or excessive consumption of menthol containing products in the form of candies, cigarettes, etc. are dangerous. Long-term exposure to menthol can also lead to cutaneous, gastrointestinal, and neurological problems. It may also cause agitation, ataxia, convulsion, coma, dizziness, and hellucinations. Thus, continuous exposure of menthol at high doses must be avoided as it can result in serious intoxication and even death (Kumar et al. 2016). Another important compound pulegone present in mint has been associated with severe hepatotoxicity and neurotoxicity. Even its metabolite menthofuran has been observed to cause liver injury. At high doses, pulegone can also lead to weight loss and atonia (Tafrihi et al. 2021).

12.8 CONCLUSION

Genus *Mentha* is extensively used in traditional medicine and culinary practices from ancient times. Due to the strong and pleasant aroma exhibited by various species of mint, its use in culinary science is inevitable as its tea is consumed globally for its peculiar taste and therapeutic properties. The bioactive compounds present in mint as EOs and phenolics have made it a popular herb of folk medicine where it was used as anti-ulcer, anorexic, anti-inflammatory, analgesic, anti-infectious, anti-microbial, antiseptic, anti-spasmodic, astringent, digestive, carminative, anti-fungal, nervine stimulant, vasoconstrictor, decongestant, and stomachic. Today also, mint is used in various industries like food, cosmetic, perfumery, and pharmaceutical. The industrially important species of mint include *M. piperita* L. (peppermint), *M. spicata* L. (spearmint), *M. arvensis* L. (cornmint), and *M. cardanensis* L. (Japanese cornmint). These species are extensively cultivated around globe because of their EO which is used in toothpastes, mouth washes, oral sprays, candies, chewing gums, and in cosmetics in body talc, etc. The major compound responsible for distinctive aroma of mint is menthol. The other important compounds include menthone, pulegone, carvone, limonene, and β-caryophyllene. The ambiguity related to various mint species has led to chaos in their distinction from each other. The different species are sometimes characterized on the basis of major constituent present in EO. Though mint species is rich in bioactive compounds having several health benefits, the work done in literature is basically focused on EO and the research related to phenolic compounds is still in its infancy. Hence, there is need to discover the benefits of phenolics as anti-microbial agents, in treatment of various types of cancers and even as anti-lipidemic and anti-cholesteromic agents. Though enough work is present in literature highlighting various pharmacological properties of mint, clinical studies involving human subjects are scarce, and hence to uncover its full efficacy, these studies need to be performed to collect enough evidence in its support as medicinal agent. Moreover, the toxic compounds present in mint EO limit its continuous use and thus methods for effective removal of these compounds should be discovered on priority.

REFERENCES

Agawane, S. B., Gupta, V. S., Kulkarni, M. J., Bhattacharya, A. K., Koratkar, S. S. 2019. Chemo-biological evaluation of antidiabetic activity of *Mentha arvensis* L. and its role in inhibition of advanced glycation end products. *Journal of Ayurveda and Integrative Medicine* 10(3):166–170. https://doi.org/10.1016/j.jaim.2017.07.003

Ahad, B., Shahri, W., Rasool, H., Reshi, Z. A., Rasool, S. and Hussain, T. 2021. Medicinal plants and herbal drugs: An overview. In: *Medicinal and Aromatic Plants*, eds. T. Aftab, and K. R. Hakeem, 1–40. Springer Nature, Switzerland. https://doi.org/10.1007/978-3-030-58975-2_1

Al Marzouqi, A. H., Rao, M. V. and Jobe, B. 2007. Comparative evaluation of SFE and steam distillation methods on the yield and composition of essential oil extracted from Spearmint (*Mentha spicata*). *Journal of Liquid Chromatography & Related Technologies* 30(4):463–475. https://doi.org/10.1080/10826070601093721

Alexa, E., Danciu, C., Radulov, I., Obistioiu, D., Sumalan, R. M., Morar, A., and Dehelean, C. A. 2018. Phytochemical screening and biological activity of *Mentha×piperita* L. and *Lavandula angustifolia* Mill. extracts. *Analytical Cellular Pathology* 2018:1–8 https://doi.org/10.1155/2018/2678924

Ali, B., Al-Wabel, N. A., Shams, S. 2015. Essential oils used in aromatherapy: A systemic review. *Asian Pacific Journal of Tropical Biomedicine* 5(8): 601–611. http://dx.doi.org/10.1016/j.apjtb.2015.05.007

Alu'datt, M. H., Rababah, T., Alhamad, M. N. et al. 2018. Pharmaceutical, nutraceutical and therapeutic properties of selected wild medicinal plants: Thyme, spearmint, and rosemary. In: *Therapeutic, Probiotic, and Unconventional Foods*, eds. Alexandru Mihai Grumezescu, Alina Maria Holban, 275–290. Academic Press. https://doi.org/10.1016/B978-0-12-814625-5.00014-5

Asfaw, D. 2019. Review on use of bioactive compounds in some spices in food preservation. *Food Science and Quality Management* 87:1–16. http://doi.org/10.7176/FSQM

AyKee, L., Shori, A. B., Hj, B. and Ahmad, S. 2017. Bioactivity and health effects of *Mentha spicata*. *Integrative Food, Nutrition and Metabolism* 5:1–2. https://doi.org/10.15761/IFNM.1000203

Banu, K. S. and Catherine, L. 2015. General techniques involved in phytochemical analysis. *International Journal of Advanced Research in Chemical Science* 2(4):25–32.

Bayania, M., Ahmadi-hamedanib, M. and Javanc, A. J. 2017. Study of hypoglycemic, hypocholesterolemic and antioxidant activities of Iranian *Mentha spicata* leaves aqueous extract in diabetic rats. *Iranian Journal of Pharmaceutical Research* 16:75–82.

Biswas, N. N., Saha, S. and Ali, M. K. 2014. Antioxidant, antimicrobial, cytotoxic and analgesic activities of ethanolic extract of *Mentha arvensis* L. *Asian Pacific Journal of Tropical Biomedicine* 4(10):792–797. https://doi.org/10.12980/APJTB.4.2014C1298

Brahmi, F., Khodir, M., Mohamed, C. and Pierre, D. 2018. Chemical composition and biological activities of *Mentha* species. In: *Aromatic and Medicinal Plants – Back to Nature*, ed. A. Hany and H. El-Shemy, IntechOpen, http://doi.org/10.5772/67291 Available from: https://www.intechopen.com/chapters/54028

Brown, N., John, J. A. and Shahidi, F. 2019. Polyphenol composition and antioxidant potential of mint leaves. *Food Production, Processing and Nutrition* 1(1). https://doi.org/10.1186/s43014-019-0001-8

Chavez Gonzalez, M. L., Leonardo, S., Deepak, K. V. et al. 2020a. Conventional and emerging extraction processes of flavonoids. *Processes* 8(4):434. https://doi.org/10.3390/pr8040434

Chavez, M. L., Sepulveda, L. and Verma, D. et al. 2020b. Conventional and emerging extraction processes of flavonoids. *Processes* 8:1–29. https://doi.org/10.3390/pr8040434

Dai, J., Orsat, V., Raghavan, G. S. V. and Yaylayan, V. 2010. Investigation of various factors for the extraction of peppermint (*Mentha piperita* L.) leaves. *Journal of Food Engineering* 96(4): 540–543. https://doi.org/10.1016/j.jfoodeng.2009.08.037

Dogan, K. and Tornuk, F. 2019. Improvement of bioavailability of bioactive compounds of medicinal herbs by drying and fermentation with *Lactobacillus plantarum*. *Functional Foods in Health and Disease* 9(12):735–748.

Farid, O., El Haidani, A. and Eddouks, M. 2018. Antidiabetic effect of spearmint in Streptozotocin-induced diabetic rats. *Endocrine, Metabolic & Immune Disorders – Drug Targets* 18(6):581–589. https://doi.org/10.2174/1871530318666180517101708

Farid, O., Zeggwagh, N. A., Ouadi, F. E. and Eddouks, M. 2019. *Mentha pulegium* aqueous extract exhibits antidiabetic and hepatoprotective effects in Streptozotocin-induced diabetic rats. *Endocrine, Metabolic & Immune Disorders – Drug Targets* 19(3): 292–301. https://doi.org/10.2174/1871530318666181005102247

Farzaei, M. H., Bahramsoltani, R., Ghobadi, A., Farzaei, F., Najafi, F. 2017. Pharmacological activity of *Mentha longifolia* and its phytoconstituents. *Journal of Traditional Chinese Medicine* 37(5):710–720. https://doi.org/10.1016/S0254-6272(17)30327-8

Grulova, D., De Martino, L., Mancini, E. et al. 2016. Phytotoxic and antibacterial activity of essential oil of new peppermint cultivar. *Natural Product Communications* 11:1721–1724. https://doi.org/10.1177/1934578X1601101124

Hadi, M., Hameed, I. and Adnan, I. 2017. *Mentha pulegium*: Medicinal uses, anti-hepatic, antibacterial, antioxidant effect and analysis of bioactive natural compounds: A review. *Research Journal of Pharmacy and Technology* 10(10):1827–1831. https://doi.org/10.5958/0974-360X.2017.00648.5

IQWiG. 2006. InformedHealth.org [Internet]. Cologne, Germany: Institute for Quality and Efficiency in Health Care, What Is an inflammation? 2010, Nov 23 [Updated 2018, Feb 22]. https://www.ncbi.nlm.nih.gov/books/NBK279298/ (accessed on: 27 July 2021).

Jothy, S. L., Zakaria, Z., Chen, Y., Lau, Y. L., Latha, L. Y., and Sasidharan, S. 2011. Acute oral toxicity of methanolic seed extract of *Cassia fistula* in mice. *Molecules* 16:5268–5282. http://doi.org/10.3390/molecules16065268

Juana, F. L., Angel, P. A. J. and Manuel, V. M. 2012. Beneficial health effects of bioactive compounds present in spices and aromatic herbs. In: *Studies in Natural Products Chemistry*, ed. Atta-ur-Rahman, 115–134, Elsevier.

Kapp, K., Tonu, P., Anne, O. et al. 2020. Chemical composition and antibacterial effect of *Mentha* spp. grown in Estonia. *Natural Product Communications* 15(12):1–14. https://doi.org/10.1177/1934578X20977615

Kaur, N. and Ahmed, T. 2021. Bioactive secondary metabolites of medicinal and aromatic plants and their disease-fighting properties. In: *Medicinal and Aromatic Plants*, eds. T. Aftab, and K. R. Hakeem, 113–142. Springer Nature, Switzerland. https://doi.org/10.1007/978-3-030-58975-2_4

Kehili, S., Boukhatem, M. A., Belkadi, A., Boulaghmen, F., Ferhat, M. A. and Setzer, W. N. 2020. Spearmint (*Mentha spicata* L.) Essential oil from Tipaza (Algeria): In vivo anti-inflammatory and analgesic activities in experimental animal models. *Acta Pharmaceutica Hungarica* 90:15–26.

Kokkini, S., Karousou, R. and Hanlidou, E. 2003. HERBS: Herbs of the Labiatae. *Encyclopedia of Food Sciences and Nutrition*, 3082–3090. http://doi.org/10.1016/B0-12-227055-X/00593-9

Kumar, A., Baitha, U., Aggarwal, P., & Jamshed, N. 2016. A fatal case of menthol poisoning. *International Journal of Applied & Basic Medical Research* 6(2):137–139. https://doi.org/10.4103/2229-516X.179015

Makkar, M. K., Sharma, S., and Kaur, H. 2018. Evaluation of *Mentha arvensis* essential oil and its major constituents for fungitoxicity. *Journal of food science and technology* 55(9):3840–3844. https://doi.org/10.1007/s13197-018-3291-y

Mamadalieva, N. Z., Hussain, H. and Xiao, J. 2020. Recent advances in genus *Mentha*: Phytochemistry, antimicrobial effects, and food applications. *Food Frontiers* 1: 435– 458. https://doi.org/10.1002/fft2.53

Marwa, C., Fikri-Benbrahim, K., Ou-Yahia, D., and Farah, A. 2017. African peppermint (*Mentha piperita*) from Morocco: Chemical composition and antimicrobial properties of essential oil. *Journal of Advanced Pharmaceutical Technology & Research* 8(3):86–90. https://doi.org/10.4103/japtr.JAPTR_11_17

Masomeh, L., Narges, M., Hassan, R., Hadi, A. 2017. Peppermint and its functionality: A review. *Archives of Clinical Microbiology* 8(4):54. http://doi.org/10.4172/1989-8436.100054

Mejdoub, K., Benomari, F., Nassim, D. et al. 2019. Antifungal and insecticidal activities of essential oils of four *Mentha* species. *Jundishapur Journal of Natural Pharmaceutical Products* 14(1):e64165. http://doi.org/10.5812/jjnpp.64165

Mikaili, P., Mojaverrostami, S., Moloudizargari, M. and Aghajanshakeri, S. 2013. Pharmacological and therapeutic effects of *Mentha longifolia* L. and its main constituent, menthol. *Ancient Science of Life* 33(2):131–138. https://doi.org/10.4103/0257-7941.139059

Mogosan, C., Vostinaru, O., Oprean, R. et al. 2017. A comparative analysis of the chemical composition, anti-inflammatory, and antinociceptive effects of the essential oils from three species of *Mentha* cultivated in Romania. *Molecules* 22(2):263. https://doi.org/10.3390/molecules22020263

Mucciarelli, M., Camusso, W., Bertea, C. M., Bossi, S. and Maffei, M. 2001. Effect of (+)-pulegone and other oil components of *Mentha×piperita* on cucumber respiration. *Phytochemistry* 57:91–98.

Mussatto, S. I. 2015. Generating biomedical polyphenolic compounds from spent coffee or silverskin. In: *Coffee in Health and Disease Prevention*, ed. Victor R. Preedy, 93–106. Academic Press. https://doi.org/10.1016/B978-0-12-409517-5.00011-5

Nandhini, S., Roy, A. and Kumar, V. J. 2018. Plants with analgesic property – A review article. *Drug Invention Today* 10(5):3847–3850.

Panda, A., Parida, A. K. and Rangani, J. 2018. Advancement of metabolomics techniques and their applications in plant science: Current scenario and future prospective. In: *Plant Metabolites and Regulation under*

Environmental Stress, ed. P. Ahmad, A. A. Mohammad, P. S. Vijay, K. T. Durgesh, A. Pravej, N. A. Mohammed, 1–36. Academic Press, Elsevier. http://dx.doi.org/10.1016/B978-0-12-812689-9.00001-7

Park, Y. J., Thanislas, B. B., Sun, K. Y. et al. 2016. Composition of volatile compounds and in vitro antimicrobial activity of nine *Mentha* spp. *Springer Plus* 5:1628. https://doi.org/10.1186/s40064-016-3283-1

Patra, N. K. and Kumar, B. 2006. Spearmint. In: *Handbook of Herbs and Spices*, ed. K. V. Peter, Vol. 3, 502–519. Woodhead Publishing Series in Food Science, Technology and Nutrition: Woodhead Publishing. https://doi.org/10.1533/9781845691717.3.503

Patti, F., Palmioli, A., Vitalini, S. et al. 2020. Anticancer effects of wild mountain *Mentha longifolia* extract in adrenocortical tumor cell models. *Frontiers in Pharmacology* 10:1647. https://doi.org/10.3389/fphar.2019.01647

Pramila, D. M., Xavier, R., Marimuthu, K. et al. 2012. Phytochemical analysis and antimicrobial potential of methanolic leaf extract of peppermint (*Mentha piperita*: Lamiaceae). *Journal of Medicinal Plants Research* 6(2):331–335.

Prasanna, X. A., Sarojini, R., GowriThilagam, T. and Susila, K. M. S. 2018. Evaluation of analgesic activity of *Mentha piperita* (aqueous extract) in Swiss male albino mice. *International Journal of Scientific Research* 7(10):68–69.

Pushpangadan, P. and Tewari, S. K. 2006. Peppermint. In: *Handbook of Herbs and Spices*, ed. K. V. Peter, 460–481. Woodhead Publishing Series in Food Science, Technology and Nutrition: Woodhead Publishing. https://doi.org/10.1533/9781845691717.3.460

Radivojac, A., Bera, O., Zekovic, Z. et al. 2021. Extraction of peppermint essential oils and lipophilic compounds: Assessment of process kinetics and environmental impacts with multiple techniques. *Molecules* 26:2879. https://doi.org/10.3390/molecules26102879

Radwan, M. N., Morad, M. M., Ali, M. M. and Wasfy, K. I. 2020. Extraction of peppermint volatile oil using a simple constructed steam distillation system. *Plant Archives* 20(2):1487–1491.

Reddy, K. M. and Manoharbabu, S. 2019. Evaluation of antioxidant and antidiabetic activity of *Mentha arvensis* Linn. *World Journal of Pharmaceutical Sciences* 7(3):199–206.

Salehi B, Stojanovic-Radic, Z., Matejic, J. et al. 2018. Plants of genus *Mentha*: From farm to food factory. *Plants* 7(3):70. https://doi.org/10.3390/plants7030070.

Saric-Kundalic, B., Silvia, F., Christoph, D. et al. 2009. Multivariate numerical taxonomy of *Mentha* species, hybrids, varieties and cultivars. *Scientia Pharmaceutica* 77(4):851–876. https://doi.org/10.3797/scipharm.0905-10

Sarkic, A. and Stappen, I. 2017. Essential oils and their single compounds in cosmetics – A critical review. *Cosmetics* 5(11). https://doi.org/10.3390/cosmetics5010011

Selles, S. M. A., Kouidri, M., Bellik, Y. et al. 2018. Chemical composition, antioxidant and in vitro antibacterial activities of essential oils of *Mentha spicata* leaf from Tiaret area (Algeria). *Dhaka University Journal of Pharmaceutical Sciences* 17(1): 87–96.

Sharma, V., Hussain, S., Gupta, M. and Saxena, A. K. 2014. In vitro anticancer activity of extracts of *Mentha* spp. against human cancer cells. *Indian Journal of Biochemistry & Biophysics* 51:416–419.

Shimotori, Y., Watanabe, T., Kohari, Y. et al. 2020. Enzyme-assisted extraction of bioactive phytochemicals from Japanese Peppermint (*Mentha arvensis* L. cv. 'Hokuto'). *Journal of Oleo Science* 69(6): 635–642.

Singh, P., Andola, H. C., Rawat, M. S. M., Pant, G. J. N. and Purohit, V. K. 2011. Fourier transform infrared (FT-IR) spectroscopy in an overview. *Research Journal of Medicinal Plants* 5(2):127–135. https://doi.org/10.3923/rjmp.2011.127.135

Singh, P. and Pandey, A. K. 2018. Prospective of essential oils of the genus *Mentha* as biopesticides: A review. *Front Plant Science* 9:1295. http://doi.org/10.3389/fpls.2018.01295

Tafrihi, M., Imran, M., Tufail, T. et al. 2021. The wonderful activities of the genus Mentha: Not only antioxidant properties. *Molecules* 26:1118. https://doi.org/10.3390/

Tambe, N. A. and Tumane, P. M. 2017. GC-MS profile of *Punica granatum*, *Mentha piperita* and *Curcuma longa* leaves. *Journal of Pharmacy Research* 11(6):639–644.

Taneja, S. C. and Chandra, S. 2012. Mint. In: *Handbook of Herbs and Spices* (2nd ed.), ed. K. V. Peter, 366–387. Woodhead Publishing Series in Food Science, Technology and Nutrition: Woodhead Publishing.

Thawkar, B. S., Jawarkar, A. G., Kalamkar, P. V., Pawar, K. P. and Kale, M. K. 2018. Phytochemical and pharmacological review of *Mentha arvensis*. *International Journal of Green Pharmacy* 10(2):71–76.

Vieira, A. J., Beserra, F. P., Souza, M. C., Totti, B. M., Rozza, A. L. 2018. Limonene: Aroma of innovation in health and disease. *Chemico Biological Interactions* 283:97–106. https://doi.org/10.1016/j.cbi.2018.02.007

Yang, C., Wang, M., Zhou, J. and Chi, Q. 2017. Bio-synthesis of peppermint leaf extract polyphenols capped nano-platinum and their in-vitro cytotoxicity towards colon cancer cell lines (HCT 116). *Materials Science and Engineering C: Materials for Biological Applications* 77:1012–1016. https://doi.org/10.1016/j.msec.2017.04.020.

Yashin, A., Yashin, Y., Xia, X. and Nemzer, B. 2017. Antioxidant activity of spices and their impact on human health: A review. *Antioxidants* 6:70. https://doi.org/10.3390/antiox6030070

Zaidi, S. and Dahiya, P. 2015. In vitro antimicrobial activity, phytochemical analysis and total phenolic content of essential oil from *Mentha spicata* and *Mentha piperita*. *International Food Research Journal* 22(6):2440–2445.

CHAPTER 13

Bioactive Compounds in Fenugreek

Prajya Arya and Pradyuman Kumar

CONTENTS

13.1 INTRODUCTION

Fenugreek (*Trigonella foenum-graecum*) is an annual crop that falls under the Fabaceae family and is recognized as the queen of forage and sometimes called queen of forage. Its major recognition belongs to its medicinal, culinary, and aromatic properties since ancient times. The fenugreek seeds, leaves, and forage have their utilization potential; thus, these parts are utilized as spice, curry, and animal feed, respectively. Fenugreek popularity also belongs to the Indian Ayurvedic system, Tibetan, and Chinese Medicinal practices for curing various human and animal health aspects of several ages (Acharya et al. 2007; Dangi et al. 2016; Zandi et al. 2011). Concerning fenugreek growth pattern and areas, fenugreek is generally cultivated in South Asia, the Middle East, Mediterranean

DOI: 10.1201/9781003215387-13

Figure 13.1 Usage of fenugreek in various forms.

Europe, and North Africa as a spice crop and as a major culinary agent in the Indian subcontinent. In India, Rajasthan is the top fenugreek-producing state and contributes approximately 80% of fenugreek production followed by Gujarat, Madhya Pradesh, Uttar Pradesh, and Bihar (Mahendra and Ritu 2019). Currently, India is one of the major producers of fenugreek with a production of 185 thousand metric tons in 2020 (Indian Spice Board 2021). Fenugreek had major usage in Indian curry as leaves, mix powder, and spices that enhance the overall flavour of the entire cuisine and are practised widely over the globe as shown in Figure 13.1 (Basu et al. 2008a, 2008b).

The usage of fenugreek seeds was quite popular in treating the low blood sugar level, impotence, prolactin, gout, diarrhoea, abdominal distension, skin inflammations, and lack of appetite, and it also enhances the sensory properties of the food in which it has been incorporated (Basch et al. 2003). The medicinal property of fenugreek includes hypolipidemic, antioxidant, antifungal, antibacterial, anti-obesity, anti-diabetic, and anti-cancer (Arya and Kumar 2021b). Fenugreek medicinal benefits are also provided by a bioactive compound that aids in more relevant features regarding health benefits towards mankind's welfare. Many studies support nutraceutical (mechanism of 4-hydroxyleucine and fibre) (Fuller and Stephen 2015), medicinal (cellular and molecular mechanism for disease prevention and molecular mechanism) (Nagulapalli et al. 2017), and bioactive (saponin rich extracts) properties of fenugreek (Navarro et al. 2020). Fenugreek, which is majorly recognized for its medicinal properties, also had different names around the globe as shown in Table 13.1.

13.2 BOTANICAL ASPECTS OF FENUGREEK SEEDS

Fenugreek is an annual legume that belongs to Fabaceae family with diploid plants ($2n = 16$) and no aneuploidy (Flammang et al. 2004; Petropoulos 2002). The genus name *Trigonella* means "little triangle" that resembles the triangular shape of its small yellowish-white flower and species name *foenum-graecum* means "Greek hay" about its initial introgression from Greece (Basu et al. 2008a) and fenugreek classification is represented in Figure 13.2. The worldwide known spice fenugreek belongs to more than 260 varieties and most used are *Trigonella foenum-graecum*, *T. calliceras*,

Table 13.1 Different Names of Fenugreek around the Globe

Country	Name	Language	Reference
India	Methi	Hindi	Adil et al. (2015)
	Methika	Sanskrit	
	Methi	Urdu	
	Meth	Kashmiri	
	Uluva	Malayalam	
	Vendayam	Tamil	
	Mentikura	Telugu	
Persia	Shanbalileh	Persian	
Arab	Hulba, Hilbeh	Arabic	
Nepal	Menthiyam	Nepali	
Burma	Penantazi	Burmese	
Germany	Bocksshornklee	German	
France	Fenugrec	French	
Japan	Koruha	Japanese	
Russia	Pazhitnik	Russian	
Ethopia	Abish	Ethiopian	Zandi et al. (2017)
Uzbekistan	Boidana	Uzbekistani	
America	Shambala	American	
England	Fenigrec	English	
Italy	Fieno Greco	Italian	

T. caerulea, *T. spicata*, *T. balansae*, *T. spinosa*, *T. polycerata*, *T. lilacina*, *T. radiata*, *T. maritima*, and *T. occulta*. Out of these, *Trigonella foenum-graecum* is the most cultivated variety of fenugreek throughout the world (Petropoulos 2002).

Generally, the fenugreek plant is aromatic erect and resembles with large clover. The stem is pinkish and long cylindrical while; its roots give appearance similarity with finger-like structure (Basu et al. 2019; Mehrafarin et al. 2011; Moradi and Moradi 2013). The leaf of fenugreek has pinnate, trifoliate, stipules triangular, long-stalked compound leaves having toothed, lanceolate, and obovate-to-oblanceolate leaflets (Srinivasan 2006). The colour of the fenugreek flower is white to

Figure 13.2 Fenugreek botanical classifications.

yellowish-white, and it has five petals as banner, wing, and keel (Mehrafarin et al. 2011). Fenugreek pods are 15-cm long with 2–8 pods, 10–20 seeds in each pod, and brownish to yellowish-brown in colour. Fenugreek seeds are hard, small, smooth, and yellow to brownish-yellow in colour (Moradi and Moradi 2013).

Fenugreek is a fast-growing plant that can grow over drylands, hillsides, planes, and field edges but it also requires a fair amount of sunlight. It needs 4–7 months to reach maturity (Petropoulos 2002). Its flowering season is midsummer (June–August), and seeds ripen in late summer (August–September). Generally, fenugreek could be grown in a tropical climate with mild winter and cool summer while its leaves and flower are temperature-dependent (McCormick et al. 2006).

13.3 COMPOSITION OF FENUGREEK SEEDS

The fenugreek seeds are cluster hub of various nutritional rich compounds that aid in betterment of human health. The fenugreek seed composition revealed that seeds contain carbohydrate (58%), galactomannan (30%), fibre (20%), protein (23.04%), ash (3%), moisture content (3.8%), and fat (3–5%) (Munshi et al. 2020; Wani and Kumar 2018; Yao et al. 2020).

Table 13.2 depicts about studies of proximate and mineral content of fenugreek seeds conducted by different researchers around the globe. The proximate content of fenugreek seeds was varied according to region, origin, climacteric conditions, variety, and availability of type and condition of soil. This chapter would enlighten different ways to utilize the hidden potential of fenugreek seeds for combating various health aspects concerning the presence of various bioactive compounds, their extraction, and utilization in curing different health issues.

The fenugreek seed matrix component reveals about its characteristic properties. Galactomannan majorly contains soluble fibres. Fenugreek seeds produce viscous solution due to mannose and galactose (Nwokocha 2021). The fibre content is non-starch polysaccharide that is capable enough in moderating human glucose metabolism. It aids in reducing sugar level by controlling blood glucose absorption and brings insulin in action. Protein lies in endosperm of fenugreek seeds and mainly contains globulin, lecithin, and albumin. Fenugreek anti-diabetic properties are linked to 4-hydroxyisoleucine and histidine that cause trigger of insulin activity (Isikli and Karababa 2005).

Table 13.2 Proximate and Mineral Content of Fenugreek Seeds

Component (%)	Yasothai (2021)	Munshi et al. (2020)	Dhawi et al. (2020)	Aljuhaimi et al. (2018)	Ali et al. (2015)
Moisture	–	3.8	5.3	6.44	9.3
Ash	3.19	9	3.4	3.93	3.1
Protein	26.65	23.04	30.7	29.09	24.9
Fat	–	4.75	4.4	6.28	5.2
Fibre	7.54	25.47	7.70	9.42	7.1
Minerals	**Yasothai (2021)**	**Pająk et al. (2019) (mg/100 g)**	**Yasothai (2018)**	**Ahmad et al. (2016) (mg/100 g)**	**–**
Calcium (%)	0.33	264	0.23	75	–
Phosphorus (%)	0.40	–	0.21	–	–
Sodium (%)	0.14	23.71	0.29	–	–
Copper (mg/kg)	13.98	1.19	54	0.9	–
Manganese (mg/kg)	9.69	162	16	0.9	–
Zinc (mg/kg)	39.94	3.82	44	2.4	–
Iron (mg/kg)	30.96	2.45	116	25.8	–

Table 13.3 Fatty Acid Profiling of Fenugreek Seed Oil by Hexane Containing Major Fatty Acids

	Fenugreek Seed Oil			
Fatty Acid Composition (%)	**Munshi et al. (2020)**	**Aljuhaimi et al. (2018)**	**Hamden et al. (2017)**	**Gu et al. (2017)**
Palmitic acid	9.89	10.64	9.76	10.00
Stearic acid	4.08	3.81	3.90	4.71
Cis-9-oleic acid	12.95	13.45	17.06	14.24
Linoleic acid	41.97	40.93	41.13	42.80
Alpha-linolenic acid	29.33	24.06	26.14	26.15
Arachidic acid	1.21	1.03	0.98	1.33

13.4 FENUGREEK SEED OIL

The fenugreek seed oil is a rich source of many fatty acids as studied by Munshi et al. (2020). Fenugreek seeds after defatted by hexane had saturated fatty acid (SFA: 15.75), monounsaturated fatty acid (MUFA: 12.95), and polyunsaturated fatty acid (PUFA: 71.30). Fenugreek seed contains neutral lipids, phospholipids, and glycolipids (85%, 10%, and 5%), respectively. Fenugreek seed oil also contains unsaturated lipids like linoleic acid, linolenic acid, and oleic acid (40%, 25%, and 14%) (Chatterjee et al. 2010). There are some fatty acids present in fenugreek seed oil as shown in Table 13.3. The presence of various chemical compounds was studied by Hamden et al. (2011) as represented in Table 13.4 for fenugreek seed oil by gas chromatography technique.

Table 13.4 Presence of Different Compounds and Their Composition in Fenugreek Seed Oil

Compound	Composition (%)	Technique Used	Reference
β-Pinene	15.05	Gas chromatography-mass spectrophotometry (GC-MS)	Hamden et al. (2011)
α-Pinene	2.61		
2,5-Dimethylpyrazine	6.14		
6-Methyl-5-hepten-2-one	4.48		
ϒ-Terpinene	2.08		
Camphor 16.32	16.32		
3-Octen-2-one	4.32		
α-Campholenal	2.63		
β-Caryophyllene	14.63		
α-Terpineol	2.77		
Neryl acetate	17.32		
α-Selinene	4.04		
Geranial	4.81		
4-Hydroxyisoleucine	4.5	High-pressure liquid chromatography-mass spectrophotometry-ultraviolet HPLC-MS-UV	Keskes et al. (2018)
Vicenin-2	18.2		
Isoschaftoside	18.6		
Isoorientin	18.7		
2-Methylpyrrolidine	0.42	Gas chromatography-mass spectrophotometry (GC-MS)	Akbari et al. (2019a)
Caryophyllene	0.32		
Trichloroacetic acid, pentadecyl ester	1.45		
Eicosane	1.61		
Quinoline	0.75		

Fenugreek seeds are used as a flavour enhancer in various cuisines due to its aromatic nature. The aroma in fenugreek seeds contributed by mien of volatile oils. A study was conducted by Murlidhar and Goswami (2012) for studying volatile compounds of fenugreek seed oil by gas chromatography and it revealed that butanoic acid, 3-isobutyl-2-methoxypyrazine, 1-octene-3-one. The olfactometry was used for the following compounds: acetic acid, diacetyl, linalool, caproic acid, and isovaleric acid. The specific metallic odour of fenugreek seeds is due to predominant substance sotolon (Ahmad et al. 2016).

13.5 BIOACTIVE COMPOUNDS IN FENUGREEK SEEDS

The fenugreek seeds are a very good source of bioactive compounds. It generally includes fibre, protein, gum, vitamin, and minerals along with steroidal sapogenin, diosgenin, trigocoumarin, trigonelline, nicotinic acid, trimethyl coumarin (Ahmad et al. 2016; Kharc 2004; Naidu et al. 2011; Sarwar et al. 2020), polyphenols, flavonoids, and alkaloids, and in the category of volatile compound, it contains stolone and anethole (Kousalya et al. 2019). Fenugreek seeds are also rich in caffeic acid and kaempferol (Benayad et al. 2014), alkaloid, galactomannan, and 4-hydroxyisoleucine (Bano et al. 2016) as shown in Table 13.5. In fenugreek seed matrices, there is presence of some major bioactive compounds depicted in Figure 13.3 such as yamogenin-based saponin (4.8%), volatile oil (1.24%), diosgenin (0.6–1.7%), trigonelline (0.2–0.38%), and C-glycosyl flavones of apigenin and luteolin (0.10%) (Yao et al. 2020). Along with this saponin, there are small amount of quercetin, vicenin, vitexincinnamate, and isovitexin (El Bairi et al. 2017).

The fenugreek seeds are rich source of polyphenols like stilbenes (desoxyrhaponticin, rhapontigenin and rhaponticin) that could be separated by modern analytical techniques (He et al. 2016). Polyphenols are compounds composed of a group of substances with various chemical structures and activities that are broadly present in nature (Pereira et al. 2009). The health-promising benefits of fenugreek and polyphenol attracted many researchers towards plant phenols like stilbenes. Resveratrol (3,5,4-trihydroxy-trans-stilbene) is one of the most popular polyphenols that work on obesity, cardiovascular preventive agents, anti-diabetic, and neuroprotective properties (Reinisalo et al. 2015).

The bioactive potential of fenugreek also cures the damage caused by day-to-day life stress. It could be rectified by the consumption of an appropriate amount of phenolic and antioxidant compounds that cause a delay in the oxidation of food as they eliminate oxygen and nitrogen reactive species that affect the DNA, protein, and lipid of the living system (Amarowicz and Fereidoon 2017). The relation of phenolic compounds concerning benefiting human health is linked to the degree of glycosylation, number, and position of the hydroxyl group along with the phenolic compound solubility, metabolism, and absorption (Singh et al. 2017).

In the case of steroidal compounds in fenugreek, diosgenin is one of the most natural bioactive compounds. It is spirostanol saponins comprising a hydrophobic steroid aglycone which is linked

Table 13.5 Fenugreek Seeds Contain Important Chemical Constituents

Type of Chemical Constituent	Chemical Constituent	Reference
Saponin	Fenugreekine, trigofoenosides A–G, fenugrin B	Kousalya et al. (2019)
Steroidal saponin	Diosgenin, apigenin, tigogenin, yamogenin, gitogenin, neogetigenin, smilagenin, sarsasapogenin, yuccagenin, hederagin	Ahmad et al. (2016)
Flavonoids	Luteloin, tricine, calycosin, orientin, quercetin, glycopyronaside, tricin-7-O-D, lilyn, kaempferol	Blumenthal et al. (2000)
Alkaloids	Trigonelline, choline, gentianine, carpaine	Rababah et al. (2011)

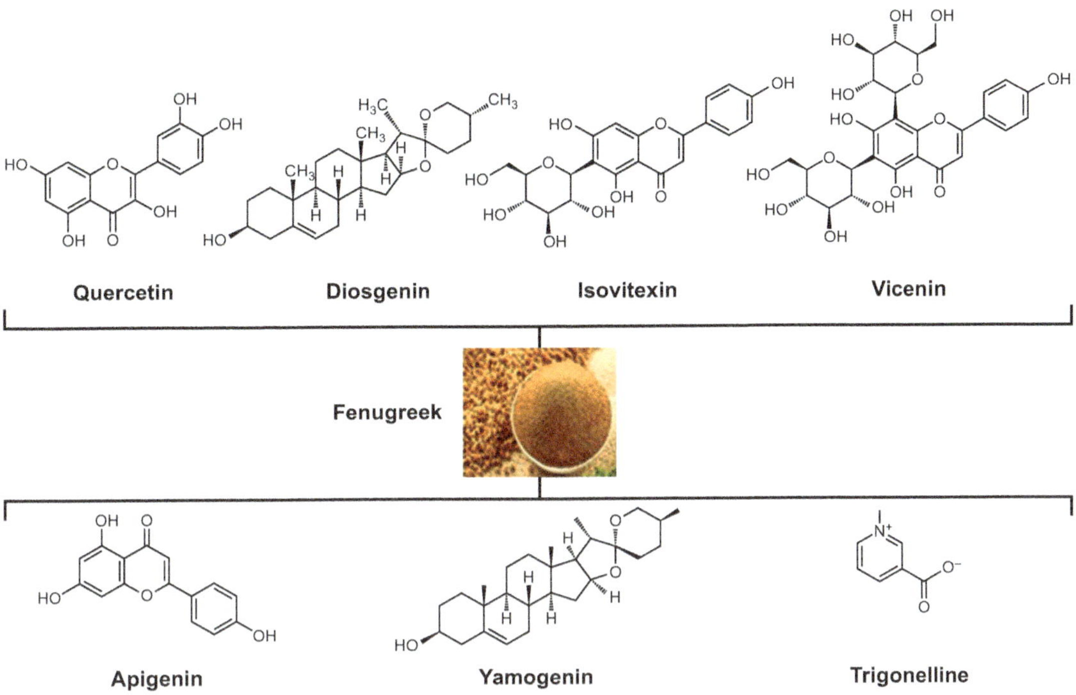

Figure 13.3 Presence of different chemical components in fenugreek seeds.

to the hydrophilic sugar moiety sapogenin that is present in fenugreek seeds. It is a major precursor used to produce synthetic steroids in the pharmaceutical industry (Raju and Chinthalapally 2012). Fenugreek seeds are recognized as a potential source of diosgenin which is a basic compound in the synthesis of the steroid drugs like cortisone and sexual hormones (Marker et al. 1947).

Steroidal sapogenin is a class of secondary metabolites that belong to the biosynthetic forerunners of sterols especially cholesterol (Dangi et al. 2016) and are found as glycosides that are known as steroidal sapogenins. Fenugreek seed diosgenin (25R-spirost-en-3b-ol) is a plant-derived steroidal sapogenin that is extensively used to produce sex hormones, steroidal drugs, and oral contraceptives due to its wide pharmaceutical properties. Diosgenin has potential against various day-to-day life disorders like cardiovascular diseases (CVD), cancer, skin irritation, oxidative stress, obesity, hyperlipidaemia, osteoporosis, and neurological disorder (Cai et al. 2020; Deshpande and Sanjivani 2014). The low solubility and less bioavailability of diosgenin made it difficult to delve into the medicinal and health promising properties of diosgenin (Cai et al. 2020).

13.6 EXTRACTION METHODS OF BIOACTIVE COMPOUNDS FROM FENUGREEK

13.6.1 Extraction of Bioactive Compound from Fenugreek

The extraction of bioactive compounds from any food matrix depends on many factors like food matrix, a technique used, extraction solvents, and modification of treatment methods (Tiwari 2015). The concept of extraction always lies between distributions of a component into two immiscible phases that aid in the separation and recovery of bioactive compounds. The extraction depends on the concentration of the solvent and solute used in the two phases. The analyte will be distributed among phases depending on temperature, the relative volume of phase, and distribution constant. It also depends on the migration kinetics, diffusion rate in two phases, and temperature (Smith 2003).

Generally, the extraction method is carved up into two methods; first is the conventional method and another one is a non-conventional method. The conventional techniques include the usage of agitation, organic solvents, temperature, and huge time consumption such as maceration, hydrodistillation, and Soxhlet extraction. In the case of non-conventional technique which is also called modern techniques, clean and green techniques, as they aid in benefiting the environment by leaving behind very less amount of residues and also require very less amount of energy as compared to the conventional techniques (Rodríguez-Pérez et al. 2015).

13.6.2 Conventional Techniques

Various studies had been conducted related to the extraction of a bioactive compound by the Soxhlet extraction method as they delve into a small sample matrix easily and require a particular set of assembly (Cravotto et al. 2011; Xhaxhiu et al. 2013). This technique is also time and energy consuming, and it also requires a large amount of solvent for extracting bioactive compounds from samples (Heleno et al. 2016). Soxhlet extraction method is majorly used for extraction of bioactive compounds with various organic solvents but it is generally harmful to the environment by releasing harmful solvent and also reduce the extract yield. Soxhlet extraction is extensively used in laboratories due to easy method of operation, has ability of multiple extraction and filtration without any aid (Siddique et al. 2020).

13.6.2.1 Maceration

Maceration is the process of breaking the sample matrix into finer particles and allowing it to react with the desired solvent for a fixed duration of time. The grinding of the sample aids in enhancing the surface area of the sample that makes extraction much easier. In maceration process, agitation helps in increasing the diffusion of the solvent into sample matrix and maintains the concentration of the solution over the sample surface (Azmir et al. 2013). The processes were performed to make the interaction more between sample matrix and solvent to extract a greater number of polyphenols, bioactive compounds (Azwanida 2015).

13.6.2.2 Hydrodistillation

Hydrodistillation is a convenient extraction method of extraction of bioactive compounds from a sample matrix. In this method, samples are packed in an allay compartment with boiling water for a fixed duration of time or direct steam could be applied to the sample matrix. The separation was carried through a mixture of oil and bioactive compounds which were cooled indirectly. After the cooling of the sample, both bioactive compound and oil get separated from the sample matrix. This method has some pros and cons such as no role of organic solvents that interact with sample matrix and no need for dehydration of the sample and it's a quick process, while it may also degrade the yield and quality of polyphenolic compounds due to the high temperature of the water during extraction (Ouzzar et al. 2015).

13.6.3 Modern Techniques

13.6.3.1 Supercritical Fluid Extraction

This method is based on the supercritical fluid which acts on the sample matrix instead of the usual liquid solvent. The supercritical fluid shows physical properties such as density, diffusivity, and viscosity (Phelps et al. 1996). Supercritical fluid extraction (SFE) is the method of extracting bioactive compounds with aid of fluid in their supercritical discipline of temperature and pressure which acquire the features of both gasses and liquids. SFE is based on the concept of mass transfer caused due to high diffusion coefficient and low viscosity (Duarte et al. 2014). It is a green technology that works with high pressure for separating the solute from sample matrix by bringing mixture in contact with fluid in a supercritical state for extraction of bioactive compounds. The plebeian solvent used in SFE for extraction purposes is carbon dioxide. CO_2 is of inert nature which enhances extraction of a bioactive compound at an economic rate and its utilization in the food, pharmaceutical, and cosmetics industries (Jesus and Meireles 2014). The bioactive compounds extracted from SFE are rich in a group of compounds like phytosterol, phenolics, carotenoids, triglycerides, and fatty acids, and no such degradative changes occurred during extraction (del Pilar et al. 2017).

13.6.3.2 Ultrasound-Assisted Extraction

Ultrasound-assisted extraction (UAE) is based on frequency (up to 20 kHz) that creates local hotspots at macroscopic scale with high temperature and shear stress which generate cavitational bubbles (Chemat et al. 2017). In case of UAE, bubble formation is accomplished due to compression and rarefaction induced by cavitation force in solvent molecule owing to change in temperature and pressure (Shirsath et al. 2012). The fragmentation of particles depends upon a collision between particle and ultrasonic waves which is caused by mass transfer and aids in the reduction of particle size. As particles get damaged, the solvent could easily interact with sample matrix by imploding bubble at the surface. Further, sonoporation and sono-capillarity can enhance alteration of the permeability of cell membrane and penetration of liquid via channel produced by bubble implosion respectively. This finally combines with the shear stress mechanism and collapse the cavitation bubble into the fluid which occurred due to oscillation phenomenon (Chemat et al. 2017). UAE method can be performed at room temperature so that it could aid in the prevention of degradation of bioactive compounds. The UAE method is gaining more importance due to its less environmental impact, simplicity, and use of less amount of solvent and can be easily conjugated with other techniques (Louie et al. 2020). Some important factors could affect the UAE method like the type of solvent, solvent composition, input power, frequency used, and extraction time. It is useful in extracting the bioactive compounds like diosgenin from fenugreek (Wani and Kumar 2018), phenolic, anthocyanin, and antioxidants from grapes (Iskalieva et al. 2012), and beta-carotene, fatty acids, and carotenoid from microalga (Cardoso et al. 2012). The extraction of fenugreek diosgenin was carried by the UAE method by Arya and Kumar (2021a) for analysing the effect of variation in solvent concentration, treatment time, and type of solvent. The study of Arya and Kumar (2021a) revealed about UAE of fenugreek diosgenin and extract yield as 40.37 mg/100 g and 21.48%, respectively. This study resulted in much better extraction of fenugreek bioactive compounds diosgenin as compared with MAE.

13.6.3.3 Microwave-Assisted Extraction

Microwave-assisted extraction functions over concept of electromagnetic waves that lies in frequency range of 0.3 to 300 GHz that corresponds to the wavelength of 1 m to 1 cm. In this setting, heat penetrates from inside to outside of the sample. In microwave-assisted extraction heating of sample matrix initiated by dipole rotation and ionic conduction of microwave energy that generate friction followed by resistance caused by solvent in the flow of ions towards sample matrix. Dipole rotation crate forced molecular movement that creates friction and causes heat generation in sample matrix (Chemat and Giancarlo 2012). As per recent grown technology, MAE has attained a big market area in the sector of extraction of bioactive compounds. During microwave radiation sample matrix breaks downs which causes evaporation of residual water and followed by more extraction

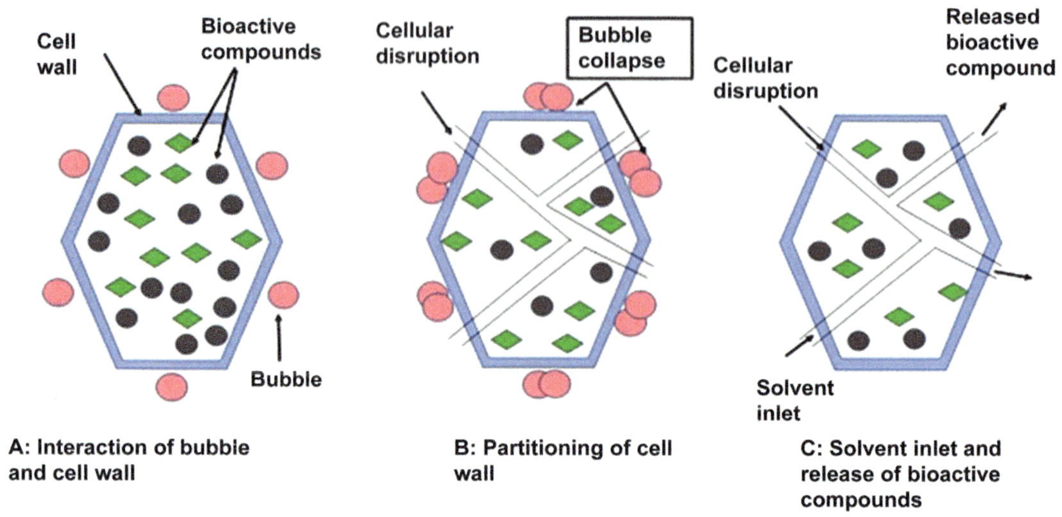

Figure 13.4 Representation of breakdown of cellular matrix and release of bioactive compound by microwave assisted extraction method.

by internal diffusion (Pimentel-Moral et al. 2018) as shown in Figure 13.4. In this method, there is no such loss of heat energy due to the execution of extraction occurred in a closed selective environment. (Reddy et al. 2020). In case of dealing with bioactive compound extraction with MAE method, it comprises three steps: (a) the come-off bioactive compounds from the active site of sample matrix due to high temperature and pressure, (b) sample matrix react with solvent via diffusion and interaction of solvent and bioactive compound occurs, and (c) release of bioactive compound in the solvent (Reddy et al. 2020). The study of Arya and Kumar (2021a) revealed about MAE of fenugreek diosgenin and extract yield as 35.50 mg/100 g and 7.83%, respectively. The MAE method produces much better recoveries of the bioactive compound and utilizes a smaller amount of solvent. The MAE procedure extracts the bioactive flavonoids with better recoveries at rapid extraction times than the conventional extraction procedures (Xiao et al. 2008). The extraction with MAE is quite cheap as its apparatus and installation cost is low and could be used in the extraction of various constituents of bioactive origin from the sample matrix (Atabani et al. 2012).

13.6.3.4 Accelerated Solvent Extraction

This method is also known as pressurized solvent extraction. It is based on the extraction of bioactive compounds at higher temperatures (50–200°C) and pressure (100–140 atm). The equipment is designed in such a way to withhold high pressure helps in extraction temperature to be raised above the boiling point of solvent utilized and high pressure maintains the solvent in the liquid stage at high temperature (Mandal et al. 2015). This aids in the quick extraction of bioactive compounds due to low viscosity, high strength of solvent and high diffusion coefficients. In the accelerated solvent extraction (ASE) method, pressure allows solvent to enter forcefully into a sample matrix and it also assists in favour desorption of analyte and good dissolution processes from a cellular matrix. Due to increased temperature at extraction time, there is an increased diffusivity of solvent that enhances the release kinetics of the bioactive compound. In ASE, sample matrix is kept in extraction cell followed by addition of solvent; a further cell is pressurized and heated up to the desired temperature and finally; a sample is extracted statically for a specific period. Ahead extract is collected; cell material is removed and washed with fresh solvent. On completion of extraction, compressed nitrogen removes all the solvent from the cell vial for further analysis. Finally, the extract is filtered before storage (Mandal et al. 2015).

13.6.3.5 Pulsed Electric Field Extraction

Pulsed electric field extraction (PEF) is a nonthermal extraction method by applying direct high voltage current pulses for a short interval of time (microsecond to the millisecond). The sample is kept between two electrodes for extracting bioactive compounds from a matrix. Due to this intermittent electric supply, the electric field gets generated, and the intensity of an electric field depends on the delivered voltage and distance between electrodes. When the electric field is strong enough, it creates pores in a cell wall of the sample matrix and a phenomenon called electroporation. This causes exchange of ions and macromolecule through cytoplasmic membrane of a sample by increasing the permeability (Martínez et al. 2020). PEF is generally affected by treatment time, specific energy, temperature of application, and electric field strength. Generally, enhancing the electric field strength and treatment time increases the release of bioactive compounds from sample matrix by heightening the electroporation but PEF is an irreversible process as the cell membrane gets ruptured entirely. Various studies supported the use of PEF for the extraction of bioactive compounds by different food matrices like cold pasteurization of fruit juices and fried potato processing (Barba et al. 2015).

13.6.3.6 Ionic Liquid Extraction

Ionic liquid extraction (ILE) method is ionic liquid interaction based which involves environment-friendly solvents, thus prove to be a substitute for the volatile organic solvents. Ionic liquids consist of ions of organic salts which have melting points equal to or lower than 100°C. Approximately 1018 ionic liquids are recognized till date; out of them mostly habituated are alkyl hydrazine, alkylammonium, and N-alkyl pyridinium (Khezeli et al. 2020). ILE extraction method is quick, high extraction efficiencies and it could be easily combined with solid–liquid extraction to boost the recovery rate for value-added constituent's biomass. The ILE method could be applied for extraction of compounds like flavonoids, phenolic compounds, saponins, alkaloids, bioactive compounds, and polycyclic aromatic compounds (Ventura et al. 2017). The different combine extraction techniques for fenugreek bioactive compounds are represented in Figure 13.5.

The recently used techniques for fenugreek extraction bioactive compounds are ultrasound-assisted, microwave-assisted, ionic liquids, and high-voltage extraction. These extraction techniques could be coupled with different techniques to make extraction much easier, better, and convenient. The extraction of bioactive compounds of fenugreek seeds and leaves from different techniques had been presented in Table 13.6.

13.7 MEDICINAL PROPERTIES OF FENUGREEK

The medicinal properties of fenugreek are quite popular from ancient times. Fenugreek medicinal properties emerge from its matrix and its related components like fibre, 4-hydroxyisoleucine, lipids, antioxidants, and bioactive compounds (Lea 2013).

It majorly includes health-promising properties towards diabetes, obesity, anti-ulcer, antibacterial, anthelmintic, rheumatoid arthritis, CVD, and anti-cancer (Adedapo et al. 2014). The major functioning of endothelial dysfunction such as blockage and toughening of arteries generate problem in normal bodily function and fenugreek seeds are having an important role in limiting cardiovascular troubles (Pal and Souvik 2020). Diabetes is one of the major deadly metabolic diseases with no cure to date. It has the highest rate of mortality with approximately 422 million people

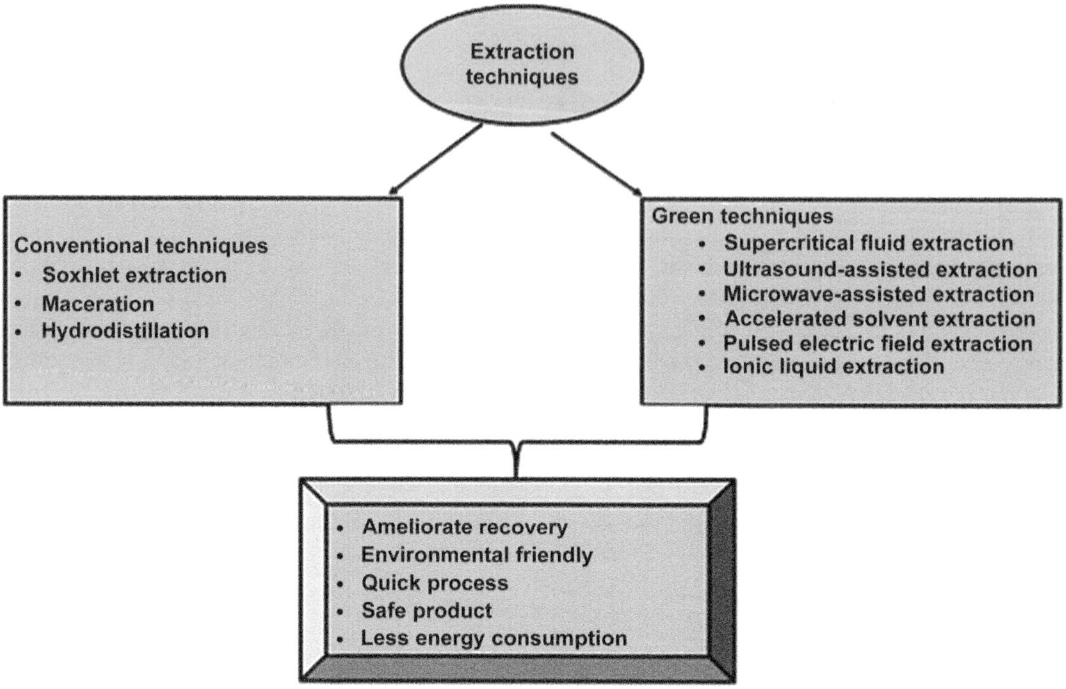

Figure 13.5 Representation of conventional and modern extraction techniques for bioactive compounds.

suffering from diabetes and 1.5 million deaths occurred due to diabetes in 2014. The blood glucose level affecting diabetes is represented in Figure 13.6. CVD cause approximately 80% mortality in diabetic patients by complications like cardiomyopathy, ischemic heart diseases, and myocardial infarction. These problems occurred by functional and structural abnormalities like mitochondrial dysfunction systolic and diastolic dysfunction, cardiac fibrosis, and cardiomyocyte hypertrophy. These types of issues are related to malfunctioning of angiotensin-converting enzyme (ACE) and receptor AT1 of the renin-angiotensin system (RAS) and it is associated with initiation and progression of diabetes and CVD (Pradeep and Krishnapura 2018). The fenugreek seeds fibre which aids in controlling diabetes by assuring improved glycaemic control, lower plasma lipid concentration and reduce hyperinsulinemia in type-2 diabetes (Chandalia et al. 2000). The fibre content of fenugreek seeds might reduce the rate of carbohydrate absorption and gastric void causing better glucose tolerance as observed in type-1 and type-2 diabetic rats (Hannan et al. 2007). Fenugreek galactomannan fibre content maintains low uptake of glucose from the jejunum and ileum segment as observed in obese and lean rats (Srichamroen et al. 2009). The CVD could be prevented by utilizing the dietary fibre potential of fenugreek seeds by blocking the RAS and it is accomplished by a down-regulated extracellular matrix. These metabolic activities aid in lowering oxidative stress by enhancing antioxidant levels in a cellular matrix and marker enzymes of cardiac damage could be restored by the dietary intervention of fenugreek. Hence, fenugreek could help in maintaining diabetes and CVD at a cellular level by altering the molecular functioning of cells (Pradeep and Krishnapura 2018).

The fenugreek seeds are rich in antioxidant potential that assists in reducing oxidative stress with help of soluble portions associated with antioxidant properties. The relation of fenugreek usage in diabetic animals was studied and it revealed the reversed commoved antioxidant level and actions of antioxidant enzymes which could be further used in the treatment of diabetic persons (Srinivasan 2019). The antioxidant potential of fenugreek has a beneficial effect on the liver and pancreas when

Table 13.6 Various Techniques Used in Extraction of Bioactive Compounds from Fenugreek

Sample	Extraction Method	Parameters	Response	References
Fenugreek seed	MAE	Treatment time: 1.5, 3.0, 4.5, 6.0 min Solvent type: acetone, ethanol, hexane, petroleum ether Solvent concentration: 40%, 60%, 80%, 100%	Extract yield: 7.83% Diosgenin content: 35.50 mg/100 g 6 min 80% ethanol concentration	Arya and Kumar (2021a)
Fenugreek seed	UAE	Treatment time: 30, 40, 50, 60 min Solvent type: acetone, ethanol, hexane, petroleum ether Solvent concentration: 40%, 60%, 80%, 100%	Extract yield: 21.48% Diosgenin content: 40.37 mg/100 g 60 min 80% ethanol concentration	Arya and Kumar (2021a)
Fenugreek seed	MAE	Extraction time: 2–12 min Ethanol concentration: 20–100% Microwave power: 300–700 W Ratio of feed to solvent: 1:8–1:16 Extraction Temperature: 40–80°C	Maximum yield: 3 min Microwave power: 600 W Ethanol concentration: 60% Feed-to-solvent ratio: 1:10 g/ml Extraction temperature: 70°C	Akbari et al. (2019b)
Fenugreek Seed	UAE	Ethanol, ethanol: water, water: 1/1 (v/v) Matric: solvent: 1:10 (w/v) Sonication amplitude: 60% Extraction temp: 75°C Saponin content Fat content Phenolic content	Maximum Extraction yield: Water: 17.16 g/100 g Saponin content: Ethanol: 12.90 g/100 g Fat content: Ethanol: 41.51 g/100 g Phenolic content: Ethanol:water: 7.14 g/100 g	Del Hierro et al. (2018)
Fenugreek seed	UAE	Extraction time: 40, 50, and 60 min Ethanol concentration: 20–100%	Maximum extract yield: 1.35 g/100 g Ethanol concentration: 80% Treatment time: 50 min	Wani et al. (2016)
Fenugreek Seed	MAE	Microwave power: Extraction time: 2, 4, 6 min Microwave power: 200 W Ethanol concentration: 20–100%	Maximum extract yield: 0.546 g/100 g Ethanol concentration: 80% Treatment time: 4 min	Wani et al. (2016)
Fenugreek seed	HVACP (high voltage atmospheric cold plasma)	Voltage: 80 kV Time: 30 min Galactomannan content	Galactomannan: extraction yield: +ce by 122%, water-binding capacity (high), swelling index (high) and viscosity (high), and melting enthalpy (low)	Rashid et al. (2020)
Fenugreek seed	HPLC	Diosgenin Limit of detection (LOD) Limit of quantification (LOQ)	Diosgenin yield and quantification LOD:0.0312 µg LOQ: 0.102 µg Quick, accurate, low cost diosgenin estimation	Chaudhary et al. (2018)
Fenugreek leaves	Ionic liquid-based ultrasonic-assisted extraction (ILUAE)	Anionic and cationic agents Five different ionic liquids	High diosgenin yield (0.77–10.24 mg/g) Five times more recovery than old methods [BMIM]HSO_4 best ionic liquid	Aminkara et al. (2018)
Fenugreek seed	UAE	Solvent concentration: 80% Time: 5 and 6 min	Maximum diosgenin yield: 13.78 g/100 g 80% 5 min	Dsouza et al. (2018)
Fenugreek seed	MAE	Solvent concentration: 80% Power: 200 W Time: 50 and 80 s	Maximum diosgenin yield: 1.49 g/100 g 80% 80 s	Dsouza et al. (2018)

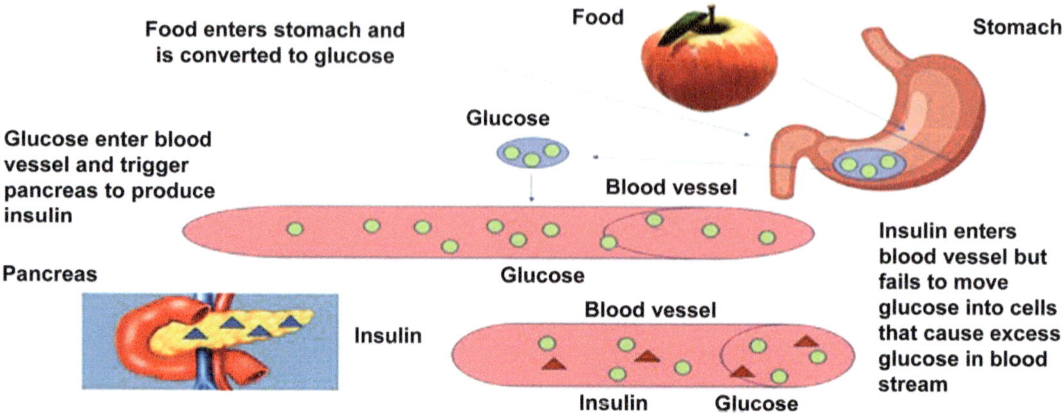

Figure 13.6 Representation of excess glucose in blood stream causes diabetes.

soaked as compared to dried seeds due to more availability of the fenugreek constituents. The more antioxidant availability was due to the presence of polyphenols and flavonoids (Murlidhar, and Goswami 2012).

Along with diabetes, fenugreek medicinal properties also cure rheumatoid arthritis, which affects joints by chronic inflammation that hampers synovial proliferation and tissue injury. The body tissue consumes more energy at the injured site and releases pro-inflammatory cytokines like tumour necrosis factor-alpha (TNF-α) and interleukin (Il-6) that caused painful movement of joints and increased soreness. The fenugreek mucilage proves to help control rheumatoid arthritis in rats by measuring paw volume on the 7th, 14th, and 21st day of conducted trial. In blood and tissue samples of rats, the analysis was conducted for inflammation caused by lipo-oxygenase, cyclo-oxygenase, arthritic ratio, rheumatoid aspect, and cytokine level (IL-6 and TNF-α). This study revealed that fenugreek mucilage helps prevent oedema in rats by giving only 75 mg/kg at dosage level (Sindhu et al. 2018). The synovial fluid structure and function showed little inflammation with synoviocytes (mesothelial cells) when supplemented with fenugreek to normalize the characteristics of these cells. Hence, fenugreek mucilage aid in suppressing oedema by reducing the activities of inflammatory enzymes due to its anti-arthritic potential (Syed et al. 2020).

Generally, fenugreek seeds are a rich source of nitrogen-containing amino acids like phenylalanine, leucine, glutamic acid, tyrosine, and aspartic acid, while sulphur-containing amino acids are quite less like cysteine and methionine (Habtemariam 2019). The fenugreek seeds are ample in 4-hydroxyisoleucine which is a nonproteinogenic amino acid that is responsible for the antidiabetic property of fenugreek seeds and can transform into a noneffective form under an acidic environment (Rangari et al. 2015). Fenugreek seeds comprise 80% of total free amino acids from 4-hydroxyisoleucine. 4-Hydroxyisoleucine contains three chiral centres and can exist in numerous enantiomeric forms but only two forms occur in fenugreek: (a) stereoisomer (2S, 3R, 4S, 90%) and (b) stereoisomeric (2R, 3R, 4S, 10%) form (Alcock et al. 1989). Studies revealed that fenugreek seed 4-hydroxyleucine aids in lowering fasting blood glucose levels in diabetic rats (Mawla and Eldien 2011).

Fenugreek role in controlling weight management is quite popular due to its galactomannan content which makes a viscous gel in the intestine and inhibits the absorption of lipid and glucose. Fenugreek consumption also led to a decrease in the production of triiodothyronine (T_3) which causes weight reduction (Syed et al. 2020). A study conducted by Kumar et al. (2014) for analysing anti-obesity activity of fenugreek seeds. It was observed that water extract of fenugreek seeds aids in lowering the digestion and absorption of fat by enhancing the glucose. This study also depicted about

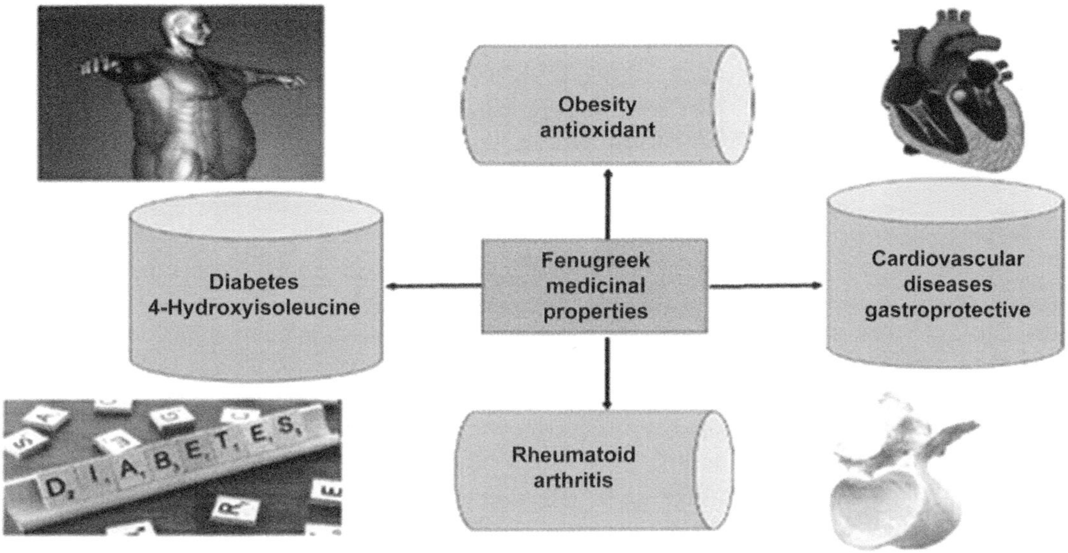

Figure 13.7 Medicinal properties of fenugreek in respect of metabolic diseases.

fat metabolism, improving insulin sensitivity, antioxidant capacity, and reduced dietary fat consumption and reducing the lipase activity in rats. Diosgenin from fenugreek seeds is helpful in controlling obesity as it raises the mRNA expression of adipose tissue at a cellular level for hepatic lipid metabolism (Fuller and Stephen 2015). Hence, fenugreek seed's galactomannan and diosgenin assist in weight management that prevents various serious forthcoming health issues in a living system.

Along with these medicinal properties, fenugreek is a promising agent in curing cancer and can be a protective complementary therapy for cancer patients for chemotherapeutic interventions as fenugreek extract revealed that shielding effect by altering cyclophosphamide caused apoptosis and free radical-mediated lipid peroxidation in the urinary bladder of mice (Bhatia et al. 2006). The fenugreek bioactive compound steroidal sapogenin is known as diosgenin that is potentially used in cancer treatments. It aids in treatment by preventing invasion, osteoclastogenesis, and suppressing proliferation via necrosis factor NF-kappa and β-regulated gene expression and enhances programmed cell death by inducing chemotherapeutic agents and cytokines (Murlidhar and Goswami 2012). Studies related to fenugreek consumption revealed its anti-cancer properties in reducing carcinoma cell, leukemic cell line HL-60, gastric cell line KATO-III, and colon cancer cell line H-29 (Raju and Bird 2006). Hence, fenugreek seeds and their bioactive compound have anti-cancerous properties against various cancer cell lines. Hence, all these medicinal properties of fenugreek are shown in Figure 13.7 with their effective organs.

13.8 MARKET AND MYTH OF FENUGREEK

The recognition of fenugreek as a whole leaf (dried-wet), its seeds are quite popular in the national and international market due to substantial development in the agriculture field, various cuisines, in different medicinal cures. Its production and utilization are in the picture for many various usages and India is an agriculture-based country that depends on the contribution of spice in its market development and recognition. India is the world's largest producer and exporter of spice which contributes to developing better statistics to investigate global spice trends (Yao et al. 2020). The fenugreek production is monitored year by year and it helps in marinating the economic perspective of financial growth of a nation. According to the Indian spice board, fenugreek production in the past 2 years had been changed significantly with ups and downs of production yield as it was 182,170 (t) in 2019–2020 and 203,360 (t) in 2020–2021, while the export data of fenugreek seeds revealed that in 2019–2020 it generated a revenue of 15,690.38 (lakhs) from 26,570 (t) and in 2020–2021 revenue of 24,642.00 (lakhs) from 38,300 (t). The difference in value and quantity in 2020 and 2021 was 57% and 47%, respectively (Indian Spice Board 2021). These data of fenugreek production and export include all the form of seeds, leaves, and extracts that provide a close look for calculating the fenugreek traded volume (Petropoulos 2002).

The excess of everything is always harmful and this quote is appropriate concerning fenugreek consumption also. Fenugreek consumption may cause some side effects like diarrhoea, allergic reactions, bloating, gas, maple syrup odour in urine, nasal congestion, coughing, facial swelling, and wheezing (Askarpour et al. 2020). When fenugreek consumption checked according to the target disease, then there are some limits of consumption like diabetes (5 g daily), obesity (392 g three times daily), and menstrual cramps (1800–2700 mg for the first three days). There are some misconceptions regarding fenugreek consumption and athletic performance (improve bench press strength in trained athletes but didn't work well), reduction in heartburn (may reduce heartburn symptoms when consumed before meals, but lack of scientific data support), improve the condition of a patient suffering from Parkinson diseases (no effect), and reduce ovarian cyst in polycystic ovary syndrome (but not supported by all research) (Di-Silvestro et al. 2011; Nathan et al. 2014; Swaroop et al. 2015). Hence, fenugreek seeds must be consumed in the belief of limits and in an amount that is commonly used in food preparation. It might cause harm when consuming in large amounts and should not be prescribed for children as supplements. If consumed in excess, it may cause an excessive drop in blood sugar level and it may also cause liver toxicity when consumed in combination with other herbs (Ouzir et al. 2016).

13.9 CONCLUSION

Fenugreek is a popular spice from ancient times that assists in curing various metabolic diseases. The medicinal properties of fenugreek include the gastric ulcer cure, anti-diabetic, anti-obesity, rich in fibre content, anti-cancerous property, and many more. This is a one-in-all spice of Indian cuisine that upholds so many health betterment properties. These properties come in frame with the presence of various bioactive compounds in fenugreek seed diosgenin. These compounds and their effective extraction could enhance the efficiency of fenugreek for a particular type of metabolic disease. The extraction of bioactive compounds could be based on conventional and modern techniques or a combination of both. The extracted compound from fenugreek seeds could be used as a curative agent against various common health issues. The utilization of fenugreek seeds is comparatively less despite their huge medicinal properties. This chapter could furnish for better extraction techniques for utilizing bioactive compounds, brief of fenugreek medical properties in protecting human body against different disorders, and employment of fenugreek seeds in day-to-day life.

REFERENCES

Acharya, S. N., Blade, S., Mir, Z., and Moyer, J. R. 2007. Tristar fenugreek. *Canadian Journal of Plant Science* 87, no. 4: 901–903. doi:10.4141/P06-047.

Adedapo, A. A., Sunday, O. O., and Kehinde, O. S. 2014. Pharmacologic and medicinal properties of fenugreek (*Trigonella foenum-graecum* L). *American Journal of Social Issues and Humanities*, Special Issue: 13–20.

Adil, S., Saim, Q., and Pattoo, R. A. 2015. A review on positive effects of fenugreek as feed additive in poultry production. *International Journal of Poultry Science* 14, no. 12: 664.

Ahmad, A., Salem, S. A., Kaiser, M., and Muhammad, A. 2016. Fenugreek a multipurpose crop: Potentialities and improvements. *Saudi Journal of Biological Sciences* 23, no. 2: 300–310. doi:10.1016/j.sjbs.2015.09.015.

Akbari, S., Nour, H. A., Rosli, M. Y., Oluwaseun, R. A., and Olalere, O. A. 2019a. Extraction, characterization and antioxidant activity of fenugreek (*Trigonella foenum-graecum*) seed oil. *Materials Science for Energy Technologies* 2, no. 2: 349–355.

Akbari, S., Nour, H. A., and Rosli, M. Y. 2019b. Optimization of saponins, phenolics, and antioxidants extracted from fenugreek seeds using microwave-assisted extraction and response surface methodology as an optimizing tool. *Comptes Rendus Chimie* 22, no. 11-12: 714–727.

Alcock, N. W., David, H. G. C., Maria, V. M. G., Edward, L., Graham, P., and Christopher, J. S. 1989. Stereochemistry of the 4-hydroxyisoleucine from *Trigonella foenum-graecum. Phytochemistry* 28, no. 7: 1835–1841.

Ali, A., Mostafa, I. W., Neeru, B., and Nadiya, A. A. 2015. Proximate and phytochemical composition and antioxidant properties of indigenous landraces of Omani fenugreek seeds. *African Journal of Traditional, Complementary and Alternative Medicines* 12, no. 2: 149–154.

Aljuhaimi, F., Şenay, Ş., Mehmet, M. Ö., Kashif, G., and Elfadıl, E. B. 2018. Effect of location on chemical properties, amino acid and fatty acid compositions of fenugreek (*Trigonella foenum-graecum* L.) seed and oils. *Journal of Food Processing and Preservation* 42, no. 4: e13569.

Amarowicz, R., and Fereidoon, S. 2017. Antioxidant activity of broad bean seed extract and its phenolic composition. *Journal of Functional Foods* 38: 656–662. doi:10.31665/JFB.2018.2146.

Aminkara, S., Shojaeiyan, A., Monfaredb, S., and Ayyari, M. 2018. Special effect of ionic liquids on extraction of diosgenin from fenugreek (*Trigonella foenum-graecum* L.) by ultrasonic assistance. *Natural Products Chemistry and Research* 6, no. 322: 1–6.

Arya, P., and Kumar, P. 2021a. Comparison of ultrasound and microwave assisted extraction of diosgenin from Trigonella foenum graceum seed. *Ultrasonics Sonochemistry* 74: 105572. doi:10.1016/j.ultsonch.2021.105572.

Arya, P., and Kumar, P. 2021b. Diosgenin a steroidal compound: An emerging way to cancer management. *Journal of Food Biochemistry 45*: e14005.

Askarpour, M., Farkhondeh, A., Marilyn, S. C., Kamesh, V., Amir, H., and Ehsan, G. 2020. Effect of fenugreek supplementation on blood lipids and body weight: A systematic review and meta-analysis of randomized controlled trials. *Journal of Ethnopharmacology* 253: 112538. doi:10.1016/j.jep.2019.112538.

Atabani, A. E., Arridina, S. S., Irfan, A. B., Mahlia, T. M. I., Masjuki, H. H., and Mekhilef, S. 2012. A comprehensive review on biodiesel as an alternative energy resource and its characteristics. *Renewable and Sustainable Energy Reviews* 16, no. 4: 2070–2093. doi:10.1016/j.rser.2012.01.003.

Azmir, J., Islam, S. M. Z., Rahman, M. M., Sharif, K. M., Mohamed, A., Sahena, F., Jahurul, M. H. A., Ghafoor, K., Norulaini, N. A. N., and Omar, A. K. M. 2013. Techniques for extraction of bioactive compounds from plant materials: A review. *Journal of Food Engineering* 117, no. 4: 426–436. doi:10.1016/j.jfoodeng.2013.01.014.

Azwanida, N. N. 2015. A review on the extraction methods use in medicinal plants, principle, strength and limitation. *Medicinal and Aromatic Plants* 4, no. 196: 2167–0412. doi:10.4172/2167-0412.1000196.

Bano, D., Heena, T., Asad, A., Abdul, M., and Iffat, Z. A. 2016. The medicinal significance of the bioactive compounds of *Trigonella foenum-graecum*: A review. *International Journal of Research in Ayurveda and Pharmacy* 7: 84–91. doi:10.7897/2277-4343.074139.

Barba, F. J., Oleksii, P., Sofia, A. P., Artur, W., Nabil, G., Nadia, B., Jorge, A. S., Raso, J., Martin-Belloso, O., Witrowa-Raichert, D., Lebovka, N., Vorobiev, E. 2015. Current applications and new opportunities for the use of pulsed electric fields in food science and industry. *Food Research International* 77: 773–798. doi:10.1016/j.foodres.2015.09.015.

Basch, E., Ulbricht, C., Kuo, G., Szapary, P., Smith, M. 2003. Therapeutic applications of fenugreek. *Alternative Medicine Review* 8, no. 1: 20–27.

Basu, S. K., Surya, N. A., and James, E. T. 2008b. Application of phosphate fertilizer and harvest management for improving fenugreek (*Trigonella foenum-graecum* L.) seed and forage yield in a dark brown soil zone of Canada. *Current Applied Science and Technology* 8, no. 1: 1–7.

Basu, S. K., Surya, N. A., and James, E. T. 2008a. Genetic improvement of fenugreek (*Trigonella foenum-graecum* L.) through EMS induced mutation breeding for higher seed yield under western Canada prairie conditions. *Euphytica* 160, no. 2: 249–258.

Basu, S. K., Peiman, Z., and William, C. 2019. Fenugreek (*Trigonella foenum-graecum* L.): distribution, genetic diversity, and potential to serve as an industrial crop for the global pharmaceutical, nutraceutical, and functional food industries. *The Role of Functional Food Security in Global Health*, 471–497. Academic Press, Kidlington.

Benayad, Z., Carmen, G., and Nour, E. E. 2014. Identification and quantification of flavonoid glycosides from fenugreek (*Trigonella foenum-graecum*) germinated seeds by LC-DAD-ESI/MS analysis. *Journal of Food Composition and Analysis* 35, no. 1: 21–29. doi:10.1016/j.jfca.2014.04.002s.

Bhatia, K., Kaur, M., Atif, F., Ali, M., Rehman, H., Rahman, S., and Raisuddin, S. 2006. Aqueous extract of *Trigonella foenum-graecum* L. ameliorates additive urotoxicity of buthioninesulfoximine and cyclophosphamide in mice. *Food and Chemical Toxicology* 44, no. 10: 1744–1750. doi:10.1016/j.fct.2006.05.013.

Blumenthal, M., Goldberg, A., Brinckmann, J. 2000. *Herbal medicine: Expanded Commission E monographs*, 130–133. American Botanical Council, Integrative Medicine Communications, Newton, MA, USA.

Cai, B., Ying, Z., Zengtao, W., Dujuan, X., Yongyan, J., Yanbin, G., Aimei, L., Gaizhi, L., Chang, J. C., and Jiansheng, L. 2020. Therapeutic potential of diosgenin and its major derivatives against neurological diseases: Recent advances. *Oxidative Medicine and Cellular Longevity* 1–16. doi:10.1155/2020/3153082.

Cardoso, L. C., Casimiro, M. S., Miguel, R., Rodríguez, E. J., de la Martínez, O., and Luis, M. L. 2012. Extraction of carotenoids and fatty acids from microalgae using supercritical technology. *American Journal of Analytical Chemistry* 3: 877. doi:10.4236/ajac.2012.312A116.

Chandalia, M., Abhimanyu, G., Dieter, L., Klaus, V. B., Scott, M. G., and Linda, J. B. 2000. Beneficial effects of high dietary fiber intake in patients with type 2 diabetes mellitus. *New England Journal of Medicine* 342, no. 19: 1392–1398. doi:10.1056/NEJM200005113421903.

Chatterjee, S., Prasad, S. V., and Sharma, A. 2010. Bioactive lipid constituents of fenugreek. *Food Chemistry* 119, no. 1: 349–353.

Chaudhary, S. A., Pooja, S. C., Bakhtiyar, A. S., Richa, M., Prashanth, G. B., Sara, V., and Marcello, I. 2018. Validation of a method for diosgenin extraction from fenugreek (*Trigonella foenum-graecum* L.). *Acta Scientiarum Polonorum Technologia Alimentaria* 17, no. 4: 1–10.

Chemat, F., and Giancarlo, C. 2012. *Microwave-assisted extraction for bioactive compounds: Theory and practice*, vol. 4: 53–68. Springer Science & Business Media, New York.

Chemat, F., Natacha, R., Anne-Gaëlle, S., Alice, M., Anne-Sylvie, F. T., and Maryline, A. V. 2017. Ultrasound assisted extraction of food and natural products: Mechanisms, techniques, combinations, protocols and applications. A review. *Ultrasonics Sonochemistry* 34: 540–560. doi:10.1016/j.ultsonch.2016.06.035.

Cravotto, G., Carlo, B., Stefano, M., Arianna, B., Valérie, T., and Farid, C. 2011. Extraction of kiwi seed oil: Soxhlet versus four different non-conventional techniques. *Natural Product Research* 25, no. 10: 974–981. doi:10.1080/14786419.2010.524162.

Dangi, R., Shubhada, T., Ritesh, K. C., and Suryaprakasa, R. 2016. Molecular phylogenetics and systematics of *Trigonella* L. (Fabaceae) based on nuclear ribosomal ITS and chloroplast trnL intron sequences. *Genetic Resources and Crop Evolution* 63, no. 1: 79–96. doi:10.1007/s10722-015-0236-4.

Del Hierro, J. N., Teresa, H., Mónica, R. G. R., Tiziana, F., Guillermo, R., and Diana, M. 2018. Ultrasound-assisted extraction and bioaccessibility of saponins from edible seeds: Quinoa, lentil, fenugreek, soybean and lupin. *Food Research International* 109: 440–447.

del Pilar, S. C., Andrea, E. I., Alejandro, C., and Miguel, H. 2017. Bioactives obtained from plants, seaweeds, microalgae and food by-products using pressurized liquid extraction and supercritical fluid extraction. *Green Extraction Techniques: Principles, Advances and Applications* 76: 27.1–24. doi:10.1016/bs.coac.2017.01.001.

Deshpande, H. A., and Sanjivani, R. B. 2014. Isolation and characterization of diosgenin from in vitro cultured tissues of *Helicteresisora* L. *Physiology and Molecular Biology of Plants* 20, no. 1: 89–94. doi:10.1007/s12298-013-0205-3.

Dhawi, F., Hossam, S. E. B., Esmat, A., and Ahmed, M. H. 2020. Antioxidant, antibacterial activities and mineral content of buffalo yoghurt fortified with fenugreek and *Moringa oleifera* seed flours. *Foods* 9, no. 9: 1157.

Di-Silvestro, R. A., Marian, A. V., and Offutt, E. J. 2011. Anti-heartburn effects of a fenugreek fiber product. *Phytotherapy Research* 25, no. 1: 88–91. doi:10.1002/ptr.3229.

Dsouza, M. R., Rufina, K., and Hana, D. 2018. Extraction of diosgenin from Fenugreek and evaluation of its pharmacological role in alleviating metabolic syndrome in vitro. *Research Journal of Biotechnology* 13, no. 12: 10–17.

Duarte, K., Justino, C. I. L., Gomes, A. M., Teresa, R. S., and Armando, C. D. 2014. Green analytical methodologies for preparation of extracts and analysis of bioactive compounds. *Comprehensive Analytical Chemistry* 65: 59–78.

El Bairi, K., Mounir, O., Najda, A., and Loubna, K. 2017. Anticancer potential of *Trigonella foenum graecum*: Cellular and molecular targets. *Biomedicine and Pharmacotherapy* 90: 479–491.

Flammang, A. M., Cifone, M. A., Erexson, G. L., and Stankowski Jr, L. F. 2004. Genotoxicity testing of a fenugreek extract. *Food and Chemical Toxicology* 42, no. 11: 1769–1775. doi:10.1016/j.fct.2004.07.003.

Fuller, S., and Stephen, J. M. 2015. Diosgenin, 4-hydroxyisoleucine, and fiber from fenugreek: Mechanisms of actions and potential effects on metabolic syndrome. *Advances in Nutrition* 6, no. 2: 189–197. doi:10.3945/an.114.007807.

Gu, L. B., Liu, X. N., Liu, H. M., Pang, H. L., and Qin, G. Y. 2017. Extraction of fenugreek (*Trigonella foenum-graceum* L.) seed oil using subcritical butane: Characterization and process optimization. *Molecules* 22, no. 2: 228.

Habtemariam, S. 2019. Medicinal foods as potential therapies for rype-2 diabetes and associated diseases. *The Chemical and Pharmacological Basis of Their Action*, 579–637. Academic Press. doi:10.1016/B978-0-08-102922-0.00017-1.

Hamden, K., Henda, K., Sahla, B., Kais, M., and Noureddine, A. 2011. Inhibitory potential of omega-3 fatty and fenugreek essential oil on key enzymes of carbohydrate-digestion and hypertension in diabetes rats. *Lipids in Health and Disease* 10, no. 1: 1–10.

Hamden, K., Henda, K., Othman, E., Abdelfattah, F., and Noureddine, A. 2017. Modulatory effect of an isolated triglyceride from fenugreek seed oil on of α-amylase, lipase and ACE activities, liver-kidney functions and metabolic disorders of diabetic rats. *Journal of Oleo Science* 66, no. 6: 633–645.

Hannan, J. M. A., Ali, L., Rokeya, B., Khaleque, J., Akhter, M., Flatt, P. R., and Abdel-Wahab, Y. H. A. 2007. Soluble dietary fibre fraction of *Trigonella foenum-graecum* (fenugreek) seed improves glucose homeostasis in animal models of type 1 and type 2 diabetes by delaying carbohydrate digestion and absorption and enhancing insulin action. *British Journal of Nutrition* 97, no. 3: 514–521. doi:10.1017/S0007114507657869.

He, Y., Xiaoyan, W., Yourui, S., Chenxu, D., and Honglun, W. 2016. Efficient protocol for isolation of rhaponticin and rhapontigenin with consecutive sample injection from fenugreek (*Trigonella foenum-graecum* L.) by HSCCC. *Journal of Chromatographic Science* 54, no. 3: 479–485. doi:10.1093/chromsci/bmv169.

Heleno, S. A., Patrícia, D., Prieto, M. A., Lillian, B., Alírio, R., Maria, F. B., and Isabel, C. F. R. F. 2016. Optimization of ultrasound-assisted extraction to obtain mycosterols from *Agaricusbisporus* L. by response surface methodology and comparison with conventional Soxhlet extraction. *Food Chemistry* 197: 1054–1063. doi:10.1016/j.foodchem.2015.11.108.

Indian Spice Board. 2021. *Estimate by Spices Board.* http://www.indianspices.com/sites/default/files/majorspicewise2021.pdf

Isikli, N. D., and Karababa, E. 2005. Rheological characterization of fenugreek paste (cemen). *Journal of Food Engineering* 69: 185–190.

Iskalieva, A., Bob, M. Y., Parag, R. G., Miklos, H., Peter, G. H., and Levente, C. 2012. Cavitation assisted delignification of wheat straw: A review. *Ultrasonics Sonochemistry* 19, no. 5: 984–993. doi:10.1016/j.ultsonch.2012.02.007.

Jesus, S. P., and Meireles, M. A. A. 2014. Supercritical fluid extraction: A global perspective of the fundamental concepts of this eco-friendly extraction technique. *Alternative Solvents for Natural Products Extraction* 39–72. doi:10.1007/978-3-662-43628-8_3.

Keskes, H., Sahla, B., El LobnaJlail, A. F., Sami, S., and Noureddine, A. 2018. LC-MS-MS and GC-MS analyses of biologically active extracts of Tunisian fenugreek (*Trigonella foenum-graecum* L.) seeds. *Journal of Food Measurement and Characterization* 12, no. 1: 209–220.

Khare, C. P. 2004. Indian herbal remedies: Rational Western therapy, Ayurvedic, and other traditional usage. *Botany*, 1–88. Springer Science & Business Media, New York.

Khezeli, T., Mehrorang, G., Ali, D., Sonia, B., Arash, A., and Mustafa, S. 2020. Ionic liquids in separation and preconcentration of organic and inorganic species. *New Generation Green Solvents for Separation and Preconcentration of Organic and Inorganic Species*, 267–318. Elsevier, Amsterdam.

Kousalya, N., Ishwarya, R., Logeshwaran, S. K., Sandhiya, S., and Arun, P. 2019. A review about bioactive compounds of fenugreek (*Trigonella foenum-graceum*). *International Journal of Engineering Applied Sciences and Technology* 4, no. 7: 332–335. doi:10.33564/IJEAST.2019.v04i07.056.

Kumar, P., Uma, B., and Shrirang, J. 2014. Fenugreek seed extract inhibit fat accumulation and ameliorates dyslipidemia in high fat diet-induced obese rats. *Biomed Research International* 1–12. doi:10.1155/2014/606021.

Lea, M. A.2013. Compounds in vegetables including okra and fenugreek of potential value in the treatment of diabetes. *Bioactive Food as Dietary Interventions for Diabetes: Bioactive Foods in Chronic Disease States* 291–306. doi:10.1016/b978-0-12-397153-1.00028-7.

Louie, K. B., Suzanne, M. K., Yuntao, H., Hiroshi, O., de Markus, R., Andrea, N. K., Nigel, J. M., Benjamin, P. B., and Trent, R. N. 2020. Mass spectrometry for natural product discovery. *Comprehensive Natural Products III. Chemistry and Biology* 263–306. doi:10.1016/b978-0-12-409547-2.14834-6.

Mahendra, K. A., and Ritu, K. 2019. Production Technology of fenugreek (Krishisewa). https://www.krishisewa.com/crop-production/vegetables/1069-production-technology-of-methee.html

Mandal, S. C., Vivekananda, M., and Anup, K. D. 2015. *Essentials of Botanical Extraction: Principles and Applications*, 83–136. Academic Press. doi:10.1016/b978-0-12-802325-9.00006-9.

Marker, R. E., Wagner, R. B., Paul, R. U., Emerson, L. Wittbecker, D. P. J. G., and Clarence, H. R. 1947. New sources for sapogenins. *Journal of the American Chemical Society* 69, no. 9: 2242–2242.

Martínez, J. M., Carlota, D., Ignacio, Á., and Javier, R. 2020. Pulsed electric field-assisted extraction of valuable compounds from microorganisms. *Comprehensive Reviews in Food Science and Food Safety* 19, no. 2: 530–552. doi:10.1111/1541-4337.12512.

Mawla, M. A. A. E. A., and Eldien, H. O. 2011. Elicitation of trigonelline and 4-hydroxyisoleucine with hypoglycemic activity in cell suspension cultures of *Trigonella foenum-graecum* L. *The Open Conference Proceedings Journal* 2, no. 1: 80–87. doi:10.2174/2210289201102010080.

McCormick, K., Norton, R., and Eagles, H. A. 2006. Fenugreek has a role in south-eastern Australian farming systems. In: *Proceedings of "Ground breaking stuff"*, 13th Annual Agronomy Conference, 639. Perth, Australia.

Mehrafarin, A., S. H. Rezazadeh, H. N. B., Noormohammadi, G. H., Zand, E., and Qaderi, A. 2011. A review on biology, cultivation and biotechnology of fenugreek (*Trigonella foenum-graecum* L.) as a valuable medicinal plant and multipurpose. *Journal of Medicinal Plants* 10, no. 37: 6–24.

Moradi, N., and Moradi. K. 2013. Physiological and pharmaceutical effects of fenugreek (*Trigonella foenum-graecum* L.) as a multipurpose and valuable medicinal plant. *Global Journal of Medicinal Plant Research* 1, no. 2: 199–206.

Munshi, M., Arya, P., Kumar, P. 2020. Physico-chemical analysis and fatty acid profiling of fenugreek (*Trigonella foenum-graecum*) seed oil using different solvents. *Journal of Oleo Sciences* 69, no. 11: 1349–1358. doi:10.5650/jos.ess20137.

Murlidhar, M., and Goswami, T. K. 2012. A review on the functional properties, nutritional content, medicinal utilization and potential application of fenugreek. *Journal of Food Processing and Technology* 3, no. 9: 1–10. doi:10.4172/2157-7110.1000181.

Nagulapalli, V., Kalyan, C., Anand, S., Debasis, B., and Anupam, B. 2017. A small plant with big benefits: Fenugreek (*Trigonella foenum-graecum* Linn.) for disease prevention and health promotion. *Molecular Nutrition and Food Research* 61, no. 6: 1600950. doi:10.1002/mnfr.

Naidu, M. Madhava, B. N., Shyamala, J. P. N., Sulochanamma, G., and Srinivas, P. 2011. Chemical composition and antioxidant activity of the husk and endosperm of fenugreek seeds. *LWT – Food Science and Technology* 44, no. 2: 451–456. doi:10. 1016/j.lwt.2010.08.013.

Nathan, J., Siddika, P., Mohan, V., Veena, J., and Prasad, A. T. 2014. Efficacy and safety of standardized extract of *Trigonella foenum-graecum* L seeds as an adjuvant to L-DOPA in the management of patients with Parkinson's disease. *Phytotherapy Research* 28, no. 2: 172–178. doi:10.1002/ptr.4969.

Navarro, D. H., Joaquín, G. R., and Diana, M. 2020. Chemical characterization and bioaccessibility of bioactive compounds from saponin-rich extracts and their acid-hydrolysates obtained from fenugreek and quinoa. *Foods* 9, no. 9: 1159. doi:10.3390/foods9091159.

Nwokocha, L. M. 2021. Galactomannans. In *Handbook of Hydrocolloids*, 273–293. Woodhead Publishing, Cambridge.

Ouzir, M., Khalid, E. B., and Saaïd, A. 2016. Toxicological properties of fenugreek (*Trigonella foenum-graecum*). *Food and Chemical Toxicology* 96, no. 2016: 145–154. doi:10.1016/j.fct.2016.08.003.

Ouzzar, M. L., Wahida, L., Ahmed, Z., and Abdeslam-Hassen, M. 2015. Comparison of the performances of hydrodistillation and supercritical CO_2 extraction processes for essential oil extraction from Rosemary (*Rosmarinus officinalis* L.). *Chemical Engineering Transactions* 43: 1129–1134. doi:10.3303/CET1543189.

Pająk, P., Robert, S., Joanna, B., Karolina, K., and Teresa, F. 2019. Antioxidant properties, phenolic and mineral composition of germinated chia, golden flax, evening primrose, phacelia and fenugreek. *Food Chemistry* 275: 69–76.

Pal, D. K., and Souvik, M. 2020. Fenugreek (*Trigonella foenum-graecum*) seeds in health and nutrition. In *Nuts and Seeds in Health and Disease Prevention*, 161–170. Academic Press. doi:10.1016/B978-0-12-818553-7.00013-9.

Pereira, D. M., Patrícia, V., José, A. P., and Paula, B. A. 2009. Phenolics: From chemistry to biology. *Molecules* 14, no. 6: 2202–2211. doi:10.3390/molecules14062202.

Petropoulos, G. A. 2002. *Fenugreek: The Genus Trigonella*, 1–8. CRC Press. doi:10.1201/9780203217474.

Phelps, C. L., Neil, G. S., and Chien, M. W. 1996. Past, present, and possible future applications of supercritical fluid extraction technology. *Journal of Chemical Education* 73, no. 12: 1163. doi:10.1021/ed073p1163.

Pimentel-Moral, S., Isabel, B. L., Jesús, L. S., David, A. R., Antonio, M. F., and Antonio, S. C. 2018. Microwave-assisted extraction for Hibiscus sabdariffa bioactive compounds. *Journal of Pharmaceutical and Biomedical Analysis* 156: 313–322. doi:10.1016/j.jpba.2018.04.050.

Pradeep, S. R., and Krishnapura, S. 2018. Alleviation of cardiac damage by dietary fenugreek (*trigonella foenum-graecum*) seeds is potentiated by onion (*Allium cepa*) in experimental diabetic rats via blocking renin–angiotensin system. *Cardiovascular Toxicology* 18, no. 3: 221–231. doi:10.1007/s12012-017-9431-1.

Rababah, T. M., Ereifej, K. I., Esoh, R. B., Al-u'datt, M. H., Alrababah, M. A., Yang, W. 2011. Antioxidant activities, total phenolics and HPLC analyses of the phenolic compounds of extracts from common Mediterranean plants. *Natural Product Research* 6: 596–605.

Raju, J., and Bird, R. P. 2006. Alleviation of hepatic steatosis accompanied by modulation of plasma and liver TNF-α levels by *Trigonella foenum-graecum* (fenugreek) seeds in Zucker obese (fa/fa) rats. *International Journal of Obesity* 30, no. 8: 1298–1307. doi:10.1038/sj.ijo.0803254.

Raju, J., and Chinthalapally, V. R. 2012. Diosgenin, a steroid saponin constituent of yams and fenugreek: Emerging evidence for applications in medicine. *Bioactive Compounds in Phytomedicine* 125: 143. doi: 10.5772/26700.

Rangari, V. D., Prachi, S., and Sachin, L. B. 2015. 4-Hydroxyisoleucine: A potential antidiabetic agent from *Trigonella foenum-graecum. Glucose Intake and Utilization in Pre-diabetes and Diabetes*, 191–198. Academic Press, London.

Rashid, F., Yiwen, B., Zaheer, A., and Jen-Yi, H. 2020. Effect of high voltage atmospheric cold plasma on extraction of fenugreek galactomannan and its physicochemical properties. *Food Research International* 138: 109776.

Reddy, A. V. B., Moniruzzaman, M., Madhavi, V., and Jafariah, J. 2020. Recent improvements in the extraction, clean up and quantification of bioactive flavonoids. *Studies in Natural Products Chemistry* 66: 197–223. doi:10.1016/b978-0-12-817907-9.00008-8.

Reinisalo, M., Anna, K., Ali, K., Kai, K., and Reijo, O. K. 2015. Polyphenol stilbenes: Molecular mechanisms of defence against oxidative stress and aging-related diseases. *Oxidative Medicine and Cellular Longevity* 9: 1–25. doi:10.1155/2015/340520.

Rodríguez-Pérez, C., Quirantes-Piné, R., Fernández-Gutiérrez, A., and Segura-Carretero, A. 2015. Optimization of extraction method to obtain a phenolic compounds-rich extract from *Moringa oleifera* Lam leaves. *Industrial Crops and Products* 66: 246–254. doi:10.1016/j.indcrop.2015.01.002.

Sarwar, S., Hanif, M. A., Ayub, M. A., Boakye, Y. D., Agyare, C. 2020. Fenugreek. In: *Medicinal Plants of South Asia*, eds. M. Hanif, H. Nawaz, M. K. Muhammad, H. J. Byrne, 257–271. Elsevier, Netherlands.

Shirsath, S. R., Sonawane, S. H., and Gogate, P. R. 2012. Intensification of extraction of natural products using ultrasonic irradiations: A review of current status. *Chemical Engineering and Processing: Process Intensification* 53: 10–23. doi:10.1016/j.cep.2012.01.003.

Siddique, M. A. B., Katerina, T., Dilip, K. R., and Nigel, B. 2020. Conventional extraction techniques for bioactive compounds from herbs and spices: Herbs, spices and medicinal plants. *Processing, Health Benefits and Safety* 69–93. doi:10.1002/9781119036685.

Sindhu, G., Shyni, G. L., Chithra, K. P., Bala, N., and Helen, A. 2018. Evaluation of anti-arthritic potential of *Trigonella foenum-graecum* L. (fenugreek) mucilage against rheumatoid arthritis. *Prostaglandins & Other Lipid Mediators* 138: 48–53. doi:10.1016/j.prostaglandins.2018.08.002.

Singh, B., Jatinder, P. S., Amritpal, K., and Narpinder, S. 2017. Phenolic composition and antioxidant potential of grain legume seeds: A review. *Food Research International* 101: 1–16. doi:10.1016/j.foodres.2017.09.026.

Smith, R. M. 2003. Before the injection – Modern methods of sample preparation for separation techniques. *Journal of Chromatography A* 1000, no. 1-2: 3–27.

Srichamroen, A., Alan, B. R. T., Catherine, J. F., and Tapan, K. 2009. In vitro intestinal glucose uptake is inhibited by galactomannan from Canadian fenugreek seed (*Trigonella foenum-graecum* L) in genetically lean and obese rats. *Nutrition Research* 29, no. 1: 49–54. doi:10.1016/j.nutres.2008.11.002.

Srinivasan, K. 2019. Fenugreek (*Trigonella foenum-graecum* L.) seeds used as functional food supplements to derive diverse health benefits. *Nonvitamin and Nonmineral Nutritional Supplements*, 217–221. Academic Press. doi:10.1016/B978-0-12-812491-8.00031-X.

Srinivasan, K. 2006. Fenugreek (*Trigonella foenum-graecum*): A review of health beneficial physiological effects. *Food Reviews International* 22, no. 2: 203–224. doi:10.1080/87559120600586315.

Swaroop, A., Amrita, S. J., Sushil, K. G., Manashi, B., Pawan, K., Harry, G. P., and Debasis, B. 2015. Efficacy of a novel fenugreek seed extract (*Trigonella foenum-graecum*, Furocyst TM) in polycystic ovary syndrome (PCOS). *International Journal of Medical Sciences* 12, no. 10: 825. doi:10.7150/ijms.13024.

Syed, Q. A., Zainab, R., Muhammad, H. A., Rizwan, S., Anum, I., Niaz, M., and Hafiz, U. U. R. 2020. Nutritional and therapeutic properties of fenugreek (*Trigonella foenum-graecum*): A review. *International Journal of Food Properties* 23, no. 1: 1777–1791. doi:10.1080/10942912.2020.1825482.

Tiwari, B. K. 2015. Ultrasound: A clean, green extraction technology. *Trends in Analytical Chemistry* 71: 100–109. doi:10.1016/j.trac.2015.04.013.

Ventura, S. P. M., Francisca. A. S., Maria, V. Q., Dibyendu, M., Mara, G. F., and Joao, A. P. C. 2017. Ionic-liquid-mediated extraction and separation processes for bioactive compounds: Past, present, and future trends. *Chemical Reviews* 117, no. 10: 6984–7052. doi:10.1021/acs.chemrev.6b00550.

Wani, S. A., and Kumar, P. 2018. Fenugreek: A review on its nutraceutical properties and utilization in various food products. *Journal of the Saudi Society of Agricultural Sciences* 17, no. 2: 97–106. doi:10.1016/j.jssas.2016.01.007.

Wani, S. A., Sonu, B., and Kumar, P. 2016. Ultrasound and microwave assisted extraction of diosgenin from fenugreek seed and fenugreek-supplemented cookies. *Journal of Food Measurement and Characterization* 10, no. 3 (2016): 527–532.

Xhaxhiu, K., Arjan, K., Altin, M., and Teodor, K. 2013. Ultrasonic and Soxhlet extraction characteristics of the orange peel from "Moro" cultivars grown in Albania. *Journal of Essential Oil-Bearing Plants* 16, no. 4: 421–428. doi:10.1080/0972060X.2013.813277.

Xiao, W., Lujia, H., and Bo, S. 2008. Microwave-assisted extraction of flavonoids from Radix Astragali. *Separation and Purification Technology* 62, no. 3: 614–618. doi:10.1016/j.seppur.2008.03.025.

Yao, D., Zhang, B., Zhu, J., Zhang, Q., Hu, Y., Wang, S., Wang, Y., Cao, H., and Xiao, J. 2020. Advances on application of fenugreek seeds as functional foods: Pharmacology, clinical application, products, patents and market. *Critical Reviews in Food Science and Nutrition* 60, no. 14: 2342–2352. doi: 10.1080/10408398.2019.1635567.

Yasothai, R. 2018. Nutritional value of fenugreek seed for livestock and poultry. *International Journal of Science, Environment and Technology* 7: 1066–1071.

Yasothai, R. 2021. Chemical composition of fenugreek (*Trigonella foenum-graecum* L.) seed and galactomannan depleted fenugreek residue. *The Pharma Innovation Journal* 10, no. 7, (2021): 1–4.

Zandi, P., Shirani-Rad, A. H., Daneshian, J., and Bazrkar-Khatibani, L. 2011. Agronomic and morphologic analysis of fenugreek (*Trigonella foenum-graecum* L.) under nitrogen fertilizer and plant density via factor analysis. *African Journal of Agricultural Research* 6, no. 5: 1134–1140. doi:10.5897/AJAR11.004.

Zandi, P., Saikat, K. B., William, C., Mojtaba, K., Shahram, K. C., and Leila, B. K. 2017. Fenugreek (*Trigonella foenum-graecum* L.): An Important Medicinal and Aromatic Crop. *Active Ingredients from Aromatic and Medicinal Plants* 207–224. InTech Open London.

CHAPTER 14

Bioactive Compounds in Cardamom

Indira Dey Paul

CONTENTS

14.1 INTRODUCTION

Spices boast of impressive wealth of bioactive compounds which apart from imparting unique and attractive flavour to the foods, can also prevent or cure several diseases. Spices comprise different parts of plants such as their seeds, stems, leaves, roots, etc., and are either used whole or in ground form to enhance the sensory characteristics of foods and beverages. Apart from amplifying the flavour of the food products, spices are traditionally well-known for their uses in the treatment of various diseases. The health-savvy consumers increasingly prefer natural remedies over synthetic medicines. Hence food technologists are exploring traditional herbs having different pharmacological effects.

DOI: 10.1201/9781003215387-14

Cardamom is such a spice that its culinary and traditional pharmaceutical applications are well-established worldwide. Basically, the seeds from the dried fruits of different genera (Elettaria, Amomum and Aframomum) of the family Zingiberaceae are identified as cardamoms (Govindarajan et al. 1982). The name "Cardamom" has been derived from the Latin word "cardamōmum". In India, this spice is known as "Elaichi". The two popular types of cardamom are as follows:

- Small cardamom or true or green cardamom (*Elettaria cardamomum* L. Maton)
- Large cardamom or black cardamom (*Amomum subulatum* Roxburgh)

This chapter covers the availability, botany, bioactive composition and medicinal properties of both types of cardamom mentioned above.

14.2 AVAILABILITY AND BOTANY OF CARDAMOM

14.2.1 Small Cardamom

Small cardamom (*Elettaria cardamomum* L. Maton) is a perennial, herbaceous, rhizomatous monocot plant, belonging to family Zingiberaceae. It is also popularly known as 'queen of spices' and is the third most expensive spice in the world after saffron and vanilla.

The original habitat of this spice is believed to be the mountains of the south-western parts of the Indian Peninsula. It was exported to Europe probably in 1214 AD. Presently India is the largest producer of small cardamom in the world, but its cultivation is restricted mainly within the states of Kerala, Karnataka and Tamil Nadu (Korikanthimath et al. 2002). However, the major quantity of Indian cardamom is utilized for internal consumption with limited exports (UNCTAD 2006). Apart from India, small cardamom is widely cultivated in Nepal, Sri Lanka, Guatemala, Thailand, Laos, Vietnam, Costa Rica, Mexico, Tanzania and El Salvador (Mehra 2001). Presently, Guatemala is the major producer of small cardamom. Out of the total world production of 55000 MT, Guatemala produces nearly 30000 MT cardamom annually. India, being the second largest producer, accounts for 15000–18000 MT cardamom annually (https://www.spicefactors.com/spice/cardamom/).

Two interfertile varieties of *Elettaria cardamomum* are available viz., *Elettaria cardamomum* var. Minor and *Elettaria cardamomum* var. Major. The var. Major denotes the wild and more primitive type of small cardamom from which var. Minor or the cultivated type of small cardamom has been derived. Cardamom is a bushy herb of about 4.5 m height, which grows well under shade at 760–1500 m above sea level on rich, moist and well-drained soil. A typical cardamom plant consists of very long lance-shaped leaves, mauve-marked orchid-like white flowers and pod-like green fruits (Figure 14.1a), which completely mature within 120 days from flowering. The green pods or capsules turn golden yellow in colour on ripening. The length of each capsule is approximately 1–2 cm. The cardamom pods are three-celled capsules containing about 12–32 seeds (Figure 14.1b), which are covered with a white mucilaginous coat; the ripe seeds appear black in colour (Murugan et al. 2016). The pods bear pungent aroma which is lost rapidly when split open.

14.2.2 Large Cardamom

Large cardamom (*Amomum subulatum* Roxburgh), a perennial herb, is majorly produced in India, recording an average annual production of 4000 MT followed by Nepal (2500 MT) and Bhutan (1000 MT). It is native to of the Eastern Himalayan region and is popularly known as greater Indian or Nepal cardamom. India exerts monopoly over large cardamom being its largest producer and exporter. Within India, Sikkim contributes about 80% of the total annual production of this crop. Thailand, Cambodia, Vietnam and China are known to produce other species of large cardamom viz., A. xanthiodes (or bastard cardamom), A. krervanh (Siam or Cambodian cardamom), A. globosum or Chinese cardamom,

(a) (b)

Figure 14.1 (a) Cured small cardamom pods or capsules; (b) small cardamom seeds.

respectively. This chapter discusses about the varieties, bioactive composition and medicinal properties of A. subulatum Roxburgh, owing to its wide commercial cultivation. Figure 14.2(a) shows cured large cardamom pods or capsules and Figure 14.2(b) shows large cardamom seeds.

Like small cardamom, large or black cardamom also grows well under shade at an elevation of 765–1675 m above mean sea level [temperature range: 6°C (min); 33°C (max)]. Among several varieties of large cardamom, Ramsey, Golsey and Sawney are most popular. The flowers of black cardamom appear during the month of April and May, and the fruits/capsules reach maturity by the month of September and October. The fruits are trilocular many-seeded (40–50 seeds) capsules whose colour ranges from reddish-brown to dark pink.

Several local varieties of large cardamom are produced in the Himalayan regions. The popular local varieties produced in Sikkim are Ramsey, Golsey, Sawney, Kopringery, Madhusay and Rhangbhang (Gautam et al. 2016; Parthasarathy et al. 2008). Among the above-mentioned varieties, only Ramsey, Golsey and Sawney are cultivated for commercial purpose (Gautam et al. 2016). Biswas et al. (1986), through his surveys, reported that Ramsey is well-cultivated at higher altitudes while Golsey is better suited for lower altitudes; the Sawney variety can be produced at different elevations. A promising variety, Barlanga, has been reported by Rao et al. (1993), which is cultivated at higher altitudes and offer desirable high-yielding properties.

The pods of black cardamom are also used as a spice. However, its flavour is distinctly different from that of small cardamom. Hence, both these spices can't be used interchangeably. Black cardamom is not used for sweet dishes owing to its strong smoky flavour.

(a) (b)

Figure 14.2 (a) Cured large cardamom pods or capsules; (b) large cardamom seeds.

14.3 GENERAL AND BIOACTIVE COMPOSITION OF CARDAMOM

The composition of any biological produce largely depends upon its variety, maturity and geographical location of cultivation. Hence, it is appropriate to discuss about the range of the different components present in the produce.

The freshly harvested small cardamom pods are green in colour and contain approximately 80% moisture. Since high moisture content hinders safe storage, fresh cardamoms are dried or cured immediately after harvesting to reduce the moisture content to 8–12% moisture content as well as retain maximum green colour. The optimum curing temperature of small cardamom is between 40–50°C; higher temperatures may lead to yellowing of the pods, split pods and heat-induced injury. The cured cardamom is mainly composed of sugars, starch, pentosans, proteins, pigments, silica, calcium oxalate, minerals, volatile oil and fixed oil (Ajit et al. 2020).

Sontakke et al. (2018), Kizhakkayil et al. (2006), Muthusamy et al. (2012), Murugan et al. (2011) and Ereifej et al. (2015) estimated the proximate composition of small cured cardamom capsules in dry basis as tabulated in Table 14.1. The leaves of small cardamom are also rich reserves of essential minerals elements such as iron (28.4 mg/100 g), copper (1.9 mg/100 g), manganese (26 mg/100 g) and zinc (0.2 mg/100 g) (Sontakke et al. 2018, Muthusamy et al. 2012, Murugan 2011).

The general chemical composition of dried large cardamom capsules has been estimated by Madhusoodanan and Rao (2001), Oli (2011), Pruthi (1993), Shankaracharya et al. (1990) and Singh (1978) (shown in Table 14.1).

Apart from the essential nutrients discussed above, both small and large cardamoms are widely known for its rich reserve of bioactive compounds in the form of volatile oils and several

Table 14.1 Proximate Composition of Cured Small Cardamom and Large Cardamom Capsules

Proximate Composition	Small Cardamom	References	Large Cardamom	References
Moisture (%)	8.30	Singh (1978)	8.49	Singh (1978); Pruthi (1993)
Carbohydrates (%)	68.2	Sontakke et al. (2018)	–	–
Starch (%)	–	–	43.21	Pruthi (1993)
Protein (%)	10.6	Sontakke et al. (2018)	6	Madhusoodanan and Rao (2001); Oli (2011); Shankaracharya et al. (1990)
Fat (%)	2.4	Sontakke et al. (2018)	–	–
Ash (%)	5.3	Sontakke et al. (2018)	4	Madhusoodanan and Rao (2001); Oli (2011)
Crude fibre (%)	16.3	Kizhakkayil et al. (2006)	22	Pruthi (1993); Shankaracharya et al. (1990)
Calcium (mg/100 g)	92.7	Sontakke et al. (2018)	666.6	Bhandari et al. (2013)
Sodium (mg/100 g)	17.0	Sontakke et al. (2018)	–	–
Phosphorus (mg/100 g)	188.5	Muthusamy et al. (2012)	61	Bhandari et al. (2013)
Potassium (mg/100 g)	124.2	Sontakke et al. (2018)	–	–
Iron (mg/100 g)	12.8	Sontakke et al. (2018); Murugan et al. (2011)	–	–
Magnesium (mg/100 g)	181.5	Ereifej et al. (2015)	412.5	Bhandari et al. (2013)
Manganese (mg/100 g)	41.7	Muthusamy et al. (2012); Murugan (2011)	–	–
Copper (mg/100 g)	0.5	Muthusamy et al. (2012); Murugan (2011)	–	–
Sulphur (mg/100 g)	100.8	Muthusamy et al. (2012)	–	–
Zinc (mg/100 g)	3.6	Muthusamy et al. (2012); Murugan (2011)	–	–
Fluoride (ppm)	–	–	14.4	Bhandari et al. (2013)

phytochemicals. The popularity of cardamoms is chiefly attributed to their essential oils and metabolites which are responsible for their unique aroma and high therapeutic value. According to Singh (1978), the average percentage of volatile oil (essential oil) present in small cardamom and large cardamom are 2.80% and 8.30%, respectively. Robert (1986) and Boiswert and Hubert (1998) described these oils as warm, sweet, spicy, lightly camphorated and citrusy.

14.3.1 Oil and Phytochemical Composition of Small Cardamom

Several researchers studied on the essential oil composition of small cardamom. It was observed that the essential oil/volatile oil content in small cardamom ranged from 0.2 – 8.7% (Kumara et al. 1985; Sontakke et al. 2018; Savan and Küçübay 2013; Menon 2000; Abbasipour et al. 2011; Leela et al. 2008). Nigam et al. (1965) reported the first detailed chemical composition of the volatile oil present in small cardamom. The oil is predominantly composed of aromatic oxygenated compounds and mono- or sesquiterpenic hydrocarbons. The major chemical components present in the volatile oil of small cardamom are presented in Figure 14.3. The trace constituents are tabulated

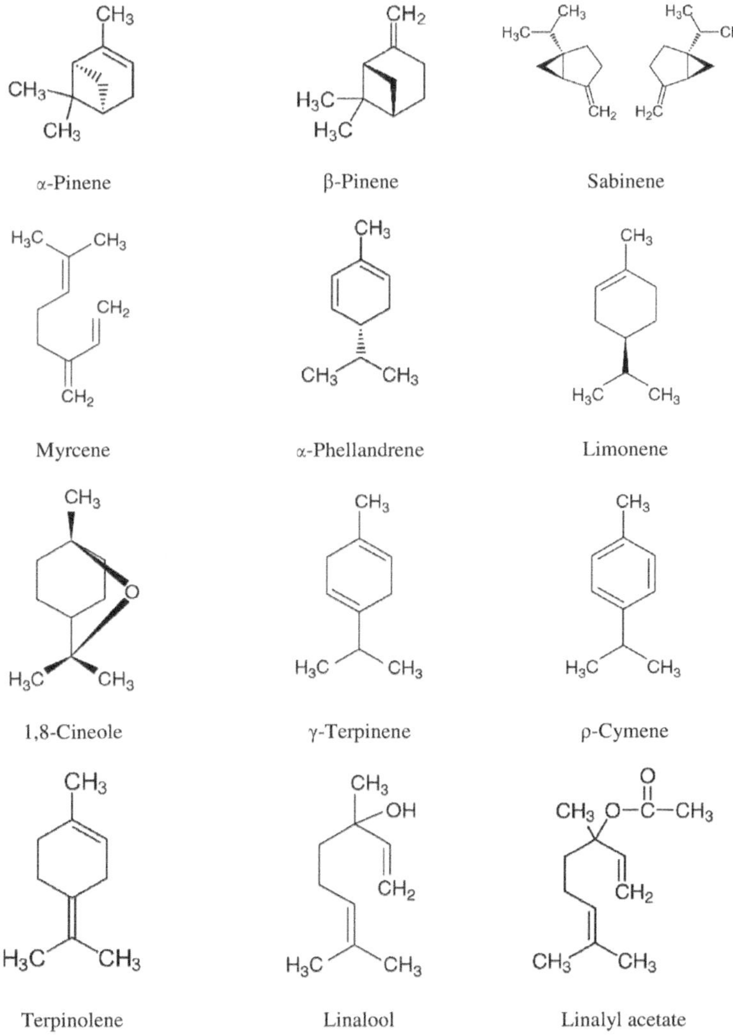

Figure 14.3 Major volatile oil components of small cardamom [*Sources:* Govindarajan et al. (1982); Lawrence (1979)]. *(Continued)*

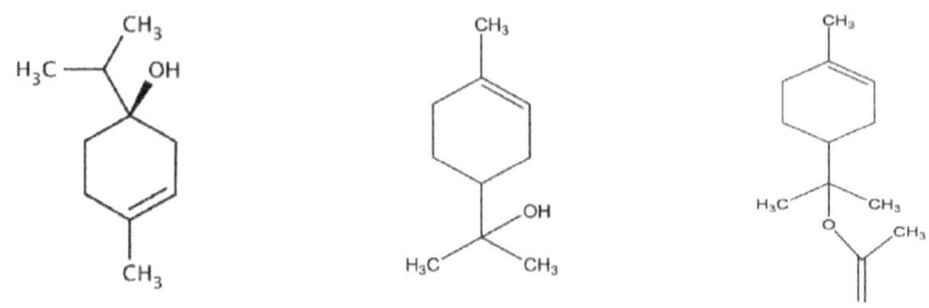

Terpinene-4-ol α-Terpineol α-Terpinyl acetate

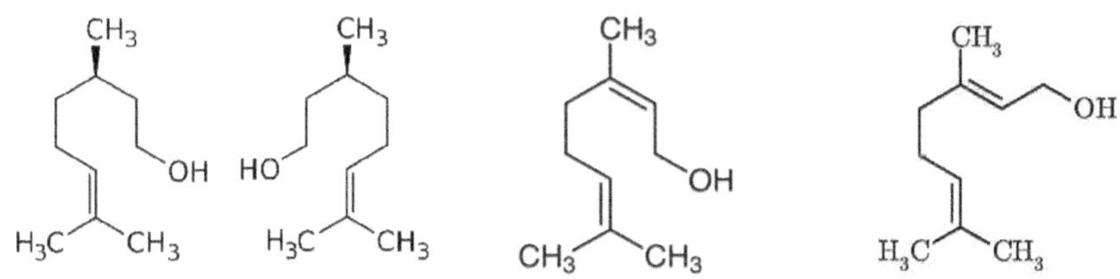

Citronellol Nerol Geraniol

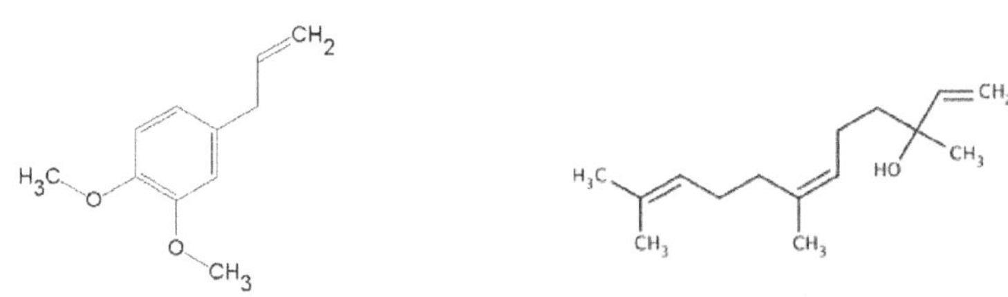

Methyl eugenol trans-Nerolidol

Figure 4.3 *(Continued)*

in Table 14.2. The concentration of different chemical compounds is greatly influenced by environmental factors such as soil conditions, altitude, latitude, climate change and seasonal variations (Heywood 2002).

Among the essential oils mentioned in Table 14.2, six oxygenated compounds viz., 1, 8-cineole (eucalyptol), linalool, α-terpinyl acetate, linalyl acetate, terpin-4-ol and α-terpineol comprise 90% of the aromatic compounds present in small cardamom. The combined effect 1, 8-cineole and α-terpinyl acetate is principally responsible for the characteristic aroma of small cardamom

Table 14.2 Minor or Trace Volatile Oil Components of Small Cardamom

Types of Trace Components	Examples
Hydrocarbon	α-Thujene, Camphene, α-Terpinene, cis-Ocimene, trans-Ocimene, Toluene, α-Copaene, α-Ylangene, ρ-Dimethylstyrene, Cyclosativene, γ-Cadinene, Δ-Cadinene
Acid	Acetic acid, Propionic acid, Butyric acid, 2-Methyl butyric acid, 3-Methyl butyric acid
Carbonyl compounds	3-Methyl butanal, 2-Methyl butanal, Pentanal, Furfural, Cuminaldehyde, Carvone, Pinole, 8-Acetoxy carvotanacetone, Terpinene-4-yl-acetate, α-Terpinyl propionate, Dihydro- α-terpinyl acetate
Alcohols and phenols	2-Methyl butanol, ρ-Methyl-3-en-l-ol, Perillyl alcohol, Cuminyl alcohol, ρ-Cresol, Thymol

Sources: Govindarajan et al. (1982) and Lawrence (1979).

(Olivero-Verbel et al. 2010; Lawrence 1979). As mentioned earlier, the volatile oil chemistry of cardamom differs greatly with their cultivation locations. The Malabar variety oozes more camphory aroma due to the presence of higher content of 1,8-cineole while the Mysore variety (commercially known as 'Alleppy green') imparts mild spicy flavour due owing to the higher content of α-terpinyl acetate. The Mysore variety also exhibit higher content of linalool and linalyl acetate, which made them the largest selling Indian cardamom grade. The oil content also varies for different parts of the cardamom plant of same cultivar. According to Krishnamurthy (1964), Krishnamurthy et al. (1967) and Korikanthimath et al. (1999), the highest oil content in the cv. Mysore and Malabar is found in their seeds (6.6–10.6%) followed by their immature capsules (4–5%) and husk (0.2%). However, other researchers have also obtained quite a deviation from the above-mentioned trend. The chemical composition of volatile oil of different parts of small cardamom plant cultivated across different geographical regions is presented in Table 14.3.

Small cardamom seeds are also composed of fixed or fatty oils, constituting primarily of oleic and palmitic acids. Verghese (1996) and Parry et al. (2006) estimated the percentage of different

Table 14.3 Chemical Composition of Volatile Oil of Different Parts of Small Cardamom Plant Cultivated Across Different Geographical Regions

Parts of Small Cardamom	Origin	Constituents	Concentration (%)	References
Capsules	South India	α-Terpinyl acetate	26.68	Ashokkumar et al. (2020a)
		1,8-Cineole	28.94	
		α-Terpineol	14.58	
		Sabinene	13.50	
		Nerol	5.01	
		α-Pinene	2.42	
	Sri Lanka	α-Terpinyl acetate	74.60	Kumara et al. (1985)
		1,8-Cineole	4.40	
		β-Terpineol	1.80	
		Linalool	12.70	
Seeds	Iran	α-Terpinyl acetate	36.61	Goudarzvand and Abbasipour (2017)
		1,8-Cineole	30.42	
		Linalyl acetate	5.79	
		Sabinene	4.85	
	Guatemala	α-Terpinyl acetate	36.80	Singh et al. (2008)
		1,8-Cineole	29.20	
		Linalyl acetate	5.20	
		Sabinene	3.90	
		Linalool	3.10	
	Costa Rica	α-Terpinyl acetate	39.31	Noleau et al. (1987)
		Sabinene	3.78	
		α-Terpineol	2.97	
		Linalool	5.91	
		Geraniol	2.66	
Leaves	Pakistan	1,8-Cineole	25.74	Mahmud (2008)
		4-Terpineol	30.26	
		α-Terpene	9.80	
		Linalool	2.67	

Source: Ashokkumar et al. (2020b).

Table 14.4 Fixed Oil/Fatty Acid Composition of Small Cardamom

Type of Fixed Oil	Fixed Oil	Concentration (%)	References
Saturated	Myristic acid	1.3–1.5	Verghese (1996); Parry et al. (2006)
	Palmitic acid	26.4–38.0	Verghese (1996); Parry et al. (2006)
	Stearic acid	2.3–3.2	Verghese (1996); Parry et al. (2006)
	Arachidic acid	0.2–2.1	Verghese (1996); Parry et al. (2006)
	Caproic acid	5.3	Verghese (1996)
	Caprylic acid	5.3	Verghese (1996)
	Capric acid	<0.1–3.8	Verghese (1996)
	Lauric acid	0.2	Verghese (1996)
	Pentadecanoic acid	0.4	Verghese 1996)
Mono-unsaturated	Palmitoleic acid	1.6–1.9	Verghese (1996); Parry et al. (2006)
	Oleic acid	42.5–49.2	Verghese (1996); Parry et al. (2006)
Poly-unsaturated	Linoleic acid	2.2–15.3	Verghese (1996); Parry et al. (2006)
	α-Linolenic acid	2.7–5.8	Verghese (1996); Parry et al. (2006)
	Eicosenoic acid	0.5	Parry et al. (2006)

fixed oils, which is tabulated in Table 14.4. The pathogen defence property of cardamom plants is mainly attributed to the presence of these fatty acids and their breakdown products. Moreover, the functional properties of small cardamom are not limited to its essential and fatty oils only. The presence of various phytochemicals in small cardamom including flavonoids and carotenoids has been reported by Ashokkumar et al. (2020c). According to review conducted by Ajit et al. (2020), the seed and leaf extracts of small cardamom are rich in phenolic compounds, flavonoids, alkaloids, tannins and saponins. Cardamom oil is also rich in tocopherols, a highly effective antioxidant. Tocopherols are widely known for their preventive effects against several types of cancers such as kidney cancer (Shen et al. 2015), bladder cancer (Wang et al. 2014) and lung cancer (Zhu et al. 2017). The phytochemical reserve of small cardamom, as estimated by Ashokkumar et al. (2020c), has been tabulated in Table 14.5.

14.3.2 Oil and Phytochemical Composition of Large Cardamom

The characteristic aroma of large cardamom also is attributed to the volatile oil content of its seeds and husk/rind. The steam distilled oil is dark brown in colour. The Golsey Dwarf variety showed highest level of volatile oil whereas the lowest oil reserve was observed in the White Ramna variety (Gupta, 1986). The pungency of large cardamom is due to the presence of 1,8-Cineole, which comprises 65–80% of its total volatile oil content. The trace amounts of α-terpinyl acetate

Table 14.5 Phytochemicals Present in Small Cardamom

Phytochemicals		Concentration (%)
Flavonoids (μg/g)	Catechin	281.8
	Quercetin	4.9
	Myricetin	18.6
	Kaempherol	8.7
Total Flavonoids (μg/g)		314.0
Carotenoids (μg/g)	Lutein	2.4
	β-carotene	0.5

Source: Ashokkumar et al. (2020c).

(≈5%) contributes to the pleasant aroma of large cardamom (Karibasappa, 1987). According to Satyal et al. (2012) and Singh et al. (2008), 1,8-Cineole is the principal component of large cardamom. This high level of cineole and low level of terpinyl acetate is probably responsible for the comparatively harsher aroma of large cardamom than small cardamom (Pruthi, 1993).

Researchers analysed the oil composition of both the seeds and husk of large cardamom. Qualitatively the oil composition of large cardamom resembles significantly with that of small cardamom. Table 14.6 presents the major volatile oil constituents of large cardamom seeds and husk

Table 14.6 Major Constituents of Volatile Oil of the Seeds and Husk of Large Cardamom

Part of Large Cardamom	Identification Method	Major Constituent(s)	Concentration (%)	References
Seed	Preparative gas chromatography (GC)	1,8-Cineole	74	Lawrence (1970)
	Packed column GC	Sabinene	9.1	Patra et al. (1982)
		1,8-Cineole	63.3	
		γ-Terpinene	16.2	
	Gas chromatography-mass spectrometry (GC-MS)	1,8-Cineole	61.3	Gurudutt et al. (1996)
		β-Pinene	8.85	
		α-Terpineol	7.92	
		α-Pinene	3.79	
		allo-Aromadendrene	3.17	
		1,8-Cineole	84.5 (fresh seed); 86.0 (dried seed)	Rout et al. (2003)
		α-Terpineol	4.6 (fresh seed); 2.9 (dried seed)	
		α-Thujene	0.1 (fresh and dried seed)	
		α-Pinene	1.7 (fresh seed); 2.3 (dried seed)	
		β-Pinene	3.2 (fresh seed); 3.4 (dried seed)	
		Camphene	0.1 (fresh and dried seed)	
		Myrcene	0.7 (fresh seed); 1.4 (dried seed)	
		1,8-Cineole	60.8	Satyal et al. (2012)
		α-Pinene	6.4	
		β-Pinene	8.3	
		α-Terpineol	9.8	
		Terpinen-4-ol	3.4	
Husk	GC-MS	1,8-Cineole	38.7	Pura Naik et al. (2004)
		β-Pinene	13.6	
		α-Terpineol	12.6	
		Spathulenol	8.3	
		4-Terpineol	4.5	
		Germacrene D	3.0	
		α-Pinene	2.8	
		β-Selinene	2.7	
		1,8-Cineole	39.0	Satyal et al. (2012)
		α-Pinene	4.8	
		β-Pinene	17.7	
		α-Terpineol	12.3	
		Terpinen-4-ol	3.2	
		Spathulenol	3.4	

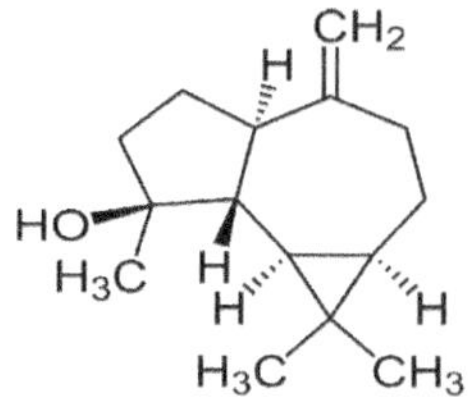

Figure 14.4 Spathulenol.

as observed in different studies. Gurudutt et al. (1996) observed that the seed oil of large cardamom grown in Sikkim contains 40 compounds, out of which seven were oxygenated monoterpenes (75.2%), 19 were monoterpene hydrocarbons (16.3%) and 11 were sesquiterpenes (6.3%). Satyal et al. (2012) identified 70 compounds in the seed oil and 64 compounds in husk oil; the component of 1,8-Cineole was found to be the major component in both seed (60.8%) and husk (39%). The compound spathulenol (Figure 14.4) was present only in the rind/husk oil. The minor or trace components of the large cardamom oleoresin have been listed in Table 14.7. The cultivation parameters and extraction method have significant influence on the yield and chemical composition of the volatile oil.

Large cardamom is also composed of non-volatile pigments. Through methanolic HCl extraction of cardamom husk, Pura Naik et al. (2004) observed presence of two pigments viz., cyanidin 3-glucoside (deep pinkish) and cyanidin 3, 5-diglucoside (red) in the ratio of 1:2.

14.4 EXTRACTION OF BIOACTIVE COMPOUNDS FROM CARDAMOM

Various techniques have been attempted for extraction of volatile oil from small and large cardamom. The traditional methods for extraction of volatile oil from spices include solvent extraction, hydro-distillation or steam distillation. However, these methods suffered from certain major disadvantages such as unwanted loss of volatiles, longer extraction time and thermal or hydrolytic degradation of unsaturated or ester compounds (Khajeh et al. 2004; Tuan and Hangantileke 1997). Since small cardamom is highly expensive and possesses rich therapeutic value, it is very important to look out for extraction techniques which can provide better yield with least damage to the volatiles. Recent studies showed that use of extraction methods such as supercritical fluid extraction, enzyme-assisted hydro distillation, solvent-free microwave extraction, cold grinding, etc. can enhance the quantity and quality of expressed oil compared to use of traditional hydro distillation.

Table 14.7 Minor or Trace Volatile Oil Components of Large Cardamom

Part of Large Cardamom	Trace Components	References
Seed	α-Phellandrene, p-Cymene, β-Ocimene, cis-Sabinene hydrate, trans-Sabinene hydrate, γ-Cadinene, δ-Cadinene, α-Fenchol, cis-Pincarveol, trans-p-Menth-2-en-1-ol, Isobomeol, trans-Piperitol, α-Muurolol, 2-Furanmethanol, Sabinene, Methyl benzoate, Perillene, trans-Thujone, cis-Carveol, α-Terpinen-7-al, Bornyl acetate, p-Cymene-7-ol, 2-Undecanone, Perilla alcohol, Eugenol, α-Copaene, β-Elemene, (E)-Caryophyllene, Caryophyllene oxide, α-Humulene, γ-Muurolene, δ-Selinene, Bicyclogermacrene, Isospathulenol	Rout et al. (2003), Satyal et al. (2012)
Husk	Sabinene, α-Phellandrene, Linalool, Nopinone, trans-Verbenol, Camphene hydrate, Sabina ketone, δ-Selinene, α-Muurolene, α-Calacorene	Satyal et al. (2012)

Baby and Ranganthan (2016) attempted enzyme pretreated steam distillation method and inferred that the enzymes help in disruption of the plant cell wall and thereby leads to better extraction of volatile oil. The use of liquid or compressed gases such as carbon dioxide for extraction eliminates chemical alteration of volatile oil (Hamdan et al. 2008). The ultrasound assisted extraction technique offers several advantages such as intensification of mas transfer, better cell disruption, enhanced solvent penetration, improved yield and shorter duration of extraction (Samaram et al. 2014; Chemat and Khan 2011; Yang and Zhang 2008). The extraction methods (both traditional and modified) used by different researchers for small and large cardamom are tabulated in Table 14.8. The identification of major oil constituents was basically conducted using gas chromatography-mass spectrometry (GC-MS) method. The extracted volatiles are highly susceptible to storage time and temperature. The hydrocarbons such as α-Pinene, sabinene and limonene (extracted using CO_2) reduced by 35–50% during storage for 90 days at 0°C. The major aroma producing volatiles viz. 1,8-cineole and terpinyl acetate also suffer reduction during storage, the loss being higher for cineole (Parthasarathy et al. 2008).

14.5 MEDICINAL PROPERTIES OF CARDAMOM

The medicinal properties of both small and large cardamom are well established since ages. The therapeutic value of these spices is mainly due to their volatile oil content and presence of certain phytochemicals as discussed in the subsection 14.3. Various researchers have demonstrated their pharmacological properties such as antioxidant, anti-microbial, anti-cancer, insecticidal, anti-inflammatory properties, etc. It has been reported by Polo et al. (2012) that α-Pinene possesses gastroprotective activity. Valente et al. (2013) observed anti-inflammatory property in sabinene and linalool. One of the major aroma producing constituents of large and small cardamom, 1,8-Cineole, can also be used for the treatment of inflammatory airway diseases such as asthma and chronic obstructive pulmonary disease (Juergens 2014).

14.5.1 Medicinal Properties of Small Cardamom

Since 4th century BC, small cardamom has been extensively used in Indian Ayurveda, ancient Chinese, Greek and Roman medical practices for treatment of various diseases as presented in Table 14.9. In Ayurveda, it helped in treating food poisoning. A specific preparation of cardamom, Eladigana, has been used for treating arthritis, itching and congestion (Ashokkumar et al. 2020b).

Govil (1998) observed that capsule powder of small cardamom, when mixed with powdered cloves, ginger and caraway, can be used for relieving digestive ailments. He also pointed out towards the anti-inflammatory activity and headache relieving effect of cardamom capsules. Chewing cardamom is a widely accepted way of avoiding bad breath. According to Singh and Singh (1996), the eyesight can be improved by daily consumption of cardamom with a tablespoon of honey.

Apart from the above-mentioned medicinal uses, small cardamom capsules are widely known for their various biological activities such as anti-inflammatory, antiseptic, anti-microbial, carminative, digestive, diuretic, stimulants, stomachic, anti-spasmodic effects.

14.5.1.1 Anti-Inflammatory Property

Many serious health issues in human such as rheumatoid arthritis, allergy, cardiovascular problems, Alzheimer's disease, inflammatory bowel disease, etc. stem from chronic inflammation. The main factors which induce inflammation include unhealthy lifestyle, poor nutrition, excessive alcohol intake, lack of physical activity and most frequently, pathogenic infection. The volatile extract as well as phytochemicals of small cardamom has been observed to have high potential of preventing inflammatory diseases through many experiments. The anti-inflammatory effect of this spice

Table 14.8 Extraction Methods Used for Extraction of Volatile Oil from Small and Large Cardamom

Raw Material	Extraction Method	Extraction Parameters	Maximum Oil Yield	Major Oil Constituents Identified	References
			Small Cardamom		
Seeds	Hydro-distillation	• Hydro distillation period: 4–5 h	8.79% (Malabar variety); 8.60% (Guatemala); 7.99% (Mysore variety); 7.90% (Vazhukka variety)	Terpinyl acetate (61.65–68.19%), 1,8-Cineole (7.23–11.76%), Linalool (0.86–5.96%); Geraniol (1.52–2.47%), Neryl acetate (1.61–2.11%), D-Limonene (1.57–2.23%). α-Terpineol (2.91–5.51%), Nerolidol (0.88–3.57%), (Terpinen-4-ol: 1.18–1.53%)	Amma et al. (2010)
	Enzyme pre-treated steam distillation [Enzymes used: Celluclast 1.5 L, Pectinex Ultra SP.L, ViscozymeL and Protease]	• Enzyme concentration: 1% • pH: 4–5.5 • Incubation period: 90 min • Temperature: 45–50°C • Steam distillation period: 4 h	7.83% (db) [with ViscozymeL]	α-Pinene (1.66%), Sabinene (4.01%), β-Pinene (2.85%), γ-Terpinene (0.75%), 1,8-Cineole (36.29%), Linalool (1.51%), Terpinen-4-ol (2.1%), α-Terpineol (2.28%), Linalyl acetate (0.75%), Terpinyl acetate (41.89%)	Baby and Ranganathan (2016)
	Ultrasound pre-treated hydro-distillation	• Seed powder: 40 g • Solid to water ratio: 1:12 (w/v) • Ultrasound extraction period: 30 min • Total extraction period: 1 h max.	7.4%	α-Pinene (0.91%), β-Myrcene (0.3%), β-Pinene (2.73%), α-Phellandrene (1.6%), ρ-Cymene (0.17%), d-Limonene (3.25%), (E)-β-Ocimene (0.18%), γ-Terpinene (0.34%), α-Terpinolene (0.55%), 1,8-Cineole (26.59%), Trans Sabinine hydrate (0.46%), 1-Octanol (0.17%), 4-Thujanol (0.28%), Linalool (5.47%), Terpinen-4-ol (1.27%), α-Terpineol (3.79%), Nerol (3.52%), Z-Citral (0.20%), δ-Terpinyl acetate (0.22%), α-Terpinyl acetate (40.56%), Neryl acetate (1.02%), Geranyl acetate (1.23%)	Morsy (2015)

(Continued)

Table 14.8 Extraction Methods Used for Extraction of Volatile Oil from Small and Large Cardamom (*Continued*)

Raw Material	Extraction Method	Extraction Parameters	Maximum Oil Yield	Major Oil Constituents Identified	References
			Small Cardamom		
	Supercritical CO_2 and subcritical propane extraction	• Raw material taken: 45 g • Pressure: 10–30 MPa (for supercritical CO_2); 2–5 MPa (for subcritical propane) • Temperature: 35–55°C (for supercritical CO_2); 25°C (for subcritical propane) • CO_2/propane flow rate: 1.0 – 1.5 L/min	6.65% (for supercritical CO_2 extraction); 7.24% (for subcritical propane extraction)	• β-Pinene: 27.6 (for supercritical CO_2 extraction); 26.9 (for subcritical propane extraction) • 1,8-Cineole: 450 (for supercritical CO_2 extraction); 386 (for subcritical propane extraction) • Linalool: 73.5 (for supercritical CO_2 extraction); 72.1 (for subcritical propane extraction) • α-Terpineol: 91.2 (for supercritical CO_2 extraction); 82.7 (for subcritical propane extraction) • Terpinyl acetate: 579 (for supercritical CO_2 extraction); 521 (for subcritical propane extraction) [In terms of 10 μL extract x 10^3]	Hamdan et al. (2008)
	Supercritical CO_2 Extraction	• Raw material taken: 280 g • Pressure: 9 MPa • Temperature: 40°C • CO_2 flow rate: 1.2 kg/h • Particle size of seeds: 250–425 μm	5.5%	α-Terpinyl acetate (42.3%), 1,8-Cineole (21.4%), Linalyl acetate (8.2%), Limonene (5.6%), Linalool (5.4%)	Marongiu et al. (2004)
	Solvent free microwave extraction	• Raw material taken: 100 g • Extraction time: 75 min • Power: 390 W • Humidity level: 67%	2.70%	1,8-Cineole (35.11–51.79%), Linalool (7.47–9.72%), Terpin-4-ol (2.30 – 4.15%), α-Terpineol (3.03–6.04%), Linalyl acetate (4.36–7.62%), α-Terpinyl acetate (19.40 – 29.22%)	Lucchesi et al. (2007)
Whole cardamom	Enzyme pre-treated hydro-distillation [Enzyme used: Lumicellulae]	• Cardamom: 100 g • Enzyme concentration: 4 mg/mL • pH: 5–5.5 • Incubation period: 2 h • Temperature: 38°C • Hydro distillation period: 4 h at 100°C	2.5%	α-Pinene (1.13%), Sabinene (1.71%), β-Myrcene (1.93%), γ-Terpinene (1.11%), 1,8-Cineole (20.9%), Sabinine hydrate (0.7%), Linalool (1.54%), Terpinen-4-ol (2.18%), α-Terpineol (4.34%), Linalyl acetate (0.93%), α-Terpinyl acetate (48.6%), Geraniol (3.69%), Geranial (0.70%), Carvacrol (0.55%), Neryl acetate (2.78%), α-Cedarene (0.75%), E-Nerolidol (1.21%), β-Caryophyllene oxide (3.86%)	Chandran et al. (2012)

(*Continued*)

Table 14.8 Extraction Methods Used for Extraction of Volatile Oil from Small and Large Cardamom (*Continued*)

Raw Material	Extraction Method	Extraction Parameters	Maximum Oil Yield	Major Oil Constituents Identified	References
	Cold grinding followed by hydro-distillation	• Temperature for cold grinding: 2–5°C • Optimum sample:ice ratio: 1:5 • Hydro distillation period: 4 h	8%	α-Terpinyl acetate (47.69%), 1,8-Cineol (34.33%), α-Pinene (1.47%), Sabinene (3.82%), α-Terpineol (2.34%), β-Elemene (0.23%), γ-Cadinene (0.01%), Nerolidol (0.09%), cis-Retinol (0.91%)	Omanakutty and Joy (2007)
			Large Cardamom		
Whole cardamom	Hydro-distillation	• Raw material taken: 50 g • Temperature: 70°C Distillation period: 3 h	1.7% (Indian variety); 1.9% (Saudi Arabian variety)	• 1,8-Cineole: 44.24% (Indian); 46.22% (Saudi Arabian) • α-Terpineol: 7.47% (Indian); 7.04% (Saudi Arabian) • Terpinen-4-ol: 5.01% (Indian); 4.83% (Saudi Arabian) • β-Pinene: 3.38% (Indian); 3.98% (Saudi Arabian) • Trans-ρ-mentha-1(7),8-dien-2-ol: 2.34% (Indian); 3.54% (Saudi Arabian) • α-Selinene: 3.14% (Indian); 2.91% (Saudi Arabian) • β-Myrcene: 2.26% (Indian); 2.53% (Saudi Arabian) • Linalool: 2.14% (Indian); 2.08% (Saudi Arabian) Geraniol D: 4.05% (Indian); 3.54% (Saudi Arabian)	Alam et al. (2021)
Seeds	Hydro-distillation	• Raw material taken: 20 g • Distillation period: 2 h	9.8–19.5 g/kg (values varied for different regions)	• 1,8-Cineole: 50.55–60.46% • α-Terpineol: 14.88–15.48% • Limonene: 5.49–11.75% • Nerolidol: 3.79–6.04% • 4-Terpineol: 2.60–5.39% • δ-Terpineol: 2.86–3.35% • δ-3-carene: 1.22–2.18% • β-Myrcene:1.16–2.36% • Germacrene D: 0.52–2.28% • α-Terpinene: 0.70–1.14%	Joshi et al. (2013)

Table 14.9 Use of Small Cardamom for Treatment of Various Health Problems

Health Problems and/or Pharmacological Effect	References
Bronchitis, asthma, constipation, anti-inflammatory agent, analgesic	Al-Zuhair et al. (1996); Bisht et al. (2011)
Diuretic, carminative, cold, cough, teeth and gum infections, urinary and kidney disorders lung congestion, pulmonary tuberculosis, and irritation of eyelids	Hamzaa et al. (2012); Jafri et al. (2001); Saeed et al. (2014); Nair and Unnikrishnan (1997)
Cataracts, nausea, vomiting, diarrhoea and cardiac problems	Khan et al. (2011); Gilani et al. (2008);
Stomach ache, bladder infection, dysentery and constipation in children	Duke et al. (2003); Kapoor (2018)

maybe attributed to its ability to inhibit NF-kB signalling pathway. NF-kB is a transcription factor which induces transcription of various pro-inflammatory cytokines, chemokines, adhesion molecules and matrix metalloproteinases, cyclo-oxygenase-2 and inducible nitric oxide (Qiblawi et al. 2020). According to Chen and Kong (2004), cardamom activates anti-inflammatory agents such as Nrf2, which in turn activates a series of enzymes having protective, antioxidative and detoxification properties. Rodrigues et al. (2009) reported that eugenol, an active constituent of small cardamom, can significantly prevent secretion of pro-inflammatory mediators IL-1b and IL-6. Choi et al. (2007) further pointed out that isoeugenol, another component of cardamom, have inhibitory effects against nitric oxide synthase and nitric oxide. Zhao et al. (2014) observed that 1,8-Cineole has ability to suppress lipo-polysaccharide-induced production of pro-inflammatory cytokine. Souissi et al. (2020) successfully examined the anti-inflammatory activity of the volatile oil of small cardamom (both seeds and fruits) through a test against periodontal pathogens. Al-Zuhair et al. (1996) performed a comparative study of the anti-inflammatory activity of volatile oil extracted small cardamom (175 and 280 μl/kg) and indomethacin (30 mg/kg) against acute carrageenan-induced planter oedema in male albino rats.

14.5.1.2 Antioxidant Property

The chief reason for induction of several chronic diseases like diabetes, cancer, Parkinson's disease and immune malfunction is considered to be oxidative stress. Cardamom pods along with their seeds are rich reserve of antioxidant elements which help in scavenging free radicals. The presence of phenolic compounds such as kaempferol, quercetin and luteolins in small cardamom capsule imparted moderate level of antioxidant activity to it (Nair et al. 1998). The essential oil of small cardamom behaves as an antioxidant and thereby helps in bringing down cell ageing through neutralization of free radicals (Saeed et al. 2014; Nanasombat and Wimuttigosol 2011). Amma et al. (2010) stated that the antioxidant properties of small cardamom can raise glutathione, a natural antioxidant, in the body. They further experimentally found out that the ethyl acetate extract of Malabar cultivar of small cardamom exhibited highest DPPH (2,2-diphenyl-1-picrylhydrazyl) radical scavenging activity (88.7 μg/mL) followed by other popular varieties viz., Vazhukka (186 μg/mL), Guatemala (223 μg/mL) and Mysore (293 μg/mL). Elguindy et al. (2018) observed that the daily oral intake of volatile/essential oil of small cardamom (100 and 200 mg/kg/day) for 26 weeks significantly reduced serum creatinine and urea, and LDH activity in diethylnitrosamine (DENA) induced oxidative stress in rat kidney and brain.

14.5.1.3 Anti-Cancerous Property

The most widely accepted reason for human mortality is cancer. Recent studies have led to increased trust in chemoprevention with the aid of food phytochemicals (Hussain et al. 2018; Najm et al. 2012). These natural phytochemicals counteract, prevent, delay or reduce risk of cancer by acting on multiple targets in the cellular signalling pathways and regulation of critical molecules (Pratheeshkumar et al. 2012). Several experimental evidences have indicated towards the potential health benefits of small cardamom against various types of cancer. Hashim et al. (1994) conducted in vitro microsomal enzyme intermediated reaction, where the essential oil of small cardamom established its anti-carcinogenic activity by impeding damage to adult DNA by aflatoxin B1. According to Bhattacharjee et al. (2007), the volatile oil constituents viz., 1,8-cineole and limonene possess protective role against development of cancer. Small cardamom antagonizes diverse chemical-induced carcinogenesis basically through activation of the detoxification enzymes, which plays vital role in influencing the ultimate destination of carcinogens/procarcinogens (Chen and Kong 2004). Elguindy et al. (2016) reported that geraniol, a cardamom extract, is significantly effective in prevention of DENA-induced hepatocellular carcinoma.

14.5.1.4 Anti-Microbial Property

The natural anti-microbials exhibit an array of advantages including reduction in antibiotic demand, inhibiting microbial contamination in food, retardation of microbial spoilage thereby enhancement of shelf-life of biological products and improvement of human immune cells. Cardamom seed extracts have exhibited robust anti-microbial activity of variable intensity against different foodborne microorganisms. In a study conducted by Shakila et al. (1996), application of the essential oil of cardamom resulted in moderate inhibition of the growth of Morgenella morganii. Several assays proved that essential oil of small cardamom is effective against various bacteria such as Staphylococcus aureus, Escherichia coli, Salmonella typhi, Streptococcus mutans, Candida albicans, Bacillus pulmilus, Listeria monocytogenes, Mycobacterium smegmatis, Klebsiella pneumoniae, Enterococcus faecalis and Micrococcus luteus (Abdullah et al. 2017; Bano et al. 2016; Ağaoğlu et al. 2005). Ağaoğlu et al. (2005) found no inhibitory effect of cardamom oil against Pseudomonas aeruginosa. Grădinaru et al. (2014) observed noteworthy anti-bacterial activity of 1,8-Cineole (dosage: 2.5 mg/mL) against target microorganism Staphylococcus aureus. Owing to its inhibitory effect against wide range of microorganisms, essential oil of small cardamom is a safe option for commercial production of antibiotics. It can also be effectively used for prevention of food-borne pathogenic and spoilage microorganism induced damage.

14.5.1.5 Gastroprotective Property

The gastroprotective effect of small cardamom, especially of its major constituent 1,8-cineole has been established in many studies. Jamal et al. (2006) investigated the anti-ulcerogenic activity of crude methanolic extract, essential oil, petroleum ether soluble and insoluble fractions of methanolic extract of small cardamom on aspirin, ethanol and pylorous ligature induced gastric lesions in rats. The researchers found that the lesions induced by ethanol and aspirin could be effectively inhibited by the cardamom fractions; the extracts could not relieve the lesion induced by pylorus ligation. Santos and Rao (2001) also pointed towards the gastroprotective property of 1,8-Cineole, whose oral administration (50–200 mg/kg) could significantly reduce ethanol-induced mucosal damage in rats. 1,8-Cineole was also found to have potential to prevent gastrointestinal ulceration (Qiblawi et al. 2020).

Amongst all these studies favouring the medicinal value of small cardamom, few negative reports have also been surfaced. According to a study conducted by Nair (2006), some sections of consumers believe that excessive intake of cardamom can lead to impotency.

14.5.2 Medicinal Properties of Large Cardamom

Like small cardamom, large cardamom is also widely known for its physiological and pharmacological properties. This spice has been used in Ayurvedic, Yunani, Tibetan and Chinese medical systems. Its bioactivity is attributed to its essential oil constituents. Large cardamom can act as stimulant, alexipharmic, stomachic and astringent. Bisht et al. (2011) reported that the above-mentioned medicinal properties of large cardamom have enabled this spice to be used for the treatment of vomiting, nausea, cough, itching, biliousness, indigestion, abdominal pain and rectal problems. It is also known to cure or prevent throat troubles, lung congestion, inflammation of eyelids and pulmonary tuberculosis (Verma et al. 2010). The spice also acts as an antidote for snake and scorpion venom (Parthasarathy et al. 2008).

Several researchers have explored the bioactivities of large cardamom viz., its anti-inflammatory, antimicrobial, antioxidant, hepatoprotective, anti-ulcerogenic, etc. properties.

14.5.2.1 Anti-Microbial Property

Apart from essential oils, large cardamom is also a rich reserve of secondary metabolites like alkaloids, flavonoids and tannins, few of which possess anti-microbial properties. Satyal et al. (2012) screened its essential oil for anti-microbial activity against Gram-positive bacteria such as Bacillus cereus and Staphylococcus aureus, and Gram-negative bacteria such as Pseudomonas aeruginosa and Escherichia coli. Hussain et al. (2011) observed that the acetonic, methanolic and ethanolic extracts of large cardamom showed inhibitory activity against two dental caries causing bacteria viz., Streptococcus mutans and Staphylococcus aureus. Tijjani et al. (2012) studied anti-microbial efficacy of the methanolic seed extract of this spice against Staphylococcus aureus, Streptococcus pneumoniae, Bacillus subtilis, Salmonella typhi, Salmonella pyrogenes, Klebsiella pneumoniae, Escherichia coli, Psudomonas aeruginosa and Candida albican sengalensis; the extract exhibited varying inhibitory activities against all the test organisms except Salmonella pyrogenes and Escherichia coli. Aneja and Joshi (2009) reported anti-fungal activity of large cardamom extracts against Candida albicans and Saccharomyces cerevisae. In a study conducted by Kapoor et al. (2008), the essential oil of large cardamom showed 100% inhibition against Aspergillus flavus at a dosage of 6 μL and moderate inhibitory effects against Aspergillus niger, A. awamori, A. oryzae, A. solani and Fusarium monoliforme. Hence, the authors recommended its usage as natural food preservatives.

14.5.2.2 Antioxidant Property

The antioxidant activity of volatile oil constituents such as 1,8-cineole, α-terpineol, protocatechualdehyde, protocatechuic acid against lipid peroxidation has been reported by Kikuzaki et al. (2001). Dhuley (1999) pointed out that the essential oil of large cardamom has ability to activate and enhance hepatic and cardiac antioxidant enzymes. According to Kapoor et al. (2008), the significant antioxidant activity of large cardamom is attributed to the presence of high level of phenolic compounds in its essential oil; the oleoresins of large cardamom were found to be a better antioxidant than butylated hydroxytoluene. Yadav and Bhatnagar (2007) reported that the polyphenol content of large cardamom act as superoxide radical scavenging compounds, which have strong reducing and inhibitory activity against iron-induced lipid peroxidation in rat liver. Prakash et al. (2012) evaluated in vitro antioxidant activity of the ethanolic and aqueous extracts of cardamom leaves; the ethanolic extracts showed higher antioxidant activity than aqueous extracts, but less activity than butylated hydroxyanisole. Kanthlal et al. (2020) observed that the aqueous extract of the crude large cardamom fruit possesses preventive activity against N^{ω}-Nitro-L-arginine methyl ester (L-NAME)-induced hypertension in rats, which indicates positively towards its high antioxidant activity.

14.5.2.3 Anti-Inflammatory Property

As discussed earlier, modern medical experts and nutritionists are getting more inclined towards herbal remedies for treatment of several health problems including inflammatory diseases. The edible parts of large cardamom have been fairly explored for their anti-inflammatory activity. Alam et al. (2011) determined in vitro anti-inflammatory activity of Indian and Saudi Arabian varieties of large cardamom using bovine albumin inhibitory assay; the Indian and Saudi Arabian varieties exhibited 69.09% and 66.81% inhibition of protein denaturation, respectively. Kumar et al. (2012) assessed anti-inflammatory activity of dried methanolic and aqueous extracts of large cardamom against carrageenan induced paw oedema in rats. It was observed that both methanolic (250 mg/kg) and aqueous (500 mg/kg) extracts showed significant and dose dependent anti-inflammatory activity; the doses were administered orally. Based on the dosage, it was clearly seen that the methanolic extract was more effective than the aqueous extract. Alam et al. (2011) also observed dose-dependent anti-inflammatory activity of ethanolic (100 mg/mL) and aqueous (200 mg/mL) extracts of large cardamom against carrageenan induced paw oedema in rats.

14.5.2.4 Anti-Ulcerogenic Property

Large cardamom has been extensively used in the Unani system of medicine for treatment of gastrointestinal disorders such as gastric ulcers. In an experiment performed by Jafri et al. (2001), the crude methanolic extract of large cardamom and its fractions namely volatile oil, petroleum ether and ethyl acetate significantly inhibited ethanol induced gastric ulcers in rats, but could not inhibit ulcers induced by pylorus ligation and aspirin. It was also observed that the ethyl acetate fraction increased the mucus wall in pylorus ligated rats, hence, exerted a protective effect on gastric mucosal wall. Supporting the above trend, Sen et al. (2009) reported that the ethanol induced gastric lesion can be significantly inhibited through treatment with methanolic fraction, petrol soluble fraction, ethyl acetate fraction, methanol insoluble fraction and essential oil of large cardamom.

14.5.2.5 Hepatoprotective Property

Large cardamom has also been proved to provide protection against liver damage. The major components of A. subulatum oil viz., 1,8-cineole, γ-terpinene, α-terpineol, sabinene, α-pinene and limonene helps in bile elimination and reduction in liver congestion; they also act as heart and liver tonic, and an appetizer (Parthasarathy et al. 2008). Parmar et al. (2009) studied the hepatoprotective activity of methanolic extract of A. subulatum seeds against 20% ethanol induced liver damage in rats; treatment with the said extract significantly prevented the ethanol induced physical, biochemical, functional and histological changes thereby leading to the recovery of hepatic cells. According to Parmar et al. (2011), the hepatoprotective activity of large cardamom can be attributed to mitochondrial protection and regulation of voltage-dependent anion channel expression.

14.6 CONCLUSION

Cardamom is one of the major spices produced in the world. Both small (*Elettaria cardamomum* L. Maton) and large (*Amomum subulatum* Roxburgh) cardamom are known for their distinct aroma and therapeutic value. The high medicinal value of cardamom can be attributed to their rich reserve of bioactive compounds viz., aromatic volatile oil, fixed oil, phytochemicals and pigments. The oil composition is highly dependent upon type of soil, altitude, latitude, climate change and seasonal variations. The major aroma producing compounds of small and large cardamom are 1,8-cineole (gives camphory aroma) and α-terpinyl acetate (gives spicy flavour). The long list of pharmacological properties of the bioactive compounds of cardamom led to extensive extraction studies by various researchers. Since the bioactive compounds are heavily influenced by processing conditions, continuous improvements are made in the extraction methods as well as storage conditions for better retention of the original oil parameters and longer shelf-life.

REFERENCES

Abbasipour, H., Mahmoudvand, M., Rastegar, F., and Hosseinpour, M. H. 2011. Fumigant toxicity and oviposition deterrency of the essential oil from the cardamom, *Elettaria cardamomum* (Maton) against three stored-product insects. *Journal of Insect Science* (167): 1.

Abdullah, A. A., Butt, MS, Shahid, M., and Huang, Q. 2017. Evaluating the antimicrobial potential of green cardamom essential oil focusing on quorum sensing inhibition of Chromobacterium violaceum. *Journal of Food Science and Technology* 54(8): 2306–2315.

Ajit, A., Brijeetha, P. D., and Varkey, P. 2020. A mini review on the bioactive compounds in dried cardamom seed and its leaf. *World Journal of Pharmaceutical Research* 9(14): 1410–1423.

Alam, A., and Singh, V. 2021. Composition and pharmacological activity of essential oils from two imported Amomum subulatum fruit samples. *Journal of Taibah University Medical Sciences* 16(2): 231–239.

Alam, K., Pathak, D., and Ansari, S. H. 2011. Evaluation of anti-inflammatory activity of Ammomum subulatum fruit extract. *International Journal of Pharmaceutical Sciences and Drug Research* 3(1): 35–37.

Al-Zuhair, H., El-Sayeh, B., Ameen, H. A., and Al-Shoora, H. 1996. Pharmacological studies of cardamom oil in animals. *Pharmacological Research* 34(1–2): 79–82.

Amma, K. P. A. P., Rani, M. P., Sasidharan, I., and Nisha, V. N. P. 2010. Chemical composition, flavonoid-phenolic contents and radical scavenging activity of four major varieties of cardamom. *International Journal of Biology Medical and Research* 1(3): 20–24.

Aneja, K. R., and Joshi, R. 2009. Antimicrobial activity of Amomum subulatum and *Elettaria cardamomum* against dental caries causing microorganisms. *Ethnobotanical Leaflets* (7): 3.

Ashokkumar, K., Murugan, M., Dhanya, M. K., and Warkentin, T. D. 2020b. Botany, traditional uses, phytochemistry and biological activities of cardamom [*Elettaria cardamomum* (L.) Maton]–A critical review. *Journal of Ethnopharmacology* 246: 112244.

Ashokkumar, K., Murugan, M., Dhanya, M. K., Raj, S., and Kamaraj, D. 2020a. Phytochemical variations among four distinct varieties of Indian cardamom *Elettaria cardamomum* (L.) Maton. *Natural Product Research* 34(13): 1919–1922.

Ashokkumar, K., Pandian, A., Murugan, M., et al. 2020c. Profiling bioactive flavonoids and carotenoids in select south Indian spices and nuts. *Natural Product Research* 34(9): 1306–1310.

Baby, K. C., and Ranganathan, T. V. 2016. Effect of enzyme pre-treatment on extraction yield and quality of cardamom (*Elettaria cardamomum* maton.) volatile oil. *Industrial Crops and Products*: 89: 200–206.

Bano, S., Ahmad, N., and Sharma, A. K. 2016. Phytochemical screening and evaluation of anti-microbial and anti-oxidant activity of *Elettaria cardamom* (Cardamom). *Journal of Applied and Natural Science* 8(4): 1966–1970.

Bhandari, A. K., Bisht, V. K., Negi, J. S., and Baunthiyal, M. 2013. 1, 8-Cineole: A predominant component in the essential oil of large cardamom (*Amomum subulatum* Roxb.). *Journal of Medicinal Plants Research* 7(26): 1957–1960.

Bhattacharjee, S., Rana, T., and Sengupta, A. 2007. Inhibition of lipid peroxidation and enhancement of GST activity by cardamom and cinnamon during chemically induced colon carcinogenesis in Swiss albino mice. *Asian Pacific Journal of Cancer Prevention* 8(4): 578–582.

Bisht, V. K., Negi, J. S., Bh, A. K., and Sundriyal, R. C. 2011. *Amomum subulatum* Roxb: Traditional, phytochemical and biological activities-An overview. *African Journal of Agricultural Research* 6(24): 5386–5390.

Biswas, A. K., Gupta, R. K., and Bhutia, D. T. 1986. Characteristics of different plant parts of large cardamom. *Cardamom* 19(3): 7–10.

Boiswert, C. and Hubert, A. 1998. *L'ABCdaire des épices*. Flammarion, Paris.

Chandran, J., Amma, K. P. P., Menon, N., Purushothaman, J., and Nisha, P. 2012. Effect of enzyme assisted extraction on quality and yield of volatile oil from black pepper and cardamom. *Food Science and Biotechnology* 21(6): 1611–1617.

Chemat, F., and Khan, M. K. 2011. Applications of ultrasound in food technology: processing, preservation and extraction. *Ultrasonics Sonochemistry* 18(4): 813–835.

Chen, C., and Kong, A. N. T. 2004. Dietary chemopreventive compounds and ARE/EpRE signaling. *Free Radical Biology and Medicine* 36(12): 1505–1516.

Choi, C. Y., Park, K. R., Lee, J. H., et al. 2007. Isoeugenol suppression of inducible nitric oxide synthase expression is mediated by down-regulation of NF-κB, ERK1/2, and p38 kinase. *European Journal of Pharmacology* 576(1–3): 151–159.

Dhuley, J. N. 1999. Anti-oxidant effects of cinnamon (*Cinnamomum verum*) bark and greater cardamom (*Amomum subulatum*) seeds in rats fed high fat diet. *Indian Journal of Experimental Biology* 37(3): 238–242.

Duke, J. A., Bogenschutz-Godwin, M. J., deCellier, J., and Duke, P. K. 2003. *Elettaria cardamomum* (L.) Maton (Zingiberaceae) Cardamon, Malabar or Mysore cardamon. *CRC Handbook of Medicinal Spices* 159.

Elguindy, N. M., Yacout, G. A., and El Azab, E. F. 2018. Amelioration of DENA-induced oxidative stress in rat kidney and brain by the essential oil of *Elettaria cardamomum*. *Beni-Suef University Journal of Basic and Applied Sciences* 7(3): 299–305.

Elguindy, N. M., Yacout, G. A., El Azab, E. F., and Maghraby, H. K. 2016. Chemoprotective effect of *Elettaria cardamomum* against chemically induced hepatocellular carcinoma in rats by inhibiting NF-κB, oxidative stress, and activity of ornithine decarboxylase. *South African Journal of Botany* 105: 251–258.

Ereifej, K. I., Feng, H., Rababah, T. M. et al. 2015. Microbiological status and nutritional composition of spices used in food preparation. *Food and Nutrition Sciences* 6(12): 1134.

Gautam, N., Bhattarai, R. R., Khanal, B. K. S., and Oli, P. 2016. Technology, chemistry and bioactive properties of large cardamom (*amomum subulatum* roxb.): an overview. *International Journal of Applied Sciences and Biotechnology* 4(2): 139–149.

Gilani, A. H., Jabeen, Q., Khan, A. U., and Shah, A. J. 2008. Gut modulatory, blood pressure lowering, diuretic and sedative activities of cardamom. *Journal of Ethnopharmacology* 115(3): 463–472.

Goudarzvand Chegini, S., and Abbasipour, H. 2017. Chemical composition and insecticidal effects of the essential oil of cardamom, *Elettaria cardamomum* on the tomato leaf miner, Tuta absoluta. *Toxin Reviews* 36(1): 12–17.

Govil, J. N. (Ed.). 1998. *Current Concepts of Multidiscipline Approach to the Medicinal Plant* Vol. 12. Today and Tomorrow Publisher.

Govindarajan, V. S., Narasimhan, S., Raghuveer, K. G., Lewis, Y. S., and Stahl, W. H. 1982. Cardamom—Production, technology, chemistry, and quality. *Critical Reviews in Food Science and Nutrition* 16(3): 229–326.

Grădinaru, A. C., Aprotosoaie, A. C., Trifan, A. et al. 2014. Interactions between cardamom essential oil and conventional antibiotics against Staphylococcus aureus clinical isolates. *Farmacia* 62: 1214–1222.

Gupta, P. N. (1986). Studies on capsule morphology of large cardamom cultivars (*Amomum subulatum* Roxb), *Journal of Plantation Crops* 16: 371–375.

Gurudutt, K. N., Naik, J. P., Srinivas, P., and Ravindranath, B. (1996). Volatile constituents of large cardamom (*Amomum subulatum* Roxb.). *Flavour and Fragrance Journal*: 11(1): 7–9.

Hamdan, S., Daood, H. G., Toth-Markus, M., and Illés, V. (2008). Extraction of cardamom oil by supercritical carbon dioxide and sub-critical propane. *The Journal of Supercritical Fluids* 44(1): 25–30.

Hamzaa, R. G., and Osman, N. N. 2012. Using of coffee and cardamom mixture to ameliorate oxidative stress induced in γ-irradiated rats. *Biochemistry and Analytical Biochemistry* 1(113): 2161–1009.

Hashim, S., Aboobaker, V. S., Madhubala, R., Bhattacharya, R. K., and Rao, A. R. 1994. Modulatory effects of essential oils from spices on the formation of DNA adduct by aflatoxin B1 in vitro. *Nutrition Cancer* 21(2): 169–175.

Heywood, V. H. 2002. *The conservation of genetic and chemical diversity in medicinal and aromatic plants. In Biodiversity* (pp. 13–22). Springer, Boston, MA.

Hussain, T., Bajpai, S., Saeed, M., et al. 2018. Potentiating effect of ethnomedicinal plants against proliferation on different cancer cell lines. *Current Drug Metabolism* 19(7): 584–595.

Hussain, T., Arshad, M., Khan, S., Sattar, H. and Qureshi, M. S. 2011. In vitro screening of methanol plant extracts for their antibacterial activity. *Pakistan Journal of Botany* 43(1): 531–538.

Jafri, M. A., Javed, K., and Singh, S. 2001. Evaluation of the gastric antiulcerogenic effect of large cardamom (fruits of *Amomum subulatum* Roxb). *Journal of Ethnopharmacology* 75(2-3): 89–94.

Jamal, A., Javed, K., Aslam, M., and Jafri, M. A. 2006. Gastroprotective effect of cardamom, *Elettaria cardamomum* Maton fruits in rats. *Journal of Ethnopharmacology* 103(2): 149–153.

Joshi, R., Sharma, P., Sharma, V., Prasad, R., Sud, R. K., and Gulati, A. 2013. Analysis of the essential oil of large cardamom (*Amomum subulatum* Roxb.) growing in different agro-climatic zones of Himachal Pradesh, India. *Journal of the Science of Food and Agriculture* 93(6): 1303–1309.

Juergens, U. R. 2014. Anti-inflammatory properties of the monoterpene 1.8-cineole: current evidence for co-medication in inflammatory airway diseases. *Drug research* 64(12): 638–646.

Kanthlal, S. K., Joseph, J., and Paul, B. 2020. Antioxidant and vasorelaxant effects of aqueous extract of large cardamom in L-NAME induced hypertensive rats. *Clinical and Experimental Hypertension* 42(7): 581–589.

Kapoor, I. P. S., Singh, B., Singh, G., Isidorov, V., and Szczepaniak, L. 2008. Chemistry, antifungal and antioxidant activities of cardamom (*Amomum subulatum*) essential oil and oleoresins. *International Journal of Essential Oil Therapeutics* 2(1): 29–40.

Kapoor, L. D. 2018. *CRC handbook of Ayurvedic medicinal plants.* CRC press.

Karibasappa, G. S. 1987. Post-harvest studies in large cardamom (*Amomum subulatum* Roxb), *Journal of Plantation Crops* 15: 63–64.

Khajeh, M., Yamini, Y., Sefidkon, F., and Bahramifar, N. 2004. Comparison of essential oil composition of Carum copticum obtained by supercritical carbon dioxide extraction and hydrodistillation methods. *Food Chemistry* 86(4): 587–591.

Khan, A. U., Khan, Q. J., and Gilani, A. H. 2011. Pharmacological basis for the medicinal use of cardamom in asthma. *Bangladesh Journal of Pharmacology* 6(1): 34–37.

Kizhakkayil, J., Thomas, E., Zachariah, T. J., Syamkumar, S., and Sasikumar, B. 2006. *A comparative quality appraisal of exported cardamoms of India*, Sri Lanka and Guatemala.

Kikuzaki, H., Kawai, Y., and Nakatani, N. 2001. 1, 1-Diphenyl-2-picrylhydrazyl radical-scavenging active compounds from greater cardamom (*Amomum subulatum* Roxb.). *Journal of Nutritional Science and Vitaminology* 47(2): 167–171.

Korikanthimath, V. S., Mulge, R., and Zachariah, T. J.1999. Variation in essential oil constituents in high yielding selections of cardamom. *Journal of Plantation Crops* 27: 230–232.

Korikanthimath, V. S., Rao, G., and Hiremath, G. M. 2002. Cultivation of cardamom (*Elettaria cardamomum*) in valley bottoms under evergreen forest shade. *Journal of Medicinal and Aromatic Plant Sciences* 24: 53–59.

Krishnamurthy, M. N. (1964). *Studies on curing aspects and utilization of cardamom.* Assoc. thesis, CFTRI, Mysore, India.

Krishnamurthy, M.N., Padma Bhai, R. and Natarajan, C.P. 1967. Chemical composition of cardamom. *Journal of Food Science and Technology* 4: 170.

Kumar, G., Chauhan, B., and Ali, M. 2012. Anti-inflammatory effect of *Amomum subulatum* Roxb. fruits extract. *International Journal of Pharmaceutical Sciences and Research* 3(12): 4722.

Kumara, S. J. S., Packiyasothy, E. V., and Jansz, E. R. 1985. Some studies on the effect of maturity and storage on the chlorophyll content and essential oils of the cardamon fruit (*Elettaria cardamomum*). *Journal of the Science of Food and Agriculture* 36(6): 491–498.

Lawrence, B. M. 1970. Terpenes in two Amomum species. *Phytochemistry* 9: 665.

Lawrence, B. M. 1979. Major tropical spices–cardamom (*Elettaria cardamomum*). *Essential oils* 104.

Leela, N. K., Prasath, D., and Venugopal, M. N. 2008. Essential oil composition of selected cardamom genotypes at different maturity levels. *Indian Journal of Horticulture* 65(3): 366–369.

Lucchesi, M. E., Smadja, J., Bradshaw, S., Louw, W., and Chemat, F. 2007. Solvent free microwave extraction of Elletaria cardamomum L.: A multivariate study of a new technique for the extraction of essential oil. *Journal of Food Engineering* 79(3): 1079–1086.

Madhusoodanan, K. J., and Rao, Y. S. 2001. Cardamom (large). *Indian Cardamom Research Institute*, Woodhead Publishing Ltd., Kerala.

Mahmud, S. 2008. Composition of essential oil of *Elettaria cardamomum* Maton leaves. *Pakistan Journal of Science* 3(4): 111–114.

Marongiu, B., Piras, A., and Porcedda, S. 2004. Comparative analysis of the oil and supercritical CO2 extract of *Elettaria cardamomum* (L.) Maton. *Journal of Agricultural and Food Chemistry* 52(20): 6278–6282.

Menon, A. N. 2000. Studies on the volatiles of cardamom (Elleteria cardamomum). *Journal of Food Science and Technology* 37(4): 406–408.

Morsy, N. F. 2015. A short extraction time of high quality hydrodistilled cardamom (*Elettaria cardamomum* L. Maton) essential oil using ultrasound as a pretreatment. *Industrial Crops and Products* 65: 287–292.

Murugan, M. 2011. Factors and patterns of pesticides usage and sustainability of cardomom (*Elettaria cardamomum* (L.) Maton) in Indian cardamom hills (Doctoral dissertation, NIAS).
Murugan, M., Dhanya, M. K., Deepthy, K. B. et al. 2016. *Compendium on Cardamom*, second ed. Kerala Agricultural University, Cardamom Research Station, Pampadumpara, pp. 66.
Muthusamy, M., Panigrahy, B. K., Shetty, P. K., Subbiah, A., and Ravi, R. 2012. Effect of heavy metal and nutrient uptake by soils in Indian Cardamom Hills. *Journal of Soil Science and Environmental Management* 3(8): 196–206.
Nair, P. R. S., and Unnikrishnan, G. 1997. *Evaluation of medicinal values of cardamom alone and in combination with other spices in ayurvedic system of medicine-project (Part I and II)*, 51. Government Ayurveda College, Trivandrum, Kerala, India, 46.
Nair, S., Nagar, R., and Gupta, R. 1998. Antioxidant phenolics and flavonoids in common Indian foods. *The Journal of the Association of Physicians of India* 46(8): 708–710.
Nair, K. P. 2006. The agronomy and economy of Cardamom (*Elettaria cardamomum* M.): the "queen of spices". *Advances in Agronomy* 91: 179–471.
Najm, M. Z., Akhtar, M. S., Ahmad, I. et al. 2012. Mutational analysis of prohibition - a highly conserved gene in Indian female breast cancer cases. *Asian Pacific Journal of Cancer Prevention* 13(10): 5113–5117.
Nanasombat, S., and Wimuttigosol, P. 2011. Antimicrobial and antioxidant activity of spice essential oils. *Food Science and Biotechnology* 20(1): 45–53.
Nigam, M. C., Handa, K. L., Nigam, I. C., and Levi, L. 1965. Essential oils and their constituents: xxix. The essential oil of marihuana: composition of genuine Indian Cannabis sativa L. *Canadian Journal of Chemistry* 43(12): 3372–3376.
Noleau, I., Toulemonde, B., and Richard, H. 1987. Volatile constituents of cardamom (*Elettaria cardamomum* Maton) cultivated in Costa Rica. *Flavour and fragrance journal* 2(3): 123–127.
Oli, P. 2011. *Post-Harvest Technology of Tea, Coffee and Spices*. First Edition, Lalitpur Valley College, Nepal.
Olivero-Verbel, J., González-Cervera, T., Güette-Fernandez, J., Jaramillo-Colorado, B., and Stashenko, E. 2010. Chemical composition and antioxidant activity of essential oils isolated from Colombian plants. *Revista Brasileira de Farmacognosia* 20(4): 568–574.
Omanakutty, M., and Joy, B. 2007. A study on the retention of fresh flavour of cardamom oil by cold grinding. *Journal of Essential Oil Bearing Plants* 10(6): 490–493.
Parmar, M. Y., Shah, P. A., Gao, J., and Gandhi, T. R. 2011. Hepatoprotection through regulation of voltage dependent anion channel expression by *Amomum subulatum* Roxb seeds extract. *Indian Journal of Pharmacology* 43(6): 671.
Parmar, M. Y., Shah, P., Thakkar, V., and Gandhi, T. R. 2009. Hepatoprotective activity of *Amomum subulatum* Roxb against ethanol-induced liver damage. *International Journal of Green Pharmacy* 3(3).
Parry, J., Hao, Z., Luther, M., Su, L., Zhou, K., and Yu, L. L. 2006. Characterization of cold-pressed onion, parsley, cardamom, mullein, roasted pumpkin, and milk thistle seed oils. *Journal of the American Oil Chemists' Society* 83(10): 847–854.
Parthasarathy, V. A., Chempakam, B., and Zachariah, T. J. (Eds.). 2008. *Chemistry of spices*. Cabi.
Patra, N.K., Siddiqui, M.S., Akhila, A., Nigam, M.C. and Naqvi, A.A. 1982. Chemical composition of the volatile oil from the pericarp (husk) of large cardamom (*Amomum subulatum* Roxb.). *Pafai Journal* 4(4): 29–31.
Polo, C. M., Moraes, T. M., Pellizzon, C. H. et al. 2012. *Gastric ulcers in middle-aged rats: The healing effect of essential oil from Citrus aurantium L. (Rutaceae)*. Evidence-Based Complementary and Alternative Medicine, 2012.
Prakash, K. D., Brajesh, K., Arshad, H., Shikhar, V., and Mala, M. 2012. Evaluation of antioxidant activity of large cardamom (leaves of *Amomum subulatum*). *International Journal of Drug Development and Research* 4(1): 175–179.
Pratheeshkumar, P., Sreekala, C., Zhang, Z. et al. 2012. Cancer prevention with promising natural products: mechanisms of action and molecular targets. *Anti-Cancer Agents in Medicinal Chemistry* 12(10): 1159–1184.
Pruthi, J. S. 1993. Major spices of India. Crop management and post-harvest technology. In: *Major spices of India*. Crop management and post-harvest technology.
Pura Naik, J., Jagan Mohan Rao, L., Mohan Kumar, T. M., and Sampathu, S. R. 2004. Chemical composition of the volatile oil from the pericarp (husk) of large cardamom (*Amomum subulatum* Roxb.). *Flavour and Fragrance Journal* 19(5): 441–444.

Qiblawi, S., Kausar, M. A., Shahid, S. M. A., Saeed, M., and Alazzeh, A. Y. 2020. Therapeutic Interventions of Cardamom in Cancer and Other Human Diseases. *Journal of Pharmaceutical Research International* 32: 74–84.

Rao, Y. S., Gupta, U., Anand, K., and Naidu, R. 1993. A note on large cardamom (*Amomum subulatum* Roxb.) germplasm collection. *Journal of Spices and Aromatic Crops* 2(1): 2.

Robert, J. L. 1986. *Le Livre des épices et des herbes*. Flammarion, Paris.

Rodrigues, T. G., Fernandes Jr, A., Sousa, J. P. B., Bastos, J. K., and Sforcin, J. M. 2009. In vitro and in vivo effects of clove on pro-inflammatory cytokines production by macrophages. *Natural Product Research* 23(4): 319–326.

Rout, P. K., Sahoo, D., Jena, K. S., and Rao, Y. R. 2003. Analysis of the oil of large cardamom (*Amomum subulatum* Roxb.,) growing in Sikkim. *Journal of Essential Oil Research* 15(4): 265–266.

Saeed, A., Sultana, B., Anwar, F., Mushtaq, M., Alkharfy, K. M., and Gilani, A. H. 2014. Antioxidant and antimutagenic potential of seeds and pods of green cardamom (*Elettaria cardamomum*). *International Journal of Pharmacology* 10(8): 461–469.

Samaram, S., Mirhosseini, H., Tan, C. P., and Ghazali, H. M. 2014. Ultrasound-assisted extraction and solvent extraction of papaya seed oil: Crystallization and thermal behavior, saturation degree, color and oxidative stability. *Industrial Crops and Products* 52: 702–708.

Santos, F. A., and Rao, V. S. N. 2001. 1, 8-cineol, a food flavoring agent, prevents ethanol-induced gastric injury in rats. *Digestive Diseases and Sciences* 46(2): 331–337.

Satyal, P., Dosoky, N. S., Kincer, B. L., and Setzer, W. N. 2012. Chemical compositions and biological activities of *Amomum subulatum* essential oils from Nepal. *Natural Product Communications* 7(9): 1934578X1200700935.

Savan, E. K., and Küçükbay, F. Z. 2013. Essential oil composition of *Elettaria cardamomum* Maton. *Journal of Applied Biological Sciences* 7(3): 42–45.

Sen, S., Chakraborty, R., De, B., and Mazumder, J. 2009. Plants and phytochemicals for peptic ulcer: An overview. *Pharmacognosy Reviews* 3(6): 270.

Shakila, R. J., Vasundhara, T. S., and Rao, D. V. 1996. Inhibitory effect of spices on in vitro histamine production and histidine decarboxylase activity of Morganella morganii and on the biogenic amine formation in mackerel stored at 30 C. *Zeitschrift für Lebensmittel-Untersuchung und Forschung* 203(1): 71–76.

Shankaracharya, N. B., Raghavan, B., Abraham, K. O., and Shankaranarayana, M. L. (1990). Large cardamom chemistry, technology and uses. *Spice India* 3(8): 17–25.

Shen, C., Huang, Y., Yi, S., Fang, Z., and Li, L. 2015. Association of vitamin E intake with reduced risk of kidney cancer: a meta-analysis of observational studies. *Medical science monitor: International Medical Journal of Experimental and Clinical Research* 21: 3420.

Singh, D. B. (1978). Large cardamom. *Cardamom* 10(5): 3–13.

Singh, G., Kiran, S., Marimuthu, P., Isidorov, V., and Vinogorova, V. 2008. Antioxidant and antimicrobial activities of essential oil and various oleoresins of *Elettaria cardamomum* (seeds and pods). *Journal of the Science of Food and Agriculture* 88(2): 280–289.

Singh, V.B., Singh, K., 1996. *Spices*. New Age International Ltd., New Delhi, India, 253

Sontakke, M. D., Syed, H. M., and Sawate, A. R. 2018. *Studies on extraction of essential oils from spices (Cardamom and Cinnamon)*. The Pharma Innovation: India.

Souissi, M., Azelmat, J., Chaieb, K., and Grenier, D. 2020. Antibacterial and anti-inflammatory activities of cardamom (*Elettaria cardamomum*) extracts: Potential therapeutic benefits for periodontal infections. *Anaerobe* 61: 102089.

Tijjani, M. A., Dimari, G. A., Buba, S. W., and Khan, I. Z. 2012. In-vitro antibacterial properties and preliminary phtytochemical analysis of *Amomum subulatum* Roxburg (Large Cardamom). *Journal of Applied Pharmaceutical Science* 2(5): 69.

Tuan, D. Q., and Ilangantileket, S. G. 1997. Liquid CO2 extraction of essential oil from star anise fruits (*Illicium verum* H.). *Journal of Food Engineering* 31(1): 47–57.

UNCTAD, I. (2006). World markets in spice trade—2000–2004.

Valente, J., Zuzarte, M., Gonçalves, M. J., et al. 2013. Antifungal, antioxidant and anti-inflammatory activities of Oenanthe crocata L. essential oil. *Food and chemical Toxicology* 62:349–354.

Verghese, J. 1996. The world of spices and herbs. *Spice India*: 12–20.

Verma, S. K., Rajeevan, V., Bordia, A., and Jain, V. (2010). Greater cardamom (*Amomum subulatum* Roxb.)– A cardio-adaptogen against physical stress. *Journal of Herbal Medicine and Toxicology* 4(2): 55–8.

Wang, Y. Y., Wang, X. L., and Yu, Z. J. 2014. Vitamin C and E intake and risk of bladder cancer: a meta-analysis of observational studies. *International Journal of Clinical and Experimental Medicine* 7(11): 4154.

Yadav, A. S., and Bhatnagar, D. 2007. Modulatory effect of spice extracts on iron-induced lipid peroxidation in rat liver. *Biofactors* 29(2, 3): 147–157.

Yang, Y., and Zhang, F. 2008. Ultrasound-assisted extraction of rutin and quercetin from *Euonymus alatus* (Thunb.) Sieb. *Ultrasonics Sonochemistry* 15(4): 308–313.

Zhao, C., Sun, J., Fang, C., and Tang, F. 2014. 1, 8-cineol attenuates LPS-induced acute pulmonary inflammation in mice. *Inflammation* 37(2): 566–572.

Zhu, Y. J., Bo, Y. C., Liu, X. X., and Qiu, C. G. 2017. Association of dietary vitamin E intake with risk of lung cancer: a dose-response meta-analysis. *Asia Pacific Journal of Clinical Nutrition* 26(2): 271–277.

CHAPTER 15

Bioactive Compounds in Fennel Seed

Upasana Yadav, Isha Kaushik, Srishti Khurana, and Karnika Prakash

CONTENTS

DOI: 10.1201/9781003215387-15

15.1 INTRODUCTION

Seed spices are the abode of bioactive compounds which give therapeutic effects on to the human health despite of their culinary concern. Fennel (*Foneiculum vulgare)* is an important biennial medicinal and aromatic spicy herb of *Umbelliferae* family and is commonly known as "*saunf*" (Khan and Musharaf 2014) with many subspecies and varieties. One is sweet fennel (*F. vulgare subsp. vulgare var. Dulce*) another is bitter fennel (*F. vulgare mill. Subsp. vulgare var. vulgare*). Mostly these two varieties are commercially available and are used medicinally. It is mostly cultivated in the temperate and subtropical regions of the world. In India, main regions of production of fennel are Gujarat, Uttar Pradesh, Maharashtra, Bihar and Punjab (Sur 2020; Akhbari et al. 2018).

All parts of the fennel i.e., the roots, stems, leaves and seeds are used in cooking (Cautela et al. 2018). Fennel seeds are useful part of plant which are like anise in aroma and used extensively as flavouring agents among meat products, baked items, fish dishes, bakery and confectionery stuffs, alcoholic beverages and also as herb mixtures. Fresh seeds are green or brown in colour and on maturity it turns into grey (Kooti et al. 2015). It is used in cooking as an important spice in the Indian subcontinent cultures of Kashmiri Pandit, Gujarati people and the Middle East (Rather et al. 2016). It is widely used as a mouth freshener and digestive aid after a meal. Many studies have found that fennel herb and seeds effectively control numerous infectious disorders of bacterial, fungal, viral, mycobacterium and protozoal origin. Such pharmacological activities are attributed due to chemical composition of herb and seed (Sur 2020). Numerous scientists revealed that biological activities of fennel seed is due to presence of the various phytochemicals like volatile compounds of essential oil, flavonoids, phenolic compounds, fatty acids and amino acids (Badgujar et al. 2014). Various extraction techniques such as solvent method, microwave-assisted method, ultrasonic method etc. are used to isolate essential oil from fennel seed. Further, identification of bioactive components such as anethole, estragole, fenchone and limonene, flavonoids and phenolic compounds are done by using different types of purification techniques (chromatographic methods and spectroscopic) (Badgujar et al. 2014; Kooti et al. 2015; Hirondart et al. 2020).

15.2 GROWTH AND PRODUCTION

Its cultivation is naturalized throughout the Northern, Eastern and Western hemispheres, specifically in Asia, North America and Europe. In India, mainly growing states are Gujarat, Rajasthan and Uttar Pradesh, whilst many other states grow it on a small scale, such as Punjab, Tamil Nadu, Bihar, Karnataka, Maharashtra and Jammu and Kashmir (Dheebisha and Vishwanath 2020). It is a *rabi* crop which grows to height of 2.5 m tall, with hollow stems. The leaf matures to 40 cm long and 0.5 mm wide. The flowers are yellow in colour and 5–15 cm wide that are produced terminally in compound umbels. The fruit is 4–10 mm long and 1.5–2 mm broad having brownish-greenish grey colour (Grover et al. 2013). Fennel seed is elliptical, slightly grooved and obtuse at ends which is harvested when colour turns yellow and light brown with 20% moisture (Saxena et al. 2018). In the year 2019–2020, 127,790 tons of fennel seed were produced in India from 75,260 ha area (Spice Board of India).

15.3 PROPERTIES

Literature accounted different properties of fennel along with its long lists of applications in food, liquor, pharma and cosmetic industry. Due to unique flavour and aroma, it finds place in culinary uses in preparation of meat, baked food, confectionary products, beverages (Diao et al. 2014; Saber and Eshra 2019). Despite kitchen place, it has also occupied space in pharma, cosmetic & perfume industry (Mohamad et al. 2011; Ahmad et al. 2018). Its oil composition shows high biological and cytotoxic activities such as antifungal, antibacterial, antitumor, antithrombotic, carminative, anti-inflammation, anti-spasmodic, analgesic etc. In addition, antioxidant or antimicrobial nature make it more useful as food additive (Mohamad et al. 2011; Dua et al. 2013; Kooti et al. 2015; Alam et al. 2019; Saber and Eshra 2019) or for active packaging system (Goncalves et al. 2020). Recently encapsulation of essential oil of fennel seed is used as a tool for sustainable management of weeds due to its bioherbicidal properties (Kaur et al. 2021).

15.4 EXTRACTION, ISOLATION, AND IDENTIFICATION OF BIOACTIVE COMPOUNDS IN FENNEL SEED

Considering the utilization of fennel seed for its different properties, there is need to have knowledge of suitable methods of extraction, isolation and identification of bioactive compounds from fennel seed. The extraction method initiated by selection of appropriate plant species. After that seeds are dried and ready for grinding process, they are ground to powder form by manual or mechanical process as per the requirements. Selection of available extraction technique (conventional & advanced) applicable for fennel govern their utilization potential in such a way that should be cost effective, less time consuming and limit the processing aids (Repajić et al. 2020; Patra et al. 2018).

15.4.1 Extraction of Bioactive Compounds Using Hydro-Distillation

It basically involves water and heat to extract material from seed, especially essential oil is extracted with different yield and composition as interval for hydro-distillation also effect the extract yield (Patra et al. 2018; Saxena et al. 2018; Sur 2020) as depicted in *Figure 15.1 hydrodistillation apparatus.*

Saxena et al. (2018) used hydrodistillation method for 3 h to extract essential oil (0.22% by weight) from seeds. The extracted essential oil showed 82.12% anethole+estragol. Whereas Ghasemian et al. (2019) reported 78.68% anethole in essential oil by hydro-distillation extraction method. Diao et al. (2014) obtained light yellow transparent liquid 1.74% (v/w) yield of essential oil with 68.53% trans-anethole with 4 h distillation method. The extraction yield of essential oil obtained from hydro-distillation during 240 min was computed as 0.82% (Akhbari et al. 2018). Garzoli et al. (2018) observed that different chemical composition of essential oil was obtained depending on the extraction time of steam distillation, with variations in their related experimental biological activity.

15.4.2 Extraction of Bioactive Compounds Using Solvents

Scientists have studied and analysed the impact of different types of solvents, such as methanol, hexane and ethyl alcohol, for the purpose of extraction of bioactive components from seed (Lourenço et al. 2019). Saber and Eshra (2019) obtained extract by using solvent 70% ethanol (1:40, w/v), for 72 h at room temperature reported 77.91% trans-anethole and 64.37 mg/100 g total phenols, whereas Allaithy (2017) stated 10.27 ppm gallic acid with 70% methanol for 72 h. Salami et al. (2016) used hexane solvent in hydro-distillation method for extracting essential oil while methanol solvent to determine total phenol. Studies concluded that to extract different bioactive compounds from seeds with a high degree of accuracy, various solvents of differing polarities must

Figure 15.1 Hydrodistillation apparatus.

be used (Altemimi et al. 2017). Beyazen et al. (2017) used four solvents chloroform, water, methanol and aqueous: methanol (20:80, v/v) and observed high extraction yield with chloroform solvent while TPC & TFC content high in methanolic extract.

15.4.3 Extraction of Bioactive Compounds Using Pressurized Liquid

This is a novel technique to substitute conventional solvent extraction method that are time consuming, maximal resources (solvent and electricity) wastage and even contribute to raise pollution. Pressurized liquid extraction (PLE) represents a novel technique with numerous advantages over conventional extraction approaches. PLE is environment friendly technique works on combination of high pressure and temperature, short time interval and low solvent level (Repajić et al. 2020). There are studies which revealed that PLE conditions, e.g., extraction temperature, time and type of solvent effects extraction yield of bioactive compounds from fennel seed (Repajic et al. 2020).

15.4.4 Extraction of Bioactive Compounds Using Ultrasonication

It is green technology which use green solvents such as ethanol and water to get extracts of seed by producing cavitation bubbles which increase the shear force to disperse compounds in the extraction solvent. The mechanism of ultrasound processing is based on physical (cavitational, mechanical effects) and/or by chemical means (formation of free radicals such as OH^{-}, H+ due to sonochemical reactions). Ultrasonication produce advantages such as less processing time, high yield and lower energy consumption while reducing thermal effects. It has been recognized for potential

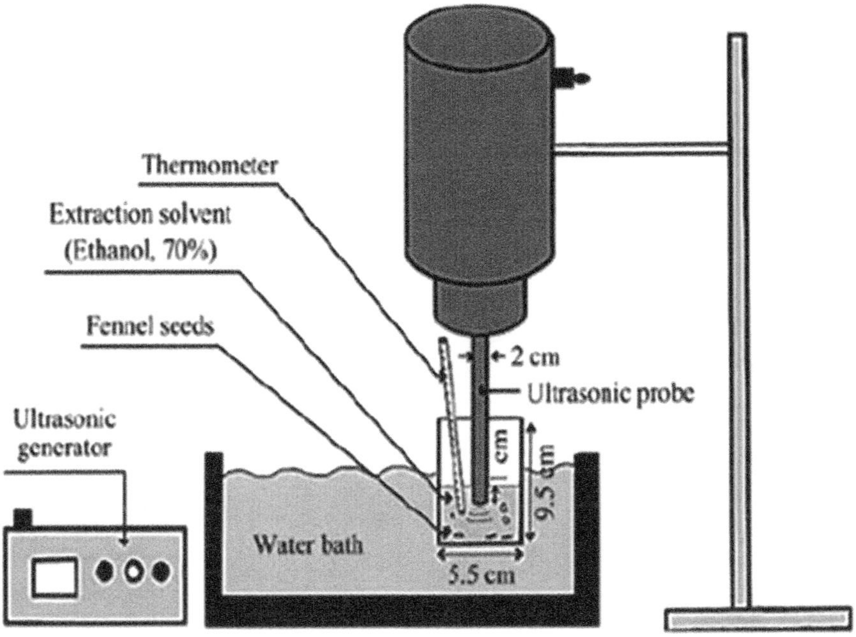

Figure 15.2 Ultrasonic wave extractor.

industrial application in the phyto-pharmaceutical extraction industry for a wide range of herbal extracts (Ghorbani et al. 2017). Ghorbani et al. (2017) observed that adoption of ultrasonication increases the extraction yield as in competition with conventional method. Its application on to the fennel seed even depresses processing time and temperature as illustrated in *Figure 15.2 ultrasonic apparatus*. Moradia et al. (2020) optimize the parameters for ultrasonication extraction of essential oil especially anethole from fennel seed.

15.4.5 Extraction of Bioactive Compounds Using Microwave

Extraction of bioactive components by use of electromagnetic waves (300 MHz to 300 GHz) is an innovative approach to minimize temperature effect on to the heat labile bioactive constituents of fennel components. In addition, there is increase in extraction efficiency with less effluents production (Akhtar et al. 2020) in microwave extractor as shown in *Figure 15.3 microwave extractor* (Benmoussa et al. 2019). With respect to high extraction yield (2.82% essential oil) and higher biological activity (antifungal, antibacterial, antioxidant & cytotoxic) of bioactive components from fennel seed, microwave technique is combined with hydro-distillation (Chen et al. 2020). This solvent free microwave assisted hydro-distillation (MHD) extracted high proportion of trans-anethole (88.28%) and oxygenated components (95.95%). In addition, enhanced solvent free microwave extraction (ESFME) using double walled microwave reactor appear to be promising tool to improve and speed up the SFME process for the extraction of essential oils from *Foeniculum vulgare* Mill. Seeds with advantages in terms of yield and low extraction time and better essential oil composition, making ESFME as simple, cheap, safe and simple technique (Benmoussa et al. 2019).

15.4.6 Extraction of Bioactive Compounds by Supercritical Fluid (SFE)

This is one of green technique of extraction in which non-toxic solvent i.e., carbon dioxide which is gas at room temperature and its gaseous nature is changed on supercritical conditions

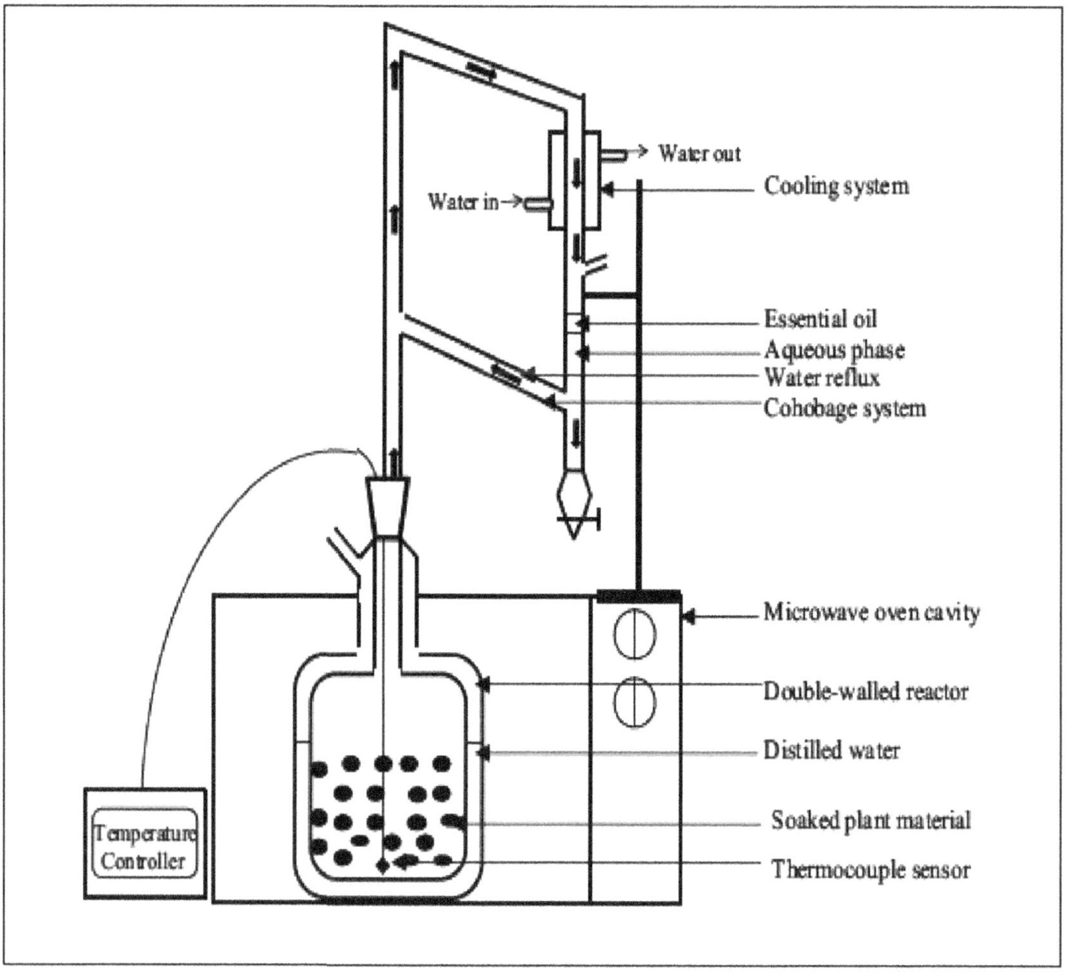

Figure 15.3 Microwave extractor.

(Michailidis et al. 2021). Mokhtari and Ghoreishi (2019) optimized the SFE parameters for extraction of trans-anethole from fennel seeds: 55.5% at 21.55 MPa, 55.59°C, 1.49 ml/min and 96.38 min through RSM and genetic algorithm. In another study, Reverchon et al. (1999) used two steps SFE of fennel seeds in which the first, 90 bar pressure at 50°C is applied in order to perform selective extraction of essential oil and in the second, vegetable oil is yield through 200 bar pressure at 40°C. Hammouda et al. (2013) noticed increased extraction yield of oil and anethole (72%) by SFE as compared to traditional methods and also concluded that process results in reduction in extraction time, high efficiency, low temperature, high process speed and full solvent recovery.

15.5 IDENTIFICATION OF BIOACTIVE COMPONENTS

Success of extraction, govern isolation and identification of bioactive components from seed extract by way of various chromatographic approaches viz., column chromatography, thin-layer chromatography, high-performance liquid chromatography along with Fourier-transform infrared spectroscopy, gas chromatography-mass spectrometry, nuclear magnetic resonance spectroscopy.

Table 15.1 Identification of Different Bioactive Components

Plant Extract	Active Components	Identification Technique	Reference
Essential oil from different parts of fennel (dry straw, fruit, leaves)	10 fractions 82.12% anethole+estragol (82.12%),4-allyl-anisole (16.86%)	GC-MS	Saxena et al. (2018)
Essential oil from seed	16 fractions trans-anethole (70.7%), fenchone (5.72%), anisketone (4.07%) and p-anisaldehyde (3.79%)	GC-MS	Ahmad et al. (2018)
Essential oil from seed	28 fractions trans-anethole (68.53%), estragole (10.42%), limonene (6.24%), fenchone (5.45%)	GC-MS	Diao et al. (2014)
Essential oil from seed	21 fractions trans-anethole (78.5%), fenchone (10.5%), limonene (6.9%), and tarragon (2.3%)	GC-MS	Ghasemian et al. (2019)
Essential oil from seed	78 fractions estragole (34–89%) anethole (5–36%)	GC-FID-MS	Napoli et al. (2010)
Methanolic seed extract	gallic acid 277.131 μg, caffeic acid 166.062 μg, ellagic acid 99.476 μg, quercetin 781.986 μg and kaempferol 92.856 μg/g	HPLC	Dua et al. (2013)
Methanolic seed extract	quercetin, gallic acid, catechin and tannin	HPLC	Allaithy (2017)
Essential oil from seed	24 fractions trans-anethole (63.41%–78.26%), estragole (44.72%–88.92%)	GC-MS	Kalleli et al. (2019)
Methanolic seed extract	27 fractions quinic acid, 4-O-caffeoylquinic acid, p-coumaric acid and 4-O-caffeoylquinic acid	HPLC-MS	Kalleli et al. (2019)
Methanolic seed extract	57 fractions trans-anethole (31.49%), 2-pentanone (25.01%), fenchone (11.68%) and benzaldehyde-4-methoxy (8.01%)	GC-MS	Alam et al. (2019)

All of these processes are performed together or in groups to identify specific bioactive compounds from aromatic herb (Patra et al. 2018). Various studies reported identification of active components from fennel seed in Table 15.1.

15.6 CHARACTERIZATION OF BIOACTIVE COMPONENTS IN FENNEL SEED

The fennel seed contain about 6.3% water, 9.5% protein, 10% fat, 13.4% minerals (calcium, potassium, magnesium, phosphorus, sodium), 18.5% fibres and 42.3% carbohydrates (Rather et al. 2016). Apart from above components, seed contain vegetable oil (fixed oil) and essential oil (volatile oil) which are rich in bioactive molecules. Many scientists have noticed variation in yield of seed oil i.e., essential oil (0.7–6%) and vegetable oil (10–20%) due to various factors such as genetic structure and environmental condition rather influenced only through previously stated extraction methods etc. (Rather et al. 2016; Saxena et al. 2018; Lal et al. 2014; Ahmad et al. 2018). Vegetable oil of seed extract of fennel mainly comprises fatty acids which are palmitic acid, petroselinic acid, oleic acid, linoleic acid (Ahmed et al. 2018). Thus, rich source of monounsaturated fatty acids (MUFAs) makes it applicable in cosmetic industry for the moisturizing, protection and healing of problematic skins (Chen et al. 2020; Ahmed et al. 2018). Many studies revealed essential oil of fennel seed is

Table 15.2 Various Components of Essential Oil, Phenols and Flavonoids

Essential Oil/Classes ↓	Components	Properties
Monoterpene hydrocarbon	α-pinene	Use in camphor, insecticides, solvents and perfumes
	Sabinene	
	β-Pinene	
	β-Myrcene	Hepatoprotective
	p-Cymene	
	D-Limonene	Hepatoprotective/dissolving agent
	γ-Terpinene	
Oxygenated monoterpenes	Fenchone	Repellent
	Camphor	
	Cuminic aldehyde	
	Carvone	
Phenylpropanoids	Estragol	Perfumery & flavouring
	Trans-anethole	Antithrombotic, Anticancer
Others	p-Anisaldehyde	Acarcidal
Phenols	3-O-caffeoylquinic acid 4-O-caffeoylquinic acid 5-O-caffeoylquinic acid (chlorogenic acid) 1,3-O-di-caffeoylquinic acid 1,4-O-di-caffeoylquinic acid 1,5-O-di caffeoylquinic acid	
Flavonoids	Eriodictyol-7-rutinoside (eriocitrin)	Antinociceptive and anti-inflammatory Activity
	Quercitin-7-rutinoside (rutin)	Immunomodulatory
	Quercetin-3-O-galactoside Kaempferol-3-O-rutinoside Quercetin-3-O-glucoside (isoquercitrin) Kaempferol-3-O-glucoside (astragalin) Quercetin-3-glucuronide Kaemferol-3-glucuronide Rosmarinic acid	

Source: Rather et al. (2016) and Badgujar et al. (2014).

rich in various types of bioactive compounds grouped in four chemical classes according to their functional groupings (Ghasemian et al. 2019; Ahmed et al. 2018; Rather et al. 2016) as depicted in Table 15.2. Among which major fractions included anethole (40–80%) and fenchone (3–20%) (Lal et al 2014; Badgujar et al. 2014; Ahmed et al. 2018; Chen et al. 2020) exhibited pharmacological activities. The main constituent of fennel plant is the trans anethole [1-methoxy-4-(1-propenyl) benzene] that attribute anise like odour (Al-Amoudi, 2017; Kooti et al. 2015), applicable in the food and liquor industries (Castro et al. 2010). It is also used in pharma industry due to inhibitory action on inflammation and carcinogenesis (Bharat et al. 2008). However, GC revealed the presence of seventeen compounds in essential oil; limonene (26.44%), fenchone (23.5%), (E)-anethole (21.68%), α-phellandrene (9.26%) and α-pinene (6.22%) were major components found in Spanish organic fennel essential oil (Marin et al. 2016). Whereas Napoli et al. (2010) identified 78 components into five groups: monoterpene hydrocarbons, oxygenated monoterpenes, sesquiterpenes, phenylpropanoids and others. In which, estragole as the main compound ranging between 34 and 89%, while (E)-anethole is the other phenylpropanoid ranging between 0.1 and 36%. Numerous studies have reported that essential oil seed extract of fennel is used in food, pharma & cosmetic industries due to typical fragrance & wide array of biological activities such as antimicrobial, anti-inflammatory, antioxidant, antitumour properties and antithrombotic etc. effects (Chen et al. 2020; Al-Amoudi 2017; Alam et al. 2019). Besides its therapeutic nature, one of its component i.e., estragole, has become a cause of concern, as the structurally similar methyl-eugenol has been recently found to be a potential carcinogen. Estragole is rapidly metabolized in humans and excreted in urine as 1'-hydroxyestragole and the HMPC (committee on herbal medicinal products) of European medicine agency recommends that the total human daily intake of estragole from all food sources should not exceed 0.07 mg/kg (Akbar 2018).

The other class of phytochemicals found in seed extract are phenols which are studied to be linked with the prevention of oxidative stress diseases such as inflammation, heart disease and cancer. Major phenolic acids include 3-O-caffeoylquinic acid, 4-O-caffeoylquinic acid, 5-O-caffeoylquinic acid, 1, 3-O-di-caffeoylquinic acid, 1, 4-O-di-caffeoylquinic acid and 1, 5-O-di-caffeoylquinicacid (Rather et al. 2016; Kooti et al. 2015; Badgujar et al. 2014) and flavonoid are eriodictyol-7-rutinoside, quercetin-3-rutinoside and rosmarinic acid (Kooti et al. 2015), luteolin-7-o-glucoside, apegenin-7-o-glucoside, apigenin, cirsiliol, cirsilineol, naringin and naringenin also found in fennel seed. Otherqunic acid, gallic acid, ferulic acid, p-coumaric acid and caffeic acid are present in minor (Anwar et al. 2009; Kalleli et al. 2019). It was evident that fennel seed contained noticeable amounts of total phenols and flavonoids which also contribute therapeutic properties ranged from 5.93-64.37 mg GAE/g and 0.35-48.8 mg/100 g respectively (Beyazen et al. 2017; Dua et al. 2013; Saber and Eshra 2019; Chatterjee et al. 2012; Mohamad et al. 2011; Salami et al. 2016). Due to these components, fennel seed have gained futuristic opportunities among nutritionists, food scientists and consumers due to their roles to uplift human health (Rather et al. 2016). As matter of fact, it is also considered that genetic and geographic factors, cultivation procedure affect the composition of essential oil, phenolics and their application (Badgujar et al. 2014).

15.7 HEALTH BENEFITS

Fennel seeds have been found to contain a good quantity of nutrients like proteins, vitamins, dietary fibre, antioxidants like phenolic compounds. They also are known for the content of essential oils (Weiping and Bookang 2011; Moghtader 2013). Fennel is also extremely rich in potassium, sodium, phosphorus and calcium (Badgujar et al. 2014). Some of the major compounds found in fennel seed oil includes anethole and 4-allyl-anisole. Other than this it contains smaller amounts of camphene, cymene, β-pinene, myrcene, α-pinene and γ-terpinene. The essential oil abstracted from fennel has been used for many potential benefits like hormone balancing and as a digestive (Saxena et al. 2018).

15.7.1 Use in Traditional Medicine

In countries like India, China and other East Asian countries the use of fennel in order to treat diseases has a long history. Fennel seeds were chewed by people in the middle ages to get rid of abdomen noise. They were used for their sedative effect in the fifth century. People from Rome thought that fennel could improve their sight. It is still used as treatment for cataract. The people from England on the other hand thought that fennel was effective in aiding digestion and relieving a bloated stomach (Kooti et al. 2015). Fennel seeds are particularly used for its digestive properties in the present time. They are used in cases of chronic diarrhoea, stomach infections and colic in

Table 15.3 Traditional and Contemporary Uses of Fennel Seeds

S. No.	Disease	Preparation of Fennel Used	Region
1.	Diuresis	Seeds	Miami, Florida, USA
2.	Abdominal Pain	Seeds	Northern Badia Jordan,
3.	Hair growth	Seed oil	Middle Navarra
4.	Diarrhoea	Seeds, Ground seeds with tuber of *Hemidesmus indicus*, paste to be consumed two times a day for three days	Northern Portugal Bhandara, Maharashtra, India
5.	Kidney diseases	Seeds	Gujranwala, Pakistan
6.	Purgative	Seeds	Gujranwala, Pakistan
7.	Laxative	Seeds	Gujranwala, Pakistan
8.	Liver ache	Seeds	Pernambuco, Northeast Brazil
9.	Leucorrhoea	Seed powder	Rajasthan, India
10.	Repeated abortions	Seed powder	Rajasthan, India
11.	Digestive issues	Seed	Balikesir, Turkey
12.	Carminative agent	Seed	South Europe North Iran
13.	Diuretic	Seed	South Europe Northern Portugal
14.	Emmenagogue	Seed	Haryana, India
15.	Blurry and itchy eye	Seed	Northern Portugal Gujranwala, Pakistan

Source: Badgujar et al. (2014).

children. Fennel is known to give relief from vomiting and nausea. Additionally, fennel is believed to be a remedy for ailments like kidney stones, diabetes and even chronic cough and fever. Fennel is known for its diuretic effect; hence it is used to treat kidney and bladder related conditions (Kooti et al. 2015). Table 15.3 shows the traditional and contemporary uses of fennel seeds.

15.7.2 Hepatoprotective Effect

The protective effect of fennel essential oil against the hepatotoxicity of Calcium tetrachoride (CCl_4) was studied at Yuzuncu Yil University. The researchers observed that the serum asparate aminotransferase (AST), alanine aminotransferase (ALT), alkaline phosphatase (ALT) and alkaline phosphatase (ALP) levels and bilirubin were reduced. This exhibited the effectiveness of fennel essential oil (FEO) as a hepatoprotective agent (Ozbek et al. 2003). In order to verify the protective effect from hepatotoxicity of fennel and dill oil against CCl_4 in rats, the rats were divided into various groups. This experiment was designed as given in Figure 15.4.

The results showed that fennel oil, dill oil and the use of fennel with dill oil, all were effective in protecting the rats against the hepatotoxic effects of CCl_4. These effects were verified by the significant ($p < 0.05$) drop in levels of serum ALT and AST and the significant ($p < 0.05$) rise in serum albumin and total protein. Treatment with fennel or dill oil can undo the rise in the MDA level with a decline in SOD enzyme activity, caused a result of hepato-damage caused by $CCl_{4.}$ Other than this the use of dill and fennel subdued the escalated ALP activity at the same time decreasing the amount of increased bilirubins. This points to potential of these oils to stabilize the biliary dysfunction taking place in the rat liver caused by $CCl_{4.}$ The benefits of fennel and dill oil in this case were attributed to their antioxidant activity. This study concluded that a combination of fennel and dill oil when used in food formulations, can benefit people suffering from liver-related dysfunctions and diseases (Rabeh and Aboraya 2014).

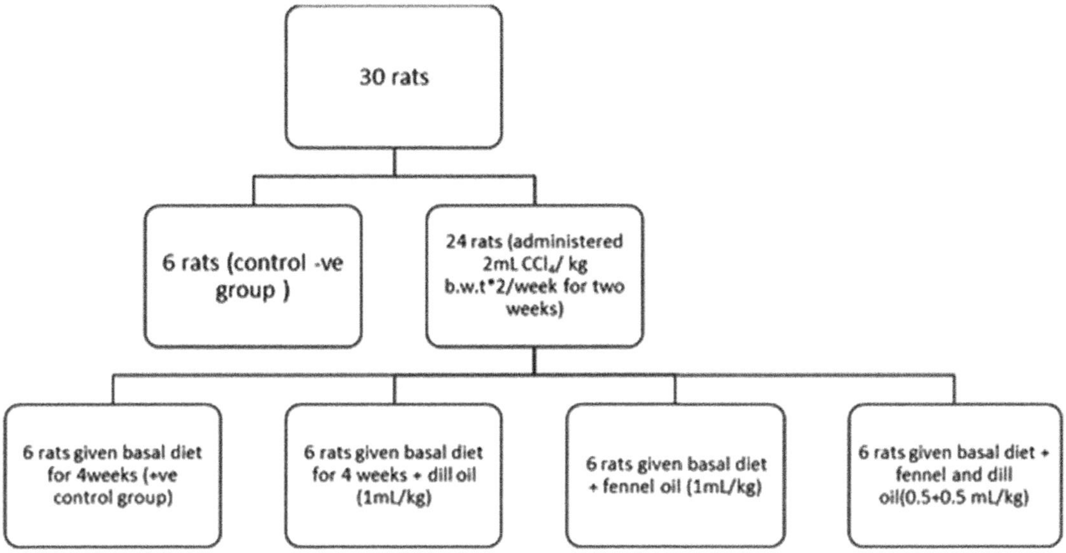

Figure 15.4 The design of experiment to study the effect of fennel and dill oil on hepatotoxicity caused by CCl_4 in rats (*Source:* Rabeh and Aboraya 2014).

15.7.3 Skincare

Fennel seed oil has been found to be good for skin health due to its ideal fatty acid composition. It has been recommended as a substitute to harmful ingredients like paraffin oil thus making cosmetics a more environment friendly option. The presence of petroselinic acid is effective for skin problems such as dryness (Sayed Ahmad et al. 2018). Fennel essential oil has also been studied as a skin penetration enhancer for cosmetics and creams (Herman and Herman 2014). Enterosorbent made with Inulin, pectin and Fennel extract proved to be a good treatment for moderate cases Atopic Dermatitis in children (Danilova et al. 2016).

15.7.4 Antioxidant Activity

Fennel contains bioactive agents such as polyphenols and flavonoids and thus acts as an antioxidant. Antioxidant activity of fennel makes it a potential treatment for a number of ailments like chronic coughs, kidney stones, bronchitis and diabetes. All of these mentioned diseases involve oxidative stress caused by radical species. When the optimization of fennel extract processing was studied it was found that the highly concentrated extracts showed greater antioxidative properties. This goes on to prove that the presence of substances such as phenolic compounds is responsible for the antioxidation exhibited by fennel. In this particular study the antioxidant capacity of fennel extract was studied. 'Antioxidant Capacity' was described in terms of the concentration of polyphenolic compounds present in the extract. The two types of extract used for the study included ethanol-water extract and water extract. Out of the two the latter was more concentrated and hence possessed a higher antioxidant capacity. The reason for this was believed to be the presence of non-volatile compounds namely phenolic compounds (Angelov and Boyadzhieva 2016).

To analyse the phenolic content and composition of plants with medicinal and herbal value in the Mediterranean region. Fennel was tested in the form of its hydroalcoholic extract, made from freeze-dried fennel. The results of the LC/MS/MS analysis showed that fennel possesses flavanol hexuronides as the major constituent from the fennel collected during spring season. It also had many hydroxycinnamoyl quinic acids, two chromone derivatives and flavanol glycosides. On the other hand, the winter sample had a different composition that showed DPPH and ABTS scavenging properties along with decreasing Fe^{3+} (Pacifico et al. 2018).

In a study conducted in Pakistan, it was found that fennel seeds native to Pakistan, *Foeniculum vulgare Mill*, had a phenolic activity of 627.21–9.67.50 GAE, mg/100 g. The flavonoid content was in the range of 374.88–681.96 CE, mg/100 g. Waste from fennel distillation has been found to be rich in antioxidant activity (Anwar et al. 2009).

15.7.5 Antibacterial Effects

The essential oil of fennel has been studied for its bactericidal and antibacterial activity. In a study conducted on various gram-positive and gram-negative bacteria it was found that the oil was effective against – *Staphylococcus albus, Bacillus subtilis, Salmonella typhimurium, Shigella dysenteriae* and *Escherichia coli.* These results were based on the result of minimum inhibitory concentration (MIC) and minimum bactericide concentration (MBC) tests. It was also found that *S. dysenteriaeis* was most sensitive to fennel essential oil. Fennel oil was successful in damaging the cell membrane of the organism resulting in leakage of electrolytes, proteins, carbohydrates etc. and thus cell lysis. (Diao et al. 2014). Another study showed the activity of *Foeniculum vulgare* on *Bacillus magterium, Bacillus pumilus, Klebsiella pneumonia* and *Candida albicans* among others (Gulfraz et al. 2008).

15.7.6 Digestive and Carminative Agent

Fennel is a well-known carminative natural agent. It releases any gas trapped in the gastric canal, due to the presence of anethole in it. It also has a good amount of fibre content that eases the movement of bowels through the system (Peter and Malhotra 2012). The carminative and anti-spasmodic effect of fennel has been utilised to treat patients having Irritable Bowel syndrome (IBS) in combination with the anti-inflammatory activity of curcumin. Researchers evaluated 121 patients with mild to moderate IBS, the desired result was a reduction in IBS-SSS (Irritable Bowel Syndrome- symptom severity score). An IBS-QoL questionnaire was also used to understand the influence of the treatment on quality of life. This treatment showed promising results, it showed symptomatic alleviation in patients. A notable difference in the IBS-SSS was observed in the course of a month from 50.05+/- 28.85% to 26.12+/- 30.62%, $p < 0.001$. Patients also experienced relief from abdominal pain and other related symptoms. The results of the IBS-QoL were in good agreement with the IBS-SSS (Portincasa et al. 2016).

15.7.7 Anti-Inflammatory

Fennel seed methanolic extract is a potent anti-inflammatory agent for diseases and allergic reactions. It also has an analgesic effect (Choi and Hwang 2004). The anti-inflammatory properties of fennel were assessed to study the prevention of radiation harm in human blood culture. The study concluded that that fennel showed anti-inflammatory against damage caused by γ-radiation (Farid et al. 2020). Table 15.4 shows the fennel seed components and their health effects. In a study conducted on rats and mice, the methanolic extract of fennel showed protective effect against acute and subacute inflammation. In this study the anti-inflammatory effect was studied using arachidonic acid induced ear oedema, carrageenan-induced paw oedema and formaldehyde-induced arthritis. In case of acute inflammation disease, the results of this experiment were as follows:

1. Methanol extract of 200 mg/kg repressed ear oedema by 70%
2. The serum transaminase, alanine aminotransferase (ALT) and aspartate aminotransferase (AST) showed an appreciable rise after administration of the methanolic extract
3. These results point to the possibility that the methanolic extract affects both cyclooxygenase and lipoxygenase pathways (Choi et al. 2004).

Table 15.4 Fennel Seed Components and Their Health Effects

S. No.	Health Benefit	Components
1.	Digestive, Carminative and antioxidant reactions	Myrcene, Fenchone, Chavicol, Cineole
2.	Promotes production of Red Blood Cells (RBCs)	Copper
3.	Promotes regular growth and development	Zinc
4.	Maintain heart rate and blood pressure, increase brain function and cognitive abilities	Potassium
5.	Treats Anaemia	Iron and Histamine
6.	Prevents oxidative stress and macular damage	Flavonoids
7.	Treats congestion, asthma and bronchitis	Cineole
8.	Anti-inflammatory effects	Anethole
9.	Anti-oxidative effects	Kaemferol and Quercetin

Source: Choudhary et al. (2018).

15.7.8 Anti-Stress and Anxiolytic Activity

Stress is a major problem in the present times. The extract of fennel has been known to exhibit significant anti-stress properties. A study conducted on rats showed that there was a notable the levels of vanillyl mandelic acid (VMA) and ascorbic acid in the urine. This is a good marker to study stress levels. The same study showed that fennel is also a good memory enhancer (Koppula and Kumar 2013). A double-blind study was conducted on post-menopausal women. It was found after analysing the patients that the symptoms of anxiety and depression were notably better. This can be attributed to the fact that fennel has phytoestrogen (Ghazanfarpour et al. 2017). Fennel can also be used as a treatment in disorders that arise as a result of oestrogen deficiency. Anxiety is one such disorder. In a study that was conducted on rats that the patients who were given fennel essence showed a lowered level of stress (Mesfin et al. 2014). When 30 Swiss albino and wistar rats were used in a study to evaluate the anti-depressant and ant amnesic effect of *Foeniculum vulgare*, it was found that fennel can enhance memory and act as an anti-depressant (Abbas et al. 2020).

15.7.9 Effect on Liver

A study in rats showed that a high amount of fennel consumption has direct influence on the liver. The rats who were fed a large amount of fennel exhibited a decrease in the weight of their liver when compared to the control group. A rise in alkaline phosphatase (ALP) serum was observed in the rats that were given fennel. Sinusoidal dilatation along with mild degenerative, inflammatory cells infiltration, necrosis and blockage in blood vessels were also observed (Luaibi et al. 2020).

15.7.10 Effect on Cardiovascular Health, Hypolipidaemia, and Diabetes

Fennel extract can also be used to lower plasma lipid levels and triglycerides. This can help treat fatty liver and cardiovascular disorders. When given intravenously, fennel extract can also lower levels of blood pressure and act as a diuretic and natriuretic (Kooti et al. 2015). Diabetes can be a comorbidity that can cause cardiovascular diseases. Fennel with its high antioxidant activity can be used to protect against cardiovascular diseases. In a study on diabetic rats, the use of fennel and senna displayed increased amelioration and thus antidiabetic and cardioprotective effects (Farid et al. 2017; Nour et al. 2017). Fennel showed slight improvement in HDL-C, LDL-C and triglyceride levels in post-menopausal women (Afiat at al. 2018).

15.7.11 Chemo Modulatory Effect

7,12-dimethylbenz(a)anthracene-(DMBA-) was used on Swiss albino mice to test the chemo modulatory properties of fennel. These rats had skin papillamagenesis and forestomach papillamagenesis. Fennel seed extract was potent in decreasing the tumours on skin and in the forestomach along with tumour multiplicity. This study established the chemoprotective properties of fennel that need to be studied further (Singh and Kale 2008).

15.7.12 Cytotoxicity

Murine fibrosarcoma L929sA cells and MDA-MB231 and MCF7 cells which cause human breast cancer were studied for the cytotoxic effect that various herbs have on them. Fennel had a IC (50) value after 24 h 500 ± 17 and 700 ± 28 μg/mL respectively for MCF7 and L929sA cells. Acetone extract of fennel was studied for its cytotoxicity in the cell line of the kidney of African green monkey and adenocarcinoma cervical cancer cells (Berrington and Lall 2012).

15.7.13 Antiaging Effect

Rasul and Akhtar (2014) studied the fennel for its antiaging effect, they used skin moisture and trans-epidermal water loss (TEWL) parameters to measure this. The formulation prepared with fennel showed effective results. Fennel hold promise as an ingredient in antiaging cosmetic formulations (Rasul and Akhtar 2014).

15.7.14 Gastrointestinal Effects

Treating rats with an aqueous extract of fennel before inducing gastric lesions in rats by the use of ethanol reduced the intensity of damage. The levels of malondialdehyde in blood also dipped and nitrate, nitrites, ascorbic acid, retinol and beta-carotene levels increased (Birdane et al. 2007).

15.8 UTILIZATION IN TREATMENT OF DISORDERS

Fennel oil has been considered a potential treatment for various disorders. One such study was conducted on albino rats to study the action of fennel oil against hepato- and nephro-toxicity of Sodium-Valoproic (SVP). They were administered 1 ml/kg b.w. fennel oil and for a duration of 6 week where 3 doses were administered per week. The results of biomarkers of renal and liver damage were used as the evaluation criteria in addition to histological examination. The population that was only given SVP underwent changes such as drastic rise in levels of enzymes namely ALP, SAT, AST, bilirubin, creatinine and urea-nitrogen along with a significant decline in total protein content. Damage to liver and kidney tissue was also observed (Al-Amoudi 2017). Chewing of fennel seeds has been shown to reduce the pH of saliva. This points towards the anticarcinogenic properties of fennel (Manohar et al. 2020).

15.8.1 As a Treatment for Cough and Sore Throat

The use of fennel tea made from fennel seeds and the aromatic oil of fennel seeds to treat cough and sore throat has also been a topic of research. Scientists found that not only do fennel seeds a rich source of nutrients such as proteins, carbohydrates, crude fibre and minerals but also bioactive substances. The methanolic extract of fennel seeds were found to have an antimicrobial and antioxidant effect. These factors make fennel a potent treatment for cough (Saber and Eshra 2019).

15.8.2 Treatment for Hirsutism

In a study conducted by Javidnia et al. (2003) ethanolic extract of *Foeniculum vulgare* was used to treat Idiopathic hirsutism. This study was based on the fact that fennel has a history of being used as an estrogenic agent. Creams containing 1% and 2% of the ethanolic extract of fennel were applied to the affected areas. It was found that the formulation containing 2% fennel ethanolic extract was more effective, both formulations performed better than the placebo. The criterion of assessment in this study was the diameter of hair. In the case of 2% fennel extract containing cream led to the decrease of 18.3% in hair diameter (Javidnia et al. 2003). In a study conducted on Iranian women having mild to moderate cases of idiopathic hirsutism, it was observed that the utilization of a fennel gel led to 3% decrement in the thickness of facial hair when compared to the control group (Akha et al. 2014).

15.8.3 As Treatment for Dysmenorrhea

Dysmenorrhea is experienced by menstruating people around the world. It causes painful cramps and discomfort. It is usually treated by the use of oral contraceptive pills or non-steroidal anti-inflammatory drugs (NSAIDs). These decrease the myometrial function. Fennel has antispasmodic properties and contains anethol, both these factors make fennel an important herbal remedy for dysmenorrhea. Researchers have used oral fennel drops in the past on patients to evaluate its efficiency in treating dysmenorrhea. They analysed the results using SPSS version 16 with a statistical significance of $p > 0.05$. Fennel proved to be reduced dysmenorrhea and its symptoms (Bokaie et al. 2013).

15.8.4 As Treatment for Dyslipidaemia

In a study conducted on dyslipidemic rats, the use of nano-volatile oil extracted from fennel resulted in a drastic reduction in the blood glucose levels. The presence of *trans*-anethole in fennel has been believed to be the reason for its anti-diabetic effect. Trans-anethole can improve the tolerance of the body to glucose, insulin production and secretion also block the pathway of kappa B. Along with this fennel nano-volatile oil also led to higher uric acid levels. Fennel oil lowering of TNF-α and MDA points to the antioxidative and anti-inflammatory effect of fennel (Al-Okbi et al. 2018).

15.8.5 As Treatment for PCOS and Other Reproductive Disorders

Anethole, a major component present in fennel has been found to be possess activity for PCOS. It was proved to be this way in an in-silico docking study done to investigate the drug target. The drug protein called androgen receptors, insulin receptors substrate 1(IRS) along with follicle stimulating hormone (FSH) receptor (Chandran et al. 2019). In female mice consumption of fennel led to a rise in serum levels of hormones such as progesterone, oestrogen and prolactin. This means that fennel can be a potential treatment for infertility (Sadeghpour et al. 2015).

People having PCOS experience a number of symptoms which include- oligomenorrhoea and amenorrhea. The use of fennel as a remedy for oligomenorrhoea was studied at random in a group of 61 patients for a period of 6 months. Fennel infusion along with dry cupping were given to a group of patients, whereas metformin was given to the other group. The results suggested that fennel can be an effective treatment. The patients in the group where fennel infusion and dry cupping were used often experienced lesser pain. This therapy is safer and doesn't lead to side-effects as is the case with metformin and hormone therapy (Mokaberinejad et al. 2019).

15.8.6 As Treatment for Staphylococcal Infection

The use of fennel in treating Staphylococcal infection in medical students was studied. Fennel essential oil (80% of *trans*-anethole) when combined with mupirocin was the ideal treatment used against S. aureus having strains with $iMLS_B$, $cMLS_B$, MS_B resistant phenotype. The use of fennel essential oil also results in a decrement in antibiotic resistance (Kwiatkowski et al. 2017).

15.8.7 Effect on Fertility

The effect of *Foeniculum vulgare* on the fertility of mice and the serum sex hormones of rats and rabbits was studied. Food containing 2% and 4% of fennel seed powder was given to mice for nine months. Rats and rabbits were given diet with fennel seed powder for two months and then their serum levels of enzymes like testosterone, progesterone, oestrogen, LH and FSH were examined. A higher number of pups were born with each generation. The study of sex hormone also pointed to a rising trend. Fertility in and ex hormones both went up on the use of fennel. The results of this study showed that fennel can be used to treat infertility and needed to be tested further in this respect (Abbas et al. 2020). In another such study the effect of dill, chicory root and fennel were studied on neonate rats. 184 (wistar rats) each having a weight of 230–250 gm were put into 4 groups. Every group had an equal number of males and females who were isolated from the opposite gender. Extract of fennel, dill and chicory root in the amounts of 45, 100 and 100 mg/kg was given to three of the four groups. After three weeks the male and females were put in close proximity and allowed to mate. The final result showed that only fennel had a pro-fertility effect on the rats (FallahHuseini and Kianbakht 2012). In another recent study cinnamon and fennel were studied for their effect on Busulfan induced infertility. Busulfan is a drug for treating cancer, one of its side-effects is infertility. In this study, 40 male rats were grouped as follows:

1. Sham group – Healthy rats without intervention
2. Control group – Rat treated with busulfan (two doses – 25 mg/kg and 15 mg/kg in a gap of 14 days)
3. Fennel group – Rats treated with busulfan and then with fennel extract intraperitoneally (100 mg/kg for 5 days)
4. Fennel and cinnamon group – Rats treated with Busulfan and then with fennel (100 mg/kg) and cinnamon (200 mg/kg) extract intraperitoneally for 5 days

The combination of fennel and cinnamon was found to have a noticeable preservative effect against infertility caused by the cancer drug. Fennel by itself surprisingly in this case further damaged the fertility of rats (Chegini et al. 2019).

15.8.8 Recovery from Injury

Use of *trans*-anethole as treatment for acute lung injury induced by lipopolysaccharide was studied by scientists. It was noticed that *trans*-anethole reduced the number of inflammatory cells. It may also be Th17/Treg response regulatory. The anti-inflammatory activity of fennel makes it ideal for use during recovery from an injury or a surgery (Zhang et al. 2018).

In a group of patients going through surgery to treat gynecological cancers, the effect of fennel tea was assessed for post-operative recovery. The group that consumed fennel tea experienced lesser complications, a briefer hospital stay and had an improved bowel mobility (Ma et al. 2015). Gastrointestinal problems are frequently experienced after many surgeries. Fennel has the potential to improve the process of recovery. When patients who had undergone a complicated abdominal surgery were given a heated fennel therapy, it was noticed that fastened the gastrointestinal motility and recovery. Heated fennel therapy also increased the serum albumin (ALB) levels and shortened the duration of hospital stay (Chen et al. 2020). The methanolic extract of fennel was tested as a natural treatment for nerve injury. The results were promising, methanolic extract of fennel exhibited signs of functional recovery. Increase in muscle mass in the group of rats treated with fennel extract was observed (Maqbool et al. 2020).

15.9 FOOD APPLICATIONS

Fennel seeds are used in savoury and sweet dishes. It is an essential part of spice mixes like the Indian panch phoran and the Chinese five spice blend. Mukhwas is a kind of mouth freshener prepared in Indian cuisine from fennel seed. These seeds may be sugar-coated or uncoated and is used to aid digestion after a meal (Badgujar et al. 2014). Fennel tea made by brewing fennel seeds in hot water is also consumed in many parts of the world. In a study conducted by the scientists at Kyung Hee University in South Korea. It was found that fennel and fenugreek tea were successful in controlling subjective appetite in overweight females. Fennel seeds are also considered to be effective in preventing weight gain (Bae et al. 2015).

15.9.1 Fennel Seed Flavour

Fennel seeds found to perform the best when analysed with an electric tongue. It scored 8.7 and 8.4 respectively for umami taste (UMS) and sourness taste (SRS) (Hong et al. 2021). This gives fennel seeds a unique scope for food application. Fennel seeds are used in both sweet and savoury food products. It used to flavour baked goods like breads and biscuits and even ice-creams and confectionery like candies. It is used in fermented dishes such as sauerkraut and even in sweet pickles. Some good flavour combination that utilizes fennel are fennel seeds with tomato, cucumber or asparagus. Fennel seeds also compliment meat dishes like stuffing, Italian sausages and lasagna. Fennel seeds were used as a deodorizer in kimchi. The kimchi sample containing fennel seed extract exhibited a superior performance during sensory analysis. Fennel seeds extract also altered the fermentation characteristics of kimchi (Choi et al. 2010).

15.9.2 Fennel as a Flavouring Agent

Fennel is known to possess an anise like aroma. The presence of volatile components in fennel have allowed its use as a flavouring agent. It is traditionally used in meat and fish preparations, alcoholic beverages and herb mixtures. It has also been used in baked goods. Fennel seeds are the preferred part of the fennel plant for cooking. In Indian Cuisine, particularly in Kashmiri Pandit and Gujarati cooking fennel seeds are important ingredient (Rather et al. 2016).

Biscuits made with the addition of fennel seeds and its aromatic oils in various quantities (25% fennel seeds, 50% fennel seeds and 25% fennel seeds +2% fennel seed essential oil) were found to have a longer shelf life. The sensory property remained unchanged during the period of their storage and no appreciable microbial growth was observed. The mode of action against microbes is to damage the microbial cell and cause the leakage of its contents, this leads to the lysis of the cell.

The flavour associated with fennel is offered by anethole, estragole and limonene. These have been described as warm aromatic compounds with a pleasing flavour. The addition of essential oil to the processing can also protect the food products from oxidation with minimal change to appearance and texture (Saber and Eshra 2019).

15.9.3 Fennel Extract in Films

Researchers are exploring to utilize fennel essential oil (FEO) in conjugation with cellulose acetate (CA) for the development of an edible film having antimicrobial activity against *Staphylococcus aureus* and *Escherichia coli* (Gonçalves et al. 2020). Similarly, another novel chitosan-based film fused with spice extracts like fennel were synthesized to increase the shelf life of chilled meat. The films had antioxidant and antibacterial activity which increased the shelf life of pork significantly (Liu et al. 2021). Pectin extracted for *Foeniculum vulgare* have also been used to produce a biopolymer film with Phaseolin proteins (Giosafatto et al. 2007).

15.10 CONCLUSION

Fennel (*Foneiculum vulgare)* is an important biennial medicinal and aromatic spicy herb commonly known as "*saunf*". All parts of the fennel i.e., the roots, stems, leaves and seeds are used in cooking. Due to unique flavour and aroma, it is used extensively as flavouring agents among meat products, baked items, fish dishes, bakery and confectionery stuffs, alcoholic beverages. Fennel contains bioactive components such as anethole, estragole, fenchone and limonene, flavonoids and phenolic compounds. Numerous studies have reported that essential oil seed extract of fennel is used in food, pharma and cosmetic industries due to typical fragrance and biological activities such as antimicrobial, anti-inflammatory, antioxidant, antitumour and antithrombotic properties. Besides its therapeutic nature, one of its component i.e., estragole, has become a cause of concern, as the structurally similar methyl-eugenol has been recently found to be a potential carcinogen and more research is required to eliminate/decrease its level. Fennel is a spice that holds immense potential that governs the healthy longevity of its users. Additional research and global funding opportunities are necessary to conduct more studies to explore the health benefits and potential of fennel for treatment of diseases and commercial product development.

REFERENCES

Abbas, A., Bukhari, Q. A., Sarfaraz, S, Hasan, F and Ikram, R. 2020. Fennel fortified diet: New perspective with regard to fertility and sex hormones. *Pakistan Journal of Pharmaceutical Sciences* 33 (6): 2595–2600.

Abbas, A., Ikram, R., Hasan, F., Adil, A., Nisar, U. and Bukhari, Q.A. 2020. "Antidepressant and Antiamnesic Potential of FoeniculumVulgare". *Journal of Advances in Medicine and Medical Research* 32 (1): 131–138.

Ahmad, B. S., Talou, T., Saad, Z., Hijazi, A., Cerny, M., Kanaan, H., Chokr, A., and Merah, O. 2018. Fennel oil and by-products seed characterization and their potential applications. *Industrial Crops and Products* 111: 92–98.

Akbar, S. 2018. Fennel (Foeniculum vulgare Mill.): A Common Spice wit Unique Medicinal Properties. *Annals of Complementary and Alternative Medicine* 1 (1): 1–9.

Akha, O., Rabiei, K., Kashi, Z., Bahar, A., Zaeif-Khorasani, E., Kosaryan, M., Saeedi, M., Ebrahimzadeh, M. A., & Emadian, O. 2014. The effect of fennel (Foeniculum vulgare) gel 3% in decreasing hair thickness in idiopathic mild to moderate hirsutism, A randomized placebo controlled clinical trial. *Caspian Journal of Internal Medicine* 5 (1): 26–29.

Akhbari, M., Kord, R., Nodooshan, S. J., and Hamedi, S. 2018. Analysis and evaluation of the antimicrobial and anticanc activities of the essential oil isolated from Foeniculum vulgare from Hamedan, Iran. *Natural Product Research* 1–5. https://doi.org/10.1080/14786419.2017.1423310

Akhtar, I., Javad, S., Ansari, M., Ghaffar, N., and Tariq, A. 2020. Process optimization for microwave assisted extraction of Foeniculum vulgare Mill using response surface methodology. *Journal of King Saud University – Science* 32: 1451–1458.

Al-Amoudi, W. M. 2017. Protective effects of fennel oil extract against sodium valproate-induced hepatorenal damage in albino rats. *Saudi Journal of Biological Sciences* 24: 915–924. https://doi.org/10.1016/j.sjbs.2016.10.021.

Al-Okbi, S.Y., Hussein, A.M.S., Elbakry, H.F.H., Fouda, K.A., Mahmoud K.F. and Hassan, M.E. 2018. Health Benefits of Fennel, Rosemary Volatile Oils and their Nano-Forms in Dyslipidemic Rat Model. *Pakistan Journal of Biological Sciences* 21: 348–358.

Alam, P., Abdel-Kader, M. S., Alqarni, M. H., Zaatout, H. H., Ahamad, S. R., and Shakeel, F. 2019. Chemical composition of fennel seed extract and determination of fenchone in commercial formulations by GC–MS method. *Journal of Food Science and Technology* 1–9. Published online. https://doi.org/10.1007/s13197-019-03695-9.

Allaithy, S. A. M. 2017. Chemical compound of cumin and fennel seed extracts against some types of pathogenic bacteria. *Iraq Medical Journal* 1 (1): 1–6.

Altemimi, A., Lakhssassi, N., Baharlouei, A., Watson, D. G., and Lightfoot, D. A. 2017. Phytochemicals: Extraction, Isolation, and Identification of Bioactive Compounds from Plant Extracts. *Plants* 6 (42): 1–23.

Angelov, G. and Boyadzhieva, S. 2016. Extraction of Fennel (Foeniculum Vulgare) Seeds: Process Optimization and Antioxidant Capacity of the Extracts. *Chemical and Biochemical Engineering Quarterly* 30 (2): 245–253. https://doi.org/10.15255/cabeq.2015.2281.

Anwar, F., Ali, M., Hussain, A. I., and Shahid, M. 2009. Antioxidant and antimicrobial activities of essential oil and extracts of fennel (Foeniculum vulgare Mill.) seeds from Pakistan. *Flavour and Fragrance Journal* 24: 170–176. doi:10.1002/ffj.1929.

Badgujar, S. B., Patel, V. V., Bandivdekar, A. H. 2014. Foeniculum vulgare Mill: A Review of Its Botany, Phytochemistry, Pharmacology, Contemporary Application, and Toxicology. *BioMed Research International*: 1–32.

Bae, J.Y., Kim, J.E., Choue, R. and Lim, H. 2015. Fennel (Foeniculum Vulgare) and Fenugreek (Trigonella Foenum-Graecum) Tea Drinking Suppresses Subjective Short-Term Appetite in Overweight Women. *Clinical Nutrition Research* 4 (3): 168. https://doi.org/10.7762/cnr.2015.4.3.168.

Benmoussa, H., Farhat, A., Romdhane, M., and Bouajila, J. 2019. Enhanced solvent-free microwave extraction of Foeniculum vulgare Mill. essential oil seeds using double walled reactor. *Arabian Journal of Chemistry* 12: 3863–3870.

Berrington, D. and Lall, N. 2012. Anticancer Activity of Certain Herbs and Spices on the Cervical Epithelial Carcinoma (HeLa) Cell Line. *Evidence-Based Complementary and Alternative Medicine*: 1–11. https://doi.org/10.1155/2012/564927.

Beyazen, A., Dessalegn, E., and Mamo, W. 2017. Phytochemical screening, antioxidant and antimicrobial activities of seeds of Foeniculum vulgare (ensilal). *World Journal of Pharmaceutical Sciences* 5 (3): 198–208.

Bharat, B. A., Ajaikumar, B. K., Kuzhuvelil, B. H., Sheeja, T. T., Sung, B., and Preetha, A. 2008. Potential of spice-derived phytochemicals for cancer prevention. *Planta Medica* 74 (13): 1560–1569.

Birdane, F. M., Cemek, M., Birdane, Y.O., Gülçin, I. and Büyükokuroğlu, M.E. 2007. Beneficial effects of Foeniculum vulgare on ethanol-induced acute gastric mucosal injury in rats. *World Journal of Gastroenterology* 13 (4): 607. https://doi.org/10.3748/wjg.v13.i4.607.

Bokaie, M., Farajkhoda, T., Enjezab, B., Khoshbin, A. and Mojgan, K.Z. 2013. Oral fennel (Foeniculum vulgare) drop effect on primary dysmenorrhea: Effectiveness of herbal drug. *Iranian Journal of Nursing and Midwifery Research* 18 (2): 128–132.

Castro, H. T., Martíne, J. R., Stashenko, and Elena, 2010. Anethole isomerization and dimerization induced by acid sites or UV irradiation. *Molecules* 15: 5012–5030.

Cautela, D., Vellab, F. M., Castaldoc, D., and Larattab, B., 2018. Characterization of essential oil recovered from fennel horticultural wastes. *Natural Product Research*. doi: 10.1080/14786419.2018.1478827.

Chandran, A., N.J., Merlin, L., Ammu, and Dharan, S.S. 2019. Fennel Treatment to PCOS: An Insilico Evaluation to explore the Therapeutic Efficacy of Anethole. *Research Journal of Pharmacy and Technology* 12 (10): 4958. doi:10.5958/0974-360x.2019.00860.6.

Chatterjee, S., Goswami, N., and Bhatnagar, P. 2012. Estimation of Phenolic Components and in vitro Antioxidant Activity of Fennel (Foeniculum vulgare) and Ajwain (Trachyspermumammi) seeds. *Advances in Bioresearch* 3 (2): 109–118.

Chegini, R., Soleimani, P., Sadeghi, M., Yosef, R.M. and Zafari, F.2019. Investigating the Effect of Fennel and Cinnamon Combined Extract on Spermatogenesis and Testis Tissues in Busulfan Induced Infertile Rats. *Journal of Applied Biotechnology Reports* 6 (3): 96–100. https://doi.org/10.29252/jabr.06.03.03.

Chen, B., He, Y., Xiao, Y., Guo, D., Liu, P., He, Y., Sun,Q., Jiang, P., Liu, Z. and Liu, Q. 2020. Heated Fennel Therapy Promotes the Recovery of Gastrointestinal Function in Patients after Complex Abdominal Surgery: A Single-Center Prospective Randomized Controlled Trial in China. *Surgery* 168 (5): 793–799. doi:10.1016/j.surg.2020.05.040.

Chen, F., Guo, Y., Kang, J., Yang, X., Zhao, Z., Liu, S., Ma, Y., Gao, W., and Luo, D. 2020. Insight into the essential oil isolation from Foeniculum vulgare Mill. Fruits using double-condensed microwave-assisted hydrodistillation and evaluation of its antioxidant, antifungal and cytotoxic activity. *Industrial Crops & Products* 144: 1–12.

Choi, A.R., Park, D.I, Son, M.H., Kim, S.Y., Jang, J.B., Kwon, S.C. and Chae, H. J. 2010. Effects of Deodorizing Plant Products on the Odor and Fermentation Characteristics of Kimchi. *Korean Journal of Food Science and Technology* 42: 63–68.

Choi, E.M. and Hwang, J.K. 2004. Antiinflammatory, analgesic and antioxidant activities of the fruit of Foeniculum vulgare. *Fitoterapia* 75 (6): 557–565. https://doi.org/10.1016/j.fitote.2004.05.005.

Choudhary, S., Sharma, R., Meena, R. S. and Verma, A. K. 2018. Molecular diversity analysis in fennel (Foeniculum vulgare Mill) genotypes and its implications for conservation and crop breeding. *International Journal of Current Microbiology and Applied Sciences* 7(3): 794–809.

Danilova, E. I., Trusova, O. Y., Roschupkin, A. N., Veterkova, Z. A., Golovacheva, E. I. and Ignatova, T. N. 2016. Efficacy of Enterosorbent Based on Pectin, Inulin and Fennel Extract in Treating Children with Atopic Dermatitis: «Case-Control» Prospective Study. *Current Pediatrics* 15 (3): 268–272. doi:10.15690/vsp.v15i3.1564.

Dheebisha, C., and Vishwanath, Y. C. 2020. Advances in cultivation of fennel. *Journal of Pharmacognosy and Phytochemistry* 9 (2): 1295–1300.

Diao, W. R., Hua, Q. P., Zhang, H, and Xu, J. G. 2014. Chemical composition, antibacterial activity and mechanism of action of essential oil from seeds of fennel (Foeniculum vulgare Mill.). *Food Control* 3: 109–116. 10.1016/j.foodcont.2013.06.056.

Dua, A., Garg, G., and Mahajan, R. 2013. Polyphenols, flavonoids and antimicrobial properties of methanolic extract of fennel (Foeniculum vulgare Miller). *European Journal of Experimental Biology* 3(4): 203–208.

FallahHuseini, H. and Kianbakht, S. 2012. Study on Effects of Chicory (Cichorium intybus L.), Fennel (Foeniculum vulgare Mill.) and Dill (Anethum graveolens L.) on Fertility and Neonatal Gender in Rats. *Journal of Medicinal Plants* 11 (42): 192–196.

Farid, A., Kamel, D., Montaser, S.B., Ahmed, M.M., El Amir, M. and El Amir, A. 2020. Assessment of Antioxidant, Immune Enhancement, and Antimutagenic Efficacy of Fennel Seed Extracts in Irradiated Human Blood Cultures. *Journal of Radiation Research and Applied Sciences* 13 (1): 260–266. doi:10.1080/16878507.2020.1728963.

Garzoli, S., Božović, M., Baldisserotto, A., Sabatino, M., Cesa, S., Pepi, F., ... and Ragno, R. 2018. Essential oil extraction, chemical analysis and anti-Candida activity of Foeniculum vulgare Miller – new approaches. *Natural Product Research* 32(11): 1254–1259. https://doi.org/10.1080/14786419.2017.1340291

Ghasemian, A., Al-Marzoqi, A., Mostafavi, S. S. K., Alghanimi, Y. K., and Teimouri, M. 2019. Chemical composition and antimicrobial and cytotoxic activities of Foeniculum vulgare Mill Essential Oils. *Journal of Gastrointestinal Cancer*: 1–7. Published online. https://doi.org/10.1007/s12029-019-00241-w.

Ghazanfarpour, M., Mohammadzadeh, F., Shokrollahi, P., Khadivzadeh, T., Najafi, M.N., Hajirezaee, H. and Afiat, M. 2017. Effect of Foeniculum vulgare (fennel) on symptoms of depression and anxiety in postmenopausal women: a double-blind randomised controlled trial. *Journal of Obstetrics and Gynaecology* 38 (1): 121–126. https://doi.org/10.1080/01443615.2017.1342229.

Ghorbani, M., Aboonajmi, M., Javid, M. G., and Arabhosseini, A. 2017. Optimization of ultrasound-assisted extraction of ascorbic acid from fennel (Foeniculum vulgare) seeds and evaluation its extracts in free radical scavenging. *Agricultural Engineering International: CIGR Journal Open Access* 19 (4): 1–11.

Giosafatto, C.V.L., Mariniello, L. and Ring, S. 2007. Extraction and Characterization of Foeniculum vulgare Pectins and Their Use for Preparing Biopolymer Films in the Presence of Phaseolin Protein. *Journal of Agricultural and Food Chemistry* 55 (4): 1237–1240. https://doi.org/10.1021/jf062725d.

Goncalves, S. M., Motta, J. F. G., Ribeiro-Santos, R., Chavez, D. W. H., and Melo, N. R. D. 2020. Functional and antimicrobial properties of cellulose acetate films incorporated with sweet fennel essential oil and plasticizers. *Current Research in Food Science* 3: 1–8. https://doi.org/10.1016/j.crfs.2020.01.001.

Grover, S., Malik, C. P., Hora, A., and Kushwaha, H. B., 2013. Botany, Cultivation, Chemical Constituents and Genetic Diversity in Fennel (Foeniculum vulgare Mill.): A Review. *LS – An International Journal of Life Sciences* 2 (2): 128–132.

Gulfraz, M. and Mehmood, S., Minhas, N., Jabeen, N., Kausar, R., Jabeen, D. and Arshad, G. 2008. Composition and antimicrobial properties of essential oil of Foeniculum vulgare. *African Journal of Biotechnology* 7 (24): 4364–4368.

Hammouda, F. M., Saleh, M. A., Abdel-Azim, N. S., Shams, K. A., Ismail, S. I., Shahat, A. A., and Saleh, I. A., 2013. Evaluation of The Essential Oil OfFoeniculum Vulgare Mill (Fennel) Fruits Extracted By Three Different Extraction Methods By Gc/Ms. *African Journal of Traditional, Complementary and Alternative Medicines* 11 (2): 277–279.

Herman, A. and Herman, A. 2014. Essential oils and their constituents as skin penetration enhancer for transdermal drug delivery: a review. *Journal of Pharmacy and Pharmacology* 67 (4): 473–485. doi:10.1111/jphp.12334.

Hirondart, M., Rombaut, N., Fabiano-Tixier, A. S., Bily, A., and Chemat, F. 2020. Comparison between Pressurized Liquid Extraction and Conventional Soxhlet Extraction for Rosemary Antioxidants, Yield, Composition, and Environmental Footprint. *Foods* 9 (584): 1–14; doi:10.3390/foods9050584.

Hong, S.J., Chang, G.B., Heo, S.U. and Shin, E.C. 2021. Comparative Study between Parts of Fennel (Foeniculum vulgare Mill.) for Taste and Flavor Properties. *Journal of the Korean Society of Food Science and Nutrition* 50 (4): 384–394. https://doi.org/10.3746/jkfn.2021.50.4.384.

Javidnia, K., Dastgheib, L., Samani, S. M. and Nasiri, A. 2003. Antihirsutism activity of Fennel (fruits of Foeniculum vulgare) extract. A double-blind placebo controlled study. *Phytomedicine* 10 (6-7): 455–458. https://doi.org/10.1078/094471103322331386.

Kalleli, F., Rebey, I. B., Wannes, W. A., lBoughalleb, F., Hammami, M., Tounsi, M. S., and M'hamdi, M. 2019. Chemical composition and antioxidant potential of essential oil and methanol extract from Tunisian and French fennel (Foeniculum vulgare Mill.) seeds. *Journal of Food Biochemistry* e12935: 1–14. doi: 10.1111/jfbc.12935.

Kaur, P., Gupta, S., Kaur, K., Kaur, N., Kumar, R., and Bhullara, M. S. 2021. Nanoemulsion of Foeniculum vulgare essential oil: A propitious striver against weeds of Triticum aestivum. *Industrial Crops and Products* 168 (15), https://doi.org/10.1016/j.indcrop.2021.113601

Khan, M., and Musharaf, S. 2014. Foeniculum vulgare Mill. A medicinal herb. *Medicinal Plant Research* 4.

Kooti, W., Moradil, M., Ali-Akbari, S., Sharafi-Ahvazi, N., Asadi-Samani, M., and Ashtary-Larky, D. 2015. Therapeutic and pharmacological potential of Foeniculum vulgare Mill: a review. *J HerbMedPharmacol* 4 (1): 1–9.

Koppula, S. and Kumar, H. 2013. Foeniculum vulgare Mill (Umbelliferae) Attenuates Stress and Improves Memory in Wister Rats. *Tropical Journal of Pharmaceutical Research* 12 (4). https://doi.org/10.4314/tjpr.v12i4.17.

Kwiatkowski, P., Mnichowska-Polanowska, M., Pruss, A., Masiuk, H., Dzięcioł, M., Giedrys-Kalemba, S. and Sienkiewicz, M. 2017. The effect of fennel essential oil in combination with antibiotics on Staphylococcus aureus strains isolated from carriers. *Burns* 43 (7): 1544–1551. doi:10.1016/j.burns.2017.04.014.

Lal, G., Saran, P. L., Devi, G., Deepak, and Raj, R. 2014. Production Technology of Fennel (Foeniculum vulgare Mill). Chapter 38: https://www.researchgate.net/publication/334416358.

Liu, T., Liu, L., Gong, X., Chi, F. and Ma, Z. 2021. Fabrication and comparison of active films from chitosan incorporating different spice extracts for shelf life extension of refrigerated pork. *LWT* 135: 110181. https://doi.org/10.1016/j.lwt.2020.110181.

Lourenço, S. C., Moldão-Martins, M., and Alves, V. D. 2019. Antioxidants of Natural Plant Origins: From Sources to Food Industry Applications. *Molecules* 24: 1–25.

Luaibi, N., Shafiq, A.A., Mohammed, N. and Ayser, L. 2020. Study of Biochemical and Histological effects of fennel seeds (Foeniculum vulgare) on Liver in Male Rats. 10.13140/RG.2.2.15963.31521.

Ma, H.W., Zhao, J.T. and Zhao, X. 2015. The Effect of Fennel Tea Drinking on Postoperative Gut Recovery after Gynecological Malignancies Operation. *Journal of Sichuan University* 46 (6): 940–943.

Manohar, R., Ganesh, A., Abbyramy, N., Abinaya, R., Balaji, S.K. and Priya, S.B. 2020. The effect of fennel seeds on pH of saliva – A clinical study. *Indian Journal of Dental Research* [serial online] [cited 2021 Jun 17]; 31: 921–923. Available from: https://www.ijdr.in/text.asp?2020/31/6/921/311647.

Maqbool, J., Anwar, H., Iqbal, J., Rasul, A., Imran, A., Malik, S.A., Shabbir, A., Ijaz, F., Sajid, F., Akram, R., Sun, T., Imran, M., Hussain, G. and Islam, S. 2020. Methanolic extract of Fennel (Foeniculum vulgare) escalates functional restoration following a compression injury to the sciatic nerve in a mouse model. *Food Science & Nutrition* 9 (2): 701–710. doi:10.1002/fsn3.2033.

Marin, I., Sayas-Barberá, E., Viuda-Martos, M., Navarro, C. and Sendra, E. 2016. Chemical composition, antioxidant and antimicrobial activity of essential oils from organic fennel, parsley, and lavender from Spain. *Foods 5*: 18. https://doi.org/10.3390/foods5010018

Mesfin, M., Asres, K. and Shibeshi, W. 2014. Evaluation of Anxiolytic Activity of the Essential Oil of the Aerial Part of Foeniculum Vulgare Miller in Mice. *BMC Complementary and Alternative Medicine* 14 (1). https://doi.org/10.1186/1472-6882-14-310.

Michailidis, D., Angelis, A., Nikolaou, P. E., Mitako, S., and Skaltsounis, A. L., 2021. Exploitation of Vitis vinifera, Foeniculum vulgare, Cannabis sativa and Punica granatum By-Product Seeds as Dermo-Cosmetic Agents. *Molecules* 26 (731): 1–20. doi.org/10.3390/molecules26030731.

Moghtader, M. 2013. Chemical composition, antimicrobial and antioxidant from the seeds and flowers of Foeniculum vulgare Mill. from Kerman Province. *Journal of Horticulture and Forestry* 5 (3): 37–40.

Mohamad, R. H., El Bastawesy, A. M., Abdel Monem, M. G., Noor, A. M., Al-Mehdar, H. A. R., Sharawy, S. M., and El-Merzabani, M. M. 2011. Antioxidant and Anticarcinogenic Effects of Methanolic Extract and Volatile Oil of Fennel Seeds (Foeniculum vulgare). *Journal of Medicinal Food* 14 (9): 986–1001.

Mokaberinejad, R., Rampisheh, Z., Aliasl, J. and Akhtari, E. 2019. The Comparison of Fennel Infusion plus Dry Cupping versus Metformin in Management of Oligomenorrhoea in Patients with Polycystic Ovary Syndrome: a Randomised Clinical Trial. *Journal of Obstetrics and Gynaecology* 39 (5): 652–58. doi:10.1080/01443615.2018.1541232.

Mokhtari, L., and Ghoreishi, S. M., 2019. Supercritical carbon dioxide extraction of trans-anethole from Foeniculum vulgare (fennel) seeds: Optimization of operating conditions through response surface methodology and genetic algorithm. *Journal of* CO_2 *Utilization* 30: 1–10.

Moradia, H., Bahmanyara, H., Azizpoura, H., Rezamandia, N., and Ashkezaria, S. M. M., 2020. Modeling and Optimization of Anethole Ultrasound-Assisted Extraction from Fennel Seeds using Artificial Neural Network, *Journal of Chemical and Petroleum Engineering* 54(1): 143–153.

Napoli, E. M., Curcuruto, G., and Ruberto, G. 2010. Screening the essential oil composition of wild Sicilian fennel. *Biochemical Systematics and Ecology* 38: 213–223.

Nour, N., Backer, W.S., Al-Ahmadi, A.M. 2017. The Role of Senna and Fennel In Ameliorating Cardiovascular Disease In Diabetic Rats. *American Journal of Pharmacy & Health Research* 5.

Ozbek, H., Uğra, S., Dülger, H., Bayram, I., Tuncer, I., Oztürk, G. and Oztürk, A. 2003. Hepatoprotective effect of Foeniculum vulgare essential oil. *Fitoterapia* 74 (3): 317–319. doi: 10.1016/s0367-326x(03)00028-5.

Pacifico, S., Galasso, S., Piccolella, S., Kretschmer, N., Pan, S.P., Nocera, P., Lettieri, A., Bauer, R. and Monaco, P. 2018. Winter Wild Fennel Leaves as a Source of Anti-Inflammatory and Antioxidant Polyphenols. *Arabian Journal of Chemistry* 11 (4): 513–524. https://doi.org/10.1016/j.arabjc.2015.06.026.

Patra, J.K., Das, G., Lee, S., Kang, S., and Shin. H. 2018. Selected commercial plants: A review of extraction and isolation of bioactive compounds and their pharmacological market value. *Trends in Food Science & Technology*: 1–78. doi: https://doi.org/10.1016/j.tifs.2018.10.001

Peter, K. V., and Malhotra, S.K. 2012. Fennel and Fennel Seed. Essay. In *Handbook of Herbs and Spices*, 2nd ed., 275–302. Oxford: Woodhead Pub.

Portincasa, P., Bonfrate, L., Scribano, M. L., Kohn, A., Caporaso, N., Festi, D., Campanale, M. C., Di Rienzo, T., Guarino, M., Taddia, M., Fogli, M. V., Grimaldi, M. and Gasbarrini, A. 2016. Curcumin and Fennel Essential Oil Improve Symptoms and Quality of Life in Patients with Irritable Bowel Syndrome. *Journal of Gastrointestinal and Liver Diseases: JGLD* 25 (2): 151–157. https://doi.org/10.15403/jgld.2014.1121.252.ccm

Rabeh, N.M. and Aboraya, A.O. 2014. Hepatoprotective Effect of Dill (Anethum Graveolens L.) and Fennel (Foeniculum Vulgare) Oil on Hepatotoxic Rats. *Pakistan Journal of Nutrition* 13 (6): 303–309. https://doi.org/10.3923/pjn.2014.303.309.

Rajic, J.R., Djordjevic, S.M., Tesevic, V.V., Zivkovic, M.B., Djordjevic, N.O., Paunovic, D.M., Nedovic, V.A. and Petrovic, T.S. 2018. The Extract of Fennel Fruit as a Potential Natural Additive in Food Industry. *Journal of Agricultural Sciences, Belgrade* 63 (2): 205–215. https://doi.org/10.2298/jas1802205r

Rasul, A. and Akhtar, N. 2014. Anti-aging potential of a cream containing milk thistle extract: Formulation and in vivo evaluation. *African Journal of Biotechnology* 11 (6): 54–58. https://doi.org/10.5897/ajb11.2678.

Rather, M. A., Dar, A. B., Sofi, S. N., Bhat, B. A., and Qurishi, M. A. 2016. Foeniculum vulgare: A comprehensive review of its traditional use, phytochemistry, pharmacology, and safety. *Arabian Journal of Chemistry* 9: S1574–S1583. https://doi.org/10.1016/j.arabjc.2012.04.011.

Repajić, M., Ekić, S., Kruk, V., and Dragović-Uzelac, V. 2020. Effect of accelerated solvent extraction conditions on the isolation of bioactive compounds from fennel (Foeniculum vulgare Mill.) seeds. *Croatian Journal of Food Technology, Biotechnology and Nutrition* 15 (3-4): 102–106.

Reverchon, J., Daghero, C., Marrone, M., Mattea, M., and Polleto, M. 1999. Supercritical fractional extraction of fennel seed oil and essential oil: experiments and mathematical modeling. *Industrial & Engineering Chemistry Research* 38: 3069–3075.

Saber, J. I., and Eshra, D. H., 2019. Using Fennel Seeds and their Oil as a Preservative and Functional Food to Produce Some Food and Drink Products to Alleviate Cough Symptoms. *Alexandria Science Exchange Journal* 40(3): 406–414. https://doi.org/10.21608/asejaiqjsae.2019.44629.

Sadeghpour, N., Khaki, A.A., Najafpour, A., Dolatkhah, H. and Montaseri, A. (2015). Study of Foeniculum vulgare (Fennel) Seed Extract Effects on Serum Level of Estrogen, Progesterone and Prolactin in Mouse. *Crescent Journal of Medical and Biological Sciences* 2 (1): 23–27. https://www.sid.ir/en/journal/ViewPaper.aspx?id=467233

Salami, M., Rahimmalek, M., Ehtemam, M. H., and Sabzalian, M. R. 2016. Variation in bioactive compounds and anatomical characteristics of different fennel (Foeniculum vulgare Mill.) populations as affected by self-pollination. *Journal of Applied Botany and Food Quality* 89: 38–48. doi:10.5073/JABFQ.2016.089.005.

Saxena, S. N., Agarwal, D., John, S., Dubey, P. N., and Lal, G. 2018. Analysis of fennel (Foeniculum vulgare) essential oil extracted from green leaves, seeds and dry straw. *International J. Seed Spices* 8 (1): 60–64.

Sayed Ahmad, B., Talou, T., Saad, Z., Hijazi, A., Cerny, M., Kanaan, H., Chokr, A. and Merah, O. 2018. Fennel Oil and by-Products Seed Characterization and Their Potential Applications. *Industrial Crops and Products* 111: 92–98. https://doi.org/10.1016/j.indcrop.2017.10.008.

Singh, B. and Kale, R.K. 2008. Chemomodulatory Action of Foeniculum Vulgare (Fennel) on Skin and Forestomach Papillomagenesis, Enzymes Associated with Xenobiotic Metabolism and Antioxidant Status in Murine Model System. *Food and Chemical Toxicology* 46 (12): 3842–3850. https://doi.org/10.1016/j.fct.2008.10.008.

Sur, A. 2020. Antimicrobial Study of Fennel (Foeniculum vulgare) Seed Oil against Bacteria and Fungi: A Comparative Study. *International Journal of Innovative Science, Engineering & Technology*. 7(2): 91–98.

Weiping, H. and Bookang, H. 2011. A review of chemistry and bioactivities of a medicinal spice: Foeniculum vulgare. *Journal of Medicinal Plants Research* 5 (16): 3539–3600.

Zhang, S., Chen, X., Devshilt, I., Yun, Q., Huang, C., An, L., Dorjbat, S. and He, X. 2018. Fennel Main Constituent, Trans-Anethole Treatment against LPS-Induced Acute Lung Injury by Regulation of Th17/Treg Function. *Molecular Medicine Reports* 18 (2): 1369–1376. doi:10.3892/mmr.2018.9149.

CHAPTER 16

Bioactive Compounds in Coriander

Shweta Suri and Anuradha Dutta

CONTENTS

16.1 INTRODUCTION

Spices are seeds, roots, fruits, or bark of plants that contribute their unique natural flavour, colour, aroma, and spiciness to the cooked, semi-cooked and raw foods (Dutta et al. 2018). In addition to imparting flavour, spices also have health benefits due to their rich phytochemical content.

DOI: 10.1201/9781003215387-16

Spices exhibit good antioxidant, anti-microbial, anti-diabetic, hypocholesterolemic, anti-cancer, and hypo-protective properties.

Coriander (*Coriandrum sativum*) is globally used as a seasoning in cooking. It is also traditionally used as herbal medicine. Coriander is a yearly herb having green lanceolate-formed leaves, articulated taproot, white/pink umbellate blossoms. The plant may go to a height of 20–70 cm (Mandal and Mandal 2015). This herb has its origin in the Eastern Mediterranean region and is widely cultivated in Asia, Northern Africa, and Central Europe for its use as medicinal and culinary purposes (Nadeem et al. 2013).

Coriander is well acknowledged owing to its therapeutic and pharmacological effects, including antioxidant, hypo-glycemic, anti-inflammatory, neuro-protective, anti-analgesic, anti-cancerous, and hypotonic action (Laribi et al. 2015). All the parts of the coriander plant, including leaves, stem, roots, and fruits, hold pleasant aroma and flavour and are used for their medicinal properties. Coriander can be said to be a worthy storehouse of functional compounds. The seeds are widely utilized for the preparation of pickles, sausages, seasonings and curry powder. It flavours numerous food varieties of savoury foods. Furthermore, the seeds are used as a traditional medicine in treating several digestion-related issues and their diuretic, carminative, anti-bilious, and refrigerant properties. Likewise, Coriander leaves, also referred to as cilantro or Chinese parsley, are broadly utilized in eastern cuisine and Indian food as a food seasoning or else to cover up the unpleasant flavours of specific food sources (Gil et al. 2002).

16.2 TAXONOMY OF CORIANDER

Coriander belongs to the family of carrot Umbelliferae (*Apiaceae*) with taxonomical description:

- Division: Angiospermae
- Class: Dicotyledonae
- Sub-class: Calyciflorae
- Order: Umbellales
- Family: Umbelliferae/Apiaceae
- Genus: Coriandrum
- Species: Umbellifera

The genus *Coriandrum L.* has two popularly known species namely the cultivated species *Coriandrum sativum* and the wild species *Coriandrum tordylium. Coriandrum sativum* is 30–100 cm in length, widely used in dishes for its fresh and pleasant flavour. Seeds of Coriander are globular, dry schizocarps having different longitudinal ridged surfaces (Mandal et al. 2015). In addition, two other varieties of Coriander exist, namely the *C. sativum* L. variety *microcarpum* DC, a small-fruited, volatile oil-rich Russian variety, and *C. sativum* L. variety *vulgare* Alef, which is large-fruited mainly grown in India, Morocco, and other Asian countries (Rajeshwari and Andallu, 2011). Coriander is known by several names in varying languages; English (*Coriander*), Hindi and Urdu (*Dhania*), Chinese (*Yuan Sui*), Greek (*Korion*), Thai (*Phak chee*), Sanskrit (*Dhanya*) and Spanish (C*ilantro*) (Singletary, 2016). Coriander is known by different names in several parts of India (Gopalan et al. 2007), as shown in Table 16.1.

16.3 TRADITIONAL USE AND BENEFITS OF CORIANDER

Literature reveals that Coriander is used as a folk remedy from ancient times since the Greek Era. Traditionally it has diverse functions, for instance, an appetite and digestive stimulant (Platel and Srinivasan 2004; Ugulu et al. 2009), lipid-lowering and glucose-lowering function (Lal et al. 2004; Tahraoui et al. 2007), diuretic (Ugulu et al. 2009), and as an anti-microbial compound

Table 16.1 Different Names of Indian Coriander

Language	Coriander
Hindi	Dhania, Dhanya
Bangla	Dhana, Dhania
Kannada	Kothambri, Kothamiribija
Gujarati	Kothmiri, Libdhaba
Marathi	Dhana
Malayalam	Kothumpalaribija
Punjabi	Dhania
Tamil	Kothamalli
Telugu	Dhaniyalu

(Chaudhry and Tariq 2006). It is also known for its role in the management of digestive diseases, for example, flatulence, jaundice, dysentery, diarrhoea, as well as vomiting; diseases relating to the central nervous system such as insomnia, seizure, anxiety; inflammatory diseases and diseases related to respiratory disorders such as cough, bronchitis, etc. (Thabrew et al. 2003; Emam and Hamedani 2006; Khan and Khatoon 2008; Ugulu et al. 2009).

Remarkably, Coriander is recognized as a natural additive to substitute the utilization of artificial antioxidants (for example, butylated hydroxyl anisole (BHA) and butylated hydroxytoluene (BHT)) (Nanditha et al. 2009) owing to its excellent antioxidant and anti-bacterial plus anti-fungal properties (Mandal et al. 2015).

Subsequently, due to the high bactericidal and fungicidal effects in coriander oil, the oil is utilized as a carminative, spasmolytic and stomachic compound. It is likewise used to treat loose bowels, sub-acidic gastritis, dyspepsia, digestive, as well as stomach-related incitement (Platel and Srinivasan 2004). Coriander also exhibits substantial lipolytic property (Leung and Foster 1996); since it has origin from the carrot family, its utilization has been recommended with caution in view of possible allergic responses from furanocoumarins (Brinker, 1998). Coriander can also be utilized in the form of seed volatile oil, oleoresin, curry powder, organic spices and flavours. Coriander essential oil is primarily utilized as a flavouring compound for cocoa, alcohol and chocolate. Coriander oil is also utilized in Pharmaceutical industries. The essential oil extracted from Coriander ranges from 0.5 to 2.5% and is utilized for flavour in cooking and production of perfumes and soaps. It enjoys being a reasonably stable compound and holding its pleasing scent longer than some other oil belonging to a similar taxonomical class (Purseglove et al. 1981).

16.4 NUTRITIONAL COMPOSITION OF CORIANDER

The nutritional composition of coriander varies depending on the geographical area, and climate of the region. As per Indian Food Composition Table, the coriander leaves contain 86.99% moisture, 3.52% protein, 1.93% carbohydrates, 2.19% total ash, 0.70% fat, 4.66% dietary fibre, 130 KJ energy, 0.09 mg/100 g thiamine, 0.05 mg/100 g riboflavin, 0.73 mg/100 g niacin, 0.63 mg/100 g pantothenic acid, 0.19 mg/100 g pyridoxine, 4.17 μg/100 g biotin, 51.01 μg/100 g total folate, 23.87 mg/100 g vitamin C, 6351 μg/100 g lutein, 28.30 μg/100 g zeaxanthin, 3808 μg/100 g β-carotene, 13808 μg/100 g total carotenoids, 146 mg/100 g calcium, 5.30 mg/100 g iron, 72.68 mg/100 g magnesium, 64.69 mg/100 g phosphorus, 37 mg/100 g sodium and 546 mg/100 g potassium. Conversely, the coriander seeds contains 8.72% moisture, 10.66% protein, 12.98% carbohydrates, 5.36% total ash, 17.47% fat, 44.81% dietary fibre, 1125 KJ energy, 0.19 mg/100 g thiamine, 0.23 mg/100 g riboflavin, 1.20 mg/100 g niacin, 0.28 mg/100 g pantothenic acid, 0.04 mg/100 g pyridoxine, 1.46 μg/100 g biotin, 22.07 μg/100 g total folate, 225 μg/100 g lutein, 10.96 μg/100 g zeaxanthin, 122 μg/100 g β-carotene, 996 μg/100 g total carotenoids, 718 mg/100 g calcium, 17.64 mg/100 g iron, 343 mg/100 g magnesium, 293 mg/100 g phosphorus, 34.41 mg/100 g sodium and 1473 mg/100 g potassium (Longvah et al. 2017).

Furthermore, a study done on herbs and spices stated that the fresh coriander leaves comprises of 87.90% moisture, 1.7% total ash, 6.5% carbohydrates, 3.3% protein, 0.01% iron, 0.14% calcium, 0.06% phosphorus, 60 mg/100 g riboflavin, 0.8 mg/100 g niacin, 135 mg/100 g ascorbic acid, and 10,460 IU/100 g vitamin A. Coriander seeds contain 11% protein, 11% starch, 30% crude fibre, 20% crude fat and 1.7% total ash (Peter, 2004). As per USDA, both the coriander seeds as well as leaf are highly nutritious. Coriander seeds contain zero cholesterol and are also a good source of micronutrients such as vitamins and minerals (USDA 2018). Different nutrients present in both coriander seeds and leaf are presented in Table 16.2.

Coriander contains reasonable amounts of essential oils that are vital for brain development and functioning. Coriander seed oil is known for its high essential oil content. Also, coriander oil attained a place in the twenty major oils category in the global market. The commercial value of coriander seed oil mainly depends on chemical composition, physical properties, and aroma plus flavour. Essential oils found in the oil glands located in the mericarp portion of Coriander impart a unique flavour and aroma to the Coriander (Diederichsen 1996). Coriander seeds as well as the aerial part of plant contain sufficient amounts of lipids (mainly petroselinic acid) and essential oils (mainly linalool).

16.5 BIOACTIVE COMPOUNDS IN DIFFERENT PARTS OF *CORIANDRUM SATIVUM* (L)

Several bioactive compounds, including polyphenols, essential oils, sterols, etc., have been isolated from different parts of coriander as shown in Table 16.3.

The bioactive components found in coriander exhibit several functions: antioxidant, anti-diabetic, cancer-protective, anti-microbial, and hepato-protective effects. In general, the aerial parts and seeds of Coriander contain reasonable amounts of essential oils, lipids, and polyphenols; on the other hand, the water-soluble constituents are merely located in aerial parts as well as seeds of Coriander (Prachayasittikul et al. 2018). Conversely, the presence and stability of bioactive compounds in coriander seeds, leaves, and stem also depends on the developmental stage of the plant, geographic area or location, season of sowing, cultivation condition, processing, and storage parameters as well as duration (Prachayasittikul et al. 2018; Msaada et al. 2017; Turek and Stintzing 2013; Mandal and Mandal 2015). Also, the extraction method may perhaps affect the bioactive compounds in terms of the type and amounts (Benyoussef and Saibi 2013). Composition of oil extracted from Coriander also varies because of the difference in the geographic area; hence, the Indian variety of Coriander is different from the European coriander variety in terms of linalool concentration found in lesser amount in the former. Additionally, in comparison to the mature coriander seeds, the fresh coriander produce exhibits a pleasant odour and flavour; this could be because of the predominant presence of the aliphatic aldehyde group compared to the linalool content in the mature coriander seed oil (Iqbal et al. 2017). The bioactive components present in coriander oil are coriandrol, jireniol and vebriniol (Bhat et al. 2014). Details of the bioactive compounds commonly found in the coriander are shown in Figure 16.1.

16.6 PHYTOCHEMICAL CHARACTERIZATION OF CORIANDER SEEDS AND HERB/LEAVES

Coriander seeds exhibit efficient therapeutic properties because of fatty acids, sterols, tocols, polyphenols, water-soluble components, and volatile compounds. The phytochemical properties of different parts of the coriander plant are discussed below.

Table 16.2 Nutritional Composition of Coriander Seeds and Leaf (per 100 g)

	Amount per 100 g	
Nutrients	**Coriander Seeds (1 Table Spoon)**	**Coriander Leaf (Dried) (1 Table Spoon)**
Moisture	8.86 g	7.30 g
Energy	298 kcal	279 kcal
Protein	12.40 g	21.90 g
Total lipid (Fat)	17.80 g	4.78 g
Ash	6.02 g	14.10 g
Carbohydrate by difference	55.00 g	52.10 g
Fibre (Total Dietary Fibre)	41.9 g	10.4 g
Calcium	709 mg	1250 mg
Iron	16.30 mg	42.50 mg
Magnesium	330 mg	694 mg
Phosphorus	409 mg	481 mg
Potassium	1270 mg	4470 mg
Sodium	35 mg	211 mg
Zinc	4.70 mg	4.72 mg
Copper	0.975 mg	1.79 mg
Manganese	1.90 mg	6.36 mg
Selenium	26.20 μg	29.30 μg
Vitamin C	21 mg	567 mg
Thiamin	0.239 mg	1.250 mg
Riboflavin	0.29 mg	1.50 mg
Niacin	2.13 mg	10.70 mg
Vitamin B6	–	0.61 mg
Folate (total)	0 μg	274 μg
Vitamin K	–	1360 μg
Vitamin E	–	1.03 mg
Lutein + Zeaxanthin	–	2430 μg
Vitamin A	–	5850 μg
β-Carotene	–	3410 μg
α- Carotene	–	31 μg
Fatty acids (Total saturated fatty acids)	0.99 g	0.115 g
SFA 14:0	0.02 g	0.003 g
SFA 16:0	0.81 g	0.096 g
SFA 18:0	0.11 g	0.005 g
Fatty acids (total monounsaturated fatty acids)	13.6 g	2.23 g
MUFA 16:1	0.1 g	0.016 g
MUFA 18:1	13.5 g	2.22 g
Fatty acids (total polyunsaturated fatty acids)	1.75 g	0.328 g
PUFA 18:2	1.75 g	0.328 g
Cholesterol	0 mg	0 mg

Source: USDA (2018).

Table 16.3 Bioactive Compounds in Different Parts of Coriander

Plant Part	Bioactive Compounds		Bioactive Action
Seeds	Essential Oils	Alcohols: Linalool Hydrocarbons: Monoterpenes Ketones: Camphor Esters: Geranyl acetate and Linalyl acetate	Antioxidant activity Anti-microbial activity Anti-carcinogenic activity Anti-diabetic activity Anti-inflammatory activity
	Lipids and Fats	Fatty acids: Petroselinic acid Polyunsaturated fatty acid (PUFA): Linoleic acids β-Carotene Sterols: Stigmasterol and β-sitosterol Tocols: Tocopherol and tocotrienol	
	Polyphenolic compounds	Phenolic acids: Caffeic acid, gallic acid, ferulic acid and chlorogenic acid Flavonoids: Luteolin, quercetin, kaempferol, tannins and anthocyanins	
	Others	Monoterpenoids and glycosides	
Leaves	Essential Oils	Aromatic acids: 2-decenoic acid, capric acid and (E)-11-tetradecenoic acid	
	Lipids and Fats	Polyunsaturated fatty acid (PUFA): α-linolenic acid β-Carotene	

Source: Adapted with permission from Prachayasittikul et al. (2018). Copyright 2018, Elsevier.

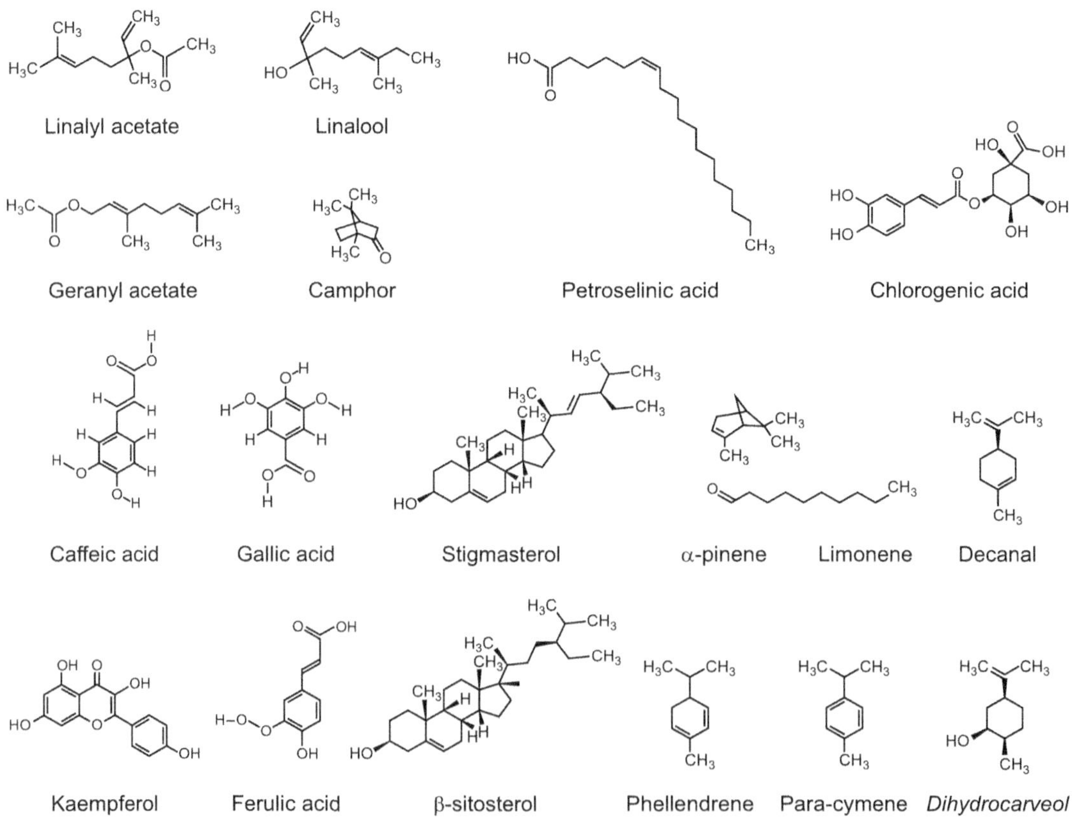

Figure 16.1 Illustration showing the bioactive compounds found in coriander (*Coriandrum sativum* L.) (*Source:* Pubchem, https://pubchem.ncbi.nlm.nih.gov/).

16.6.1 Seeds

16.6.1.1 Polyphenolic Compounds

Polyphenols are a large group of organic compounds derived from plant-based sources. These possess excellent antioxidant properties. Several studies have been conducted in the past to check the total phenolic content of the coriander plant. Eleven phenolic compounds, namely ferulic, vanillic, gallic, rosmarinic, salicylic, chlorogenic, trans-hydroxycinnamic, caffeic, o-coumaric, p-coumaric, and trans-cinnamic acids and ten flavonoids, namely luteolin, kaempferol, resorcinol, quercetin dihydrate, quercetin-3-rhamnoside, rutin trihydrate, apigenin, naringin, coumarin, and flavone are observed in different varieties of Coriander (Iqbal et al. 2017). Variation among the levels and types of polyphenolics availability in coriander is because of its geographical distribution differences. Extract of Coriander grown in Syria exhibits polyphenol content of 1.09 mg GAE/g dry weight, while Tunisian and Egyptian coriander extract exhibits 1.00 and 0.94 mg GAE/g polyphenol content (Msaada et al. 2017). The antioxidant potential of coriander seeds was analysed in a study done in India. They reported good total phenol content (189.61 mg GAE/100 g), total flavonoid content (310 mg RE/100 g), total antioxidant activity by FRAP (951.61 mg TE/100 g) and total antioxidant activity by DPPH (24.28 mg TE/100 g) in coriander seeds (Dutta et al. 2018). The phenolic content of the methanolic extract of coriander seeds grown in Canada was 15.16 mg GAE/g, while that of Tunisian coriander seeds was 12.10 mg GAE/g (Neffati et al. 2011). The ethyl acetate extract of coriander seeds grown in Norway showed a total phenol content of 1.89 g GAE/100 g. It should be noted that a wide variation in the total phenolic proportion also exist owe to varied extraction solvents utilized and differences in adopted procedure as well (Wangensteen et al. 2004).

16.6.1.2 Essential Oils

16.6.1.2.1 Yield of Coriander Essential Oil

The yield and quantity of essential oil extracted from coriander seeds varies depending on the geographical conditions, climate, and cultivar's origin. Hence it's not uniform across nations; for instance, the yield of essential coriander oil extracted from the Indian variety of Coriander varied from 0.18 to 0.39% (Ravi et al. 2007); the yield attained from Tunisian Coriander was 0.35% (Msaada et al. 2009). Iranian coriander variety was reported to have an oil yield of 0.1 to 0.36% (Nejad Ebrahimi et al. 2010). On the other hand, the seed oil yield from the Bangladesh variety was nearly 0.42% (Bhuiyan et al. 2009). Likewise, diverse cultivars of Canadian Coriander grown in cold/wet settings were reported to have oil concentrations ranging from 0.8 to 2.2%. Besides, a study done on two different varieties of Coriander, namely *vulgare* and *microcarpum* grown in Turkey, reported having essential oil ranging from 0.15 to 0.25% in *vulgare*, while it was 0.31–0.43% in *microcarpum* (Iqbal et al. 2017).

16.6.1.2.2 Composition of Coriander Essential Oil

Various parts of Coriander contain compounds such as linalool (a volatile oil compound), α-pinene, citronellol, limpnene, geraniol, borneol, p-cymene, monoterpenes, camphor, coriandrin, coriandrons A-E, dihydrocoriandrin, γ-terpinene, flavonoids, and other essential oils (Pathak et al. 2011). Key components of essential oil found in coriander seeds are linalool (57.57%), β-Caryophyellene (3.26%), geranylacetate (15.9%), camphor (3.02%), and p-cymene (2.5%) (Khani and Rahdari, 2012) (Table 16.4). Coriander's chief volatile bioactive compound is the alcoholic monoterpene, predominantly the (S) – (þ) – linalool moiety. Linalool accounts for above 50% of the total essential oil obtained from coriander seeds. Coriander seeds comprise of 18% oil in the form of fatty acids or triglycerides;

Table 16.4 Constituents of Essential Oil Derived from Coriander Seeds

Compounds	Composition %	Compounds	Composition %	Compounds	Composition %
Heptanol	0.78	γ-Terpinene	0.11	Nerol	1.98
α-Thujene	0.19	Cis-linalool oxide	0.07	Neral	0.29
α-Pinene	0.12	Trans-linalool oxide	0.83	Carvone	1.14
Sabinene	0.58	Terpinolene	0.17	Geraniol	0.24
β-Pinene	0.14	Linalool	57.57	Geranial	1.03
l³-Carene	0.11	β-Citronella	0.39	Anethole	0.21
α-Terpinene	0.39	Camphor	3.02	Thymol	0.12
p-Cymene	2.52	Borneol	1.27	Carvacrol	0.29
Limonene	0.62	Menthol	0.54	δ-Elemene	0.56
1,8-Cineol	0.97	Terpinene-4-ol	0.14	Eugenol	0.76
(z)-β-Ocimene	0.41	p-Cymen-8-ol	0.29	Eugenyle acetate	0.49
β-Caryophyellene	3.26	α-Terpineol	0.18	Geranyl acetate	15.09
α-Humulene	0.19	Cis-Dihydrocarvone	0.35	Other compounds	1.60
Germacrene D	1.01	Neryle acetate	0.01		

Source: Adapted in modified form from Khani and Rahdari (2012). (Copyright © 2012 A. Khani and T. Rahdari. This is an open access article distributed under the Creative Commons Attribution License, which permits unrestricted use, distribution, and reproduction in any medium, provided the original work is properly cited.)

and nearly 0.84% as essential oils. For instance, the other class of fatty acids, for example oleic, linolenic acids, and petroselinic, account for 2/3rd of the oil present in coriander seed (Gil et al. 2002).

Coriander seed oil extracted from the Tunisian variety was reported to have a linalool content of 87.54% and cis-dihydrocarvone content of 2.36% (Msaada et al. 2007). The biochemical analysis of the seed oil extracted from Algerian variety showed the linalool content to be 73.1%. In comparison, other components recognized were 6.51% of *p*-mentha-1, 4-dien-7-ol, 3.41% of α-pinene, and 3.22% of neryl acetate (Zoubiri and Baaliouamer 2010). Likewise, the linalool content ranging from 40.9 to 79.9%, neryl acetate content of 2.3 to 14.2%, γ-terpinene content of 0.1 to 13.6%, and α-pinene content of 1.2 to 7.1% was observed in the volatile coriander seed oil extracted from Iranian coriander variety (Nejad et al. 2010) while the coriander seed oil extracted from Bangladeshi variety showed linalool (37.65%) followed by geranyl acetate (17.57%) and γ-terpinene (14.42%), respectively (Bhuiyan et al. 2009).

The concentration of linalool in the essential oil extracted from coriander seed grown in Pakistan was 69.60%. In comparison, other components such as 4.99% of geranyl acetate, 4.17% of γ-terpinene, 1.63% of α-pinene, 1.15% of anethole, and 1.12% of p-cymene were also identified (Anwar et al. 2011). Additionally, the biochemical content of the coriander seed oil obtained from the Indian coriander variety was similar to the Pakistani variety. Coriander essential oil from the Indian variety comprised of a higher percentage of linalool (75.30%), followed by geranyl acetate (8.12%), as well as α-pinene (4.09%). However, various essential oil constituents from Indian varieties were dissimilar (Singh et al. 2006; Bankar et al. 2011).

Atlantic Canadian coriander variety showed a wide range in the proportion of linalool (64.00–84.60%). In contrast, camphor (3.40–6.20%), linalyl acetate (2.40–3.30%), limonene (0.70–1.80%), phellandrene (1.7–4.1%), geranyl acetate (0.90–1.60%), α-pinene (1.20–3.20%), and para-cymene (0.50–1.30%) were the additional components identified in seed oil (Zheljazkov et al. 2008). The key compounds identified in the coriander seed oil from Brazil were linalool, γ-terpinene, α-pinene, limonene, geraniol and (E)-2-decenal (de Figueiredo et al. 2004). Besides, the two Turkish coriander varieties, viz. *microcarpum* and *vulgare* showed the linalool content of 63.50 to 71.00% and 42.10 to 52.70%, respectively. In contrast, other components identified in the coriander oil extract were γ-terpinene, geraniol, p-cymene, (Z)-isoapiole dillapiole, and geranyl acetate (Telci et al. 2006). On the contrary, the immature Tunisian coriander seeds showed high geranyl acetate content (46.27%) as compared to linalool content (10.96%); yet, upon maturation, the linalool content increased to 87.54%, indicating that the difference in the essential oil content may be linked with alterations in secondary metabolism (Msaada et al. 2009). Dissimilarities in the biochemical constituents of coriander seed oil from region to region can be attributed to the factors like the chemical configuration, genetic, seasonal, and climatic factors. Several studies investigating the chemical composition of Coriander focused on the influence of developmental stage (Neffati and Marzouk, 2008), sowing period (Zheljazkov et al. 2008), maturity stage (Msaada et al. 2009), interaction amongst the growing area with abiotic stress, for instance, salinity (Neffati and Marzouk, 2008), drought, etc. on the biochemical configuration and yield of essential oil. Numerous other studies have focused on the extraction techniques such as hydro-distillation (Benyoussef and Saibi 2013), microwave-assisted extraction (Kosar et al. 2005), and supercritical carbon dioxide fluid extraction (Grosso et al. 2008) used for coriander seed oil extraction to check their influence on the yield as well as constituents of coriander seed oil.

16.6.1.3 Water-Soluble Constituents

Scanty studies report about the water-soluble constituents of Coriander; only few studies stated that around thirty-three different compounds are present in the hydrophilic part of methanolic extract of coriander seeds. These comprise two aromatic glycosides, four monoterpenoid glycosides, as well as two monoterpenoid glucoside sulphates (Kitajima et al. 2003).

16.6.1.4 Fatty Acids

Current researches have evaluated the fatty acid composition of coriander seeds. They alluded that it contains a variety of fatty acids, including palmitic acid, petroselinic acid, linoleic acid, oleic acid plus stearic acid. Some other fatty acids such as arachidic acid (0.10–0.25%), α-linolenic acid (0.15–0.50%), and palmitoleic acid (0.41–1.1%) are also observed in coriander seeds. The total oil yield of coriander seeds has been reported to range from 19.24 to 28.4% on a dry weight basis. The neutral lipids cover coriander seed oil's significant quantity, comprising triacylglycerols, free fatty acids, and diacylglycerol (Laribi et al. 2015).

16.6.1.4.1 Sterols

Oil extracted from coriander seed contains a good amount of sterols that show inhibitory action towards the absorption of dietary cholesterol. Stigmasterol and β-sitosterol are the major sterol markers found in coriander seeds. Total sterol content in coriander seed oil varies from 36.93–51.86 mg/g oil. Stigmasterol and β-sitosterol are the predominantly occurring sterol markers in the coriander seeds, while Δ^7-stigmasterol, Δ^5, 24-stigmastadienol, and campestrol are some representative sterols. Smaller quantities of other sterols such as Δ^7-avenasterol and Δ^5-avenasterol are also present in coriander seeds (Laribi et al. 2015).

16.6.2 Herb/Leaves

There is scanty information regarding the bioactive compounds found in the coriander leaves/herb compared to seeds. Polyphenolic compounds, flavonoids, and essential oils are the major compounds identified in the coriander leaves.

16.6.2.1 Essential Oils

16.6.2.1.1 Yield of Coriander Essential Oil

Coriander leaves contain smaller amounts of essential oil compared to the coriander seeds (Bhuiyan et al. 2009). For example, Air-dried Tunisian coriander variety consists of essential oil (0.12%).

16.6.2.1.2 Essential Oil Composition

A variation in coriander leaves' essential oil composition was observed compared to oil extracted from seeds. Although the composition varies, most of the research quoted aldehydes and alcohols as the key component in coriander leaf oil. A study done on the essential oil obtained through extraction from Coriander leaves grown in Bangladesh reported the presence of 44 moieties, mainly consisting of aromatic acids for example 2-decenoic acid, capric acid and E-11-tetradecenoic acid (Bhuiyan et al. 2009). Similarly, the essential oil extraction from coriander leaves grown in Kenya revealed the predominance of aldehydes (56.10%) followed by alcohols (46.30%). The key compounds found were (E)-2-decenal, decanal, (E)-2-decen-1-ol and n-decanol in concentrations of 15.90%, 14.30%, 14.2%, and 13.6%, respectively. However, the compounds isolated in lesser quantity were *(E)*-2-tridecen-1-al, undecanol, *(E)*-2-dodecenal, undecanal, dodecanal, and alkanes (Matasyoh et al. 2009).

Extraction of essential oils from the leaves of a Fiji coriander variety demonstrated the predominance of (E)-2-decen-1-ol (26.00%) (Eyres et al. 2005). The chief compounds recognized in the oil isolated from coriander leaves of Brazil were of alcoholic origin, for instance, 1-decanol (24.20%), (E)-2-decenol (18.00%) and (Z)-2-dodecenol (17.60%) together with aldehydes (89%) (Begnami et al. 2010). The Indian coriander was seen to contain bioactive components such as (E)-2-decenal, dec-9-en-1-ol, dodecanal, n-tetradecanol, decanal, (E)-2-dodecenal and decanol (Padalia et al. 2011).

16.6.2.2 Lipids

The fatty acid constituents present in the coriander leaves were expounded upon interestingly for the first time by Neffati and Marzouk (2008). They explored the influence of salinity on the plasma membrane of fats present in Tunisian coriander leaves cultivated in hydroponic culture. The study reported that the highest total fats are located in the basal leaves (61.21 mg/g) compared to the leaves from the upper part (41.80 mg/g) on the dry weight. Also, the analysis of coriander leaf fats showed a high proportion of polyunsaturated fats (PUFA). Additionally, it was revealed that α-linolenic was the utmost predominant unsaturated fat found in the basal along with upper leaves at a level of approximately 41.1% and 39.4%, respectively. As far as mg/g dry weight was concerned, it was 17.1 and 24.1 mg/g, individually.

Coriander leaves also exhibited the predominant presence of linoleic, heptadecenoic, and palmitic acid. Some fatty acids, for example, stearic acid, oleic acid, cis/trans-palmitoleic acid and stearidonic acid were present in smaller amounts in both basal as well as upper leaves of the coriander plant; collectively accounting for almost 9.6% and 4.7% of total fats content, individually. The study also stated that salinity decreases the total fat content in both basal as well as upper leaves. Thus, with an increase in the NaCl concentration, there is a decrease in unsaturated fats to saturated fats, leading to a rigid membrane (Neffati and Marzouk 2008).

Five bioactive components, namely β-carotene, neoxanthin, lutein-5,6-epoxide, violaxanthin, plus β-cryptoxanthin epoxide were reported to be obtained through ether extraction of coriander leaves (de Almeida et al. 2005). The coriander leaves also exhibited reasonable quantities of β-carotene (a well-known vitamin A precursor), signifying about 61.14%.of the total carotenoids identified in the ether extract. Additionally, it was reported that at the pre-flowering stage, there is an elevation in the β-carotene plus total carotenoids content (Divya et al. 2012).

16.6.2.3 Polyphenolic Compounds

In an investigation, the methanolic extract of the green part of Coriander was examined for the total phenols plus total flavonoid content; the results depicted the presence of 1013.95 mg/kg of total phenols and 5259.52 mg/kg of total flavonoid content. The major polyphenols present in the green part of coriander comprises of quercetin-3-O-rutinoside (3296.16 mg/kg), quercetin 3-O-glucuronide (1237.13 mg/kg), dimethoxycinnamoyl hexoside (406.39 mg/kg), quercetin-3-O-glucoside (405.36 mg/kg) as well as kaempferol-3-O-rutinoside (320.86 mg/kg). Green portion of the Coriander exhibited four major flavonoids, namely caffeoylquinic acid (7.92 mg/kg), ferulic acid glucoside (122.29 mg/kg), 3-O-caffeoylquinic acid (173.51 mg/kg), and *p*-coumaroylquinic acid (303.83 mg/kg). The green section of coriander grown in Portugal demonstrated the presence of quercetin derivatives (Barros et al. 2012).

In the past, around 21 phenolic constituents were documented to be present in the green part of Coriander. These constituents were primarily phenolcarboxylic acids, flavonoids, as well as coumarins. Other compounds such as apigenin, hesperidin, esculetin, luteolin, dihydroquercetin, hyperoside, vicenin, dicoumarin, chrysoeriol, gallic acid, diosmin, orientine, maleic acid, catechin, ferulic acid, arbutin, 4-hydroxycoumarin, salicylic acid, esculin, and tartaric acid were initially recognized in Coriander (Oganesyan et al. 2007).

It is also stated that the chief phenols recognized in an Indian variety of coriander leaves were *p*-coumaric, vanilic, and *cis/trans*-ferulic acids. Flavonoids such as quercetin, acacetin, kaempferol, 3′-OMe quercetin, and 4′-OMe quercetin were also present in leaf portion of coriander. Conversely, glycoflavones were not identified in coriander leaves. Hence quercetin is identified as a chief bioactive component present in coriander leaves of Indian origin (Nambiar et al. 2010).

The phenols comprising of protocatechinic acid (6.43 μg/ml), caffeic acid (4.34 μg/ml), plus glycitin (3.27 μg/ml) were found through the aqueous extraction of Coriander leaves grown in Brazil (de Almeida et al. 2005). Also, the coriander leaves were found to possess anthocyanin, whose biosynthesis might be boosted by microelements, predominantly by applying zinc and salicylic acid. On the other hand, nitrogen, phosphorus, and potassium negatively affected the anthocynanin content.

A comprehensive *in-vitro* study is done to analyse the polyphenolic content of the Coriander leaves samples. The study showed that coriander leaves possess various polyphenols. C-glycosylated apigenin was found as the vital component having a 2,983 mg/kg concentration. However, the clones A and B exhibited a similar flavonoid content; anthocyanins were identified individually in clone A, which relates to its purple pigment. Peonidin-3-O-feruloylglucoside-5-O-glucoside was found to be the main anthocyanin with a concentration of 1.70 μg/kg. Hence, more in-vitro culture studies may possibly be carried out to discover the potential of medicine, industry, and pharmaceuticals in producing secondary metabolites such as flavonols, flavones, and anthocyanins (Barros et al. 2012).

16.7 THERAPEUTIC PROPERTIES OF CORIANDER

Different parts of the coriander plant show varied biological activities. Some of the properties are listed below:

16.7.1 Analgesic Property

The aqueous extract of Coriander possesses analgesic properties which perhaps is mediated through the inhibition of central pain receptors (Pathan et al. 2011). Bioactive compounds, mainly linalool, a monoterpene alcohol, is frequently known to be the most crucial constituent of essential oil of Coriander that exhibits good analgesic activity (Guimaraes et al. 2013).

Various past studies reported the antinociceptive property of linalool (Peana et al. 2003); some supplementary studies reported the impact of the glutamatergic system in the antinociception stimulated through linalool in the mice (Batista et al. 2008). In addition, other studies reported that β-cyclodextrin-complexed linalool exhibits antinociceptive properties, which were comparatively superior to linalool in experimental pain trials (Quintans-Junior et al. 2013).

16.7.2 Anti-Microbial Property

Coriander seed extract exhibits anti-microbial action against gram-negative bacteria, namely *Klebsiella pneumonia and Pseudomonas aeruginosa* and gram-positive bacteria such as *Staphylococcus aureus* PTCC1431 (Zardini et al. 2012). A peptide viz. "*Plantaricin CS*" sequestered from the extract of coriander leaf exhibits great anti-bacterial activity against the *Staphylococcus aureus* strain (MIC = 1.3 mg/mL). Another peptide displayed active germicidal action on *K. pneumonia* strain (MIC = 2.65 mg/mL). Nonetheless, the *Pseudomonas aeruginosa* was examined to be the most resistant strain (MIC = 3.2 mg/mL). Hence, the *Plantaricin CS* peptide has been observed to have a lower impact of anti-microbial action on Gram-negative bacteria; this may be a consequence of cell wall polysaccharides, which may stop the active complexes from reaching the bacterial cytoplasmic membrane (Zare-Shehneh et al. 2014).

Also, the anti-bacterial activity of the extract is equivalent to the standard antibiotic drugs, for example, gentamicin sulphate, ciprofloxacin, bramycin, and ofloxacin studied in similar settings (Reddy et al. 2012). The anti-bacterial action against both Gram-positive as well as Gram-negative bacteria is attributed to the permeability of the membrane of coriander essential oil (Silva et al. 2011). Moreover, the long-chain (C_6–C_{10}) alcoholic groups along with aldehydes are the main reason behind the action of coriander leaf oil against *Listeria monocytogenes* (Delaquis et al. 2002). Consequently, it may be said that the synergistic effect of active compounds of Coriander leads to its significant anti-bacterial activity (Reddy et al. 2012).

The synergy amongst coriander essential oil constituents plus antibiotics has been studied to emphasize the potential of the essential oil as the possible resistance modifying agent (Langeveld et al. 2014; Yap et al. 2014). Combined action of essential oil constituents in contrast to the effects of individual constituents of the coriander oil appear unanswered. Also, the whole coriander oil exhibits better anti-bacterial activity in contrast to the vital oil constituents individually (Burt, 2004). Consequently, it is much more practical to examine the whole coriander essential oil instead of few components because they are synergistically related (Bassole and Juliani 2012).

16.7.3 Antioxidant Property

All parts of Coriander have been studied for their antioxidant potential and are stated to possess excellent antioxidant effects. Antioxidant activity in Coriander is attributed to its redox potential, significant adsorption plus neutralization of free radicals, quenching of reactive oxygen species, and chelation of metal, particularly iron as well as copper cations (Balasundram et al. 2006). Antioxidant activity of coriander essential oil is due to its active constituents. It may perhaps be attributable to the large fraction of key constituents along with the occurrence of other elements in smaller amounts or to the synergy between them (Reddy et al. 2012). The terpene and terpenoid compounds such as limonene, camphor, α-pinene and geraniol may also contribute to the antioxidant properties of the coriander essential oil (Darughe et al. 2012). Coriander leaf essential oil is also known to contain a good amount of natural antioxidants, which are a potent replacer of artificial antioxidants in the food applications like preservation (Darughe et al. 2012).

In a study, *in-vivo* and *in-vitro* developed Coriander was comparatively examined for both lipophilic (tocopherols, chlorophylls, and carotenoids) and hydrophilic (ascorbic acid, sugars, flavonols, phenolics, and anthocyanins) antioxidants. The *in-vitro* samples were clone A by purple pigmentation in stems and leaves and clone B green. The *in-vivo* samples exhibited higher antioxidant activity because of the high levels of hydrophilic constituents. On the other hand, *in-vitro* samples, primarily clone A, presented greater lipophilic compounds concentration yet a diverse profile in contrast to the *in-vivo* sample. Clones A as well as B showed decline in α-, β- and δ-tocopherols along with the β-carotene, and a rise in γ-tocopherol and chlorophylls in contrast to the *in-vivo* sample. The study reported that *in-vitro* culture might be beneficial to discover the potential of plants in industrial uses, monitoring environmental friendly settings for production of more significant quantities of certain bioactive goods (Dias et al. 2011).

16.7.4 Anxiolytic Activity

Some studies advise the anxiolytic properties of coriander seeds. These seeds perhaps show sedative and muscle relaxing properties (Emamghoreishi et al. 2005; Mahendra and Bisht 2011), still more rigorous pharmacological and biochemical researches are required to explain the action mechanism of the coriander extract and quantify the bioactive constituents causing such effects. Further, essential oils and flavonoids are reported to be linked with the anxiolytic properties of the Coriander (Mahendra and Bisht 2011).

16.7.5 Hypo-Lipidemic Property

Coriander exhibits hypolipidemic properties. Several studies were conducted in experimental animals to check the effect of coriander seed administration on various aspects of lipid metabolism (Lal et al. 2004; Dhanapakiam et al. 2007). Some studies reported the hypolipidemic activity of coriander seeds on account of the synergy between the bioactive components in the seed (Rajeshwari and Andallu 2011). Fatty acids found in Coriander, such as oleic acid, linoleic acid, palmitic acid, and stearic acid, helped lower cholesterol levels in the blood. Also, these fats lessen the chances of deposition of cholesterol in the inner walls of arteries and veins (Ertas et al. 2005). The reduction in serum and tissue cholesterol level by administering coriander seeds may also be facilitated by its improved frequency of degradation to bile acids as well as neutral sterols (Dhanapakiam et al. 2007). Thus, Coriander appears to have immense potential to be recognized as an effective domestic herbal remedy for preventing/curing hyperlipidemia (Lal et al. 2004).

16.7.6 Hypo-Glycemic Property

Coriander extract exhibits anti-diabetic or hypo-glycemic property. Nonetheless, few studies are available on the influence of coriander essential oil administration on diabetic experimental animals (Naquvi et al. 2004; Waheed et al. 2006; Aissaoui et al. 2011; Rajeshwari and Andallu 2011). Therefore the hypoglycemic action of Coriander essential oil is suggested to be associated with the synergistic effect of the bioactive components such as linalool, γ-terpinene, and geranyl acetate (El-Soud et al. 2012).

Numerous investigators have harmonized that the counter hyperglycemic activity of Coriander is related to stimulation of insulin secretion and boosting glucose uptake plus digestion via muscle, imitating the effect of different bioactive component of Coriander. Thus, Coriander may be considered as a potent pool of dietary supplement for regulating the blood glucose level and forestalling long-term complications associated with non-insulin dependent diabetes mellitus (Gallagher et al. 2003).

16.7.7 Anti-Inflammatory Property

Coriander also exhibits potential for anti-inflammatory effects. Coriander extract, as well as the essential oil obtained from Coriander, seems to possess anti-inflammatory properties (Reuter et al. 2008; Sonika et al. 2010; Wu et al. 2010). Remarkably, a Poly-herbal Ayurveda formulation containing the Coriander along with certain other herbs has exhibited action against inflammatory bowel disease (Jagtap et al. 2004).

16.7.8 Anti-Cancerous Property

The linalool component present in coriander essential oil contributes to its anti-cancer properties. The coriander natural components moderately inhibited cell proliferation and, therefore, may expand the therapeutic index of anthracyclines in the managing breast cancer, particularly in multidrug-resistant tumours (Ravizza et al. 2008). The *in-vitro* anti-tumour and immune-modulating property of both coriander leaf and seed extract, similar to the counter proliferative effect of coriander root extract on the breast cancer cell line, propose its command in malignancy counteraction and restraint of metastasis (Gomez-Flores et al. 2010; Tang et al. 2013).

Anti-tumour causing effects of Coriander has also been recognized through its defensive role against the harmful impact of lipid metabolism linked with this malignancy in experimental colon cancer. Therefore, Coriander and the conventional drug would be beneficial in enhancing the treatment of harmful diseases like cancer (Chithra and Leelamma 2000).

16.8 EFFECT OF PROCESSING ON NUTRITIONAL COMPOSITION OF CORIANDER SEEDS

Coriander is chiefly consumed in its natural form. It is also processed to facilitate the global exchange/trade, profit and palatability (Bhat et al. 2014). Various studies focused on the nutritional aspects of coriander processing. Vacuum cooling of Coriander leads to change in physicochemical properties. Vacuum cooling is usually a part of the processing of Coriander to prolong the shelf life after harvesting. It is done to pre-cool the harvest before transportation. Vacuum cooling exhibits a considerable effect on the vitamin C content of the Coriander (Apichart et al. 2012).

Moreover, a study compared the retention of bioactive compounds in fresh coriander leaves paste using high-pressure processing at 50 MPa, 200 MPa, 400 MPa, and 600 MPa, and thermal processing at 90°C. The results showed that phenolic content of Coriander improved by 0.99% and 1.10% at 400 MPa and 600 MPa, correspondingly upon high-pressure processing. Flavonoids present in high pressure processed (400 MPa) Coriander were considerably more significant than thermally processed Coriander. Untreated Coriander showed higher vitamin C content (1.32 mg/g) as compared to high pressure processed (600 Mpa) coriander (1.17 mg/g) and thermally processed coriander sample (0.50 mg/g). Chlorophyll (natural colorant in Coriander) declined from 1.68 mg/g to 1.58 and 1.21 mg/g in high pressure processed (600 MPa) and thermally processed Coriander, individually. Thermal processing also led to overall enzyme inactivation (Polyphenol Oxidase (PPO) and Peroxidase (POD)), while high-pressure processing at 400 MPa showed a considerable reduction in the enzyme activity. Also, the lightness value (L*) of Coriander showed a significant (a = 0.05) decline with pressure processing. Microbial growth evaluated through total plate count also showed a decline from 3.00×10^3 CFU/g to nil and 1.9×10^1 CFU in thermally processed and high pressure processed (600 MPa) coriander sample. Consequently, it may be concluded that the high pressure processing (400 MPa) is more efficient for withholding bioactive constituents than thermal processing (Nath et al. 2016).

Likewise, the effect of blanching on the stability of polyphenols and antioxidant activity of coriander paste prepared using fresh coriander leaves was assessed. Steam blanching of coriander paste for 1 minute showed no change in the phenols and antioxidant activity. Water blanching and steam blanching for an extended duration showed increased levels in contrast to the unheated coriander sample. Water blanching for a short period of time leads to higher bioactive levels than continued heat treatment. Based on these results, water blanching for a short time period is suggested as the preferred procedure in preparing coriander leaves into novel pasty items (Kaiser et al. 2013).

16.9 CONCLUSION

The present chapter gives an in-depth account of various qualities of coriander such as its nutritional composition, bioactive compounds, phytochemicals, traditional use, therapeutic potential, and processing effects on its composition. It sums up various reports on the chemical morphology and bioactive constituents of different portions of the coriander plant, including seeds, leaves and roots. It can be suggested that Coriander has excellent potential to be used in both the nutraceutical and pharmaceutical sector. Each part of the coriander plant, including leaves, stem, roots, and fruits, holds a pleasant aroma and flavour and is used for its medicinal effects. Coriander acts as a worthy storehouse for the functional compounds in view of the richness of the bioactive components. Coriander seeds and herbs contain enormous amounts of bioactive compounds that exhibit various biological activities for instance analgesics, anxiolytic, anti-microbial, hypoglycemic, antioxidant, hypolipidemic, anti-inflammatory, and anti-cancerous activities.

Lists of therapeutic potential existing among coriander seeds is in direct concern with its bioactive constituents. All this govern its enormous utilization as defensive cum preventive traits against various diseases along with health promotion. Thus in nutshell, coriander is regarded as "spice of joy". Keeping in mind, the above headed attributes this chapter will serve as a comprehensive guide to researchers in various health, medicine, and agriculture fields.

REFERENCES

Aissaoui, A., S. Zizi, Z.H. Israili and B. Lyoussi. 2011. Hypoglycemic and hypolipidemic effects of Coriandrum sativum L. in Meriones shawi rats. *Journal of Ethnopharmacology 137*:652–661.

Anuradha, D., S. Pushpa, T. Soni, T. and S. Shweta. 2018. Antioxidant properties of traditional spices of Uttarakhand. *International Journal of Basic and Applied Agricultural Research 16*:180–184.

Anwar, F., M. Sulman, A.I. Hussain, N. Saari, S. Iqbal and U. Rashid. 2011. Physicochemical composition of hydro-distilled essential oil from coriander (Coriandrum sativum L.) seeds cultivated in Pakistan. *Journal of Medicinal Plants Research 5*:3537–3544.

Apichart, S., B. Danai, B. Pichaya. 2012. Effect of vacuum cooling on physico-chemical properties of organic Coriander. *Asian Journal of Food and Agro-Industry* 5:96–103.

Balasundram, N., K. Sundram and S. Samman. 2006. Phenolic compounds in plants and agri-industrial by-products: Antioxidant activity, occurrence, and potential uses. *Food Chemistry 99*:191–203.

Bankar, R., A. Kumar, S. Puri, I. Ali, A. Sharma and I.A. Khan. 2011. Chemical Composition and Antimicrobial Activity of Essential Oil from Seed of Coriandrum sativum L. *Analytical Chemistry Letters* 1:189–193.

Barros, L., M. Duenas, M.I. Dias, M.J. Sousa, C. Santos-Buelga and I.C. Ferreira. 2012. Phenolic profiles of in vivo and in vitro grown Coriandrum sativum L. *Food Chemistry 132*:841–848.

Bassole, I.H.N. and H.R. Juliani. 2012. Essential oils in combination and their antimicrobial properties. *Molecules 17*:3989–4006.

Batista, P.A., M.F. de Paula Werner, E.C. Oliveira, L. Burgos, P. Pereira, L.F. da Silva Brum and A.R.S. Dos Santos. 2008. Evidence for the involvement of ionotropic glutamatergic receptors on the antinociceptive effect of (–)-linalool in mice. *Neuroscience Letters 440*:299–303.

Begnami, A.F., M.C.T. Duarte, V. Furletti and V.L.G. Rehder. 2010. Antimicrobial potential of Coriandrum sativum L. against different Candida species in vitro. *Food Chemistry 118*:74–77.

Benyoussef, E.H. and S. Saibi. 2013. Influence of essential oil composition on water distillation kinetics. *Flavour and Fragrance Journal 28*:300–308.

Bhat, S., P. Kaushal, M. Kaur and H.K. Sharma. 2014. Coriander (Coriandrum sativum L.): Processing, nutritional and functional aspects. *African Journal of Plant Science 8*:25–33.

Bhuiyan, M.N.I., J. Begum and M. Sultana. 2009. Chemical composition of leaf and seed essential oil of Coriandrum sativum L. from Bangladesh. *Bangladesh Journal of Pharmacology 4*:150–153.

Brinker, F.J. 1998. *Herb contraindications and drug interactions: with appendices addressing specific conditions and medicines*. Eclectic medical publications.

Burt, S. 2004. Essential oils: their antibacterial properties and potential applications in foods—a review. *International Journal of Food Microbiology 94*:223–253.

Chaudhry, N.M. and P. Tariq. 2006. Bactericidal activity of black pepper, bay leaf, aniseed and coriander against oral isolates. *Pakistan Journal of Pharmaceutical Sciences 19*:214–218.

Chithra, V. and S. Leelamma. 2000. Coriandrum sativum—effect on lipid metabolism in 1, 2-dimethyl hydrazine induced colon cancer. *Journal of Ethnopharmacology 71*:457–463.

Darughe, F., M. Barzegar, and M. A. Sahari. 2012. Anti-oxidant and anti-fungal activity of Coriander (*Coriandrum sativum* L.) essential oil in cake. *International Food Research Journal 19*: 1253–1260.

de Almeida M.E., J. M. Filho, and N. B. Guerra. 2005. Characterization of anti-oxidant compounds in aqueous coriander extract (*Coriandrum sativum* L.). *LWT-Food Science and Technology 38*: 15–19.

de Figueiredo, R.O., M.O.M. Marques, J. Nakagawa and L.C. Ming. 2004. Composition of coriander essential oil from Brazil. *Acta Horticulturae 629*:135–138.

Delaquis, P.J., K. Stanich, B. Girard and G. Mazza. 2002. Antimicrobial activity of individual and mixed fractions of dill, cilantro, coriander and eucalyptus essential oils. *International Journal of Food Microbiology 74*:101–109.

Dhanapakiam, P., J.M. Joseph, V.K. Ramaswamy, M. Moorthi and A.S. Kumar. 2007. The cholesterol lowering property of coriander seeds (Coriandrum sativum): mechanism of action. *Journal of Environmental Biology 29*:53.

Dias, M.I., L. Barros, M.J. Sousa and I.C. Ferreira. 2011. Comparative study of lipophilic and hydrophilic antioxidants from in vivo and in vitro grown Coriandrum sativum. *Plant Foods for Human Nutrition 66*:181–186.

Diederichsen, A. 1996. *Coriander (Coriandrum sativum L.) promoting the conservation and use of underutilized and neglected crops*. International Plant Genetic Resources Institute (IPGRI), Rome.

Divya, P., B. Puthusseri and B. Neelwarne. 2012. Carotenoid content, its stability during drying and the antioxidant activity of commercial coriander (Coriandrum sativum L.) varieties. *Food Research International 45*:342–350.

Dutta, A., P. Shukla, S. Tilara and S. Suri. 2018. Antioxidant properties of traditional spices of Uttarakhand. *International Journal of Basic and Applied Agricultural Research 16*(2): 180–184.

El-Soud, N.H.A., N.A. El-Lithy, G.S.M. El-Saeed, M.S. Wahby, M.Y. Khalil, L.T.A. El-Kassem., F. Morsy and N. Shaffie. 2012. Efficacy of Coriandrum sativum L. essential oil as antidiabetic. *Journal of Applied Sciences Research* 8:3646–3655.

Emam, G.M. and H.G. Hamedani. 2006. Sedative-hypnotic activity of extracts and essential oil of coriander seeds. *Iranian Journal of Medical Sciences 31*: 22–27.

Emamghoreishi, M., M. Khasaki and M.F. Aazam. 2005. Coriandrum sativum: evaluation of its anxiolytic effect in the elevated plus-maze. *Journal of Ethnopharmacology 96*:365–370.

Ertas, O.N., T. Guler, M. Ciftci, B. Dalkilic and O. Yilmaz. 2005. The effect of a dietary supplement coriander seeds on the fatty acid composition of breast muscle in Japanese quail. *Revue de Medecine Veterinaire 156*:514.

Eyres, G., J.P. Dufour, G. Hallifax, S. Sotheeswaran and P.J. Marriott. 2005. Identification of character-impact odorants in coriander and wild coriander leaves using gas chromatography-olfactometry (GCO) and comprehensive two-dimensional gas chromatography–time-of-flight mass spectrometry (GC× GC–TOFMS). *Journal of Separation Science 28*:1061–1074.

Gallagher, A.M., P. R. Flatt, G. A. W. Y. Duffy, and Y. H. A. Abdel-Wahab. 2003. The effects of traditional antidiabetic plants on in vitro glucose diffusion. *Nutrition Research 23*:413–424.

Gil, A., E.B. De La Fuente, A.E. Lenardis, M. Lopez Pereira, S.A. Suarez, A. Bandoni, C. Van Baren, P.L. Di Leo and C.M. Ghersa. 2002. Coriander essential oil composition from two genotypes grown in different environmental conditions. *Journal of Agricultural and Food Chemistry 50*:2870–2877.

Gomez-Flores, R., H. Hernandez-Martinez, P. Tamez-Guerra, R. Tamez-Guerra, R. Quintanilla-Licea, E. Monreal-Cuevas and C. Rodriguez-Padilla. 2010. Antitumor and immunomodulating potential of Coriandrum sativum. *Journal of Natural Products 3*:54–63.

Gopalan, C, B.V. RamaSari and S.C. Balasubramanian. 2007. *Nutritive Value of Indian Foods* India/National Institute of Nutrition.

Grosso, C., V. Ferraro, A.C. Figueiredo, J.G. Barroso, J.A. Coelho and A.M. Palavra. 2008. Supercritical carbon dioxide extraction of volatile oil from Italian coriander seeds. *Food Chemistry 111*:197–203.

Guimaraes, A.G., J.S. Quintans and L.J. Quintans-Junior. 2013. Monoterpenes with analgesic activity—a systematic review. *Phytotherapy Research 27*:1–15.

Iqbal, M. J., S. B. Masood and A.R.S. Hafiz. 2017. *Bioactive molecules in food*. Berlin, Germany/Springer.

Jagtap, A.G., S.S. Shirke and A.S. Phadke. 2004. Effect of polyherbal formulation on experimental models of inflammatory bowel diseases. *Journal of Ethnopharmacology 90*:195–204.

Kaiser, A., D.R. Kammerer and R. Carle. 2013. Impact of blanching on polyphenol stability and antioxidant capacity of innovative coriander (Coriandrum sativum L.) pastes. *Food Chemistry 140*:332–339.

Khan, S.W. and S. Khatoon. 2008. Ethnobotanical studies on some useful herbs of Haramosh and Bugrote valleys in Gilgit, northern areas of Pakistan. *Pakistan Journal of Botany 40*:43–58.

Khani, A. and T. Rahdari. 2012. Chemical composition and insecticidal activity of essential oil from Coriandrum sativum seeds against Tribolium confusum and Callosobruchus maculatus. *International Scholarly Research Notices 2012*:1–5.

Kitajima, J., T. Ishikawa, E. Fujimatu, K. Kondho and T. Takayanagi. 2003. Glycosides of 2-C-methyl-D-erythritol from the fruits of anise, coriander and cumin. *Phytochemistry 62*:115–120.

Kosar, M., T. Ozek, F. Goger, M. Kurkcuoglu and C.B.K Husnu. 2005. Comparison of microwave-assisted hydrodistillation and hydrodistillation methods for the analysis of volatile secondary metabolites. *Pharmaceutical Biology 43*:491–495.

Lal, A., T. Kumar, P.B. Murthy and K.S. Pillai. 2004. Hypolipidemic effect of Coriandrum sativum L. in triton-induced hyperlipidemic rats. *Indian Journal of Experimental Biology 42*: 909–912.

Langeveld, W.T., E.J. Veldhuizen and S.A. Burt. 2014. Synergy between essential oil components and antibiotics: a review. *Critical Reviews in Microbiology 40*:76–94.

Laribi, B., K. Kouki, M. M'Hamdi and T. Bettaieb. 2015. Coriander (Coriandrum sativum L.) and its bioactive constituents. *Fitoterapia 103*:9–26.

Leung, A.Y. and S. Foster. 1996. Encyclopedia of common natural ingredients used in food, drugs, and cosmetics. *Journal of the American Chemical Society 118*: 8988.

Longvah, T., A. Irajacinkam, K. Bhaskarachary, K. Venkaiah, and T. Longvah. 2017. *Indian food composition tables*. Hyderabad: National Institute of Nutrition, Indian Council of Medical Research.

Mahendra, P. and S. Bisht. 2011. Anti-anxiety activity of Coriandrum sativum assessed using different experimental anxiety models. *Indian Journal of Pharmacology 43*: 574.

Mandal, S. and M. Mandal. 2015. Coriander (Coriandrum sativum L.) essential oil: Chemistry and biological activity. *Asian Pacific Journal of Tropical Biomedicine 5*:421–428.

Matasyoh, J.C., Z.C. Maiyo, R.M. Ngure, and R. Chepkorir. 2009. Chemical composition and antimicrobial activity of the essential oil of Coriandrum sativum. *Food Chemistry 113*:526–529.

Msaada, K., K. Hosni, M.B. Taarit, T. Chahed, M.E. Kchouk and B. Marzouk. 2007. Changes on essential oil composition of coriander (Coriandrum sativum L.) fruits during three stages of maturity. *Food Chemistry 102*:1131–1134.

Msaada, K., M.B. Jemia, N. Salem, O. Bachrouch, J. Sriti, S. Tammar, I. Bettaieb, I. Jabri, S. Kefi, F. Limam and B. Marzouk. 2017. Antioxidant activity of methanolic extracts from three coriander (Coriandrum sativum L.) fruit varieties. *Arabian Journal of Chemistry 10*:S3176–S3183.

Msaada, K., M.B. Taarit, K. Hosni, M. Hammami and B. Marzouk. 2009. Regional and maturational effects on essential oils yields and composition of coriander (Coriandrum sativum L.) fruits. *Scientia Horticulturae 122*:116–124.

Nadeem, M., F.M. Anjum, M.I. Khan, S. Tehseen, A. El-Ghorab and J.I. Sultan. 2013. Nutritional and medicinal aspects of coriander (Coriandrum sativum L.): A review. *British Food Journal 115*: 743–755.

Nambiar, V.S., M. Daniel and P. Guin. 2010. Characterization of polyphenols from coriander leaves (Coriandrum sativum), red amaranthus (A. paniculatus) and green amaranthus (A. frumentaceus) using paper chromatography and their health implications. *Journal of Herbal Medicine and Toxicology 4*:173–177.

Nanditha, B.R., B.S. Jena and P. Prabhasankar. 2009. Influence of natural antioxidants and their carry-through property in biscuit processing. *Journal of the Science of Food and Agriculture 89*:288–298.

Naquvi, K.J., M.O.H.D., Ali and J. Ahmad. 2004. Antidiabetic activity of aqueous extract of Coriandrum sativum L. fruits in streptozotocin induced rats. *Indian Journal of Experimental Biology 42*:909–12.

Nath, P., S. J. Kale., O.P. Chauhan. and R.K. Gupta. 2016. High pressure processing induced changes in bioactive compounds, antioxidant activity, microbial safety and color attributes of coriander paste. *Agricultural Research 5*:182–192.

Neffati, M. and B. Marzouk. 2008. Changes in essential oil and fatty acid composition in coriander (Coriandrum sativum L.) leaves under saline conditions. *Industrial Crops and Products 28*:137–142.

Neffati, M., J. Sriti, G. Hamdaoui, M.E. Kchouk and B. Marzouk. 2011. Salinity impact on fruit yield, essential oil composition and antioxidant activities of Coriandrum sativum fruit extracts. *Food Chemistry 124*:221–225.

Nejad, E. S., J. Hadian and H. Ranjbar. 2010. Essential oil compositions of different accessions of Coriandrum sativum L. from Iran. *Natural Product Research 24*:1287–1294.

Oganesyan, E.T., Z.M. Nersesyan and A. Y. Parkhomenko. 2007. Chemical composition of the above-ground part of Coriandrum sativum. *Pharmaceutical Chemistry Journal 41*:149–153.

Padalia, R.C., N. Karki, A.N. Sah and R.S. Verma. 2011. Volatile constituents of leaf and seed essential oil of Coriandrum sativum L. *Journal of Essential Oil Bearing Plants 14*:610–616.

Pathak, N. L., S.B. Kasture, M. Bhatt Nayna and J.D. Rathod. 2011. Phytopharmacological properties of Coriander sativum as a potential medicinal tree: an overview. *Journal of Applied Pharmaceutical Science 1*:20–25.

Pathan, A.R., K.A. Kothawade and M.N. Logade. 2011. Anxiolytic and analgesic effect of seeds of Coriandrum sativum Linn. *International Journal of Research in Pharmacy and Chemistry 1*:1087–1099.

Peana, A.T., S.D. Paolo, M.L. Chessa, M.D. Moretti, G. Serra and P. Pippia. 2003. (–)-Linalool produces antinociception in two experimental models of pain. *European Journal of Pharmacology 460*:37–41.

Peter, K. V. 2004. *Hand Book of Herbs and Spices*. Woodhead Publishing Ltd/Cambridge.

Platel, K. and K. Srinivasan. 2004. Digestive stimulant action of spices: a myth or reality?. *Indian Journal of Medical Research 119*:167–179.

Prachayasittikul, V., S. Prachayasittikul, S. Ruchirawat and V. Prachayasittikul. 2018. Coriander (Coriandrum sativum): A promising functional food toward the well-being. *Food Research International 105*:305–323.

Purseglove, J. W., E.G. Brown, C.L. Green and S.R.J. Robbins. 1981. *Spices 2*. Longman: New York.

Quintans-Junior, L.J., R.S. Barreto, P.P. Menezes, J.R. Almeida, A.F.S. Viana, R.C. Oliveira, A.P. Oliveira, D.P. Gelain, J.W. de Lucca and A.A. Araujo. 2013. β-Cyclodextrin-complexed (–)-linalool produces antinociceptive effect superior to that of (–)-linalool in experimental pain protocols. *Basic and Clinical Pharmacology and Toxicology 113*:167–172.

Rajeshwari, U. and B. Andallu. 2011. Medicinal benefits of coriander (Coriandrum sativum L). *Spatula DD 1*:51–58.

Ravi, R., M. Prakash and K.K. Bhat. 2007. Aroma characterization of coriander (Coriandrum sativum L.) oil samples. *European Food Research and Technology 225*:367–374.

Ravizza, R., M.B. Gariboldi, R. Molteni and E. Monti. 2008. Linalool, a plant-derived monoterpene alcohol, reverses doxorubicin resistance in human breast adenocarcinoma cells. *Oncology Reports 20*:625–630.

Reddy, L.J., D.J. Reshma, B. Jose and S. Gopu. 2012. Evaluation of antibacterial and DPPH radical scavenging activities of the leaf extracts and leaf essential oil of Coriandrum sativum Linn. *World Journal of Pharmaceutical Research 1*:705–716.

Reuter, J., C. Huyke, F. Casetti, C. Theek, U. Frank, M. Augustin and C. Schempp. 2008. Anti-inflammatory potential of a lipolotion containing coriander oil in the ultraviolet erythema test. *JDDG: Journal of the German Society of Dermatology 6*:847–851.

Silva, F., S. Ferreira, J.A. Queiroz and F.C. Domingues. 2011. Coriander (Coriandrum sativum L.) essential oil: its antibacterial activity and mode of action evaluated by flow cytometry. *Journal of Medical Microbiology 60*:1479–1486.

Singh, G., S. Maurya, M.P. De Lampasona and C.A. Catalan. 2006. Studies on essential oils, Part 41. Chemical composition, antifungal, antioxidant and sprout suppressant activities of coriander (Coriandrum sativum) essential oil and its oleoresin. *Flavour and Fragrance Journal 21*:472–479.

Singletary, K. 2016. Coriander: overview of potential health benefits. *Nutrition Today 51*:151–161.

Sonika, G., R. Manubala and J. Deepak. 2010. Comparative studies on anti-inflammatory activity of Coriandrum sativum, Datura stramonium and Azadirachta indica. *Asian Journal of Experimental Biological Sciences 1*:151–154.

Tahraoui, A., J. El-Hilaly, Z.H. Israili and B. Lyoussi. 2007. Ethnopharmacological survey of plants used in the traditional treatment of hypertension and diabetes in south-eastern Morocco (Errachidia province). *Journal of Ethnopharmacology 110*:105–117.

Tang, E.L., J. Rajarajeswaran, S.Y. Fung and M.S. Kanthimathi. 2013. Antioxidant activity of Coriandrum sativum and protection against DNA damage and cancer cell migration. *BMC Complementary and Alternative Medicine 13*:1–13.

Telci, I., O.G. Toncer and N. Sahbaz. 2006. Yield, essential oil content and composition of Coriandrum sativum varieties (var. vulgare Alef and var. microcarpum DC.) grown in two different locations. *Journal of Essential Oil Research 18*:189–193.

Thabrew, M.I., M.G. Dharmasiri and L. Senaratne. 2003. Anti-inflammatory and analgesic activity in the polyherbal formulation Maharasnadhi Quathar. *Journal of Ethnopharmacology 85*:261–267.

Turek, C. and F.C. Stintzing. 2013. Stability of essential oils: a review. *Comprehensive Reviews in Food Science and Food Safety 12*:40–53.

Ugulu, I., S. Baslar, N. Yorek and Y. Dogan. 2009. The investigation and quantitative ethnobotanical evaluation of medicinal plants used around Izmir province, *Turkey. Journal of Medicinal Plants Research 3*:345–367.

USDA. 2018. National Nutrient Database for Standard Reference (All Nutrients). Nutrient data for 2018, Spices, Coriander seed. https://fdc.nal.usda.gov/fdc-app.html#/food-details/170921/nutrients (Accessed May 12, 2021).

Waheed, A., G.A. Miana, S.I. Ahmad and M.A. Khan. 2006. Clinical investigation of hypoglycemic effect of Coriandrum sativum in type-2 (NIDDM) diabetic patients. *Pakistan Journal of Pharmacology 23*:7–11.

Wangensteen, H., A.B. Samuelsen and K.E. Malterud. 2004. Antioxidant activity in extracts from coriander. *Food Chemistry 88*:293–297.

Wu, T. T., C.W. Tsai, H.T. Yao, C.K. Lii, H.W. Chen, Y.L. Wu, P.Y. Chen and K.L. Liu. 2010. Suppressive effects of extracts from the aerial part of Coriandrum sativum L. on LPS-induced inflammatory responses in murine RAW 264.7 macrophages. *Journal of the Science of Food and Agriculture 90*:1846–1854.

Yap, P. S. X., B.C. Yiap, H.C. Ping, and S.H.E. Lim. 2014. Essential oils, a new horizon in combating bacterial antibiotic resistance. *The Open Microbiology Journal 8*:6–14.

Zardini, H. Z., B. Tolueinia, Z. Momeni, Z. Hasani and M. Hasani. 2012. Analysis of antibacterial and antifungal activity of crude extracts from seeds of *Coriandrum sativum. Gomal Journal of Medical Sciences 10*:167–1771.

Zare-Shehneh, M., M. Askarfarashah, L. Ebrahimi, N.M. Kor, H. Zare-Zardini, H. Soltaninejad, Z. Hashemian, and F. Jabinian. 2014. Biological activities of a new antimicrobial peptide from Coriandrum sativum. *International Journal of Biosciences 4*:89–99.

Zheljazkov, V.D., K.M. Pickett, C.D. Caldwell, J.A. Pincock, J.C. Roberts and L. Mapplebeck 2008. Cultivar and sowing date effects on seed yield and oil composition of coriander in Atlantic Canada. *Industrial Crops and Products 28*:88–94.

Zoubiri, S. and A. Baaliouamer. 2010. Essential oil composition of Coriandrum sativum seed cultivated in Algeria as food grains protectant. *Food Chemistry 122*:1226–1228.

Index

For Product Safety Concerns and Information please contact our EU representative GPSR@taylorandfrancis.com Taylor & Francis Verlag GmbH, Kaufingerstraße 24, 80331 München, Germany

Batch number: 10732790

Printed by Printforce, the Netherlands